COMPUTING AND VISUALIZATION FOR INTRAVASCULAR IMAGING AND COMPUTER-ASSISTED STENTING

The Elsevier and MICCAI Society Book Series

Advisory board

COMPUTING AND VISUALIZATION FOR INTRAVASCULAR IMAGING AND COMPUTER-ASSISTED STENTING

Edited by

SIMONE BALOCCO

MARIA A. ZULUAGA

GUILLAUME ZAHND

SU-LIN LEE

STEFANIE DEMIRCI

AMSTERDAM • BOSTON • HEIDELBERG • LONDON
NEW YORK • OXFORD • PARIS • SAN DIEGO
SAN FRANCISCO • SINGAPORE • SYDNEY • TOKYO
Academic Press is an imprint of Elsevier

Academic Press is an imprint of Elsevier
125 London Wall, London EC2Y 5AS, United Kingdom
525 B Street, Suite 1800, San Diego, CA 92101-4495, United States
50 Hampshire Street, 5th Floor, Cambridge, MA 02139, United States
The Boulevard, Langford Lane, Kidlington, Oxford OX5 1GB, United Kingdom

Notices
Knowledge and best practice in this field are constantly changing. As new research and experience broaden our understanding, changes in research methods, professional practices, or medical treatment may become necessary.

Practitioners and researchers must always rely on their own experience and knowledge in evaluating and using any information, methods, compounds, or experiments described herein. In using such information or methods they should be mindful of their own safety and the safety of others, including parties for whom they have a professional responsibility.

To the fullest extent of the law, neither the Publisher nor the authors, contributors, or editors, assume any liability for any injury and/or damage to persons or property as a matter of products liability, negligence or otherwise, or from any use or operation of any methods, products, instructions, or ideas contained in the material herein.

Library of Congress Cataloging-in-Publication Data
A catalog record for this book is available from the Library of Congress

British Library Cataloguing-in-Publication Data
A catalogue record for this book is available from the British Library

ISBN: 978-0-12-811018-8

For information on all Academic Press publications visit our website at https://www.elsevier.com/

Publisher: Joe Hayton
Acquisition Editor: Tim Pitts
Editorial Project Manager: Charlotte Kent
Production Project Manager: Lisa Jones
Cover Designer: Greg Harris

Typeset by SPi Global, India

CONTENTS

CONTRIBUTORS

T. Adriaenssens
University Hospitals Leuven; KU Leuven, Leuven, Belgium

S. Beier
The University of Auckland, Auckland, New Zealand

P. Berg
University of Magdeburg "Otto von Guericke", Magdeburg, Germany

J.-L. Bigras
University of Montreal, Montreal, QC, Canada

O. Bonnefous
Medisys-Philips Research, Paris, France

D. Burgner
Murdoch Childrens Research Institute, Parkville; University of Melbourne, Parkville; Monash University, Clayton, VIC, Australia

S. Carlier
University of Mons, Mons, Belgium

J. Cater
The University of Auckland, Auckland, New Zealand

K.Y.H. Chen
Murdoch Childrens Research Institute; University of Melbourne, Parkville, VIC, Australia

S. Conjeti
Technical University of Munich, Munich, Germany

B. Cowan
The University of Auckland, Auckland, New Zealand

N. Dahdah
University of Montreal, Montreal, QC, Canada

L.B. Daniels
University of California San Diego, La Jolla, CA, United States

L. Daróczy
University of Magdeburg "Otto von Guericke", Magdeburg, Germany

S. Demirci
Technical University of Munich, Munich, Germany

C. Doblado
Vicomtech-IK4 Foundation; Biodonostia Health Research Institute, San Sebastián, Spain

P. Fallavollita
University of Ottawa, Ottawa, ON, Canada

L. Flórez-Valencia
Pontifical Xavierian University, Bogotá, Colombia

A. Gastounioti
National Technical University of Athens, Athens, Greece

R. Ghotbi
Helios Klinikum Munich West Academic Hospital of Ludwig-Maximilian University, Munich, Germany

S. Golemati
National Kapodistrian University of Athens, Athens, Greece

K. Houissa
University of Mons, Mons, Belgium

N. Idris
University of Indonesia/Cipto Mangunkusumo General Hospital, Jakarta, Indonesia

G. Janiga
University of Magdeburg "Otto von Guericke", Magdeburg, Germany

L. Kabongo
Vicomtech-IK4 Foundation; Biodonostia Health Research Institute, San Sebastián, Spain

A. Katouzian
IBM Almaden Research Center, San Jose, CA, United States

A. Kermani
Iran University of Science and Technology (IUST), Tehran, Iran

M. Kowarschik
Siemens Healthineers, Forchheim, Germany

J.H. Legarreta
Vicomtech-IK4 Foundation; Biodonostia Health Research Institute, San Sebastián, Spain

K. López-Linares
Vicomtech-IK4 Foundation; Biodonostia Health Research Institute, San Sebastián, Spain

I. Macía
Vicomtech-IK4 Foundation; Biodonostia Health Research Institute, San Sebastián, Spain

R. Mansour
Helios Klinikum Munich West Academic Hospital of Ludwig-Maximilian University, Munich, Germany

R.L. Maurice
University of Montreal, Montreal, QC, Canada

P. Medrano-Gracia
The University of Auckland, Auckland, New Zealand

P. Mermigkas
National Technical University of Athens, Athens, Greece

H.G. Morales
Medisys-Philips Research, Paris, France

N. Navab
Technical University of Munich, Munich, Germany; Johns Hopkins University, Baltimore, MD, United States

S. Norris
The University of Auckland, Auckland, New Zealand

K.S. Nikita
National Technical University of Athens, Athens, Greece

M. Orkisz
Univ Lyon, CNRS UMR5220, Inserm U1206, INSA-Lyon, Université Lyon 1, CREATIS, F-69621, Lyon, France

J. Ormiston
Auckland Heart Group, Auckland, New Zealand

A. Pourmodheji
Azad University, Dubai, UAE

M. Prevenios
National Technical University of Athens, Athens, Greece

S.M. Ranjbarnavazi
Tehran Azad University of Medical Science, Tehran, Iran

J. Rigla
Barcelona Perceptual Computing Laboratory, Universitat de Barcelona, Barcelona, Spain

A.G. Roy
Technical University of Munich, Munich, Germany; Indian Institute of Technology Kharagpur, Kharagpur, West Bengal, India

D. Sheet
Indian Institute of Technology Kharagpur, Kharagpur, West Bengal, India

T. Syeda-Mahmood
IBM Almaden Research Center, San Jose, CA, United States

A. Taki
Technical University of Munich (TUM), Munich, Germany

G.J. Ughi
Massachusetts General Hospital and Harvard Medical School, Boston, MA, United States

L. Vaujois
University of Montreal, Montreal, QC, Canada

M. Webster
Auckland City Hospital, Auckland, New Zealand

A. Young
The University of Auckland, Auckland, New Zealand

ABOUT THE EDITORS

Simone Balocco is associate professor at the Department of Mathematics and Informatics, University of Barcelona, Spain, and is a senior researcher at the Computer Vision Center, Bellaterra. He obtained a PhD degree in Acoustics at the CREATIS laboratory, Lyon, France, and in Electronic and Telecommunication in MSD Lab, University of Florence, Italy. He performed postdoctoral research at the laboratory CISTIB, at Universitat Pompeu Fabra, Spain. Balocco's main research interests are pattern recognition and computer vision methods for the computer-aided detection of clinical pathologies. In particular, his research focuses on ultrasound and magnetic imaging applications and vascular modeling.

Maria A. Zuluaga In 2011, Maria A. Zuluaga obtained her PhD degree from Université Claude Bernard Lyon 1, France, investigating automatic methods for the diagnosis of coronary artery disease. After a year as a postdoctoral fellow at the European Synchrotron Radiation Facility (Grenoble, France), she joined University College London, UK, in March 2012, as a research associate to work on cardiovascular image analysis and computer-aided diagnosis of cardiovascular pathologies. Since August 2014, she has been part of the Guided Instrumentation for Fetal Therapy and Surgery (GIFT-Surg) as a senior research associate.

Guillaume Zahnd received his engineering degree from the National Institute of Applied Science (INSA-Lyon, France) in 2007, and obtained his PhD from the CREATIS Laboratory, University of Lyon, France, in 2012. In 2013, he joined the Biomedical Imaging Group Rotterdam, Erasmus MC, The Netherlands, as a postdoctoral researcher. From 2016, he is a research fellow in the Imaging-based Computational Biomedicine Laboratory at the Nara Institute of Science and Technology, Japan. His work focuses on image-processing methodologies toward cardiovascular risk assessment. His fields of interest include vascular imaging, image-based biomarkers, ultrasound, intracoronary optical coherence tomography, motion tracking, contour segmentation, and machine learning.

Su-Lin Lee received her MEng in Information Systems Engineering and PhD from Imperial College London, UK, in 2002 and 2006, respectively, for her work on statistical shape modeling and biomechanical modeling. She is currently a lecturer at the Hamlyn Centre for Robotic Surgery and the Department of Computing, Imperial College London. Her current research focuses on machine learning and shape modeling with application to guidance in cardiovascular interventions. Of particular interest to her are improved navigation and decision support for safer and more efficient robotic-assisted minimally invasive cardiovascular procedures.

Stefanie Demirci is a postdoctoral researcher and research manager at the Technical University of Munich (TUM), Germany. She received her PhD from the same institution in 2011 for her work on novel approaches to computer-assisted endovascular procedures. After being a postdoctoral fellow at the SINTEF Medical Technology Laboratory in Trondheim, Norway, she returned back to the TUM where she is currently teaching Interventional Imaging and Image Processing, and managing the Computer Aided Medical Procedures (CAMP) Laboratory. Her current research focuses on multimodal imaging and image processing, machine learning, and biomedical gamification, with particular interest in crowd sourcing for biomedical ground truth creation.

PREFACE

Cardiovascular disease is the primary cause of mortality worldwide, necessitating clinical and technical improvements to cardiovascular disease prediction, prevention, and treatment. To address this major public health issue, a myriad of techniques have been developed that offer increasingly useful information regarding vascular anatomy and function and are poised to have dramatic impact on the diagnosis, analysis, modeling, and treatment of vascular diseases. Today, scientific research in the field of computer-assisted technological advances in diagnostic and intraoperative vascular imaging and stenting benefits from an ever-growing interest.

In 2006, a workshop was organized to bring together researchers, clinicians, and industry in the field of intravascular imaging. This event, created by Gözde Ünal, Ioannis Kakadiaris, Greg Slabaugh, and Allen Tannenbaum, was the first International Workshop on Computer Vision for Intravascular and Intracardiac Imaging (CVII), organized in conjunction with MICCAI (Medical Image Computing and Computer-Assisted Intervention) in Copenhagen, Denmark. The aim was to provide a platform to present state-of-the-art techniques in computing and visualization for vascular applications.

Technological advances in intravascular imaging such as B-mode ultrasound (US), intravascular ultrasound (IVUS), and optical coherence tomography (OCT), as well as more traditional vascular imaging methods such as computed tomography angiography (CTA), X-ray angiography, and fluoroscopy, have had a dramatic impact on the diagnosis, analysis, modeling, and treatment of vascular diseases. Computer vision methods applied to these images have received tremendous interest, allowing for improved modeling, simulation, visualization, classification, and assessment. This first workshop addressing these issues was a success with 24 papers presented; it was followed by workshops in New York City, USA (2008) and Toronto, Canada (2011).

Following these footsteps, the first International MICCAI Workshop on Computer-Assisted Stenting was organized by Stefanie Demirci, Gözde Ünal, Su-Lin Lee, and Petia Radeva, and was held in Nice, France, in 2012. This workshop brought together researchers in the field of endovascular stenting procedures and covered research dealing with cerebral, coronary, carotid, and aortic stenting, crossing anatomical boundaries. It was also, to the best of our knowledge, the first technical workshop dealing with this difficult, minimally invasive procedure. The program of this first workshop included 16 papers and a plenary talk by Dr. Reza Ghotbi, a vascular surgeon, describing his experience of stenting and the current clinical challenges. A second workshop was held the following year—2013—in Nagoya, Japan.

With the aim of these two different workshops targeting improved vascular procedures, it seemed like the most natural step to combine both workshops. The Joint MICCAI-Workshop on Computing and Visualization for Intravascular Imaging and Computer-Assisted Stenting (CVII-STENT) was launched in 2014 in Boston, USA and since then has become the main annual technical workshop dedicated to computer assistance for vascular procedures and vascular imaging. The workshop was held again in 2015 in Munich, Germany as well as in 2016 in Athens, Greece.

After the second edition of the CVII-STENT workshop (2015), we were presented with the opportunity to edit a book providing a snapshot of the state-of-the-art methods and techniques in both the clinical and technical domains. This is the book that you are holding in your hands now. We have invited the top researchers in the field of computer-assisted endovascular stenting and intravascular imaging to contribute a chapter describing their research. Both the clinical and technical communities have provided material on the existing clinical challenges and context, as well as the latest in technical innovations. Their views on the future research perspectives in the field are a tremendous resource to any academic, clinical, and industrial researchers working in the area of endovascular imaging and interventions.

This book is divided into four main sections. Section I covers the *Clinical Introduction* to the book. Here the authors provide a clinical overview of vessel pathologies, which can be treated endovascularly, and the imaging technologies involved in their assessment and treatment. In Chapter 1, we compare IVUS and OCT in the assessment of coronary artery disease and review their role in percutaneous coronary interventions. A broader review of IVUS and OCT, not limited to just the heart but to plaque progression and atherosclerosis in general, is then presented (Chapter 2). Afterward, we address abdominal aortic aneurysms (AAA) and the key role imaging has played in all the aspects of the clinical workflow (Chapter 3). Finally, given that IVUS and OCT are the most common imaging modalities in plaque assessment, the section is concluded by providing a more global overview of all the modalities involved in this task (Chapter 4).

Section II covers the *Vascular and Intravascular Analysis of Plaque*. This part focuses on imaging and modeling techniques aimed at plaque analysis; it includes segmentation, vessel reconstruction, and movement quantification methods applied to several imaging modalities (B-mode US, CTA, IVUS, and OCT). We first explore the implications of the kinematic activity of the atherosclerotic plaque applied to B-mode US images of the carotid artery (Chapter 5). We then present a right generalized cylinder model which can be used both for 3D vessel reconstruction and for vascular segmentation of CTA images (Chapter 6). Then we present a technique for in vivo IVUS tissue characterization which exploits a domain-adapted model (Chapter 7). Finally, in Chapter 8, we cover different imaging techniques for plaque segmentation, stent analysis, and tissue characterization applied to OCT sequences.

Vascular Biomechanics and Modeling is presented in Section III. This section focuses on numerical simulations of arterial parameters to characterize relevant biomarkers. We first investigate a large-scale approach to model and quantify coronary flow with computational fluid dynamics (CFD) (Chapter 9). We then address the challenge of CFD-based numerical simulations of aneurysms and hemodynamics in cerebral arteries (Chapter 10). To conclude this section, we focus on methods to identify at-risk subjects in the early stage of the pathology for improvement of preventive care, by means of an US elastography approach devised to detect functional alterations in the vascular wall (Chapter 11).

Finally, Section IV covers *Computer-Assisted Stenting*. The chapters here cover computing methods for both preoperative and intraoperative stages of endovascular procedures. For the former, we explore the careful sizing required for aortic aneurysm stenting (Chapter 15) and also the simulation of stenting for intracranial aneurysms (Chapter 14). For the latter, we review the navigation support for live endovascular procedures (Chapter 12) as well as the live quantification of blood flow during a procedure (Chapter 13). Research into effective intraoperative guidance of stenting is starting to take off and we hope this section is a good starting point for further investigations.

Editing this book has been a pleasure. We hope that it will constitute a very valuable introduction to readers new to the field of intravascular imaging and stenting. Likewise, we hope that experts in the field will be exposed to different research approaches to known problems. Finally, our sincerest thanks go to the authors of all chapters for their dedication to this project and to Elsevier for their support.

Su-Lin Lee
Guillaume Zahnd
Maria A. Zuluaga
Stefanie Demirci
Simone Balocco

SECTION I

Clinical Introduction

CHAPTER 1

Intravascular Imaging to Assess Coronary Atherosclerosis and Percutaneous Coronary Interventions

S. Carlier, K. Houissa
University of Mons, Mons, Belgium

Chapter Outline

1. INTRAVASCULAR IMAGING DEVELOPMENT

In clinical practice today, two intravascular imaging modalities are mainly used: intravascular ultrasound (IVUS) and optical coherence tomography (OCT). They are based on ultrasound and near-infrared light, two technologies that measure very different physical properties, with varying resolution and penetration range. This gives each methodology advantages and drawbacks, and these are important to understand in order to use each method accordingly for optimal use.

1.1 Intravascular Ultrasound

The first medical ultrasound application was described in 1953 by Edler and Hertz [1], who introduced the recording of the motion pattern of cardiac structures along a single

Computing and Visualization for Intravascular Imaging and Computer-Assisted Stenting
http://dx.doi.org/10.1016/B978-0-12-811018-8.00001-1

sound beam. Beside the development of hand-held probes offering 2D, and now 3D, imaging widely used in cardiology today, an intracardiac scanner was already described by Bom and Lancée in 1972 [2]. Further miniaturization allowed the development of a mechanically rotated intravascular system that Paul Yock patented in 1991 [3]. The technique of intravascular ultrasound (IVUS) is based on a flex-shaft mechanically rotated single element or an electronically steered phased array system (Fig. 1.1). The systems that have been used for the last 10 years are running at frequencies between 20 and 40 MHz to produce cross-sectional vascular real-time images of the lumen, plaque, and arterial wall with an axial resolution of ~120 μm and a lateral resolution of ~200 μm [4]. The higher the central frequency and the bandwidth, the higher the resolution is. Fig. 1.2 shows the evolution in the quality of IVUS systems by looking at a stent with a solid-state 20 MHz system (Philips Volcano) with a synthetic aperture image formation from a multielement transducer array, and single unfocused ultrasound transducer of 30 and 40 MHz (Boston Scientific) and 45 MHz (Philips Volcano), up to the latest available broadband element with 60 MHz central frequency (ACIST HDi High-Definition IVUS System). ACIST's claim on this system that received CE marking in May 2016 is a 40 μm axial resolution and 90 μm lateral resolution. The circumferential resolution was improved by high-speed imaging at 60 Hz frame rate with 2048 vectors per frame, up from 30 frames/s of 512 vectors on older systems. The drawback of higher-frequencies IVUS systems is the fact that the intensity of the blood speckle

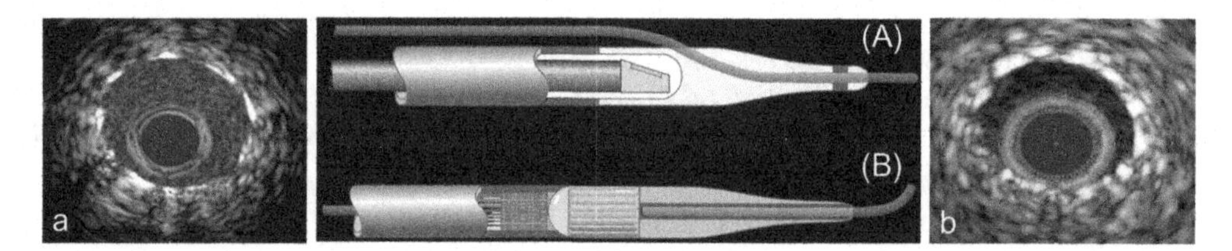

Fig. 1.1 The two types of intravascular ultrasound (IVUS) imaging systems. (A) Mechanical system with a rotating element; a: Cross-sectional image given by a mechanical system (Atlantis SR Pro Catheter iLab Ultrasound Imaging System). (B) Electronic system with a multielement array; b: Cross-sectional image given by an electronic system (Eagle Eye Gold Catheter S5 System).

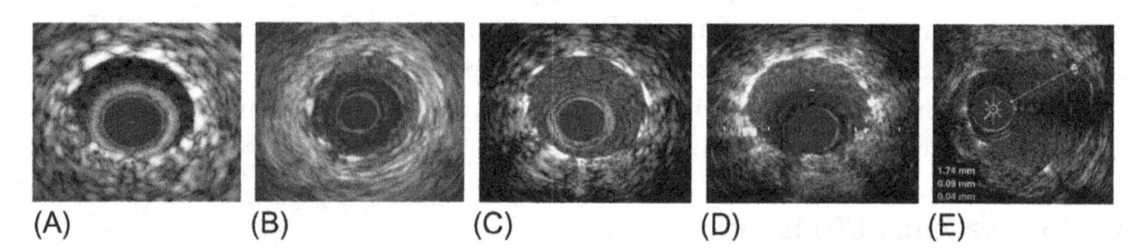

Fig. 1.2 Image resolution: influence of bandwidth. Metallic stent struts as seen by (A) 20 MHz, (B) 30 MHz, (C) 40 MHz, (D) 45 MHz, and (E) 60 MHz IVUS catheters.

increases to the fourth power of the transducer frequency so that the echodensity of blood might become as high as the plaque, if not higher in the presence of blood stasis with rouleaux red-cell formation. Electronic filtering and image processing can reduce this phenomenon [5].

The normal appearance of a coronary artery by IVUS depends on the differential scattering of ultrasound by the different wall component. The strongest reflection of ultrasound comes from collagen. The adventitia of the coronary arteries is very rich in collagen and appears as the brightest structure. The external elastic lamina (EEL) lies between the adventitia and the media, and is mostly muscular and typically echolucent (dark). In normal, nonatherosclerotic arteries, the thickness of the media is typically 200 μm. The internal elastic lamina (IEL) separates the media from the most inner structure of the artery, the intima that is covered by a single layer of endothelial cells. Intimal thickness increases with age and it is typically 200 μm at 40 years of age [6]; this will produce the classical three-layer appearance of a normal coronary artery by IVUS. Intimal thickening is the first pathophysiological change related to atherosclerosis. With accumulation of plaque, intima and IEL tend to merge, and the separation from the media is difficult to assess. IVUS studies use the EEL and lumen borders, delimiting the plaque and media histologically, as a surrogate of plaque measurement. The media layer is thus added to the plaque volume in the plaque progression/regression studies. An IVUS image of such an atherosclerotic artery is given in Fig. 1.3, as well as one with a plaque disruption (dissection) after a balloon angioplasty (upper right panel). These high-resolution images obtained at 60 MHz (ACIST HDi® High-Definition IVUS System) compare well to the higher-resolution OCT examples (lower row) recorded, respectively, with a Lunawave (Terumo Corporation) on the left and C7-XR FD-OCT (LightLab Imaging, now St. Jude Medical) on the right.

The profile of the IVUS imaging catheters nowadays offers compatibility with 5 French (Fr) guiding catheter. Imaging is recorded during manual pullback or automatic pullback at a speed of 0.5–1 mm/s.

1.2 Optical Coherence Tomography

Intravascular optical coherence tomography (OCT) is a catheter-based method for high-resolution imaging of coronary arteries. It was developed in the early 1990s at the Massachusetts Institute of Technology for both retinal and vascular imaging, as described initially by Huang et al. [7]. Current OCT systems incorporate advanced near-infrared light sources and optical components that operate in a wavelength band centered on 1310 nm. As with the IVUS, OCT measures the depth of reflections from tissue according to the round-trip propagation time of reflected energy. However, as the speed of light is much faster than that of sound, an interferometer is required to measure the backscattered light. OCT uses low-coherence interferometry to produce

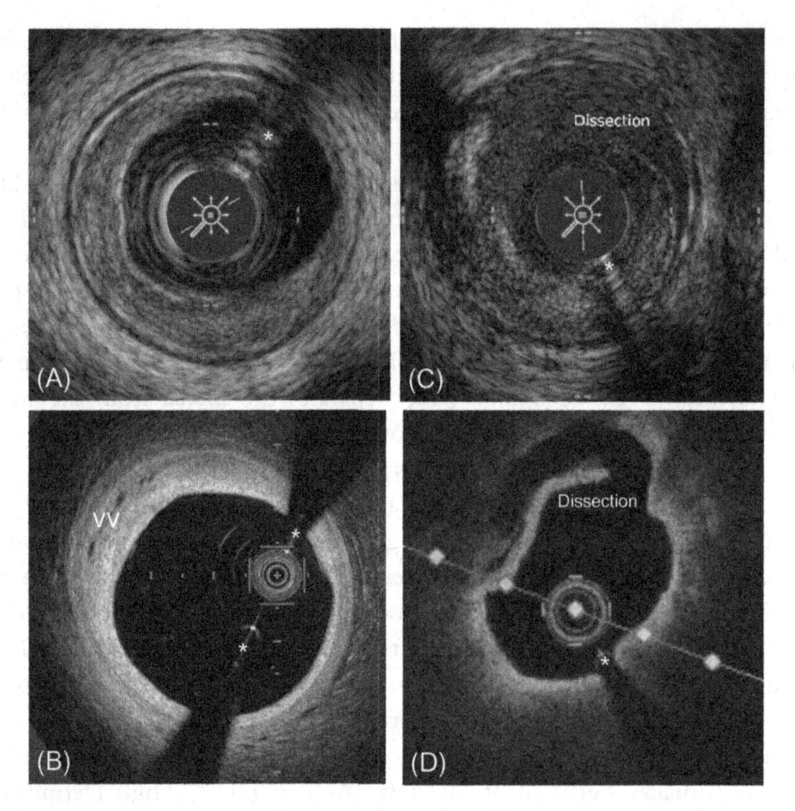

Fig. 1.3 (A) Mild coronary atherosclerotic plaque seen with a 60 MHz IVUS while flushing the lumen to reduce blood speckle artifact; (B) similar mild plaque seen by OCT; (C) a dissection seen with a 60 MHz IVUS; and (D) a similar dissection seen by OCT. Of note in panel (B), the higher resolution of OCT allows the visualization of the vasa vasorum (vv) in the plaque (image courtesy of Dr. Satoru Sumitsuji, Osaka University, Japan). The wires, and their acoustic/optical shadow, used to advance the IVUS or OCT catheters are shown by the *.

a two-dimensional image of optical scattering from internal tissue microstructures in a way that is analogous to ultrasonic pulse-echo imaging. Clinically available intravascular OCT systems have longitudinal and lateral spatial resolutions 10 times higher than IVUS (15 μm axially, 25 μm laterally), as can be appreciated on Fig. 1.3. However, the penetration depth is only 0.5–2 mm. Moreover, as near-infrared light penetrates only a short distance through blood, temporary blood clearance is required for OCT imaging.

The early OCT systems were based on a technology coined time domain (TD) OCT in which the optical probe in the catheter could not be pulled back faster than an IVUS catheter, at a speed of 1 mm/s. It was then necessary to use an occlusion balloon proximally inflated in the coronary artery to be imaged and flush the lumen with some saline solution. Instead of the broadband light source used in TD-OCT, nowadays, ultrafast imaging is available using Fourier-domain OCT (FD-OCT) based on a novel wavelength-swept laser as a light source. Today it is no more necessary to occlude the artery: the injection for a few seconds of the X-ray contrast media that has a high viscosity is enough to clear the coronary lumen, while the optical fiber of the FD-OCT can be pulled back at a maximal speed of 20 mm/s with 100 frames of 450 lines acquired by second [8]. Similar to modern IVUS catheters, the FD-OCT system is designed for rapid-exchange delivery over a 0.014-in guidewire and is compatible with a 6 Fr guide catheter. Inside the catheter, there is a fiber-optic imaging core threaded through a hollow torque cable that rotates at 100 rotations/s, three times faster than in an IVUS catheter, and pulls back within a transparent plastic sheath.

While IVUS has been around for the last 25 years, with well-described guidelines for acquisition and measurements [9, 10], clinical experience of FD-OCT has been gained in less than 10 years. However, important enthusiasm and interest from the interventional community allowed the building of strong clinical evidence of its usefulness to assess coronary atherosclerosis [11], and imaging recommendations were more quickly published [12].

2. SAFETY OF INTRAVASCULAR IMAGING

Epicardial coronary arteries and venous or arterial grafts can be imaged after intubation of their ostium with a guiding catheter and the insertion of a 0.014-in guidewire on which the IVUS or OCT catheter will be advanced. The patient must be anticoagulated, for example, with heparin, and preferably nitroglycerin will be administered through the guiding catheter in order to avoid catheter-induced spasm. Once these simple recommendations have been followed, many reports have demonstrated that IVUS and OCT can be performed safely. There is, on average, 3% of coronary spasms that can occur. In an early multicenter report in more than 2000 patients, the most common complications other than spasms were acute occlusion, embolism, dissection, and thrombus formation with some patients presenting major events such as myocardial infarction or emergency coronary artery bypass surgery [13]. The complication rate was higher (2.1%) in the case of acute coronary syndrome (ACS) compared with patients with stable angina pectoris and asymptomatic patients (0.8% and 0.4%, respectively). Similar rates of complication were reported from another multicenter registry [14]. More recently, three-vessel imaging by IVUS and/or OCT has been performed in longitudinal prospective studies. There were 1.6% of events (10 coronary dissections and 1 perforation) among the 697 ACS patients of the IVUS PROSPECT trial [15]. In the 103 patients with ST-segment elevation myocardial infarction (STEMI) who underwent three-vessel coronary imaging during primary percutaneous coronary intervention (PCI) for the IBIS4 study [16] conducted at five European centers, imaging of the noninfarct-related vessels was successful in approximately 90% of them. Periprocedural complications occurred in less than 2.0% of OCT procedures. There were no IVUS-related complication. More recent data from the Thoraxcentre team (Rotterdam, NL) are even more reassuring: in more than 1000 OCT and nearly 2500 IVUS procedures, imaging-related complications rates were low, without a difference between the two modalities (OCT: 0.6%; IVUS: 0.5%). Complications were self-limiting after retrieval of the imaging catheter or easily treatable in the catheterization laboratory, and no major adverse events occurred [17].

These small event rates support the safety of IVUS and OCT imaging procedures; nevertheless, they should not be performed without a good clinical indication.

3. INTRAVASCULAR IMAGING VERSUS CORONARY ANGIOGRAPHY

IVUS was developed to alleviate the limitations of coronary angiography, which shows only the lumen of an artery filled transiently by X-ray contrast media. Although angiography remains the gold standard for the evaluation of symptomatic patients with coronary artery disease (CAD), there are numerous limitations to these projected luminograms unable to show the arterial wall [18]. Atherosclerosis is in essence a disease of the arterial wall [19]. The lumen will only be compromised late on, as first described by Glagov from histologic sections of heart obtained at autopsy [20], and later by others using IVUS [21]. A (coronary) atherosclerotic lesion can evolve over several years without any clinical symptom or flow limitation. The plaque may reach several millimeters in thickness. Ultrasound waves from IVUS catheters running from 20 to 60 MHz can travel deep enough to image completely a thickened atherosclerotic vessel wall. Calcifications will produce a strong reflection of the US beam hampering the evaluation of deeper atherosclerosis, to be restricted to regions with enough signal confidence [22]. The near infra-red light of the OCT systems interacts differently with the arterial wall components. Light travels through calcifications with a low backscattering. Signal-poor regions with diffuse border correspond to the presence of a lipid pool or necrotic core [23].

There is a large variability in the evaluation of intermediate coronary stenosis from angiography. Interventional cardiologists tend to overestimate the degree of a stenosis before an intervention, while underestimating it after [24]. When arterial wall irregularities appear on angiography, up to 80% of the coronary tree is already diseased and will show atherosclerosis [25]. An IVUS or OCT pullback in a coronary artery will then always teach us better the amount of atherosclerotic burden and help us in the understanding of ambiguous lesions with intraluminal filling defects (thrombus, dissection, etc.), coronary aneurysms, ostial or bifurcation lesions, calcifications, and left main disease.

4. INTRAVASCULAR IMAGING ASSESSMENT OF PLAQUE PROGRESSION/REGRESSION

CAD progression is clinically very difficult to assess and to visualize from coronary angiography. Attempts to quantify the amount of plaque on coronary computed tomography angiography [26] have been described. However, only IVUS and OCT offer a resolution high enough to study coronary atherosclerosis progression accurately. IVUS and OCT offer important insights but are invasive and therefore limited to high-risk patients. Pullback IVUS data files contain thousands of cross-sectional images that allow a 3D volumetric reconstruction of the lumen, vessel, and plaque as depicted in Fig. 1.4. Automatic extraction of the media/adventitia and lumen borders has been the topic of numerous research efforts that we reviewed recently [27] and were benchmarked during the MICCAI 2011 Computing and Visualization for (Intra)Vascular Imaging (CVII) workshop, comparing the results of eight teams that participated [28]. It remains a challenging image processing problem: automated segmentation methods have seen limited success due to the presence of guide wire, the presence of an arc of calcified plaques, the motion of the catheter as well as the heart, and the appearance of side branches. Other researchers are still using manual tracing of the lumen and vessel contours to study plaque progression in a coronary segment defined at the baseline using clear landmarks such as side branches or the ostium of the coronary artery investigated. Measurements will be repeated after 6 to 24 months of follow-up, between the same landmarks.

Several studies have demonstrated the value of IVUS in evaluating plaque volume regression over time using different treatments. There have been important differences in the methodologies between some limited observational studies and large randomized, placebo-controlled trials so that a consensus document was necessary [29]. For example,

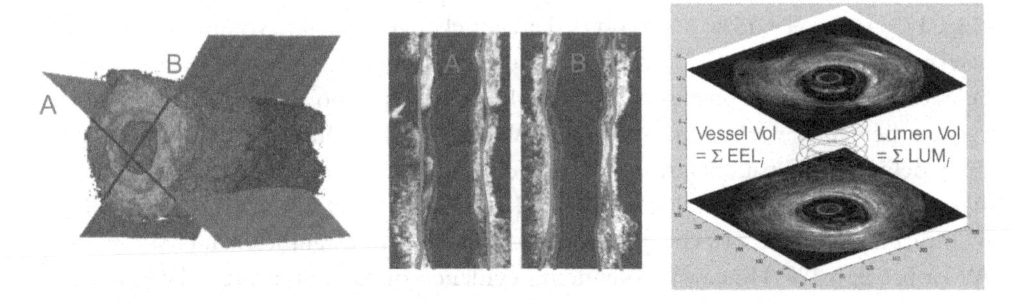

Fig. 1.4 A stack of IVUS cross-sections from a motorized pullback at 0.5 mm/s is cut by two orthogonal planes A and B (*left*). The lumen contours are seen on the sagittal views in the middle panel in *red* and the EEL in *green*. The volume of the plaque will correspond to the volume between all the *green* contours and the *red* contours (= lumen). *For interpretation of the color in this figure, please see the online version.*

Jensen et al. demonstrated that an ECG-gated three-dimensional IVUS-derived plaque volume significantly decreased by 6.3% ($p = 0.002$) after 12 months of simvastatin treatment (40–80 mg, titrated to reach a low-density lipoprotein (LDL) cholesterol level <120 mg/dL). Although the reported changes in mean volume are highly significant, such an uncontrolled study is methodologically inadequate, lacking a control group. The observed changes could have been due to a variety of other reasons besides the simvastatin therapy. Nevertheless, some of these early reports are worth mentioning, because the hypothesis-generating observations were further confirmed in larger randomized trial and metaanalyses. For example, von Birgelen et al. reported in 2003 that an LDL level lower than 75 mg/dL might be associated with a freezing of the progression of coronary atherosclerosis. He followed the progression of mild atherosclerotic plaques located in the left main coronary artery (LMCA) of 60 patients using IVUS. Over a follow-up period of 18 ± 9 months, plaque progression was correlated with LDL level, and inversely correlated with HDL cholesterol values. Regression analysis predicted no annual plaque cross-sectional increase, on average, below an LDL cutoff value of 75 mg/dL. The REVERSAL (Reversal of Atherosclerosis with Aggressive Lipid Lowering) trial was the first one to demonstrate, in a significantly large number of patients, in a double-blind randomized multicenter trial, a difference in the effects of two statins (atorvastatin and pravastatin) administered for 18 months [30]. The IVUS-derived percentage change in atheroma volume showed a significantly lower progression rate in the atorvastatin group ($p = 0.02$). Progression of coronary atherosclerosis—the primary endpoint—occurred in the pravastatin group (2.7%; 95% confidence interval [CI]: 0.2–4.7; $p = 0.001$) compared with the baseline. Progression did not occur in the atorvastatin group (−0.4%; CI: −2.4% to 1.5%; $p = 0.98$) compared with the baseline. A smaller number of patients ($n = 70$) presenting with an ACS were studied in the ESTABLISH trial, with an IVUS performed in a nonculprit lesion [31]. In the intensive lipid-lowering therapy ($n = 35$, atorvastatin), plaque volume decreased significantly compared to a control group ($n = 35$) at the 6-month follow-up: $-13 \pm 13\%$ versus $9 \pm 15\%$, $p < 0.0001$. Percent change in plaque volume was positively correlated with follow-up LDL cholesterol level, even in patients with low baseline LDL cholesterol. Also in the setting of an ACS, Nissen et al. reported a significant decrease in mean percentage atheroma volume (PAV) of $-1.1 \pm 3.2\%$ ($p = 0.02$ compared with baseline) in 57 patients treated for 5 weeks with infusions of recombinant ApoA-I Milano/phospholipids complexes (ETC-216) [32]. Conversely, in this double-blind, randomized study, the placebo group demonstrated a nonsignificant increase in mean PAV of $0.1 \pm 3.1\%$. However, convincing evidence of regression using PAV, the most rigorous IVUS measure of disease progression and regression, was only seen later in the ASTEROID trial [33]. At the baseline, 507 patients with an ACS or stable CAD had a baseline IVUS examination and all received 40 mg of rosuvastatin daily. After 24 months, 349 patients had evaluable serial IVUS examinations. Their baseline LDL-C level decreased from 130 ± 34 to 61 ± 20 mg/dL (a mean reduction of 53%). The mean $\pm$ standard deviation change in PAV for the entire vessel was $-1 \pm 3\%$, with a median of -0.8% (97.5% CI: -1.2 to -0.5) ($p < 0.001$ vs. baseline), in agreement with the early observation of von Birgelen and colleagues. It would be beyond the scope of this chapter to go through all plaque regression/progression studies, and we will not review the trials that used advanced plaque characterization algorithms such as IB-IVUS [34], RF-IVUS [35], or iMAP [36] because practically, these methods are not really used in a daily clinical practice. These methods are very interesting research tools, but they must be further validated by comparing the derived quantitative results to histological validation. It has been attempted by several investigators with more or less fortune [37, 38] since many challenges remain for accurate atherosclerotic plaque characterization with IVUS [39].

It appears that the modest changes seen in plaque progression rates following high-dose statin administration versus conventional dose might be a surrogate for hard clinical endpoints that can only be derived from mega-trials such as the TNT study that randomized 10,001 patients with clinically evident CAD and LDL-C levels <130 mg/dL to 10 or 80 mg of atorvastatin per day [40]. After a median follow-up of 4.9 years, a primary major adverse cardiac event (MACE, defined as death from CAD, nonfatal nonprocedure-related MI, resuscitation after cardiac arrest, or fatal or nonfatal stroke) occurred in 434 patients (8.7%) receiving 80 mg of atorvastatin, compared with 548 patients (10.9%) receiving 10 mg. This difference represented a 22% relative reduction in the rate of MACE (hazard ratio [HR]: 0.78; 95% CI: 0.69–0.89; $p < 0.001$). There was no difference between the two treatment groups in overall mortality.

IVUS plaque regression has been investigated with other therapies such as amlodipine or enalapril versus a placebo in the CAMELOT study [41], or more recently with ezetimibe in the PRECISE-IVUS [42]. The IVUS-derived rate of progression of atherosclerotic burden might be a surrogate endpoint that reflects the beneficial clinical impact of the investigated therapies. However, the lack of conclusive data led to metaregression analysis investigating the association between plaque changes and clinical events during follow-up. Out of 11 studies with 7864 patients, it appears that plaque volume regression is significantly associated with the incidence of MI or revascularization ($p = 0.006$), but not with MACE (a composite of death, MI, or revascularization; $p = 0.208$) [43].

The major limitation of OCT is its inability to assess the full depth of advanced atherosclerosis with plaques showing a thickness of several millimeters. It is then impossible to measure plaque burden, remodeling index, and to plan studies of plaque progression/regression. It is possible to measure coronary intima-media thickness [44] and confirm the absence of significant atherosclerosis. Serial measurements might be able to monitor early structural changes over time, but the potentially young, nearly disease-free candidates do not clinically require a catheterization, nor serial OCT assessment. However, in the specific setting of transplant coronary artery vasculopathy, it has been shown that OCT revealed that patients with a history of high-grade cellular rejection, compared with those with none/mild rejection, had more coronary artery intimal thickening with macrophage infiltration, involving all coronary segments and side branches [45]. It remains to be proven that with its higher resolution, OCT might evaluate more precisely than IVUS the rate of progression of intimal thickening in order to predict more accurately the outcome of heart transplant recipients. A multicenter study has already demonstrated that an increase of more than 0.5 mm in the IVUS derived maximal intimal thickness from baseline to 1 year was associated with a higher incidence of death or graft loss, more nonfatal major adverse cardiac events, and/or more newly occurring angiographic luminal irregularities [46].

The accurate diagnosis of vulnerable plaques might be more important than the total plaque burden. What makes atherosclerosis one of the deadliest diseases is not stenosis alone but failure in detection and proper treatment of vulnerable plaques. They have been characterized in postmortem pathological studies to be mainly lipid-rich, voluminous, and outwardly remodeled plaques covered by attenuated and inflamed fibrous caps. Today, more than ever, there is a need for reliable, reproducible, clinically approved atherosclerotic plaque characterization algorithms. For this, OCT clearly has many advantages because of its ability to assess the thickness of a thin-cap fibroatheroma and to identify and quantify macrophage presence using tissue property indexes coined normalized standard deviation, signal attenuation, and granulometry index [47]. The origins of many qualitative OCT features can be explained by an investigation that we performed of both the attenuation and the backscattering properties of different plaque components of postmortem human cadaver coronary arteries [48]. The artery samples were examined both from lumen surface using a catheter and from transversely cut surface using an OCT microscope, where OCT images could be matched to histology exactly. Light backscattering coefficient μ_b and attenuation coefficient μ_t were measured for three basic plaque types based on a single-scattering physical model. For calcification $\mu_b = 4.9 \pm 1.5\,\text{mm}^{-1}$, $\mu_t = 5.7 \pm 1.4\,\text{mm}^{-1}$, fibrotic tissue $\mu_b = 18.4 \pm 6.4\,\text{mm}^{-1}$, $\mu_t = 6.4 \pm 1.2\,\text{mm}^{-1}$, and lipid pool $\mu_b = 28.1 \pm 8.9\,\text{mm}^{-1}$, $\mu_t = 13.7 \pm 4.5\,\text{mm}^{-1}$. In summary, compared to fibrous plaque, the lipid plaque exhibits both higher backscattering and higher attenuation. The lipid has a higher signal strength than the fibrous plaque at the top surface. However, the signal attenuates very fast and drops below fibrotic tissue after 50–100 μm. The calcified plaque showed both low backscattering and low attenuation, except in the shallow specular reflection region. Colormapping combining both the backscattering and attenuation measurements offered a new tool for contrast enhancing and better tissue characterization.

IVUS remains well behind any of these possibilities, although early findings suggested sonographic differences for visual discrimination among plaque constituents [49]. Next to the already mentioned methods based on the processing of the raw radiofrequency (RF) signal from the IVUS, variations of intensities attributed to repetitive tissue microstructure patterns have motivated us and other researchers to develop texture-based algorithms on IVUS images to differentiate tissue types [50–52]. However, none of these methods made it to the clinical arena to be used in day-to-day practice; the same was true for the RF-based methods. The main reason is that in the meantime, several longitudinal imaging studies in humans have demonstrated that plaque morphology changes over a few months, gaining or losing "vulnerable" characteristics [53–55]. The management of patients at risk of an ACS is based nowadays more on the atherosclerotic disease burden rather than on features of individual plaques considering the complex relationship between plaque rupture and acute coronary events [56].

5. INTRAVASCULAR IMAGING ASSESSMENT OF LESIONS TO BE REVASCULARIZED, OR NOT

In a clinical setting, we will not try to measure the lumen and vessel area every millimeter of the pullback to derive volumes as described in Section 4. As shown in Fig. 1.5, we will look for the IVUS/OCT slice showing the minimal lumen cross-sectional area (MLA). It has been shown that MLA is the parameter that has the best correlation with physiological assessment of the severity of a lesion and induced ischemia.

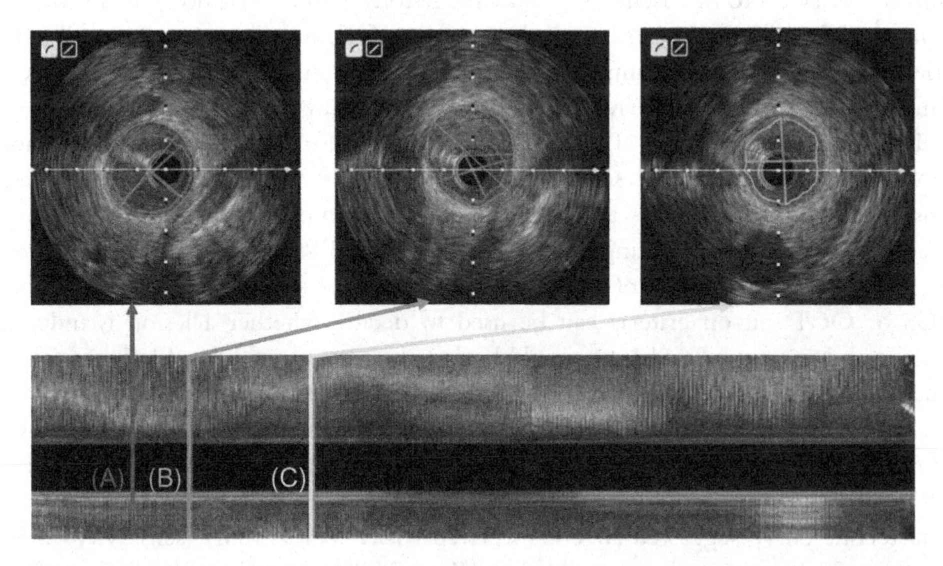

Fig. 1.5 Lesion evaluation. (A) Distal reference; (B) minimal luminal area; and (C) proximal reference images on *top*: cross-sectional view; *image below*: corresponding position seen on full pullback long view.

5.1 Coronary Artery Stenosis Excluding the LMCA

Abizaid described initially that an MLA > 4 mm^2 was associated with a coronary flow reserve ≥2 [57]. Nishioka found that an MLA < 4 mm^2 was associated with ischemia on scintigraphy [58]. In patients in whom a revascularization was deferred on the basis of an MLA > 4 mm^2, at 1 year, the only independent predictors of an MACE were IVUS MLA and area stenosis (AS) [59]. The only independent predictors of target lesion revascularization (TLR) were diabetes mellitus, MLA, and AS. In the 248 lesions with an MLA > 4.0 mm^2, the event rate was only 4.4% and the TLR rate 2.8%. A more recent study evaluated fractional flow reserve (FFR) or IVUS in 167 consecutive patients with intermediate coronary lesions [60]. When the FFR was >0.8, a previously well-defined criterion used for the last 15 years [61, 62], a PCI could be deferred. A threshold of 4 mm^2 was used for the MLA. A total of 94 lesions were in the IVUS guidance arm, and 83 in the FFR group. The primary outcome was defined as MACE at 1 year after the index procedure. The IVUS-guided group underwent revascularization therapy significantly more often (91.5% vs. 33.7%, $p < 0.001$) but at 1 year, no significant difference was found in MACE rates between the two groups (3.6% in FFR-guided PCI vs. 3.2% in IVUS-guided PCI).

In the first comparison between IVUS criteria and FFR, an MLA < 4.0 mm^2 had a sensitivity of 92% but only a specificity of 56% for a measured FFR < 0.75, while a lesion length > 10 mm had a sensitivity of 41% and a specificity of 80% [63]. Many more comparisons in the studies summarized in Table 1.1 have shed further light on the limitation of IVUS to find ischemia-producing lesions with criteria down to an MLA < 2.1 mm^2. Of course, it appears logical that the size of the reference vessel, as well as the patients under investigation, must be taken into account and that smaller cut-offs were found in Asian patients who have smaller coronary arteries.

The plaque burden or 3D criteria have been also investigated [64] without much higher specificity and sensitivity. Today, FFR is the gold standard to guide revascularization, with studies that demonstrated a prognostic value and an improved outcome for patients by treating only lesions with an FFR < 0.8 [65, 66]. It is surely safe to defer an intervention of a lesion with an MLA > 4 mm^2. However, no simple IVUS or OCT cut-off criteria can be used to decide whether a lesion is inducing ischemia: using a cut-off <4 mm^2 would lead to the treating of 50% of lesions without ischemia [67]. There is a good agreement between the MLA measured by IVUS and by OCT in a coronary artery lesion [68]. There is no reason to believe that any OCT MLA cut-off would have a better sensitivity and specificity than IVUS ones to find FFR positive lesions. Out of 62 intermediate lesions in 59 patients, a receiver-operating characteristic curve suggested an OCT-derived MLA < 1.91 mm^2 (sensitivity 93.5%, specificity 77.4%) and percentage lumen AS > 70.0% (sensitivity 96.8%, specificity 83.9%) as the best cut-off values for an FFR < 0.75.

Table 1.1 Summary of the different studies comparing IVUS MLA and ischemia testing

Study	Number of lesions	Ischemia assessed by	Threshold MLA
Abizaid 1998 [57]	86	CFR ≤ 2	4 mm^2
Nishioka 1999 [74]	70	Myocardial scintigraphy	4 mm^2
Takagi 1999 [75]	51	FFR < 0.75	3 mm^2
Briguori 2001 [63]	53	FFR < 0.75	4 mm^2
Lee 2010 [76]	94	FFR < 0.75	2 mm^2 (diameter < 3 mm)
Kang 2011 [77]	236	FFR < 0.8	2.4 mm^2
Ahn 2011 [78]	170	Myocardial scintigraphy	2.1 mm^2
Koo 2011 [79]	267	FFR < 0.8	3 mm^2 (prox IVA), 2.75 mm^2 (mid IVA)
Ben Dor 2012 [80]	205	FFR < 0.8	3.09 mm^2
Kwan 2012 [81]	169	FFR < 0.8	3.03 mm^2 (IVA)
Koh 2012 [82]	77	FFR < 0.8	3.5 mm^2 (ostial lesion)
Gonzalo 2012 [83]	61	FFR < 0.8	2.36 mm^2
Chen 2013 [84]	323	FFR < 0.8	2.97 mm^2
FIRST 2013 [85]	367	FFR < 0.8	3.07 mm^2
VERDICT 2013 [86]	312	FFR < 0.8	2.9 mm^2
Yang 2013 [87]	206	FFR < 0.8	3.2 mm^2 (prox IVA), 2.5 mm^2 (mid IVA)
Kang 2013 [88]	700	FFR < 0.8	2.5 mm^2 (IVA)

5.2 Evaluation of the LMCA

Abizaid has reported that among 122 patients with LMCA not treated at the time of the IVUS assessment, the 1-year event rate was 14% and the IVUS minimal lumen diameter (MLD) was the most important quantitative predictor of cardiac events [69]. In angiographically silent LMCA disease detected by IVUS, MLA was an independent predictor of cardiac events in another series of 107 consecutive patients [70]. Data from registries of patients with intermediate LMCA lesions show similar long-term results when a cut-off for FFR > 0.80 [71] or an the IVUS MLA was used to defer revascularization, compared with patients with an FFR < 0.80 treated with revascularization. The initial Mayo Clinic MLA cut-off [72] was 7.5 mm^2, but the multicentric LITRO study validated an MLA cut-off of 6 mm^2 in 354 patients included in 22 centers. LMCA was revascularized in 91% (152 of 168) of patients with an MLA $< 6\,mm^2$ while it was deferred in 96% (179 of 186) of the patients with an MLA $\geq 6\,mm^2$. After 2 years of follow-up, cardiac death-free survival was 98% in the deferred group versus 95% in the revascularized group ($p = 0.5$), and event-free survival was 87% versus 81%, respectively ($p = 0.3$). This was similar to the 88% event-free survival at 3 years reported by the Mayo Clinic in the patient with a deferred revascularization.

Jasti has demonstrated in Western patients that an IVUS MLA < $6\,mm^2$ in the LMCA is in agreement with an FFR < 0.80 [73]. More recently, a cut-off MLA of $4.8\,mm^2$ was reported in Korean patients, which suggests again smaller MLA cut-offs, and arteries, in Asian patients.

6. INTRAVASCULAR IMAGING ASSESSMENT OF PERCUTANEOUS INTERVENTIONS

6.1 Early Experience After Balloon Angioplasty

We learned from early studies that IVUS preintervention provides accurate determination of vessel size, severity of the lesion, and extension of the atherosclerotic process. The derived lesion composition (amount of calcification and localization, superficial or deep) and lesion length allowed interventional cardiologists to understand/predict the extension of dissections [89], which was important in the era of provisional stenting. Reports showed that IVUS modified the treatment strategy in up to 20% of cases [90]. IVUS could improve angiographic results by safely upsizing the largest balloon for angioplasty once vessel remodeling was taken into account, as demonstrated in the landmark CLOUT registry [91]. Using balloons sized to the EEL diameter, some advocated aggressive PTCA instead of systematic stent implantation [92]. Nowadays, with the advent of drug-eluting stents (DES) that solved the issues of (1) late lumen loss secondary to negative remodeling postballoon angioplasty, as demonstrated by IVUS [93] and (2) the in-stent restenotic process, such provisional angioplasty strategies based on IVUS, or physiological measurements [94], are no longer considered. On the other side, we recognize better the importance of optimal lesion preparation before stenting, using rotational atherectomy, cutting balloons, or newer devices, and IVUS and OCT are very important guidance tools to assess the results of these techniques [95, 96].

6.2 Assessment of Bare Metal Stent

6.2.1 Randomized Studies and Registries

When Colombo reported that he could achieve optimal stent expansion in 96% of his patients with IVUS guidance, and that only antiplatelet therapy could be administered, without any more anticoagulation, it was a game changer, and the first demonstration of the utility of IVUS after bare metal stent (BMS) implantation, with a decrease in hospital stay, vascular complications, and ST [97]. An analysis of 35 trials including more than 14,000 patients revealed that in-stent restenosis related to the proliferation of neointimal hyperplasia was still responsible of approximately 20% TLR at 1 year when using "modern" BMS [98]. To accommodate the intrastent proliferation of "scar" tissue, implantation guidelines were promoting the concept of "the bigger the better." IVUS was the optimal tool to achieve such results, and the so-called MUSIC criteria were introduced with the corresponding registry, requesting a complete apposition of the stent, symmetrically well open, and with a minimal stent area >90% of the average area of the proximal and distal references, or ≥100% of the smallest reference. Fig. 1.5 shows a typical IVUS pullback with the slice showing the MLA and the proximal and distal references. By definition, the reference will be chosen as the most normal looking cross-sections before, and after, the MLA, with ideally a plaque burden less than 40%. The MUSIC criteria were used, for example, in the Strategy for IVUS-Guided PTCA and Stenting (SIPS) trial that randomized consecutive patients to either a strategy of initial IVUS-guided treatment or use of angiographic guidance alone [99]. Optimal angioplasty following CLOUT was attempted with a specific device combining a semicompliant balloon and an IVUS catheter, a possibility offered by the phased array system (Endosonics, now Volcano Philips). Stenting was needed in 50% of the procedures. Although angiographic MLD did not differ significantly after 6 months, IVUS-guided provisional stenting improved 2-year clinical results with a decrease in clinically driven TLR in the IVUS group compared with the standard guidance group (17% vs. 29%, respectively; $p < 0.02$). The different trials that investigated IVUS guidance for BMS are summarized in Table 1.2. Not all of them used the MUSIC criteria. For example, the multicenter CRUISE (Can Routine Ultrasound Influence Stent Expansion) study used an optimization strategy that was not standardized but left to the discretion of the individual institutional practice [100]. Nevertheless, in this study, the incidence of IVUS-guided additional therapy was 36% and led to a decrease in clinically driven target vessel revascularization at the 9-month follow-up in the IVUS-guided group (8.5% vs. 15.3%, $p < 0.05$). The multicenter RESIST (REStenosis after IVUS-guided Stenting) used "only" in the IVUS guidance arm ($n = 79$) a threshold of 80% for the ratio of intrastent CSA to the average of the proximal and distal reference lumen CSA, and yet they could demonstrate at the 18-month follow-up a near significant trend for a higher revascularization rate in the control group (odds ratio [OR]: 1.9 (95% CI: [0.97; 2.4]). On the other side, there was an important disappointment when the results

Table 1.2 Clinical benefits of an IVUS-guided strategy for BMS implantation

Study	Number of patients	TVR	Death/myocardial infarct
SIPS 2002 [99]	43	+	–
CRUISE 2000 [100]	525	+	–
Choi and Al 2001 [101]	278	+	–
AVID 2009 [102]	744	+	–
DIPOL 2007 [103]	163	+	–
RESIST 1997 [104, 105]	155	+	–
TULIP 2003 [106]	144	+	–
Gaster 2003 [107]	103	+	–
OPTICUS 2001 [108]	548	–	–

of the large OPTICUS trial, based on the MUSIC criteria, did not show a significant difference between the angiographic- and IVUS-guidance groups.

6.2.2 Metaanalysis of Available Studies

Because there were only a few underpowered studies comparing the clinical efficacy of IVUS-guided versus angiographically guided BMStenting and it was not clear whether the significant lower restenosis rate for the vast majority of them (see Table 1.2) would translate into a substantial clinical advantage, a metaanalysis was conducted by Casella et al. [109]. Five randomized trials and four registries with a total of 2972 patients were analyzed. At 6 months, the primary endpoint (occurrence of death or nonfatal MI) was similar for both strategies (4.1% with angio-guided compared to 4.5% with IVUS-guided stenting; OR = 1.13; 95% CI: 0.79–1.61; $p = 0.5$). Pooled data of individual cardiac endpoints show a 38% reduced probability of target vessel revascularization (OR = 0.62; 95% CI: 0.49–0.78; $p < 0.00003$) in favor of IVUS-guided stenting, while death, nonfatal MI, or CABG were equally distributed in both groups. These results were confirmed by a more recent analysis, including only 7 studies and 2193 patients [110].

6.3 Assessment of DES

The antiproliferative efficacy of DES offers a dramatic decrease of the incidence of instent restenosis and secondary TLR: 5% on average from the network analysis mentioned previously [98]. However, a risk of late and very late ST remains a major safety concern.

6.3.1 Randomized Studies and Registries

Fewer randomized studies comparing IVUS versus angiographic guidance have been conducted, but many large registries are available, as summarized in Table 1.3.

In order to minimize stent implantation failures, the lessons learned from more than two decades of IVUS can be summarized as: (1) "the bigger, the better" regarding MLA and (2) "the lower, the better" regarding the stent-edge plaque burden. However, the lack of robust randomized trials, some controversial results among the trials available, and immediate additional procedural cost (the IVUS catheter is not reimbursed in Europe) might explain why IVUS guidance is used in less than 5% of PCI [131]. Earlier registries, conducted with first-generation DES, suggested that IVUS guidance may reduce the need for repeat revascularization [112] and even late cardiac death in PCI of LMCA [113]. The disappointing results of the first randomized trial HOME DES IVUS [115] in 2010 and later of the AVIO trial [123] published by Colombo's group in 2013, who failed to prove clinical benefit of IVUS-guided PCI in complex lesions, relaunched the debate about the real added value of routine IVUS guidance. In the latest randomized trial IVUS-XPL [129] IVUS-guided everolimus long stent implantation was associated with a lower rate of MACE, primarily due to a lower risk of TLR

Table 1.3 Studies of DES with IVUS guidance

Study	Type	Sample size[a]	Follow-up (months)	Specificities	IVUS benefit
Agostini 2005 [111]	Registry	24/34	14	Left main	–
Roy 2008 [112]	Registry	884/884	12	–	+
MAIN-COMPARE 2009 [113]	Registry	145/145	36	Left main	+
Kim 2010 [114]	Registry	473/285	48	Bifurcation	+
HOME-DES 2010 [115]	Randomized	105/105	18	Type B2/C	–
MATRIX 2011 [116]	Registry	548/548	24	–	+
COBIS 2011 [117]	Registry	487/487	36	Bifurcation	+
Youn 2011 [118]	Registry	125/216	36	Primary PCI	–
ADAPT-DES 2012 [119]	Registry	3349/5234	12	–	+
Chen 2013 [120]	Registry	324/304	12	Bifurcation	+
EXCELLENT 2013 [121]	Registry	463/463	12	–	–
Hur 2013 [122]	Registry	2765DES/1816	36	–	+
AVIO 2013 [123]	Randomized	142/142	24	Complex lesions	–
RESET 2013 [124]	Randomized	269/274	12	Long lesions	+[b]
IRIS-DES 2013 [125]	Registry	1616/1628	24	–	+ [c]
Yoon 2013 [126]	Registry	662/912	12	Short lesions	–[d]
IVUS-TRONCO-ICP 2014 [127]	Registry	505/505	36	Left main	+
Magalhaes [128]	Registry	1315/910 and 1270/804[e]	12	–	+
IVUS-XPL 2015 [129]	Randomized	700/700	12	Long lesions	+
CREDO-Kyoto AMI 2016 [130]	Registry	932/2096	60	Primary PCI	–[f]

[a] The number of patients with IVUS/with angiography alone.
[b] In the intention-to-treat analysis, there was no statistically significant difference, but well in a per-protocol analysis according to actual IVUS usage.
[c] In PCI with long stent implantation.
[d] Subanalysis of the RESET study.
[e] Respectively for the two subgroups: complete and incomplete revascularizations.
[f] IVUS-guided PCI was associated with significantly lower incidences of TVR (22% vs. 27%, $p < 0.001$) and definite ST (1.2% vs. 3.1%, $p = 0.003$), but no more after adjusting for confounders.

(2.5% vs. 5%; $p = 0.02$). IVUS criteria for stent optimization after PCI were defined as a minimal lumen cross-sectional area greater than the lumen cross-sectional area at the distal reference segments. In the post hoc analysis, among the patients within the IVUS-guided stent group, the patients who did not meet the IVUS criteria had a significantly higher incidence of MACE compared with those meeting the IVUS criteria for stent optimization (4.6% vs. 1.5%, respectively; HR: 0.31 [95% CI: 0.11–0.86], $p = 0.02$). These data support strongly IVUS guidance in such complex lesions.

The Assessment of Dual Antiplatelet Therapy with Drug-Eluting Stents (ADAPT-DES) study was a large-scale, prospective, multicenter study initiated after the tsunami created by the reported increased of ST rate with DES implantation. The relationship between platelet reactivity and other clinical and procedural variables with subsequent ST and adverse clinical events was the main focus. A prespecified substudy evaluated IVUS guidance ($n = 3349$) versus angiographic guidance ($n = 5234$) [119]. IVUS was associated with reduced 1-year rates of ST (0.6% vs. 1.0%; $p = 0.003$), myocardial infarction (2.5% vs. 3.7%; $p = 0.004$), and MACE (3.1% vs. 4.7%; $p = 0.002$). These benefits were especially evident in patients with ACSs, contrasting with the latest results from the Japanese registry CREDO-Kyoto AMI [130], where no significant improvement were observed in the 932 IVUS-guided primary PCI compared to the 2096 angiography-guided ones after adjustment for confounders. Recent European registries confirmed the lower rate of TLR with IVUS guidance [128] and reported in addition improvement in cardiac death and MI rates when IVUS guidance is performed in LMCA PCI [127]. IVUS guidance was also associated with reduced in hospital mortality (OR: 0.65, 9% CI: 0.52–0.83; $p < 0.001$) compared to conventional angiography-guided PCI in the largest American registry as of today: 401,571 PCIs were identified between 2008 and 2011, of which 377,096 were angiography-guided and 24,475 were IVUS-guided [132]. These results were observed even in subgroup of patients with AMI and/or shock and those with a higher comorbidity burden.

With a much higher resolution, OCT has been the optimal research tool to compare in clinical studies the minimal amount of in-stent neointimal thickness between DES, or the incidence of strut malapposition and uncovered struts at follow-up [133]. Fewer OCT-guidance studies are available. Prati reported the feasibility of performing FD-OCT imaging in patients with ambiguous lesions and/or addressing the adequacy of stent deployment [134]. Similar to IVUS studies, OCT findings led to additional interventions in 24 out of 74 patients (32%). The same group reported in 2012 the results of the CLI-OPCI study where 335 patients undergoing PCI with angiographic plus OCT guidance were compared with 335 matched patients undergoing PCI with angiographic only guidance [135]. OCT guidance was associated with a significantly lower risk of cardiac death or MI at 1 year by multivariable analysis adjusting for baseline and procedural differences between the groups (OR: 0.49 [0.25–0.96], $p = 0.037$) and at propensity-score adjusted analyses. Based on that study, a retrospective analysis of 1002 lesions (832 patients) with end-procedural OCT assessment revealed that MACE were independently increased by: (1) suboptimal stent implantation (in-stent minimum lumen area $< 4.5\,\text{mm}^2$ (HR: 1.64; $p = 0.040$), (2) dissection $> 200\,\mu\text{m}$ at the distal stent edge (HR: 2.54; $p = 0.004$), and (3) reference lumen area $< 4.5\,\text{mm}^2$ at either distal (HR: 4.65; $p < 0.001$) or proximal (HR: 5.73; $p < 0.001$) stent edges. Conversely, in-stent minimum lumen area/mean reference lumen area $< 70\%$, stent malapposition $> 200\,\mu\text{m}$, intrastent plaque/thrombus protrusion $> 500\,\mu\text{m}$, and dissection $> 200\,\mu\text{m}$ at the proximal stent edge were not associated with worse outcomes [136].

IVUS guidance versus OCT guidance was studied in a total of 70 patients by Habara [137]. Device and clinical success rates were similar. However, because of a significantly lower visibility of vessel border with OCT, minimum and mean stent area and focal and diffuse stent expansion were smaller, and the frequency of significant residual reference segment stenosis at the proximal edge was higher in the OCT group ($p < 0.05$). Incomplete apposed struts in both groups were similar. The results of the OPINION study presented by Dr. T Kubo from Wakayama University (Japan) at the EuroPCR 2016 meeting (Paris, May 17, 2016) show that optimal frequency domain imaging OCT (OFDI) to guide PCI with second-generation DES in 800 patients achieves equivalent clinical and angiographic outcomes to IVUS-guided PCI at 12 months. The noninferiority analysis demonstrates a similarly low rate of target vessel revascularization, cardiac death, target vessel-related MI, and clinically driven target vessel revascularization at 12 months in patients undergoing OFDI-guided PCI or IVUS-guided PCI (5.2% and 4.9%, respectively). At the same meeting, ILUMIEN I was also presented by Dr. Wijns from Aalst (Belgium). FFR was performed in 418 subjects, followed by OCT. Physicians changed their strategy based on the pre-PCI OCT in more than 50% of the cases (for the stent length, stent diameter, etc.). After PCI, OCT influenced optimization in 27% of cases. Post hoc analysis revealed a lower rate of periprocedural myocardial infarct with combined OCT guidance pre- and post-PCI. Overall, 30-day and 1-year MACE were similar, with very few MI and ST.

Recently, a specific study analyzed the rate and improvement of strut apposition acutely, in order to improve at 6-month stent strut coverage [138]. Jensen et al. demonstrated that in patients with non-ST-segment-elevation myocardial infarction, at 6-month follow-up, OCT guidance decreases the proportion of uncovered struts compared to angiographic guidance (4.3% vs. 9.0%, respectively, $p < 0.01$). OCT-guided patients had also significantly more completely covered stents (17.5% vs. 2.2%, $p = 0.02$). The randomized ILUMIEN III trial is currently comparing the potential for angiography, IVUS, and OCT to guide stent implantation.

While DES is today the standard-of-care therapy, two major modes of failures remain: ST (<1% at 1 year and ~0.2–0.4% per year thereafter) and in-stent restenosis (~5–10%). Patient-, stent-type-, and procedure-related factors each play a role [139]. We, and others, have demonstrated the issues of underexpansion, incomplete lesion coverage, and stent fracture that can be better imaged by IVUS or OCT than on an angiogram [140–143]. Not all DES are made the same and the latest ones fade away: they are self-degrading after their useful function has been served, healing the dissections, reducing intimal proliferation, and blocking negative remodeling. OCT has been extensively used to study these stents [144]. Some have advocated that these bioresorbable scaffolds were the moonshot cure to treat coronary artery stenosis. Although they were noninferior to new generation metallic DES in large randomized

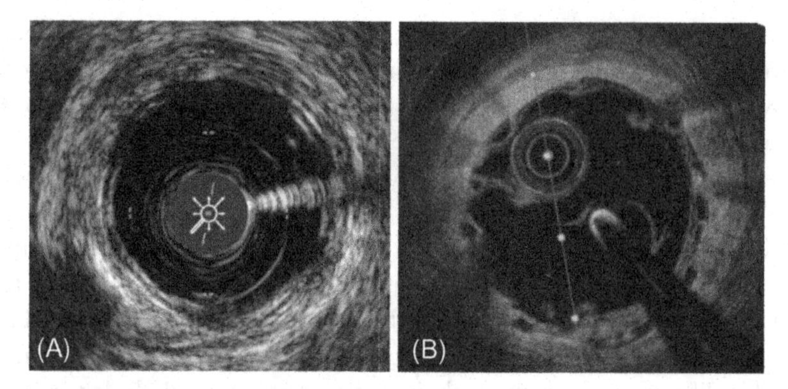

Fig. 1.6 In vivo cross-sectional view of a bioresorbable stent. (A) 60 MHz IVUS during lumen flushing with contrast. The struts are seen as small double-layer hyperechoic structure without ultrasound shadowing since the strut is made of polylactic acid. (B) Corresponding cross-section from the OCT pullback where struts appear as *black squares.*

trials [145], they appear to have an increased risk of subacute thrombosis in recent registries and metaanalysis [146–148]. Improved procedural technique with more aggressive plaque modification before implantation, routine high-pressure noncompliant balloon postdilatation to ensure adequate scaffold expansion, and more frequent use of intravascular imaging to optimize lesion coverage and scaffold dimensions are recommended to reduce thrombosis rates. Fig. 1.6 shows that OCT offers a very detailed assessment of such bioresorbable stent, with a unique appearance of the struts compared to a metallic stent. On the left side, a 60 MHz IVUS cross-section at the same level in the stent shows that high-resolution imaging of a bioresorbable stent is also possible with a high-frequency broadband ultrasound system.

6.3.2 Metaanalysis of Available Studies

Several metaanalysis of IVUS versus angiography DES-guided PCIs reported a reduced rate of cardiac death, myocardial infarction, ST, and TLR [149–152]. In the latest analysis of Steinvil [153], 25 studies were included with 31,283 patients, of whom 3192 patients were enrolled in 7 RCTs (including 2 CTO randomized trials [154, 155]). Results were significantly in favor of IVUS-guided group with lower rate of MACE, death, myocardial infarction, ST, target lesion, and vessel revascularization. However, in a separate analysis of RCTs, a favorable result for IVUS-guided DES implantation was found only for MACE (OR: 0.66, 95% CI: 0.52–0.84, $p = 0.001$), essentially due to TLR and TVR lower rates.

7. FUTURE DEVELOPMENTS AND FINAL WORD

The reviewed data support the evidence that intravascular guiding is valuable to decrease iterative revascularization procedures, but also confirm the need for new larger randomized trials to understand in which lesions and for which patients and clinical situations IVUS and/or OCT should be used to improve PCI outcomes. Until that day, image guidance for PCI remains recommended only in "selected" patients without clear criteria of selection [156].

We have tried to show the respective advantages of IVUS and OCT, and we believe that these two imaging methods should be complementary, not competing modalities. The very high resolution offered by OCT should be combined with the larger field of view obtained by IVUS, as demonstrated in Fig. 1.7. It is indeed impractical in a busy cath lab and without any reimbursement to use sequentially an IVUS, then an OCT catheter, and issues for image coregistration would have to be solved. However, the in vitro proof of concept of a combined catheter with both OCT and IVUS modalities has been reported as early as 2012 [157], and more recently, another catheter for OCT and near-infrared autofluorescence imaging was successfully used in patients [158]. Multimodality imaging must be the way forward for better plaque characterization and progression/regression studies, as well as interventional guidance.

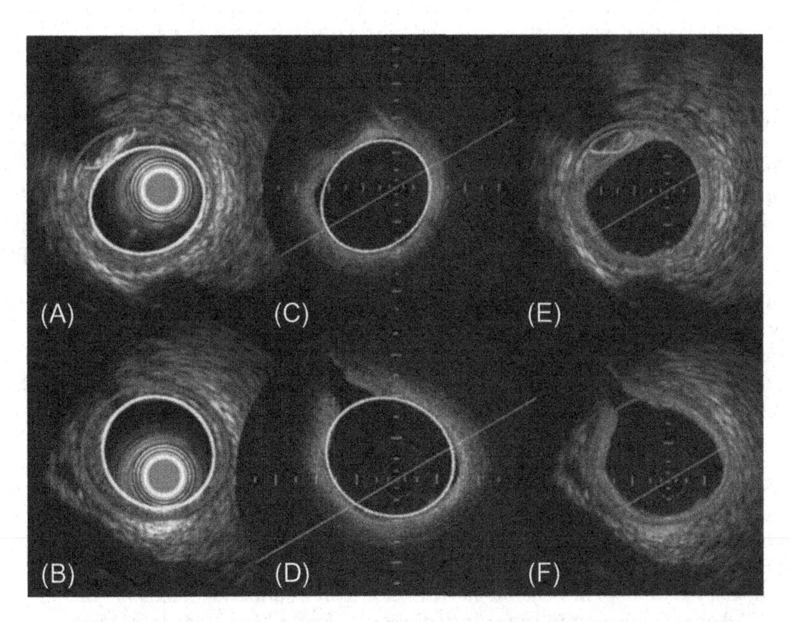

Fig. 1.7 In vitro acquisition of a human coronary artery with a 40 MHz IVUS at a level of a small calcified plaque (A) and side branch (B) and corresponding OCT cross-section extracted from a second pullback (C) and (D). (E) Image fusion of A and C using an algorithm developed in collaboration with G Unal (Sabanci University, Turkey). (F) Image fusion of B and D.

REFERENCES

[1] Edler I, Hertz CH. The use of ultrasonic reflectoscope for the continuous recording of the movements of heart walls. Clin Physiol Funct Imaging 2004;24:118–36.
[2] Bom N, Lancée CT, van Egmond FC. An ultrasonic intracardiac scanner. Ultrasonics 1972;10:72–6.
[3] Yock P. Catheter apparatus, system and method for intravascular two-dimensional ultrasonography. U.S. Patent 5,000,185, issued March 19, 1991.
[4] Bom N, Carlier SG, van der Steen AF, Lancée CT. Intravascular scanners. Ultrasound Med Biol 2000;26(Suppl. 1):S6–9.
[5] Hibi K, Takagi A, Zhang X, Teo T-J, Bonneau HN, Yock PG, et al. Feasibility of a novel blood noise reduction algorithm to enhance reproducibility of ultra-high-frequency intravascular ultrasound images. Circulation 2000;102:1657–63.
[6] Velican D, Velican C. Comparative study on age-related changes and atherosclerotic involvement of the coronary arteries of male and female subjects up to 40 years of age. Atherosclerosis 1981;38:39–50.
[7] Huang D, Swanson EA, Lin CP, Schuman JS, Stinson WG, Chang W, et al. Optical coherence tomography. Science 1991;254:1178–81.
[8] Barlis P, Schmitt JM. Current and future developments in intracoronary optical coherence tomography imaging. EuroIntervention 2009;4:529–33.
[9] Di Mario C, Görge G, Peters R, Kearney P, Pinto F, Hausmann D, et al. Clinical application and image interpretation in intracoronary ultrasound. Study Group on Intracoronary Imaging of the Working Group of Coronary Circulation and of the Subgroup on Intravascular Ultrasound of the Working Group of Echocardiography of the European Society of Cardiology. Eur Heart J 1998;19:207–29.
[10] Mintz GS, Nissen SE, Anderson WD, Bailey SR, Erbel R, Fitzgerald PJ, et al. American College of Cardiology clinical expert consensus document on standards for acquisition, measurement and reporting of intravascular ultrasound studies (IVUS). A report of the American College of Cardiology task force on clinical expert consensus documents. J Am Coll Cardiol 2001;37:1478–92.
[11] Takarada S, Imanishi T, Liu Y, Ikejima H, Tsujioka H, Kuroi A, et al. Advantage of next-generation frequency-domain optical coherence tomography compared with conventional time-domain system in the assessment of coronary lesion. Catheter Cardiovasc Interv 2010;75:202–6.
[12] Prati F, Guagliumi G, Mintz GS, Costa M, Regar E, Akasaka T, et al. Expert review document part 2: methodology, terminology and clinical applications of optical coherence tomography for the assessment of interventional procedures. Eur Heart J 2012;33:2513–20.
[13] Hausmann D, Erbel R, Alibelli-Chemarin MJ, Boksch W, Caracciolo E, Cohn JM, et al. The safety of intracoronary ultrasound. A multicenter survey of 2207 examinations. Circulation 1995;91:623–30.
[14] Batkoff BW, Linker DT. Safety of intracoronary ultrasound: data from a multicenter European registry. Cathet Cardiovasc Diagn 1996;38:238–41.
[15] Stone GW, Maehara A, Lansky AJ, de Bruyne B, Cristea E, Mintz GS, et al. A prospective natural-history study of coronary atherosclerosis. N Engl J Med 2011;364:226–35.
[16] Taniwaki M, Radu MD, Garcia-Garcia HM, Heg D, Kelbæk H, Holmvang L, et al. Long-term safety and feasibility of three-vessel multimodality intravascular imaging in patients with ST-elevation myocardial infarction: the IBIS-4 (integrated biomarker and imaging study) substudy. Int J Cardiovasc Imaging 2015;31:915–26.
[17] van der Sijde JN, Karanasos A, van Ditzhuijzen NS, Okamura T, van Geuns R-J, Valgimigli M, et al. Safety of optical coherence tomography in daily practice: a comparison with intravascular ultrasound. Eur Heart J Cardiovasc Imaging 2016. http://dx.doi.org/10.1093/ehjci/jew037 [Epub ahead of print].
[18] Topol EJ, Nissen SE. Our preoccupation with coronary luminology the dissociation between clinical and angiographic findings in ischemic heart disease. Circulation 1995;92:2333–42.
[19] Ross R. Atherosclerosis—an inflammatory disease. N Engl J Med 1999;340:115–26.
[20] Glagov S, Weisenberg E, Zarins CK, Stankunavicius R, Kolettis GJ. Compensatory enlargement of human atherosclerotic coronary arteries. N Engl J Med 1987;316:1371–5.
[21] Ge J, Erbel R, Gerber T, Görge G, Koch L, Haude M, et al. Intravascular ultrasound imaging of angiographically normal coronary arteries: a prospective study in vivo. Br Heart J 1994;71:572–8.

[22] Sheet D, Karamalis A, Eslami A, Noël P, Virmani R, Nakano M, et al. Hunting for necrosis in the shadows of intravascular ultrasound. Comput Med Imaging Graph 2014;38:104–12.
[23] Otsuka F, Joner M, Prati F, Virmani R, Narula J. Clinical classification of plaque morphology in coronary disease. Nat Rev Cardiol 2014;11:379–89.
[24] Zir LM, Miller SW, Dinsmore RE, Gilbert JP, Harthorne JW. Interobserver variability in coronary angiography. Circulation 1976;53:627–32.
[25] Nissen SE. Pathobiology, not angiography, should guide management in acute coronary syndrome/non-ST-segment elevation myocardial infarction: the non-interventionist's perspective. J Am Coll Cardiol 2003;41:103S–12S.
[26] Sandfort V, Lima JAC, Bluemke DA. Noninvasive imaging of atherosclerotic plaque progression: status of coronary computed tomography angiography. Circ Cardiovasc Imaging 2015;8:e003316.
[27] Katouzian A, Angelini ED, Carlier SG, Suri JS, Navab N, Laine AF. A state-of-the-art review on segmentation algorithms in intravascular ultrasound (IVUS) images. IEEE Trans Inf Technol Biomed 2012;16:823–34.
[28] Balocco S, Gatta C, Ciompi F, Wahle A, Radeva P, Carlier S, et al. Standardized evaluation methodology and reference database for evaluating IVUS image segmentation. Comput Med Imaging Graph 2014;38:70–90.
[29] Mintz GS, Garcia-Garcia HM, Nicholls SJ, Weissman NJ, Bruining N, Crowe T, et al. Clinical expert consensus document on standards for acquisition, measurement and reporting of intravascular ultrasound regression/progression studies. EuroIntervention 2011;6(9):1123–30.
[30] Nissen SE, Tuzcu E, Schoenhagen P, Brown B, Ganz P, Vogel R, et al. Effect of intensive compared with moderate lipid-lowering therapy on progression of coronary atherosclerosis: a randomized controlled trial. JAMA 2004;291:1071–80.
[31] Okazaki S, Yokoyama T, Miyauchi K, Shimada K, Kurata T, Sato H, et al. Early statin treatment in patients with acute coronary syndrome demonstration of the beneficial effect on atherosclerotic lesions by serial volumetric intravascular ultrasound analysis during half a year after coronary event: the ESTABLISH study. Circulation 2004;110:1061–8.
[32] Nissen S, Tsunoda T, Tuzcu E, Schoenhagen P, Cooper C, Yasin M, et al. Effect of recombinant ApoA-I Milano on coronary atherosclerosis in patients with acute coronary syndromes: a randomized controlled trial. JAMA 2003;290:2292–300.
[33] Nissen SE, Nicholls SJ, Sipahi I, Libby P, Raichlen J, Ballantyne C, et al. Effect of very high-intensity statin therapy on regression of coronary atherosclerosis: the asteroid trial. JAMA 2006;295:1556–65.
[34] Kawasaki M, Sano K, Okubo M, Yokoyama H, Ito Y, Murata I, et al. Volumetric quantitative analysis of tissue characteristics of coronary plaques after statin therapy using three-dimensional integrated backscatter intravascular ultrasound. J Am Coll Cardiol 2005;45:1946–53.
[35] Räber L, Taniwaki M, Zaugg S, Kelbæk H, Roffi M, Holmvang L, et al. Effect of high-intensity statin therapy on atherosclerosis in non-infarct-related coronary arteries (IBIS-4): a serial intravascular ultrasonography study. Eur Heart J 2015;36:490–500.
[36] Sathyanarayana S, Carlier S, Li W, Thomas L. Characterisation of atherosclerotic plaque by spectral similarity of radiofrequency intravascular ultrasound signals. EuroIntervention 2009;5:133–9.
[37] Haruta H, Hiro T, Mitsumata M, Takayama T, Sudo M, Li Y, et al. Stabilization of atherosclerotic plaque by pitavastatin in watanabe heritable hyperlipidemic rabbits: a serial tissue-characterizing intravascular ultrasound study. J Cardiol 2016;67:205–11.
[38] Thim T, Hagensen MK, Wallace-Bradley D, Granada JF, Kaluza GL, Drouet L, et al. Unreliable assessment of necrotic core by virtual histology intravascular ultrasound in porcine coronary artery disease. Circ Cardiovasc Imaging 2010;3:384–91.
[39] Katouzian A, Sathyanarayana S, Baseri B, Konofagou EE, Carlier SG. Challenges in atherosclerotic plaque characterization with intravascular ultrasound (IVUS): from data collection to classification. IEEE Trans Inf Technol Biomed 2008;12:315–27.
[40] LaRosa JC, Grundy SM, Waters DD, Shear C, Barter P, Fruchart J-C, et al. Intensive lipid lowering with atorvastatin in patients with stable coronary disease. N Engl J Med 2005;352:1425–35.
[41] Nissen SE, Tuzcu E, Libby P, Thompson PD, Ghali M, Garza D, et al. Effect of antihypertensive agents on cardiovascular events in patients with coronary disease and normal blood pressure: the CAMELOT study: a randomized controlled trial. JAMA 2004;292:2217–25.

[42] Tsujita K, Sugiyama S, Sumida H, Shimomura H, Yamashita T, Yamanaga K, et al. Impact of dual lipid-lowering strategy with ezetimibe and atorvastatin on coronary plaque regression in patients with percutaneous coronary intervention: the multicenter randomized controlled PRECISE-IVUS trial. J Am Coll Cardiol 2015;66:495–507.
[43] D'Ascenzo F, Agostoni P, Abbate A, Castagno D, Lipinski MJ, Vetrovec GW, et al. Atherosclerotic coronary plaque regression and the risk of adverse cardiovascular events: a meta-regression of randomized clinical trials. Atherosclerosis 2013;226:178–85.
[44] Kume T, Akasaka T, Kawamoto T, Watanabe N, Toyota E, Neishi Y, et al. Assessment of coronary intima-media thickness by optical coherence tomography: comparison with intravascular ultrasound. Circ J 2005;69:903–7.
[45] Dong L, Maehara A, Nazif TM, Pollack AT, Saito S, Rabbani LE, et al. Optical coherence tomographic evaluation of transplant coronary artery vasculopathy with correlation to cellular rejection. Circ Cardiovasc Interv 2014;7:199–206.
[46] Kobashigawa JA, Tobis JM, Starling RC, Tuzcu EM, Smith AL, Valantine HA, et al. Multicenter intravascular ultrasound validation study among heart transplant recipients: outcomes after five years. J Am Coll Cardiol 2005;45:1532–7.
[47] Di Vito L, Agozzino M, Marco V, Ricciardi A, Concardi M, Romagnoli E, et al. Identification and quantification of macrophage presence in coronary atherosclerotic plaques by optical coherence tomography. Eur Heart J Cardiovasc Imaging 2015;16:807–13.
[48] Xu C, Schmitt JM, Carlier SG, Virmani R. Characterization of atherosclerosis plaques by measuring both backscattering and attenuation coefficients in optical coherence tomography. J Biomed Opt 2008;13:34003.
[49] Tobis JM, Mallery J, Mahon D, Lehmann K, Zalesky P, Griffith J, et al. Intravascular ultrasound imaging of human coronary arteries in vivo. Analysis of tissue characterizations with comparison to in vitro histological specimens. Circulation 1991;83:913–26.
[50] Zhang X, McKay CR, Sonka M. Tissue characterization in intravascular ultrasound images. IEEE Trans Med Imaging 1998;17:889–99.
[51] Seabra JC, Ciompi F, Pujol O, Mauri J, Radeva P, Sanches J. Rayleigh mixture model for plaque characterization in intravascular ultrasound. IEEE Trans Biomed Eng 2011;58:1314–24.
[52] Katouzian A, Karamalis A, Sheet D, Konofagou E, Baseri B, Carlier SG, et al. Iterative self-organizing atherosclerotic tissue labeling in intravascular ultrasound images and comparison with virtual histology. IEEE Trans Biomed Eng 2012;59:3039–49.
[53] Kubo T, Maehara A, Mintz GS, Doi H, Tsujita K, Choi S-Y, et al. The dynamic nature of coronary artery lesion morphology assessed by serial virtual histology intravascular ultrasound tissue characterization. J Am Coll Cardiol 2010;55:1590–97.
[54] Motreff P, Rioufol G, Finet G. Seventy-four-month follow-up of coronary vulnerable plaques by serial gray-scale intravascular ultrasound. Circulation 2012;126:2878–9.
[55] Uemura S, Ishigami K, Soeda T, Okayama S, Sung JH, Nakagawa H, et al. Thin-cap fibroatheroma and microchannel findings in optical coherence tomography correlate with subsequent progression of coronary atheromatous plaques. Eur Heart J 2012;33:78–85.
[56] Arbab-Zadeh A, Fuster V. The myth of the "vulnerable plaque": transitioning from a focus on individual lesions to atherosclerotic disease burden for coronary artery disease risk assessment. J Am Coll Cardiol 2015;65:846–55.
[57] Abizaid A, Mintz GS, Pichard AD, Kent KM, Satler LF, Walsh CL, et al. Clinical, intravascular ultrasound, and quantitative angiographic determinants of the coronary flow reserve before and after percutaneous transluminal coronary angioplasty. Am J Cardiol 1998;82:423–8.
[58] Nishioka T, Amanullah AM, Luo H, Berglund H, Kim CJ, Nagai T, et al. Clinical validation of intravascular ultrasound imaging for assessment of coronary stenosis severity: comparison with stress myocardial perfusion imaging. J Am Coll Cardiol 1999;33:1870–8.
[59] Abizaid AS, Mintz GS, Mehran R, Abizaid A, Lansky AJ, Pichard AD, et al. Long-term follow-up after percutaneous transluminal coronary angioplasty was not performed based on intravascular ultrasound findings: importance of lumen dimensions. Circulation 1999;100: 256–61.

[60] Nam CW, Yoon HJ, Cho YK, Park HS, Kim H, Hur SH, et al. Outcomes of percutaneous coronary intervention in intermediate coronary artery disease. JACC Cardiovasc Interv 2010;3:812–7.
[61] Zimmermann FM, Ferrara A, Johnson NP, van Nunen LX, Escaned J, Albertsson P, et al. Deferral vs. performance of percutaneous coronary intervention of functionally non-significant coronary stenosis: 15-year follow-up of the DEFER trial. Eur Heart J 2015;36:3182–8.
[62] Johnson NP, Tóth GG, Lai D, Zhu H, Açar G, Agostoni P, et al. Prognostic value of fractional flow reserve: linking physiologic severity to clinical outcomes. J Am Coll Cardiol 2014;64:1641–54.
[63] Briguori C, Anzuini A, Airoldi F, Gimelli G, Nishida T, Adamian M, et al. Intravascular ultrasound criteria for the assessment of the functional significance of intermediate coronary artery stenoses and comparison with fractional flow reserve. Am J Cardiol 2001;87:136–41.
[64] Takayama T, Hodgson JM. Prediction of the physiologic severity of coronary lesions using 3D IVUS: validation by direct coronary pressure measurements. Catheter Cardiovasc Interv 2001;53:48–55.
[65] van Nunen LX, Zimmermann FM, Tonino PAL, Barbato E, Baumbach A, Engstrøm T, et al. Fractional flow reserve versus angiography for guidance of PCI in patients with multivessel coronary artery disease (FAME): 5-year follow-up of a randomised controlled trial. Lancet 2015;386:1853–60.
[66] De Bruyne B, Fearon WF, Pijls NHJ, Barbato E, Tonino P, Piroth Z, et al. Fractional flow reserve-guided PCI for stable coronary artery disease. N Engl J Med 2014;371:1208–17.
[67] Mintz GS. Clinical utility of intravascular imaging and physiology in coronary artery disease. J Am Coll Cardiol 2014;64:207–22.
[68] Bezerra HG, Attizzani GF, Sirbu V, Musumeci G, Lortkipanidze N, Fujino Y, et al. Optical coherence tomography versus intravascular ultrasound to evaluate coronary artery disease and percutaneous coronary intervention. JACC Cardiovasc Interv 2013;6:228–36.
[69] Abizaid AS, Mintz GS, Abizaid A, Mehran R, Lansky AJ, Pichard AD, et al. One-year follow-up after intravascular ultrasound assessment of moderate left main coronary artery disease in patients with ambiguous angiograms. J Am Coll Cardiol 1999;34:707–15.
[70] Ricciardi MJ, Meyers S, Choi K, Pang JL, Goodreau L, Davidson CJ. Angiographically silent left main disease detected by intravascular ultrasound: a marker for future adverse cardiac events. Am Heart J 2003;146:507–12.
[71] Hamilos M, Muller O, Cuisset T, Ntalianis A, Chlouverakis G, Sarno G, et al. Long-term clinical outcome after fractional flow reserve-guided treatment in patients with angiographically equivocal left main coronary artery stenosis. Circulation 2009;120:1505–12.
[72] Fassa A, Wagatsuma K, Higano S, Mathew V, Barsness G, Lennon R, et al. Intravascular ultrasound-guided treatment for angiographically indeterminate left main coronary artery disease: a long-term follow-up study. J Am Coll Cardiol 2005;45:204–11.
[73] Jasti V. Correlations between fractional flow reserve and intravascular ultrasound in patients with an ambiguous left main coronary artery stenosis. Circulation 2004;110:2831–6.
[74] Nishioka T, Amanullah AM, Luo H, Berglund H, Kim CJ, Nagai T, et al. Clinical validation of intravascular ultrasound imaging for assessment of coronary stenosis severity: comparison with stress myocardial perfusion imaging. J Am Coll Cardiol 1999;33:1870–8.
[75] Takagi A, Tsurumi Y, Ishii Y, Suzuki K, Kawana M, Kasanuki H. Clinical potential of intravascular ultrasound for physiological assessment of coronary stenosis: relationship between quantitative ultrasound tomography and pressure-derived fractional flow reserve. Circulation 1999;100:250–5.
[76] Lee C-H, Tai B-C, Soon C-Y, Low AF, Poh K-K, Yeo T-C, et al. New set of intravascular ultrasound-derived anatomic criteria for defining functionally significant stenoses in small coronary arteries (results from Intravascular Ultrasound Diagnostic Evaluation of Atherosclerosis in Singapore [IDEAS] study). Am J Cardiol 2010;105:1378–84.
[77] Kang S-J, Lee J-Y, Ahn J-M, Song HG, Kim W-J, Park D-W, et al. Intravascular ultrasound-derived predictors for fractional flow reserve in intermediate left main disease. JACC Cardiovasc Interv 2011;4:1168–74.
[78] Ahn J-M, Kang S-J, Mintz GS, Oh J-H, Kim W-J, Lee J-Y, et al. Validation of minimal luminal area measured by intravascular ultrasound for assessment of functionally significant coronary stenosis comparison with myocardial perfusion imaging. JACC Cardiovasc Interv 2011;4:665–71.
[79] Koo B-K, Yang H-M, Doh J-H, Choe H, Lee S-Y, Yoon C-H, et al. Optimal intravascular ultrasound criteria and their accuracy for defining the functional significance of intermediate coronary stenoses of different locations. JACC Cardiovasc Interv 2011;4:803–11.

[80] Ben-Dor I, Torguson R, Deksissa T, Bui AB, Xue Z, Satler LF, et al. Intravascular ultrasound lumen area parameters for assessment of physiological ischemia by fractional flow reserve in intermediate coronary artery stenosis. Cardiovasc Revasc Med 2012;13:177–82.
[81] Kwan TW, Yang S, Xu B, Chen J, Xu T, Ye F, et al. Optimized quantitative angiographic and intravascular ultrasound parameters predicting the functional significance of single de novo lesions in the left anterior descending artery. Chin Med J (Engl) 2012;125:4249–53.
[82] Koh J-S, Koo B-K, Kim J-H, Yang H-M, Park K-W, Kang H-J, et al. Relationship between fractional flow reserve and angiographic and intravascular ultrasound parameters in ostial lesions. JACC Cardiovasc Interv 2012;5:409–15.
[83] Gonzalo N, Gonzalo N, Escaned J, Alfonso F, Nolte C, Rodriguez V, et al. Morphometric assessment of coronary stenosis relevance with optical coherence tomography: a comparison with fractional flow reserve and intravascular ultrasound. J Am Coll Cardiol 2012;59:1080–9.
[84] Chen S-L, Xu B, Chen JB, Xu T, Ye F, Zhang J-J, et al. Diagnostic accuracy of quantitative angiographic and intravascular ultrasound parameters predicting the functional significance of single de novo lesions. Int J Cardiol 2013;168:1364–9.
[85] Waksman R, Legutko J, Singh J, Orlando Q, Marso S, Schloss T, et al. First: fractional flow reserve and intravascular ultrasound relationship study. J Am Coll Cardiol 2013;61:917–23.
[86] Vascular evaluation for revascularization: defining the indications for coronary therapy: a pilot study. Available from: https://clinicaltrials.gov/ct2/show/NCT01158053 [accessed 05.06.16].
[87] Yang H-M, Tahk S-J, Lim H-S, Yoon M-H, Choi S-Y, Choi B-J, et al. Relationship between intravascular ultrasound parameters and fractional flow reserve in intermediate coronary artery stenosis of left anterior descending artery: intravascular ultrasound volumetric analysis: FFR and IVUS. Catheter Cardiovasc Interv 2014;83:386–94.
[88] Kang S-J, Ahn J-M, Han S, Lee J-Y, Kim W-J, Park D-W, et al. Sex differences in the visual-functional mismatch between coronary angiography or intravascular ultrasound versus fractional flow reserve. JACC Cardiovasc Interv 2013;6:562–8.
[89] Fitzgerald PJ, Ports TA, Yock PG. Contribution of localized calcium deposits to dissection after angioplasty. An observational study using intravascular ultrasound. Circulation 1992;86:64–70.
[90] Mintz GS, Pichard AD, Kovach JA, Kent KM, Satler LF, Javier SP, et al. Impact of preintervention intravascular ultrasound imaging on transcatheter treatment strategies in coronary artery disease. Am J Cardiol 1994;73:423–30.
[91] Stone GW, Hodgson JM, Goar FGS, Frey A, Mudra H, Sheehan H, et al. Improved procedural results of coronary angioplasty with intravascular ultrasound-guided balloon sizing the CLOUT pilot trial. Circulation 1997;95:2044–52.
[92] Haase KK, Athanasiadis A, Mahrholdt H, Treusch A, Wullen B, Jaramillo C, et al. Acute and one year follow-up results after vessel size adapted PTCA using intracoronary ultrasound. Eur Heart J 1998;19:263–72.
[93] Kimura T, Kaburagi S, Tamura T, Yokoi H, Nakagawa Y, Yokoi H, et al. Remodeling of human coronary arteries undergoing coronary angioplasty or atherectomy. Circulation 1997;96:475–83.
[94] Serruys PW, de Bruyne B, Carlier S, Sousa JE, Piek J, Muramatsu T, et al. Randomized comparison of primary stenting and provisional balloon angioplasty guided by flow velocity measurement. Circulation 2000;102:2930–7.
[95] de Ribamar Costa J, Mintz GS, Carlier SG, Mehran R, Teirstein P, Sano K, et al. Nonrandomized comparison of coronary stenting under intravascular ultrasound guidance of direct stenting without predilation versus conventional predilation with a semi-compliant balloon versus predilation with a new scoring balloon. Am J Cardiol 2007;100:812–7.
[96] Barbato E, Carrié D, Dardas P, Fajadet J, Gaul G, Haude M, et al. European expert consensus on rotational atherectomy. EuroIntervention 2015;11:30–6.
[97] Colombo A, Hall P, Nakamura S, Almagor Y, Maiello L, Martini G, et al. Intracoronary stenting without anticoagulation accomplished with intravascular ultrasound guidance. Circulation 1995;91:1676–88.
[98] Stettler C, Allemann S, Wandel S, Kastrati A, Morice MC, Schömig A, et al. Drug eluting and bare metal stents in people with and without diabetes: collaborative network meta-analysis. BMJ 2008;337:a1331.

[99] Frey AW, Hodgson JM, Müller C, Bestehorn H-P, Roskamm H. Ultrasound-guided strategy for provisional stenting with focal balloon combination catheter results from the randomized strategy for intracoronary ultrasound-guided PTCA and stenting (SIPS) trial. Circulation 2000;102: 2497–502.
[100] Fitzgerald PJ, Oshima A, Hayase M, Metz JA, Bailey SR, Baim DS, et al. Final results of the can routine ultrasound influence stent expansion (CRUISE) study. Circulation 2000;102:523–30.
[101] Choi JW, Goodreau LM, Davidson CJ. Resource utilization and clinical outcomes of coronary stenting: a comparison of intravascular ultrasound and angiographical guided stent implantation. Am Heart J 2001;142:112–8.
[102] Russo RJ, Silva PD, Teirstein PS, Attubato MJ, Davidson CJ, DeFranco AC, et al. A randomized controlled trial of angiography versus intravascular ultrasound-directed bare-metal coronary stent placement (the AVID trial). Circ Cardiovasc Interv 2009;2:113–23.
[103] Gil RJ, Pawłowski T, Dudek D, Horszczaruk G, Zmudka K, Lesiak M, et al. Comparison of angiographically guided direct stenting technique with direct stenting and optimal balloon angioplasty guided with intravascular ultrasound. The multicenter, randomized trial results. Am Heart J 2007;154:669–75.
[104] Schiele F, Meneveau N, Vuillemenot A, Gupta S, Mercier M, Danchin N, et al. Impact of intravascular ultrasound guidance in stent deployment on 6-month restenosis rate: a multicenter, randomized study comparing two strategies—with and without intravascular ultrasound guidance. J Am Coll Cardiol 1998;32:320–8.
[105] Schiele F, Meneveau N, Seronde M-F, Caulfield F, Pisa B, Arveux P, et al. Medical costs of intravascular ultrasound optimization of stent deployment. Results of the multicenter randomized "REStenosis after Intravascular ultrasound STenting" (RESIST) study. Int J Cardiovasc Intervent 2000;3:207–13.
[106] Oemrawsingh PV. Intravascular ultrasound guidance improves angiographic and clinical outcome of stent implantation for long coronary artery stenoses: final results of a randomized comparison with angiographic guidance (TULIP study). Circulation 2003;107:62–7.
[107] Gaster AL, Skjoldborg US, Larsen J, Korsholm L, von Birgelen C, Jensen S, et al. Continued improvement of clinical outcome and cost effectiveness following intravascular ultrasound guided PCI: insights from a prospective, randomised study. Heart 2003;89:1043–9.
[108] Mudra H, di Mario C, de Jaegere P, Figulla HR, Macaya C, Zahn R, et al. Randomized comparison of coronary stent implantation under ultrasound or angiographic guidance to reduce stent restenosis (OPTICUS study). Circulation 2001;104:1343–9.
[109] Casella G, Klauss V, Ottani F, Siebert U, Sangiorgio P, Bracchetti D. Impact of intravascular ultrasound-guided stenting on long-term clinical outcome: a meta-analysis of available studies comparing intravascular ultrasound-guided and angiographically guided stenting. Catheter Cardiovasc Interv 2003;59:314–21.
[110] Parise H, Maehara A, Stone GW, Leon MB, Mintz GS. Meta-analysis of randomized studies comparing intravascular ultrasound versus angiographic guidance of percutaneous coronary intervention in pre-drug-eluting stent era. Am J Cardiol 2011;107:374–82.
[111] Agostoni P, Valgimigli M, van Mieghem CAG, Rodriguez-Granillo GA, Aoki J, Ong ATL, et al. Comparison of early outcome of percutaneous coronary intervention for unprotected left main coronary artery disease in the drug-eluting stent era with versus without intravascular ultrasonic guidance. Am J Cardiol 2005;95:644–7.
[112] Roy P, Steinberg DH, Sushinsky SJ, Okabe T, Slottow TLP, Kaneshige K, et al. The potential clinical utility of intravascular ultrasound guidance in patients undergoing percutaneous coronary intervention with drug-eluting stents. Eur Heart J 2008;29:1851–7.
[113] Park S-J, Kim Y-H, Park D-W, Lee S-W, Kim W-J, Suh J, et al. Impact of intravascular ultrasound guidance on long-term mortality in stenting for unprotected left main coronary artery stenosis. Circ Cardiovasc Interv 2009;2:167–77.
[114] Kim S-H, Kim Y-H, Kang S-J, Park D-W, Lee S-W, Lee CW, et al. Long-term outcomes of intravascular ultrasound-guided stenting in coronary bifurcation lesions. Am J Cardiol 2010;106: 612–8.

[115] Jakabčin J, Špaček R, Bystroň M, Kvašňák M, Jager J, Veselka J, et al. Long-term health outcome and mortality evaluation after invasive coronary treatment using drug eluting stents with or without the IVUS guidance. randomized control trial. HOME DES IVUS: mortality evaluation after using DES with or without the IVUS. Catheter Cardiovasc Interv 2010;75:578–83.
[116] Claessen BE, Mehran R, Mintz GS, Weisz G, Leon MB, Dogan O, et al. Impact of intravascular ultrasound imaging on early and late clinical outcomes following percutaneous coronary intervention with drug-eluting stents. JACC Cardiovasc Interv 2011;4:974–81.
[117] Kim J-S, Hong M-K, Ko Y-G, Choi D, Yoon JH, Choi S-H, et al. Impact of intravascular ultrasound guidance on long-term clinical outcomes in patients treated with drug-eluting stent for bifurcation lesions: data from a Korean multicenter bifurcation registry. Am Heart J 2011;161:180–7.
[118] Youn YJ, Yoon J, Lee J-W, Ahn S-G, Ahn M-S, Kim J-Y, et al. Intravascular ultrasound-guided primary percutaneous coronary intervention with drug-eluting stent implantation in patients with ST-segment elevation myocardial infarction. Clin Cardiol 2011;34:706–13.
[119] Witzenbichler B, Maehara A, Weisz G, Neumann F-J, Rinaldi MJ, Metzger DC, et al. Relationship between intravascular ultrasound guidance and clinical outcomes after drug-eluting stents: the assessment of dual antiplatelet therapy with drug-eluting stents (ADAPT-DES) study. Circulation 2014;129:463–70.
[120] Chen S-L, Ye F, Zhang J-J, Tian N-L, Liu Z-Z, Santoso T, et al. Intravascular ultrasound-guided systematic two-stent techniques for coronary bifurcation lesions and reduced late stent thrombosis. Catheter Cardiovasc Interv 2013;81:456–63.
[121] Park KW, Kang S-H, Yang H-M, Lee H-Y, Kang H-J, Cho Y-S, et al. Impact of intravascular ultrasound guidance in routine percutaneous coronary intervention for conventional lesions: data from the EXCELLENT trial. Int J Cardiol 2013;167:721–6.
[122] Hur S-H, Kang S-J, Kim Y-H, Ahn J-M, Park D-W, Lee S-W, et al. Impact of intravascular ultrasound-guided percutaneous coronary intervention on long-term clinical outcomes in a real world population. Catheter Cardiovasc Interv 2013;81:407–16.
[123] Chieffo A, Latib A, Caussin C, Presbitero P, Galli S, Menozzi A, et al. A prospective, randomized trial of intravascular-ultrasound guided compared to angiography guided stent implantation in complex coronary lesions: the AVIO trial. Am Heart J 2013;165:65–72.
[124] Kim J-S, Kang T-S, Mintz GS, Park B-E, Shin D-H, Kim B-K, et al. Randomized comparison of clinical outcomes between intravascular ultrasound and angiography-guided drug-eluting stent implantation for long coronary artery stenoses. JACC Cardiovasc Interv 2013;6:369–76.
[125] Ahn J-M, Han S, Park YK, Lee WS, Jang JY, Kwon CH, et al. Differential prognostic effect of intravascular ultrasound use according to implanted stent length. Am J Cardiol 2013;111:829–35.
[126] Yoon Y-W, Shin S, Kim B-K, Kim J-S, Shin D-H, Ko Y-G, et al. Usefulness of intravascular ultrasound to predict outcomes in short-length lesions treated with drug-eluting stents. Am J Cardiol 2013;112:642–6.
[127] de la Torre Hernandez JM, Baz Alonso JA, Gómez Hospital JA, Alfonso Manterola F, Garcia Camarero T, Gimeno de Carlos F, et al. Clinical impact of intravascular ultrasound guidance in drug-eluting stent implantation for unprotected left main coronary disease: pooled analysis at the patient-level of 4 registries. JACC Cardiovasc Interv 2014;7:244–54.
[128] Magalhaes MA, Minha S, Torguson R, Baker NC, Escarcega RO, Omar AF, Lipinski MJ, et al. The effect of complete percutaneous revascularisation with and without intravascular ultrasound guidance in the drugeluting stent era. EuroIntervention 2015;11:625–33.
[129] Hong S-J, Kim B-K, Shin D-H, Nam C-M, Kim J-S, Ko Y-G, et al. Effect of intravascular ultrasound-guided vs angiography-guided everolimus-eluting stent implantation: the IVUS-XPL randomized clinical trial. JAMA 2015;314:2155–63.
[130] Nakatsuma K, Shiomi H, Morimoto T, Ando K, Kadota K, Watanabe H, et al. Intravascular ultrasound guidance vs. angiographic guidance in primary percutaneous coronary intervention for ST-segment elevation myocardial infarction—long-term clinical outcomes from the CREDO-Kyoto AMI registry. Circ J 2016;80:477–84.
[131] Moschovitis A, Cook S, Meier B. Percutaneous coronary interventions in Europe in 2006. EuroIntervention 2010;6:189–94.

[132] Singh V, Badheka AO, Arora S, Panaich SS, Patel NJ, Patel N, et al. Comparison of inhospital mortality, length of hospitalization, costs, and vascular complications of percutaneous coronary interventions guided by ultrasound versus angiography. Am J Cardiol 2015;115:1357–66.
[133] Iannaccone M, D'Ascenzo F, Templin C, Omedè P, Montefusco A, Guagliumi G, et al. Optical coherence tomography evaluation of intermediate-term healing of different stent types: systemic review and meta-analysis. Eur Heart J Cardiovasc Imaging 2016. http://dx.doi.org/10.1093/ehjci/jew070 [Epub ahead of print].
[134] Imola F, Mallus MT, Ramazzotti V, Manzoli A, Pappalardo A, Di Giorgio A, et al. Safety and feasibility of frequency domain optical coherence tomography to guide decision making in percutaneous coronary intervention. EuroIntervention 2010;6:575–81.
[135] Prati F, Di Vito L, Biondi-Zoccai G, Occhipinti M, La Manna A, Tamburino C, et al. Angiography alone versus angiography plus optical coherence tomography to guide decision-making during percutaneous coronary intervention: the Centro per la Lotta contro l'Infarto-Optimisation of Percutaneous Coronary Intervention (CLI-OPCI) study. EuroIntervention 2012;8:823–9.
[136] Prati F, Romagnoli E, Burzotta F, Limbruno U, Gatto L, La Manna A, et al. Clinical impact of OCT findings during PCI: the CLI-OPCI II study. JACC Cardiovasc Imaging 2015;8:1297–305.
[137] Habara M, Nasu K, Terashima M, Kaneda H, Yokota D, Ko E, et al. Impact of frequency-domain optical coherence tomography guidance for optimal coronary stent implantation in comparison with intravascular ultrasound guidance. Circ Cardiovasc Interv 2012;5:193–201.
[138] Antonsen L, Thayssen P, Maehara A, Hansen HS, Junker A, Veien KT, et al. Optical coherence tomography guided percutaneous coronary intervention with Nobori Stent implantation in patients with non-ST-segment-elevation myocardial infarction (OCTACS) trial difference in strut coverage and dynamic malapposition patterns at 6 months. Circ Cardiovasc Interv 2015;8:e002446.
[139] Byrne RA, Joner M, Kastrati A. Stent thrombosis and restenosis: what have we learned and where are we going? The Andreas Grüntzig Lecture ESC 2014. Eur Heart J 2015;36(47):3320–31.
[140] Fujii K, Mintz GS, Kobayashi Y, Carlier SG, Takebayashi H, Yasuda T, et al. Contribution of stent underexpansion to recurrence after sirolimus-eluting stent implantation for in-stent restenosis. Circulation 2004;109:1085–88.
[141] Fujii K, Carlier SG, Mintz GS, Yang Y, Moussa I, Weisz G, et al. Stent underexpansion and residual reference segment stenosis are related to stent thrombosis after sirolimus-eluting stent implantation: an intravascular ultrasound study. J Am Coll Cardiol 2005;45:995–8.
[142] Takebayashi H, Kobayashi Y, Mintz GS, Carlier SG, Fujii K, Yasuda T, et al. Intravascular ultrasound assessment of lesions with target vessel failure after sirolimus-eluting stent implantation. Am J Cardiol 2005;95:498–502.
[143] van Werkum JW, Heestermans AA, Zomer AC, Kelder JC, Suttorp M-J, Rensing BJ, et al. Predictors of coronary stent thrombosis: the Dutch Stent Thrombosis Registry. J Am Coll Cardiol 2009;53:1399–409.
[144] Serruys PW, Ormiston JA, Onuma Y, Regar E, Gonzalo N, Garcia-Garcia HM, et al. A bioabsorbable everolimus-eluting coronary stent system (ABSORB): 2-year outcomes and results from multiple imaging methods. Lancet 2009;373:897–910.
[145] Stone GW, Gao R, Kimura T, Kereiakes DJ, Ellis SG, Onuma Y, et al. 1-year outcomes with the absorb bioresorbable scaffold in patients with coronary artery disease: a patient-level, pooled meta-analysis. Lancet 2016;387:1277–89.
[146] Capodanno D, Gori T, Nef HM, Latib A, Mehilli J, Lesiak M, et al. Percutaneous coronary intervention with everolimus-eluting bioresorbable vascular scaffolds in routine clinical practice: early and midterm outcomes from the European multicentre GHOST-EU registry. EuroIntervention 2015;10:1144–53.
[147] Cassese S, Byrne RA, Ndrepepa G, Kufner S, Wiebe J, Repp J, et al. Everolimus-eluting bioresorbable vascular scaffolds versus everolimus-eluting metallic stents: a meta-analysis of randomised controlled trials. Lancet 2016;387:537–44.
[148] Kang S-H, Chae I-H, Park J-J, Lee HS, Kang D-Y, Hwang S-S, et al. Stent thrombosis with drug-eluting stents and bioresorbable scaffolds: evidence from a network meta-analysis of 147 trials. JACC Cardiovasc Interv 2016;9:1203–12.

[149] Zhang Y, Farooq V, Garcia-Garcia HM, Bourantas CV, Tian N, Dong S, et al. Comparison of intravascular ultrasound versus angiography-guided drug-eluting stent implantation: a meta-analysis of one randomised trial and ten observational studies involving 19,619 patients. EuroIntervention 2012;8:855–65.

[150] Klersy C, Ferlini M, Raisaro A, Scotti V, Balduini A, Curti M, et al. Use of IVUS guided coronary stenting with drug eluting stent: a systematic review and meta-analysis of randomized controlled clinical trials and high quality observational studies. Int J Cardiol 2013;170:54–63.

[151] Jang J-S, Song Y-J, Kang W, Jin H-Y, Seo J-S, Yang T-H, et al. Intravascular ultrasound-guided implantation of drug-eluting stents to improve outcome. JACC Cardiovasc Interv 2014;7:233–43.

[152] Ahn J-M, Kang S-J, Yoon S-H, Park HW, Kang SM, Lee J-Y, et al. Meta-analysis of outcomes after intravascular ultrasound-guided versus angiography-guided drug-eluting stent implantation in 26,503 patients enrolled in three randomized trials and 14 observational studies. Am J Cardiol 2014;113:1338–47.

[153] Steinvil A, Zhang Y-J, Lee SY, Pang S, Waksman R, Chen S-L, et al. Intravascular ultrasound-guided drug-eluting stent implantation: an updated meta-analysis of randomized control trials and observational studies. Int J Cardiol 2016;216:133–9.

[154] IVUS-CTO: Intravascular Ultrasound-Guided vs. Angiography-Guided DES Intervention in Coronary CTO. Available from: http://www.acc.org/latest-in-cardiology/articles/2014/09/11/16/04/ivus-cto-intravascular-ultrasound-guided-vs-angiography-guided-des-intervention-in-coronary-cto [accessed 08.06.16].

[155] Tian N-L, Gami S-K, Ye F, Zhang J-J, Liu Z-Z, Lin S, et al. Angiographic and clinical comparisons of intravascular ultrasound- versus angiography-guided drug-eluting stent implantation for patients with chronic total occlusion lesions: two-year results from a randomised AIR-CTO study. EuroIntervention 2015;10:1409–17.

[156] Kolh P, Windecker S, Alfonso F, Collet JP, Cremer J, Falk V, et al. 2014 ESC/EACTS Guidelines on myocardial revascularization: the task force on myocardial revascularization of the European Society of Cardiology (ESC) and the European Association for Cardio-Thoracic Surgery (EACTS) developed with the special contribution of the European Association of Percutaneous Cardiovascular Interventions (EAPCI). Eur Heart J 2014;35:2541–619.

[157] Li J, Li X, Jing J, Mohar D, Raney A, Mahon S, et al. Integrated intravascular optical coherence tomography (OCT)—ultrasound (US) catheter for characterization of atherosclerotic plaques in vivo. Conf Proc IEEE Eng Med Biol Soc 2012:3175–8.

[158] Ughi GJ, Wang H, Gerbaud E, Gardecki JA, Fard AM, Hamidi E, et al. Clinical characterization of coronary atherosclerosis with dual-modality OCT and near-infrared autofluorescence imaging. JACC Cardiovasc Imaging 2016. http://dx.doi.org/10.1016/j.jcmg.2015.11.020 [Epub ahead of print].

CHAPTER 2

Atherosclerotic Plaque Progression and OCT/IVUS Assessment

J. Rigla
Barcelona Perceptual Computing Laboratory, Universitat de Barcelona, Barcelona, Spain

Chapter Outline

1. AIMS AND INTRODUCTION

1.1 Aim of this Chapter

Because of its high morbidity and mortality, atherosclerosis disease remains an unsolved problem and an intellectual challenge for the current investigators. The aim of this chapter is to provide new investigators with a background information in atherosclerosis disease and also an introduction in the optical coherence tomography (OCT) and intravascular ultrasound (IVUS) as atherosclerosis research tools. We believe the acquaintance between the OCT and IVUS visualization capacities and the atherosclerosis disease challenges will drive further advances in research.

Computing and Visualization for Intravascular Imaging and Computer-Assisted Stenting
http://dx.doi.org/10.1016/B978-0-12-811018-8.00002-3

1.2 Introduction

Arteries are elastic conducts responsible for the blood distribution throughout the body. Each arterial vessel consists of three concentric layers:

- the intima, the more internal layer, constituted by the endothelium (simple plane epithelium);
- the media, composed of smooth muscle fibers arranged concentrically; and
- the adventitia, the more external layer, mainly composed of fibroblasts and collagen.

Atherosclerosis lesions are a consequence of a subintimal infiltration of lipoproteins. There are three classical pathogenic theories for the origin of atherosclerosis: the theory of incrustation by Rokitansky where the trigger of atherosclerosis is the deposition of a protein into the intima, the theory of infiltration by Anitschkow where the trigger is a lipid infiltration into the intima, and the theory by Virchow where the trigger is the cell degeneration and proliferation [1]. Up today it is accepted that the key factor in the genesis of atherosclerosis is the endothelial dysfunction, which increases the endothelial permeability to the lipids across the endothelium. In reaction to the lipid infiltration, the inflammation process stimulates the migration and proliferation of smooth muscle fibers and the fibrosis healing the plaque and sealing the lipid entrance to the subintimal space, and also stimulating the macrophages to digest the lipids. This is a dynamic process, where atherosclerosis plaque progression is a consequence of the balance between the endothelial permeability to the lipids into the intima and the plaque healing and cleaning process. When the lipid pool into the subintimal space increases, atherosclerosis plaque grows [2].

2. DESCRIPTION OF ATHEROSCLEROSIS LESIONS IN CHILDREN, ADULTS, AND ELDERLY POPULATION

The first atherosclerosis lesions appear early in life in large arteries [3]. Stimulated by risk factors, atherosclerosis progresses during the lifetime.

2.1 Atherosclerosis in Children

As shown in the necropsy registries, the atherosclerosis process begins in the luminal surface of the arteries as spots of an irregular yellow-white discoloration, referred to as fatty streaks. Frequently visible to the naked eye, a fatty streak consists of aggregates of foam cells which are lipoprotein-loaded macrophages located below the endothelial cells in the subintimal space [4]. Preatherosclerosis changes of the arteries are already detectable in the prenatal period and they are significantly associated with the maternal smoking habit [5]. Children with familial hypercholesterolemia (a genetic disorder characterized by high-cholesterol level in the blood and early cardiovascular disease) have endothelial dysfunction and also an increase of carotid intima-media thickness, a consequence of the subintimal infiltration of lipoproteins [6]. Despite the risk factors, most children over the age of 3 years have some degree of aortic fatty streaks [1], and almost all children older than 10 have the presence of aortic fatty streaks [7]. In the thoracic aorta, the fatty streaks have the same prevalence during the first and second decade of life, and females show larger fatty streaks than men [8]. The fatty streaks progress to complex lesions is more frequent in the abdominal aorta [9]. No direct relation has been established between aortic fatty streaks and the future clinically relevant lesions of atherosclerosis [10–12]. In the coronary arteries, fatty streaks begin to appear a decade later than in the aorta [8, 10–12]. However, in contrast with the aorta, the localization of fatty streak in the coronary arteries has a close correspondence with the further appearance of raised lesions in adults [13].

2.2 Atherosclerosis in Adults

Fatty streaks (defined as spots of an irregular yellow-white discoloration in the luminal surface of the arteries) are the characteristic atherosclerotic lesion in children. Raised lesions (generally defined as fibrotic plaques raised on the luminal surface of the arteries) are the characteristic atherosclerotic lesion in adults. Fatty streaks and raised lesions contain lipids, collagen, and macrophages, and differ only in the proportions of each component: fatty streaks are early lesions, raised lesions are more evolved. Raised lesions appear most frequently in the second decade of life, and increases in prevalence for every additional decade. Although females show larger fatty streaks than men [8], men and woman have the same incidence of raised lesions. The PDAY is the most important study in prevalence and progression of atherosclerosis lesions from necropsy of juveniles and young adults, and concludes that the progress of fatty streaks to raised plaques, it should differ according to the arterial sites, with a higher prevalence and extent of raised lesions in the abdominal aorta than in the thoracic aorta [14, 15]. In aorta, the fatty streak areas are more prevalent in the proximal region, and the raised lesions are more prevalent in the distal segment (left in the image) [15]. In right coronary artery (RCA), the fatty streak areas are more prevalent in the proximal region, and the raised lesions have more prevalence in the proximal segment. In coronary arteries, raised lesions in RCA begin in the 20–24 year age group and become more extensive in older age groups. Men have more raised lesions in RCA than women [15]. Although raised lesions appear later in the coronaries than in the aorta, the proximal segment of the left anterior descending (LAD) is especially predisposed to plaque formation, followed by left main stem (LMS) and circumflex (LCX). The lesions within the LAD progressed faster to more advanced lesions [15]. Fig. 2.1 represents the distribution and progression of disease between the groups of 15–24 age and 25–34 years old [14]. See the American Heart Association (AHA) classification in Section 3 of this chapter.

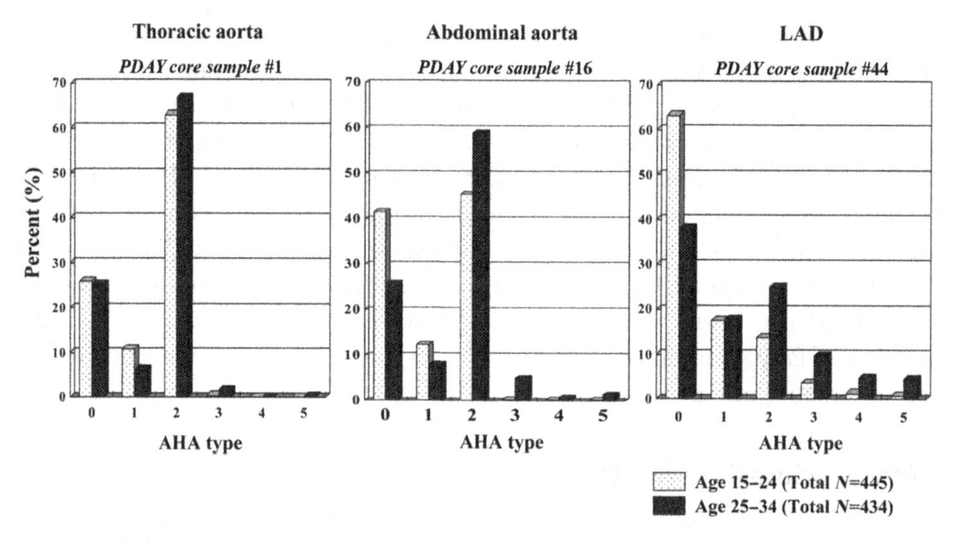

Fig. 2.1 Distribution (percent) of AHA grades within the thoracic aorta, abdominal aorta, and left anterior descending coronary artery by 10 year age groups ($N = 879$) [14].

2.3 Atherosclerosis in the Elderly

In elderly patients is characteristic the disease progression affecting to the peripheral arteries (peripheral artery disease, PAD) and to the carotids (in particular to the internal carotid, IC) [16–22]. The Rotterdam study [23] is the most important epidemiological registry in elderly population and concludes that the prevalence of PAD is high whereas the prevalence of reported IC is relatively low. PAD and IC prevalence increase sharply in advanced age (Fig. 2.2). Almost all eastern men aged 70–89 years have ultrasonically detectable atherosclerosis lesions in the IC arteries and the majority of elderly people have carotid calcification [24]. In cases of acute coronary syndrome (ACS), the odds for in-hospital death increase by 70% for each 10-year greater in age [25, 26].

3. ATHEROSCLEROSIS HISTOLOGIST CLASSIFICATION

The atherosclerosis histologist classification [9, 27–29] has been formulated from a selection of histology frames obtained from different necropsies, and sequenced from simple to complex. The atherosclerosis histologist classification distinguishes lesion types from I to VIII (where type V has three subtypes: Va, Vb, and Vc). These types may also be grouped in three progressive stages; early lesions, transitional lesions, and complex plaques. The AHA adopted and improved these classical definitions in a consensus document [30–33].

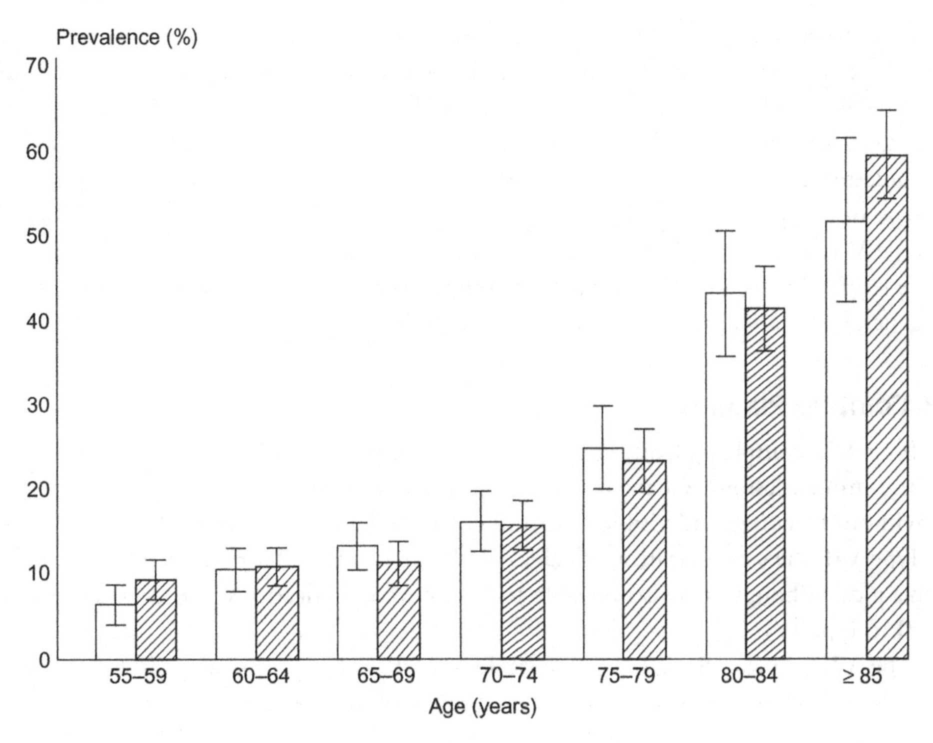

Fig. 2.2 The prevalence of PAD per age and sex-specific according to age for men (*white bars*) and women (*shaded bars*) and 95% CI [23].

3.1 Early Lesions

- Type I: initial lesion, consists of foam cells from macrophages, containing lipids. Only identifiable biochemically and microscopically.
- Type II: fatty streak with multiple foam cell layers and occasionally accompanied by vascular smooth muscle cells. They have a higher proportion of lipid-laden macrophages than type I lesions. Macroscopically visible as fatty points minimally raised in the lumen of the arteries as fatty streaks.

3.2 Transitional Lesions

- Type III: preatheroma, associated with multiple and diffuse accumulation of extra-cellular lipids.
- Type IV: atheroma, with a confluent extra-cellular lipid core. Progress to a single extra-cellular lipids accumulation, forming the characteristic lipid core. These lesions contain calcium granules, localized extra cellular in the lipid core or within some vascular smooth muscle cells.

- Type V: atheroma, where vascular smooth muscle cells migrate and proliferate producing collagen fibers over the lipid core, forming a layer on the luminal side. Usually the plaque increases its size by the lipid core through an accumulation of pasty wastes including macrophages and dead cells with abundant crystals of free cholesterol. When the new tissue is part of a lesion with a lipid core (type IV), this type of morphology may be referred to as fibroatheroma (FA) or type Va lesion. A type V lesion in which the lipid core and other parts of the lesion are calcified may be referred to as type Vb. A type V lesion in which a lipid core is absent and lipid in general is minimal may be referred to as type Vc.

3.3 Complex Plaques

- Type VI: complex plaque referred to as complicated fibroatheromatous plaques or complex lesions. Characterized by possible surface defect, hemorrhage with or without bleeding, and visible lipid and collagen deposits plus thrombotic materials.
- Type VII: the categorization of calcified plaque is reserved for advanced and calcified plaques, although it is also possible to find microcalcifications in earlier stages (type III to type IV).
- Type VIII: finally, the atherosclerotic lesions that consist almost entirely of fibrillar collagen. These lesions may be a consequence of lipid core regression.

4. PHASES OF PROGRESSION OF ATHEROSCLEROSIS DISEASE

According to the criteria published by the AHA, and based on the previous histologist classification, the description of atherosclerosis disease progression defines a road-map and a time-lined evolution to move from one stage to the consecutive, through progressive stages [9, 29–31, 34–36]. The progression of the plaque can be subdivided into five phases as follows (Fig. 2.3).

- Phase 1 (early). Lesions are small, usually seen in the young, and includes type I lesions, II, and III.
- Phase 2 (advanced). These plaques are morphological classified into one of two types: type IV lesions, extra-cellular lipid core; or type Va lesions, with an extra-cellular lipid core covered by a fibrous capsule. Lesions may not be necessarily stenosis, and tend to break due to its high-lipid content, increased inflammation, and the presence of a thin fibrous layer. The plaques of phase 2 can evolve into acute phases 3 and 4.
- Phase 3 (acute complication causing thrombosis). Type VI is a possible evolution of plaques type IV or Va eroded or broken. Acute complication of type VI plaques may be silent [37].
- Phase 4. These lesions are characterized by acute complication of type VI lesions with recurrent thrombosis and/or permanent occlusion. This process can be manifested

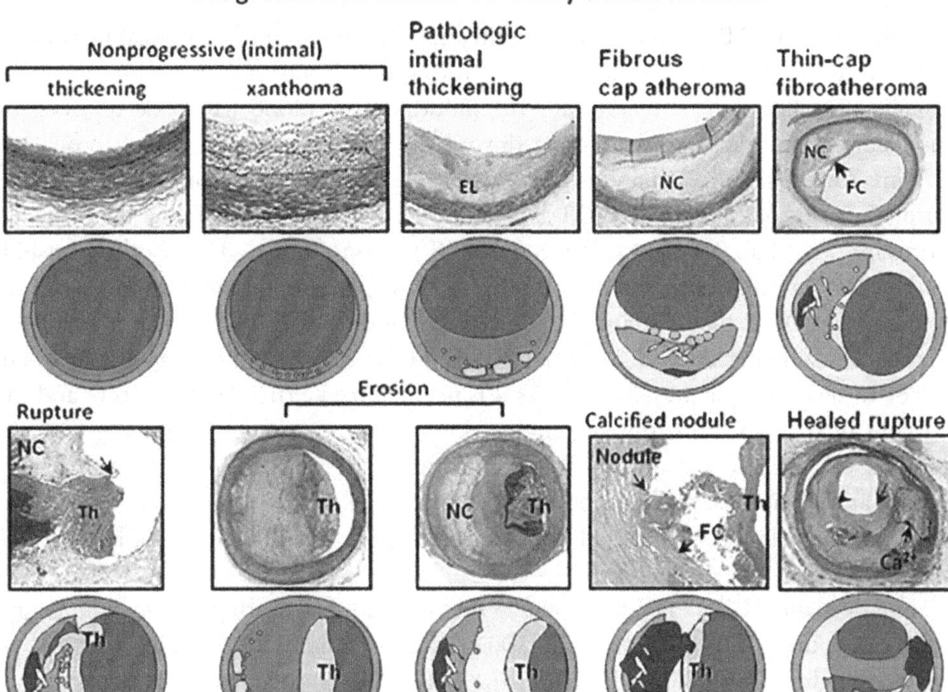

Fig. 2.3 Spectrum of representative coronary lesions seen in sudden death population, forming the basis for the modified AHA descriptive classification. The two nonprogressive lesions are intimal thickening and intimal xanthomas (foam cell collections known as fatty streaks, AHA type II). Pathological intimal thickening (PIT) (AHA type III transitional lesions) marks the first of the progressive plaques, as they are the assumed precursors to more advanced FA. Thin-cap fibroatheromas (TCFAs) are considered precursors to plaque rupture. Essentially missing from the AHA consensus classifications are alternative entities that give rise to coronary thrombosis, namely erosion and the calcified nodule. Erosion can occur on a substrate of PIT or FA, whereas calcified nodules depict eruptive fragments of calcium that protrude into the lumen, causing a thrombotic event. Lastly, healed plaque ruptures are lesions with generally smaller necrotic cores and focal areas of calcification where the surface generally shows areas of healing rich in proteoglycans. Multiple healed plaque ruptures are thought responsible for progressive luminary narrowing. *Ca^{2+}*, calcium; *EL*, extra-cellular lipid; *FC*, fibrous cap; *NC*, necrotic core; *Th*, luminal thrombus [43].

as an ACS [38, 39]. An occlusive thrombosis on a broken plaque causes about 66% of ACS, but the other 33% occurs on the surface of an eroded plaque [40].

- Phase 5. This phase corresponds to type Vb (calcified) lesions or type Vc (fibroid) lesions; regardless, it can cause angina, be silent or be clinically unapparent [41, 42].

5. OCT

The principle of OCT is based on tissue reflection of white light. An optical light beam is directed to the tissue, and a small portion of this light is reflected from the subsurface and collected. The light is opaque for the blood, so the blood has to be replaced by translucent media (radiologic contrast) during the OCT examination. In coronaries, OCT images are acquired in around 3 s, during 12–16 cc injection of contrast through a normal guiding catheter. The OCT images are recorded frame by frame. With the aid of software (LightLab (SJM) or QIVUS (MEDIS)), a longitudinal display of the vessel is available. OCT resolution is about 15 μm, with a low penetration range from 1 to 2 mm into the tissue. OCT allows seeing the intima layer of the vessel wall in greater detail and the intraluminal materials such the presence of thrombus, whether it is attached to stent struts or to ulcerated plaques. Sometimes it is possible to see the plaque rupture and the lipid core in patients with unstable angina. Due to the fact that the guiding catheter cannot be occlusive in the ostium, there is no way to clear all the blood coming from the aorta at the entrance to the right coronary ostium or to the left coronary main stem, so it is not possible to obtain OCT images at this level. There are two different OCT systems available:

1. Time domain OCT (TD-OCT) where the light is path length of a reference length for calibration. TD-OCT records 100 fps for 50 mm, with a pullback speed of 20 mm/s. In TD-OCT the length of the pullback is limited, and it is very difficult to reconstruct an artery from two different consecutive pullbacks, because the low penetration of OCT does not provide clear land-marker images as the references from the adventitia.
2. Frequency domain OCT (FD-OCT) where the broadband interference is acquired with spectral separated detectors. This feature improves imaging speed dramatically, while reducing losses. In contrast with TD-OCT, FD-OCT allows a single pullback for imaging all the interest area in the same pullback. FD-OCT records 158 fps on 15 cm, with a pullback speed of 40 mm/s.

In Fig. 2.4 you can find examples of the OCT appearance of basic tissues, vulnerable plaques, and complex lesions. OCT tissue characterization is fundamental in atherosclerosis plaque research, and need some training to a correct tissue identification:

- fibrous: homogeneous signal-rich (Fig. 2.4A);
- calcified: signal-poor with well-delimited borders (Fig. 2.4B); and
- lipid-rich: signal-poor with diffuse borders (Fig. 2.4C).

Blood artifacts are inherent to the OCT technique as may be difficult to remove the blood in the vessel totally. The blood retained may led to errors in the images interpretation [44]. In spite of the blood retention, the main pitfalls in OCT plaque characterization are the "superficial shadowing" and the "tangential signal dropout," which can produce images with the appearance of thin-cap fibroatheroma (TCFA) due

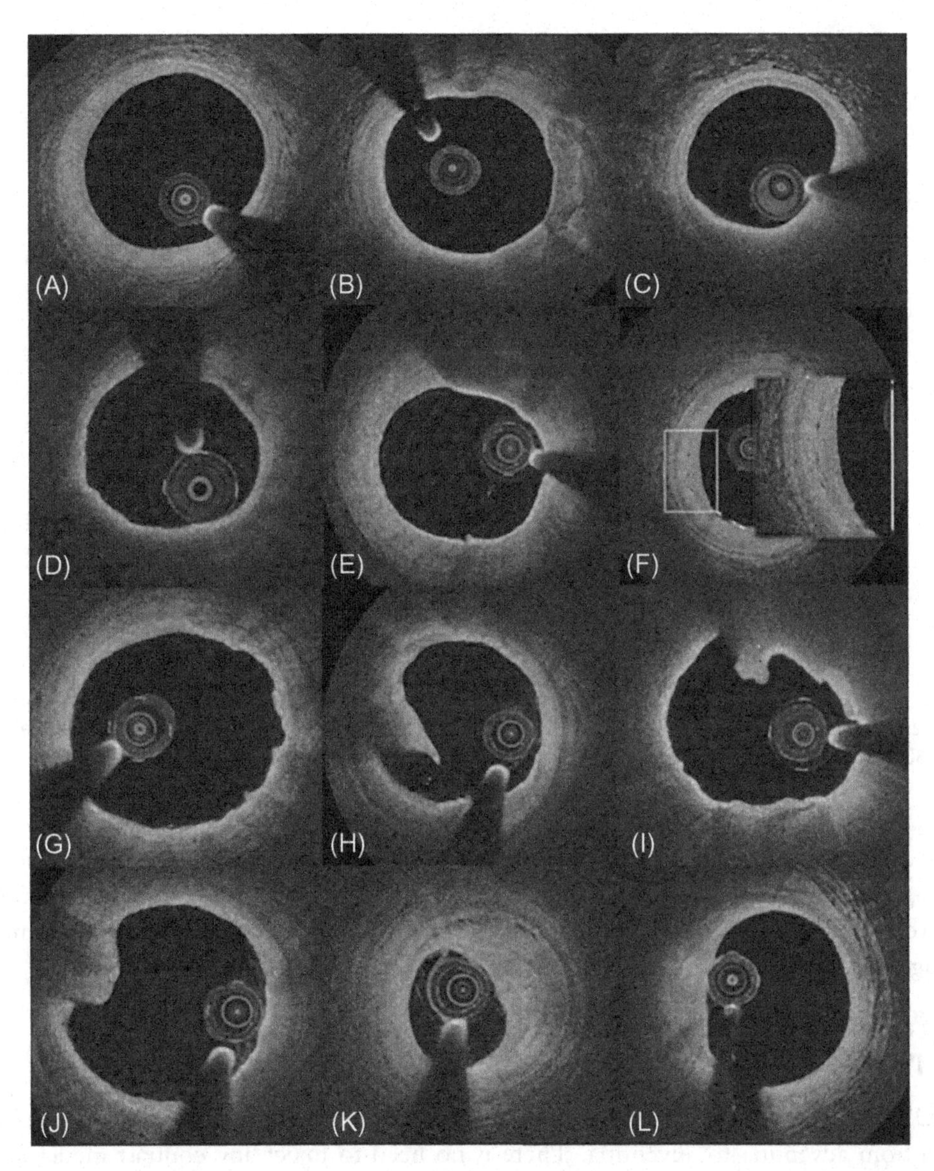

Fig. 2.4 OCT images revealing atherosclerosis plaque characteristics. (A) Eccentric intimal hyperplasia; (B) calcification; (C) lipid pool with vulnerable plaque features; (D) thin-cap fibroatheroma; (E) macrophage image; (F) microchannels; (G) intimal laceration; (H) plaque rupture; (I,J) intraluminal thrombus; and (K,L) layered complex plaque [44].

to a diffracted attenuation of the signal [45]. There is also a discrepancy between diametric measures obtained by TD-OCT (as it is significantly lower) than that obtained with FD-OCT and IVUS, which are similar [46–49]. For experts in OCT interpretation, and using histology as a gold standard, the sensitivity of OCT for

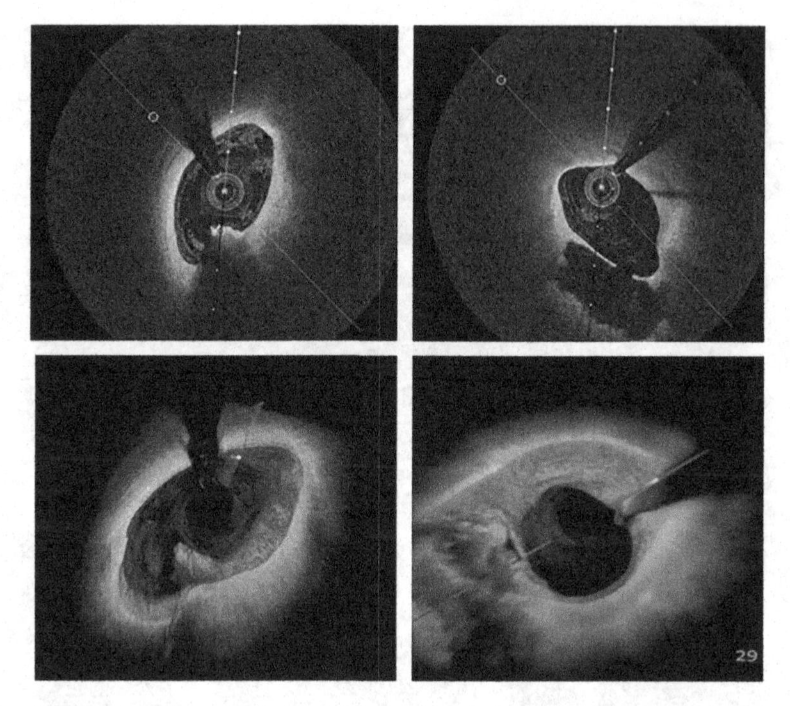

Fig. 2.5 Cross-sectional OCT images (*gray scale*) of a ulcerated plaque with thrombotic content. Below the 3D OCT corresponding images.

the characterization of superficial calcification, fibrosis, and lipid pool was 100%, 98%, and 95%, respectively (but note this is only in ex vivo OCT registration) [50]. Nevertheless, OCT 3D reconstruction is extremely clear, and allows illustrative lumen images (QAngio-OCT from MEDIS) (Fig. 2.5).

6. IVUS

IVUS is a unique technique that allows in real time the transverse views of the arterial wall from adventitia-to-adventitia. There is no need to inject any contrast media, but sometimes the injection of physiologic solution of Cl-Na or contrast media improves the visibility of intraluminal materials. Vessel penetration can be up to 8 mm, depending on plaque composition and transducer frequency (lower MHz provides reduced definition and increased penetration). IVUS consists of an ultrasound pulse emitted by the transducer into the vessel, where the ultrasound reflections are collected and translated to radio frequency (RF) and processed. The gray-scale IVUS is a reconstruction of the RF power signal (envelope) at 30 frames/s. Pullback allows 15 cm at 0.5 mm/s speed [51]. The gray-scale image use only the RF envelope but the back-scattering tissue

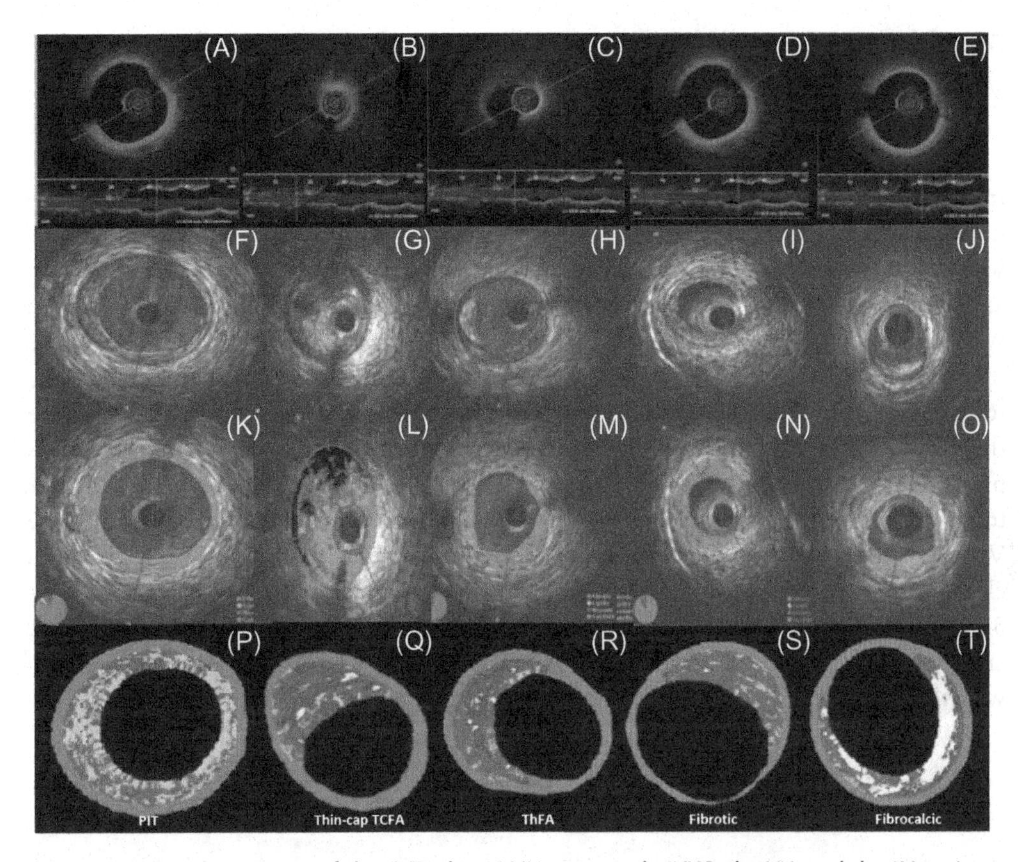

Fig. 2.6 Comparative images of the OCT, the 40 MHz gray-scale IVUS, the VH, and the iMap tissue characterization in pathological intimal thickening (PIT), thin-cap fibroatheroma (TCFA), thick-cap fibroatheroma (ThFA), fibrotic, and fibro-calcific. *For interpretation of the color in this figure, please see the online version.*

characterization use the full radio-frequency signal for selected frames (one per systole or per second). Based on their visual appearance, the atheroma have been classified in four categories in gray-scale IVUS [52]:

- soft plaque (lesion echogenicity less than the surrounding adventitia) (Fig. 2.6F);
- fibrous plaque (intermediate echogenicity between soft (echolucent) atheroma and highly echogenic calcified plaques) (Fig. 2.6I);
- calcified plaque (echogenicity higher than the adventitia with acoustic shadowing) (Fig. 2.6J);
- mixed plaques (unique acoustical subtype) (Fig. 2.6G and H).

There are two different major IVUS system developers, Volcano and Boston Scientific (BSC), both obtain and process the images in a very different manner to characterize the plaque components: Volcano with VH and BSC with iMap.

6.1 Volcano IVUS and VH

Volcano IVUS uses a 20 MHz ultrasound in a phase array system with 64 microtransducers shot sequentially. The first three emitters are shot beginning at 12 o'clock and their reflection is collected in 16 beams. Another three emitters are then shot at 6 o'clock in the same manner, both in a clockwise direction, so consecutive radial bands are obtained with a difference of 180 and reconstructed together. Once the RF data is sorted, it is filtered at the transducer frequency. Each frame image has a depth of 9.6 mm and a catheter width of 1.2 mm, with an image size fixed of 200 × 200 pixels. The system resolution is approximately 150 μm in this 20 MHz phase array catheter. The blood is not visible at 20 MHz and is filtered by the software which may mask the intraluminal materials as thrombus or dissections. A Cromaflow feature may visualize the blood speckle. At the sample rate of one frame per cardiac pulse, the RF data is processed to obtain the VH. The VH process the RF signal through an algorithm tree to classify the vessel tissues in four categories: lipid, fibroid, calcified, and necrotic. The VH algorithm classify according eight spectral characteristics, and also considering a classification tree which includes information from surrounding regions (Fig. 2.7) [53–55]. In VH the lipid plaque is yellow (Fig. 2.6P), the fibroid plaque is green (Fig. 2.6S), and the calcified plaque is white (Fig. 2.6T). The VH cannot be displayed simultaneously with their gray scale.

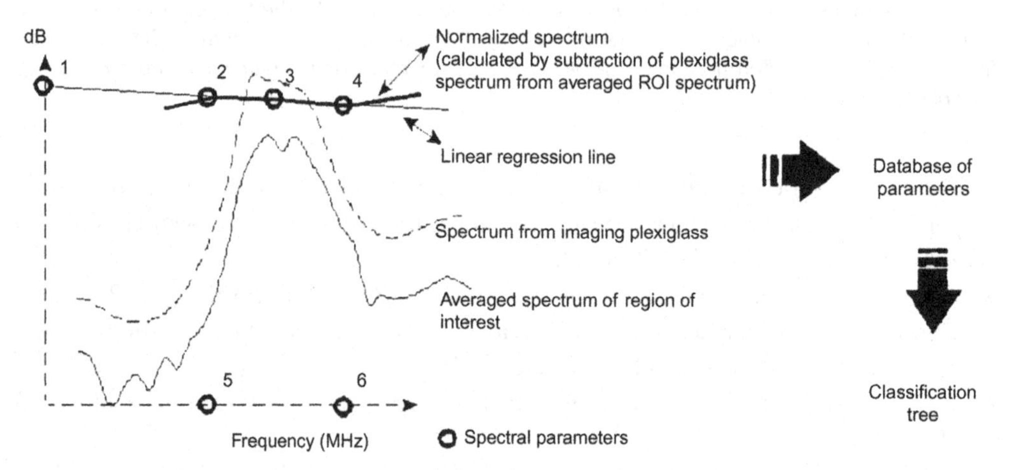

Fig. 2.7 VH algorithm uses eight spectral characteristics to identify signals. These eight spectral characteristics are evaluated using a complex decision tree to result in a determined tissue. This algorithm tree classify in four categories: lipid, fibroid, calcified, and necrotic and is decided according the surrounding tissues [53–55].

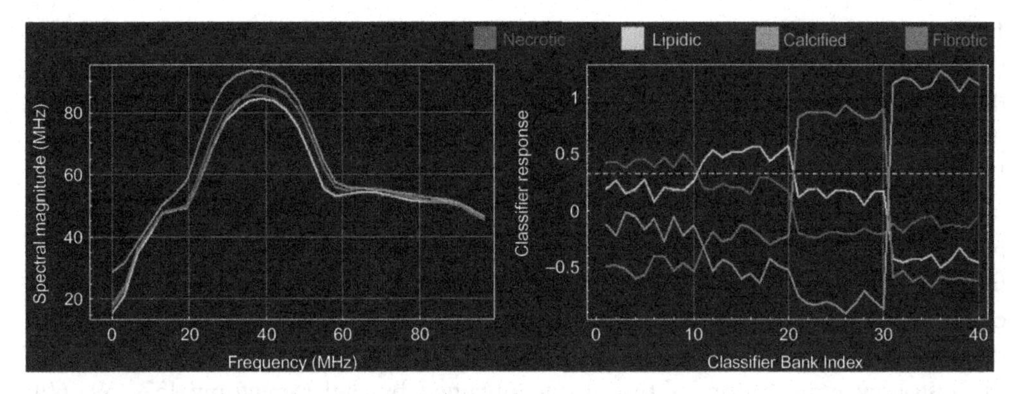

Fig. 2.8 iMap system consists of a direct matching for every single pixel of the sample RF and the RF patterns of tissue signatures (for lipid, fibrotic, and calcified) [56].

6.2 BSC-IVUS and Tissue Classification

BSC-IVUS obtains the signals with a single rotational transducer at 40 MHz and a sound speed in tissue of 1565 m/s, with a focal distance of 1.2 mm. The diameter of the transducer is 0.6 mm, and the total image depth captured is 9.6 mm. The gray-scale RF signals are acquired using a 12-bit acquisition card, ranging from −4096 to +4096 with a sampling rate of fs = 200 MHz and with a radial distance of 4.9 cm. The frame rate of the machine is 30 frames/s. The system provides 256 real angles, each collected from the rotational catheter beginning at 12 o'clock. BSC-IVUS resolution is approximately 50 μm with 40 MHz rotational catheters. In the vessel lumen the blood speckle is visualized in contrast of any other transluminal content as thrombus or dissection flaps. The BSC iMap tissue characterization system consists of a direct matching with a library of RF patterns of tissue signatures (for lipid, fibrotic, calcified, and necrotic) (Fig. 2.8). The lipid plaques appear in iMap in yellow (Fig. 2.6K), the fibrotic tissue in green (Fig. 2.6N), the calcified in blue (Fig. 2.6O), and the necrotic in pink (Fig. 2.6L). The matching between the sample RF waves and the pattern signatures is measured with a confidence level, giving a feedback on the characterization quality in every frame.

6.3 BSC-IVUS and iMap Versus Volcano 20 MHz and VH Comparison

Several techniques of computer vision have been applied in order to compare the quality in images and RF signals from the two IVUS systems in terms of plaque classification. The Barcelona Perception Laboratory (Karla L. Caballero, Joel Barajas, Oriol Pujol, Petia Radeva) measured the resolution and capacity of discrimination among tissues. Volcano images and RF signals showed less resolution than the BSC signal since the range used to capture the information is lower ([−128, 127] 8 bits for Volcano versus [−2048, 2047]

12 bits for BSC). In addition, the number of real beams is higher for BSC (256 beams) than for Volcano (64 beams) reducing the angular resolution of Volcano images. From the analysis performed for the images of both IVUS equipment, the conclusion was that for the entire image, BSC-IVUS has more resolution to characterize texture. BSC-IVUS tissue characterization is presented in iMap as a combination of IVUS gray-scale images plus colorization from the RF classification. For Volcano VH, the images are presented in a pure color scale, without the gray scale image represented in the same frame. Nevertheless, the IVUS system will be extremely useful for the classification of different types of coronary components in atherosclerosis research [48, 57, 58]. In the studies, there is strong correlation between VH IVUS plaque characterization and the histology examination of the plaque obtained by endarterectomy [57, 59, 60]. The measurements of plaque composition obtained by VH and iMap are acceptably reproducible [49, 58, 59, 61, 62].

7. VULNERABLE PLAQUES

A plaque rupture is the most important mechanism of coronary thrombosis and ACS [35, 36, 63]. The ruptured plaque use to consist in a necrotic core with an overlying thin-ruptured cap, infiltrated by macrophages and lymphocytes in the absence of smooth muscle cells within the fibroid cap [27, 28, 64]. The thin fibroid cap near the rupture site measures $23 \pm 19\,\mu m$, with 95% of caps measuring $<65\,\mu m$ (TCFA) [65]. These rupture-prone plaques are called "vulnerable plaques" [48, 49, 66–68]. In unselected autopsy subjects it is common to find TCFAs, clustering in the proximal segments of the three major epicardial coronary arteries [67], so it is more likely that clinical events occurs on a plaque rupture or over a TCFA, but not every TCFA or every plaque rupture generates a clinical event. In patients who died suddenly because of a different plaque rupture, healed ruptures are frequently visible. Furthermore, silent plaque rupture may be a form of wound healing [69]. During the past decades, much effort has been put toward accurately detecting the presence of vulnerable plaques [60, 70]. The OCT-derived TCFA was defined as the presence of a thin fibrous cap ($<65\,\mu m$) overlying a lipid-rich plaque [48, 49, 67, 68] (Fig. 2.9). VH-IVUS assessment of plaque necrosis area $>4\,mm^2$ has a high positive predictive value and correctly identifies OCT-TCFA [61, 71]. Also with iMap-IVUS larger absolute necrotic area in coronary plaque was closely associated with OCT-TCFA [48]. With OCT, IVUS-VH, and IVUS-iMap are useful to identify pathological intimal thickening (PIT), TCFA, and thick-cap fibroatheroma (ThFA); observe examples in Fig. 2.6.

8. SUMMARY

Atherosclerosis begins very early in life in the aorta and usually progresses during the lifetime to coronaries and peripheral vessels. The atherosclerosis classification is a

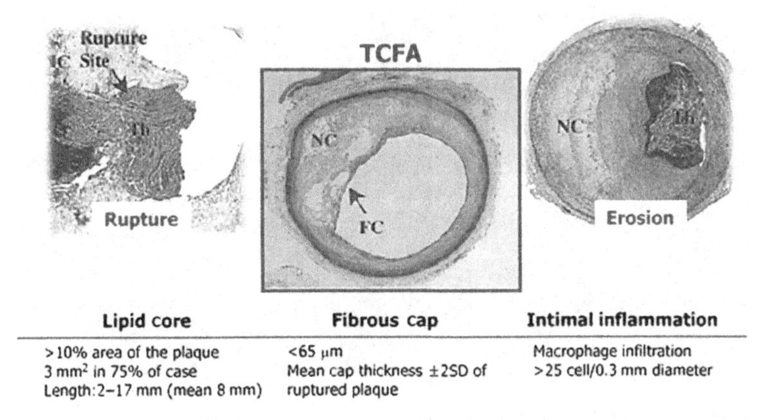

Fig. 2.9 The OCT-derived TCFA was defined as a presence of thin fibrous cap (<65 μm) overlying a lipid-rich plaque.

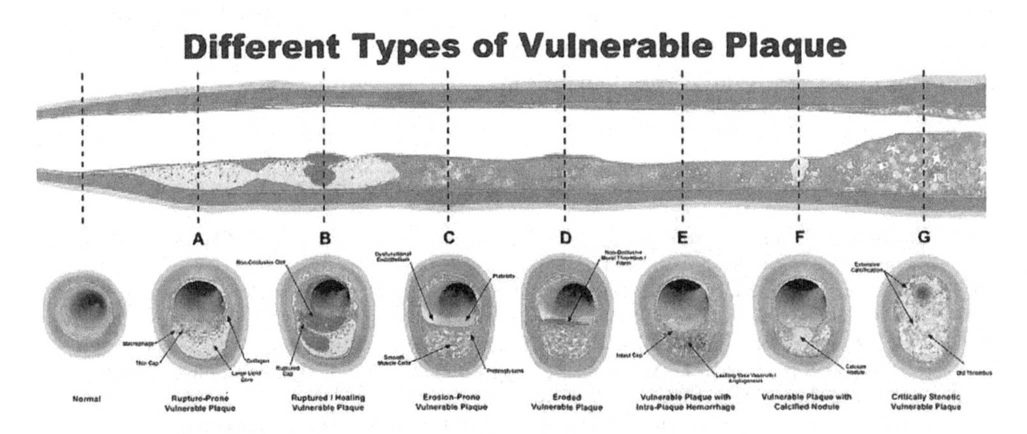

Fig. 2.10 Histological description of the vulnerable plaque evolution.

consequence of the study of atherosclerosis plaques in necropsy from different ages. The disease classification defines a sequential time-lined evolution of the atherosclerosis histology patterns. The plaque components can be characterized in vivo by IVUS and more recently with OCT, and also documented prospectively (Fig. 2.10). In contrast with the well-defined histology patterns, OCT and IVUS show high heterogeneity in atherosclerosis plaques in adults, including the simultaneous presence of multiple histology patterns in the same atherosclerosis plaque. The IVUS and OCT images are predictive for plaque evolution. In atherosclerotic plaques, OCT and IVUS may identify vulnerable areas containing a necro-lipid deposit covered by a fibroid thin cap. OCT and IVUS allow us to consider atherosclerosis plaque as a complex and dynamic structure. If

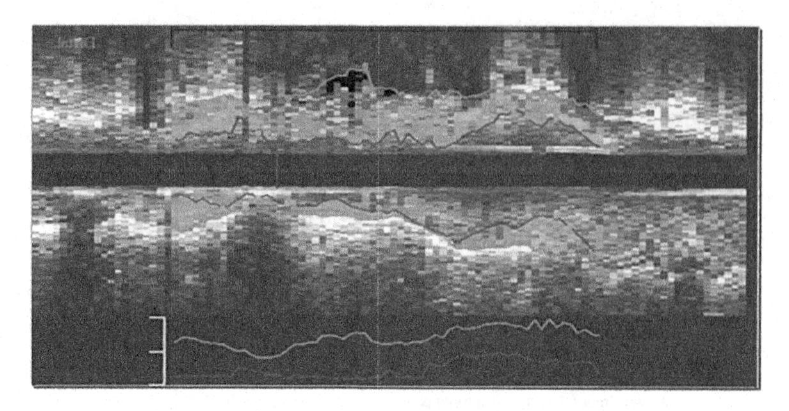

Fig. 2.11 In vivo long-view IVUS iMap tissue characterization example.

we compare the classical description of the vulnerable plaque evolution with the in vivo long-view IVUS/OCT tissue characterization (Fig. 2.11), the inverse correlation of the images is very evident, where to the right of the image is proximal, older and calcified lesions and to the left is distal, younger and lipo-fibroid lesions.

REFERENCES

[1] Holman RL, McGill Jr H, Strong JP, Geer JC. Filtration versus local formation of lipids in pathogenesis of atherosclerosis. J Am Med Assoc 1959;170(4):416–20.
[2] Tegos TJ, Kalodiki E, Sabetai MM, Nicolaides AN. The genesis of atherosclerosis and risk factors: a review. Angiology 2001;52(2):89–98.
[3] Lusis AJ. Atherosclerosis. Nature 2000;407(6801):233–41.
[4] Guyton AC, Hall JE, Zocchi L, Aicardi G. Fisiología médica, vol. 11. Rio de Janeiro: Elsevier; 2006. ISBN: 8821429369, 9788821429361.
[5] Matturri L, Lavezzi AM, Ottaviani G, Rossi L. Intimal preatherosclerotic thickening of the coronary arteries in human fetuses of smoker mothers. J Thromb Haemost 2003;1(10):2234–38.
[6] Wiegman A, Hutten BA, de Groot E, Rodenburg J, Bakker HD, Büller HR, et al. Efficacy and safety of statin therapy in children with familial hypercholesterolemia: a randomized controlled trial. J Am Med Assoc 2004;292(3):331–7.
[7] Kumar V, Abbas AK, Fausto N, Aster JC. Robbins and Cotran pathologic basis of disease. Elsevier Health Sciences; 2014. ISBN: 978-1-4377-1781-5.
[8] McGill H. Fatty streaks in the coronary arteries and aorta. Lab Invest 1968;18(5):560–4.
[9] Strong JP, McGill Jr H. The natural history of coronary atherosclerosis. Am J Pathol 1962;40:37–49.
[10] Stehbens WE. The epidemiological relationship of hypercholesterolemia, hypertension, diabetes mellitus and obesity to coronary heart disease and atherogenesis. J Clin Epidemiol 1990;43(8): 733–41.
[11] Olson RE. The dietary recommendations of the American Academy of Pediatrics. Am J Clin Nutr 1995;61(2):271–3.
[12] Newman TB, Garber AM, Holtzman NA, Hulley SB. Problems with the report of the expert panel on blood cholesterol levels in children and adolescents. Arch Pediatr Adolesc Med 1995;149(3):241–7.
[13] Montenegro MR, Eggen DA. Topography of atherosclerosis in the coronary arteries. Lab Invest 1968;18(5):586–93.

[14] Homma S, Troxclair DA, Zieske AW, Malcom GT, Strong JP, Pathobiological Determinants of Atherosclerosis in Youth (PDAY) Research Group. Histological topographical comparisons of atherosclerosis progression in juveniles and young adults. Atherosclerosis 2008;197(2):791–8.
[15] McGill Jr H, McMahan CA, Herderick EE, Tracy RE, Malcom GT, Zieske AW, et al. Effects of coronary heart disease risk factors on atherosclerosis of selected regions of the aorta and right coronary artery. PDAY Research Group. Pathobiological Determinants of Atherosclerosis in Youth. Arterioscler Thromb Vasc Biol 2000;20(3):836–45.
[16] Newman AB, Siscovick DS, Manolio TA, Polak J, Fried LP, Borhani NO, et al. Ankle-arm index as a marker of atherosclerosis in the cardiovascular health study. Cardiovascular Heart Study (CHS) Collaborative Research Group. Circulation 1993;88(3):837–45.
[17] Fowkes FG, Housley E, Cawood EH, Macintyre CC, Ruckley CV, Prescott RJ. Edinburgh artery study: prevalence of asymptomatic and symptomatic peripheral arterial disease in the general population. Int J Epidemiol 1991;20(2):384–92.
[18] McKenna M, Wolfson S, Kuller L. The ratio of ankle and arm arterial pressure as an independent predictor of mortality. Atherosclerosis 1991;87(2–3):119–28.
[19] Criqui MH, Langer RD, Fronek A, Feigelson HS, Klauber MR, McCann TJ, et al. Mortality over a period of 10 years in patients with peripheral arterial disease. N Engl J Med 1992;326(6):381–6.
[20] Smith GD, Shipley MJ, Rose G. Intermittent claudication, heart disease risk factors, and mortality. The Whitehall Study. Circulation 1990;82(6):1925–31.
[21] Stoffers HE, Rinkens PE, Kester AD, Kaiser V, Knottnerus JA. The prevalence of asymptomatic and unrecognized peripheral arterial occlusive disease. Int J Epidemiol 1996;25(2):282–90.
[22] Vogt MT, Cauley JA, Kuller LH, Hulley SB. Prevalence and correlates of lower extremity arterial disease in elderly women. Am J Epidemiol 1993;137(5):559–68.
[23] Meijer WT, Hoes AW, Rutgers D, Bots ML, Hofman A, Grobbee DE. Peripheral arterial disease in the elderly: The Rotterdam Study. Arterioscler Thromb Vasc Biol 1998;18(2):185–92.
[24] Salonen R, Tervahauta M, Salonen JT, Pekkanen J, Nissinen A, Karvonen MJ. Ultrasonographic manifestations of common carotid atherosclerosis in elderly eastern Finnish men. Prevalence and associations with cardiovascular diseases and risk factors. Arterioscler Thromb 1994;14(10): 1631–40.
[25] Alexander KP, Newby LK, Armstrong PW, Cannon CP, Gibler WB, Rich MW, et al. Acute coronary care in the elderly, part II: ST-segment-elevation myocardial infarction: a scientific statement for healthcare professionals from the American Heart Association Council on Clinical Cardiology: in collaboration with the Society of Geriatric Cardiology. Circulation 2007;115(19):2570–89.
[26] Alexander KP, Newby LK, Cannon CP, Armstrong PW, Gibler WB, Rich MW, et al. Acute coronary care in the elderly, part I: non-ST-segment-elevation acute coronary syndromes: a scientific statement for healthcare professionals from the American Heart Association Council on Clinical Cardiology: in collaboration with the Society of Geriatric Cardiology. Circulation 2007;115(19):2549–69.
[27] Fuster V, Badimon L, Badimon JJ, Chesebro JH. The pathogenesis of coronary artery disease and the acute coronary syndromes (2). N Engl J Med 1992;326(5):310–8.
[28] Fuster V, Badimon L, Badimon JJ, Chesebro JH. The pathogenesis of coronary artery disease and the acute coronary syndromes (1). N Engl J Med 1992;326(4):242–50.
[29] McGill Jr HC, Geer JC, Strong JP. Natural history of human atherosclerotic lesions. In: Sandler M, Bourne GH, editors. Atherosclerosis and its origin. New York: Academic Press; 1963. p. 39–65.
[30] Stary HC, Chandler AB, Glagov S, Guyton JR, Insull Jr W, Rosenfeld ME, et al. A definition of initial, fatty streak, and intermediate lesions of atherosclerosis. A report from the Committee on Vascular Lesions of the Council on Arteriosclerosis, American Heart Association. Circulation 1994;89(5): 2462–78.
[31] Stary HC. Changes in components and structure of atherosclerotic lesions developing from childhood to middle age in coronary arteries. Basic Res Cardiol 1994;89(Suppl. 1):17–32.
[32] Stary HC, Chandler AB, Dinsmore RE, Fuster V, Glagov S, Insull Jr W, et al. A definition of advanced types of atherosclerotic lesions and a histological classification of atherosclerosis. A report from the Committee on Vascular Lesions of the Council on Arteriosclerosis, American Heart Association. Arterioscler Thromb Vasc Biol 1995;15(9):1512–31.

[33] Stary HC, Chandler AB, Dinsmore RE, Fuster V, Glagov S, Insull Jr W, et al. A definition of advanced types of atherosclerotic lesions and a histological classification of atherosclerosis. A report from the Committee on Vascular Lesions of the Council on Arteriosclerosis, American Heart Association. Circulation 1995;92(5):1355–74.
[34] Palac RT, Hwang MH, Meadows WR, Croke RP, Pifarre R, Loeb HS, et al. Progression of coronary artery disease in medically and surgically treated patients 5 years after randomization. Circulation 1981;64(2 Pt 2):II17–21.
[35] Davies MJ, Thomas AC. Plaque fissuring—the cause of acute myocardial infarction, sudden ischaemic death, and crescendo angina. Br Heart J 1985;53(4):363–73.
[36] Davies MJ. Anatomic features in victims of sudden coronary death. Coronary artery pathology. Circulation 1992;85(1 Suppl.):I19–24.
[37] Davies MJ. Stability and instability: two faces of coronary atherosclerosis. The Paul Dudley White Lecture 1995. Circulation 1996;94(8):2013–20.
[38] Canto JG, Shlipak MG, Rogers WJ, Malmgren JA, Frederick PD, Lambrew CT, et al. Prevalence, clinical characteristics, and mortality among patients with myocardial infarction presenting without chest pain. J Am Med Assoc 2000;283(24):3223–9.
[39] Sheifer SE, Arora UK, Gersh BJ, Weissman NJ. Sex differences in morphology of coronary artery plaque assessed by intravascular ultrasound. Coron Artery Dis 2001;12(1):17–20.
[40] Falk E, Fuster V. Angina pectoris and disease progression. Circulation 1995;92(8):2033–5.
[41] Pohl T, Seiler C, Billinger M, Herren E, Wustmann K, Mehta H, et al. Frequency distribution of collateral flow and factors influencing collateral channel development. Functional collateral channel measurement in 450 patients with coronary artery disease. J Am Coll Cardiol 2001;38(7): 1872–78.
[42] Werner GS, Ferrari M, Betge S, Gastmann O, Richartz BM, Figulla HR. Collateral function in chronic total coronary occlusions is related to regional myocardial function and duration of occlusion. Circulation 2001;104(23):2784–90.
[43] Virmani R, Kolodgie FD, Burke AP, Farb A, Schwartz SM. Lessons from sudden coronary death: a comprehensive morphological classification scheme for atherosclerotic lesions. Arterioscler Thromb Vasc Biol 2000;20(5):1262–75.
[44] Cassar A, Matsuo Y, Herrmann J, Li J, Lennon RJ, Gulati R, et al. Coronary atherosclerosis with vulnerable plaque and complicated lesions in transplant recipients: new insight into cardiac allograft vasculopathy by optical coherence tomography. Eur Heart J 2013;34(33):2610–7.
[45] van Soest G, Regar E, Goderie TPM, Gonzalo N, Koljenović S, van Leenders GJLH, et al. Pitfalls in plaque characterization by oct: image artifacts in native coronary arteries. JACC Cardiovasc Imaging 2011;4(7):810–3.
[46] Bezerra HG, Attizzani GF, Sirbu V, Musumeci G, Lortkipanidze N, Fujino Y, et al. Optical coherence tomography versus intravascular ultrasound to evaluate coronary artery disease and percutaneous coronary intervention. JACC Cardiovasc Interv 2013;6(3):228–36.
[47] Kawasaki M, Bouma BE, Bressner J, Houser SL, Nadkarni SK, MacNeill BD, et al. Diagnostic accuracy of optical coherence tomography and integrated backscatter intravascular ultrasound images for tissue characterization of human coronary plaques. J Am Coll Cardiol 2006;48(1):81–8.
[48] Koga S, Ikeda S, Miura M, Yoshida T, Nakata T, Koide Y, et al. iMap-intravascular ultrasound radiofrequency signal analysis reflects plaque components of optical coherence tomography-derived thin-cap fibroatheroma. Circ J 2015;79(10):2231–7.
[49] Kubo T, Nakamura N, Matsuo Y, Okumoto Y, Wu X, Choi SY, et al. Virtual histology intravascular ultrasound compared with optical coherence tomography for identification of thin-cap fibroatheroma. Int Heart J 2011;52(3):175–9.
[50] Yabushita H, Bouma BE, Houser SL, Aretz HT, Jang IK, Schlendorf KH, et al. Characterization of human atherosclerosis by optical coherence tomography. Circulation 2002;106(13):1640–5.
[51] von Birgelen C, Mintz GS, Nicosia A, Foley DP, van der Giessen WJ, Bruining N, et al. Electrocardiogram-gated intravascular ultrasound image acquisition after coronary stent deployment facilitates on-line three-dimensional reconstruction and automated lumen quantification. J Am Coll Cardiol 1997;30(2):436–43.

[52] Mintz GS, Nissen SE, Anderson WD, Bailey SR, Erbel R, Fitzgerald PJ, et al. American College of Cardiology Clinical Expert Consensus Document on Standards for Acquisition, Measurement and Reporting of Intravascular Ultrasound Studies (IVUS). A report of the American College of Cardiology Task Force on Clinical Expert Consensus Documents. J Am Coll Cardiol 2001;37(5): 1478–92.
[53] Nair A, Kuban BD, Tuzcu EM, Schoenhagen P, Nissen SE, Vince DG. Coronary plaque classification with intravascular ultrasound radiofrequency data analysis. Circulation 2002;106(17):2200–6.
[54] Nair A, Klingensmith JD, Vince DG. Real-time plaque characterization and visualization with spectral analysis of intravascular ultrasound data. Stud Health Technol Inform 2005;113:300–20.
[55] Nair A, Kuban BD, Obuchowski N, Vince DG. Assessing spectral algorithms to predict atherosclerotic plaque composition with normalized and raw intravascular ultrasound data. Ultrasound Med Biol 2001;27(10):1319–31.
[56] Sathyanarayana S, Carlier S, Li W, Thomas L. Characterisation of atherosclerotic plaque by spectral similarity of radiofrequency intravascular ultrasound signals. EuroIntervention 2009;5(1):133–9.
[57] Nasu K, Tsuchikane E, Katoh O, Vince DG, Virmani R, Surmely JF, et al. Accuracy of in vivo coronary plaque morphology assessment: a validation study of in vivo virtual histology compared with in vitro histopathology. J Am Coll Cardiol 2006;47(12):2405–12.
[58] Heo JH, Brugaletta S, Garcia-Garcia HM, Gomez-Lara J, Ligthart JMR, Witberg K, et al. Reproducibility of intravascular ultrasound iMap for radiofrequency data analysis: implications for design of longitudinal studies. Catheter Cardiovasc Interv 2014;83(7):E233–42.
[59] Ivanovíc M, Rancić M, Rdzanek A, Filipjak KJ, Opolski G, Cvetanović J. Virtual histology study of atherosclerotic plaque composition in patients with stable angina and acute phase of acute coronary syndromes without st segment elevation. Srp Arh Celok Lek 2013;141(5–6):308–14.
[60] Vancraeynest D, Pasquet A, Roelants V, Gerber BL, Vanoverschelde JLJ. Imaging the vulnerable plaque. J Am Coll Cardiol 2011;57(20):1961–79.
[61] Liu J, Wang Z, Wang Wm, Li Q, Ma Yl, Liu Cf, et al. Feasibility of diagnosing unstable plaque in patients with acute coronary syndrome using iMap-IVUS. J Zhejiang Univ Sci B 2015;16(11):924–30.
[62] de Souza CF, Maehara A, Lima E, Guimarães LdFC, Carvalho AC, Alves CM, et al. Morphological and tissue characterization of culprit lesions in patients with ST-segment elevation myocardial infarction after thrombolytic therapy. Analysis with grayscale intravascular ultrasound and iMap technology. Rev Bras Cardiol Invasiva 2014;22(3):225–32.
[63] Falk E, Shah PK, Fuster V. Coronary plaque disruption. Circulation 1995;92(3):657–71.
[64] Fuster V. Elucidation of the role of plaque instability and rupture in acute coronary events. Am J Cardiol 1995;76(9):24C–33C.
[65] Burke AP, Farb A, Malcom GT, Liang YH, Smialek J, Virmani R. Coronary risk factors and plaque morphology in men with coronary disease who died suddenly. N Engl J Med 1997;336(18):1276–82.
[66] Finn AV, Nakano M, Narula J, Kolodgie FD, Virmani R. Concept of vulnerable/unstable plaque. Arterioscler Thromb Vasc Biol 2010;30(7):1282–92.
[67] Kume T, Okura H, Yamada R, Kawamoto T, Watanabe N, Neishi Y, et al. Frequency and spatial distribution of thin-cap fibroatheroma assessed by 3-vessel intravascular ultrasound and optical coherence tomography: an ex vivo validation and an initial in vivo feasibility study. Circ J 2009;73(6):1086–91.
[68] Kubo T, Maehara A, Mintz GS, Doi H, Tsujita K, Choi SY, et al. The dynamic nature of coronary artery lesion morphology assessed by serial virtual histology intravascular ultrasound tissue characterization. J Am Coll Cardiol 2010;55(15):1590–7.
[69] Burke AP, Kolodgie FD, Farb A, Weber DK, Malcom GT, Smialek J, et al. Healed plaque ruptures and sudden coronary death: evidence that subclinical rupture has a role in plaque progression. Circulation 2001;103(7):934–40.
[70] Mintz GS, Maehara A. Serial intravascular ultrasound assessment of atherosclerosis progression and regression. State-of-the-art and limitations. Circ J 2009;73(9):1557–60.
[71] Stone GW, Maehara A, Lansky AJ, de Bruyne B, Cristea E, Mintz GS, et al. A prospective natural-history study of coronary atherosclerosis. N Engl J Med 2011;364(3):226–35.

CHAPTER 3

AAA Treatment Strategy Change Over Time

R. Ghotbi, R. Mansour
Helios Klinikum Munich West Academic Hospital of Ludwig-Maximilian University, Munich, Germany

Chapter Outline

Computing and Visualization for Intravascular Imaging and Computer-Assisted Stenting
http://dx.doi.org/10.1016/B978-0-12-811018-8.00003-5

Chapter Points

- The effectiveness of the elective AAA-surgery implies that most cases of death caused by an AAA are preventable.
- The open surgical procedure remains the procedure of choice for young patients with high life expectancy and a low operative, anesthesiologic risk.
- Endovascular aneurysm repair (EVAR) has revolutionized the morbidity of aneurysm-treatment and remains a high technical-dependent procedure.
- Imaging plays a crucial role in the workflow and has a high impact on all aspects of diagnosis and risk stratification, as well as on the selection of the appropriate therapy modality for individual patients.

1. INTRODUCTION

The 20th century has seen great advances in surgical techniques in which the focus changed from resective emergency surgery to reconstructive and prophylactic procedures. Especially in the early part of the 20th century, surgeons gained scholarly acceptance as operations became safer. Continued advancement in pathophysiological understanding combined with milestones in anesthesia and the implementation of surgical asepsis were all fundamental requirements for surgical innovations of this period.

Two surgeons were awarded the Nobel Prize in the 20th century: Kocher for his work on the thyroid gland, and Carel in 1911 for the first successful suture of a blood vessel and the technique of the vascular anastomosis. Despite these steps forward, the aortic aneurysm still remained a fatal diagnosis until the mid-century. It was not until 1952 that Dubost [1] (Fig. 3.1), a French surgeon in Paris, could realize new findings and possibilities by successfully performing the first open surgery of an abdominal aortic aneurysm (AAA). His technique was to replace an AAA with a segment of the thoracic aorta of a recently deceased 20-year-old man. At first, the operation could exclusively be performed by using a homograft (Fig. 3.2). It took a few years until a synthetic prosthesis was available. With this prosthesis, an anastomosis could be performed proximal and distal of the AAA to maintain an antegrade blood flow. Two years later, Blakemor published a series of 17 operative procedures with aortic replacement that were carried out with the first synthetic prosthesis (Vignion N.), thus signaling the beginning of the modern age of synthetic replacement of blood vessels. Apart from new developments of prosthesis materials, the technique has remained largely unchanged until the present day.

In the same year that Dubost died (1991), Juan Parodi introduced and published a dramatic change in therapeutic management with the first endovascular AAA-repair (EVAR). This more recent kind of therapy has the same goal, namely to avoid rupture

Fig. 3.1 Charles Dubost (1914–91).

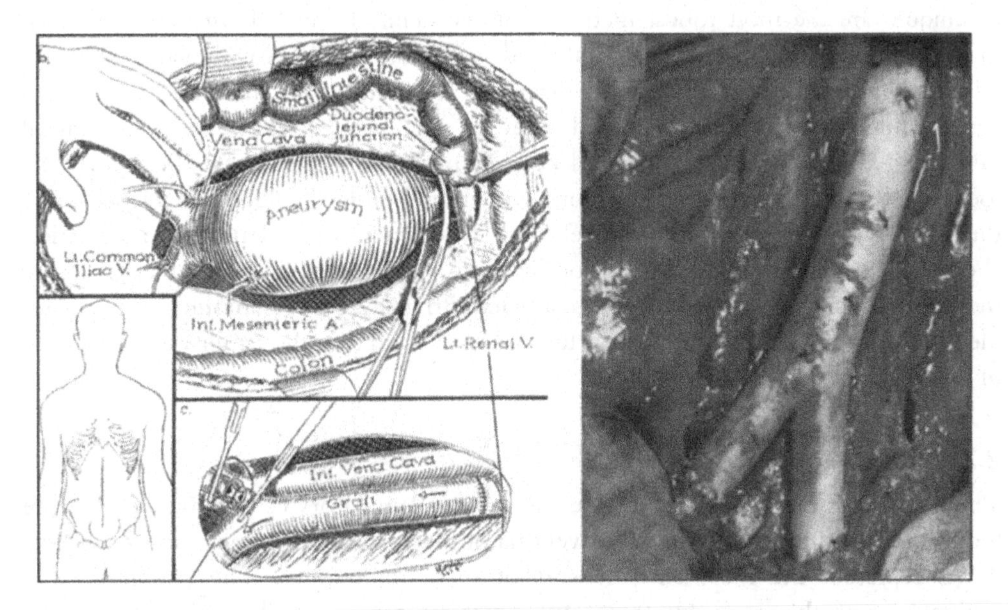

Fig. 3.2 Early preoperative planning for open surgery/homograft implant.

and eliminate the risk of death due to AAA, but follows different principles than open surgical repair.

Moving from a purely clinical/operative perspective, recent advancements in imaging technology and the implementation of operative workflow have shifted the framework in which operative procedures are conducted. No longer seen as the solution to a long-standing problem, procedures are part of an overarching treatment that includes the latest technology and workflows to allow surgeons to increasingly perform with a preventative focus.

This chapter describes a clinical perspective of the aneurysmatic disease, the present-day clinical workflow, and the basic technical challenges for a successful endovascular therapy of the AAA; the procedure that still represents the essence of vascular surgery.

2. PATHOGENESIS

2.1 Genetics

The pathogenesis of the AAA seems to have a significant genetic component. Frequent occurrence in families was first described in 1977 [2] and is nowadays recognized as an established fact by vascular surgeons. The prevalence of AAA is twice as high as that in the general population if an aneurysmatic disease exists among first-degree relatives. Different segregation and twin studies have corroborated evidence for the genetic influence on the development of an AAA [3]. Considering genetic aspects, these techniques are the most robust methods for the delineation of the genetic influence in human population. So, the probability of contracting an AAA is similar to the age-related macular degeneration or type 2 diabetes. However, the search for a corresponding gene segment that can be identified with an evident association with the clinical outcome has not been successful until now. Remarkable studies in this field currently focus on the matrix protein regulation, with most studies being relatively small and the statistical power being, respectively, too weak to provide definitive information about corresponding specific gene localizations [4, 5]. It is expected that by using modern informatics with its enormous capacity to analyze large amounts of data, this identification will succeed in the near future. Apart from this, the familiar accumulation of AAAs is a fact.

2.2 Inflammation

The histologic examinations of AAA tissue show inflammatory cell infiltrations that are localized in the intima, media, and adventitia of the aorta. The so-called inflammatoric AAAs represent the most extreme sort of these infiltrations. The reason for the infiltrations of the aorta by these inflammatory cells as well as the consecutive degeneration of the matrix is unknown [6]. Infections with *Chlamydia pneumoniae* might play a role. Different studies show that an augmented amount of chlamydia antibodies occurs significantly more often in patients with AAA than in the control groups [7]. On the other hand, the prevalence of the chlamydia infection among the general population is incomparably higher with respect to the prevalence of AAA, so there is still a lack of evidence that a chlamydia infection initiates an AAA or rather influences the progression of the aneurysm. The regularly existing inflammation of the vessel wall and the corresponding immune response that is regulated by cytokines continually leads to a decrease of the smooth muscle in the aortic media and to a significantly higher rate of apoptosis in the AAA tissue. The significant decrease of the smooth muscle mass of the media-wall inevitably leads to a reduction of the wall thickness that consecutively leads to a dilatation and further degeneration of the aorta [8].

2.3 Biomechanical Aspects

Biomechanical factors have an extensive influence on the development and progression of aneurysms of the terminal aorta. The anatomical formation of the aortic bifurcation, the large-lumen renal arteries, lumbar arteries, and the hypogastric arteries result in a unique disturbance of blood-flow in the infrarenal aorta (Fig. 3.3). The rigidity of this part of the aorta [9] increases in proportion to the decrease of the elastin/collagen-proportion. Additional intramural thrombus, which regularly appear in aortic aneurysms, as well as the absence of any vasa vasorum of the media of the normal abdominal aorta, are factors that reduce the nutritive blood-transport to the

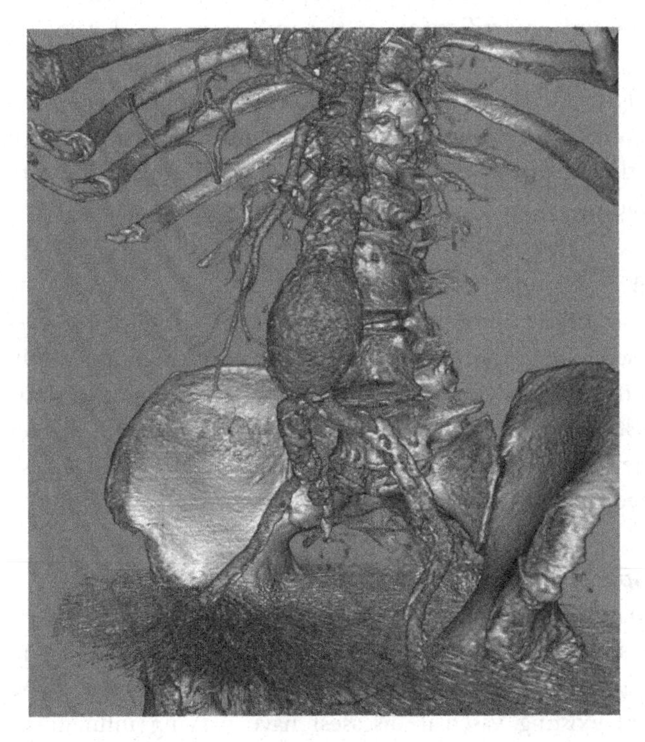

Fig. 3.3 Arteriosclerotic AAA.

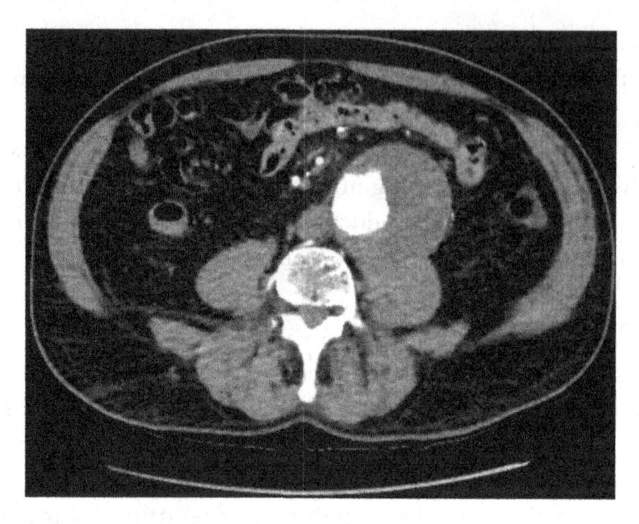

Fig. 3.4 AAA with an excentric thrombus.

aortic wall. The coexistence of these factors causes significant stress for the aortic wall that can be regarded as a key factor for the rupture (Fig. 3.4). More and more methods for the evaluation and objectification of the wall stress are developed. These aspects can influence the therapeutic decisions of the future, together with the pathogenetic factors. Accordingly, different studies have shown that the analysis of wall stress in the 3D CT-angiography (CTA) describes a better prognosis factor of the rupture risk than the diameter of the aneurysm that has been used until now as the only factor for the therapeutic decision.

2.3.1 Small AAA

Today there is substantially better understanding of the complex pathology of the aorta. Combined with screening programs, these factors can greatly improve the therapy of the small aneurysms to establish a more effective medical treatment to slow the increase of aneurysms. A 50% reduction of the progression rate of aneurysm diameter theoretically prolongs the time before surgery becomes necessary by 10 years. These 10 years often raise the life expectancy for many patients that have a small aneurysm [10, 11].

2.3.2 Clinical Aspects and Epidemiology of AAA

Most AAAs remains asymptomatic until rupture (Fig. 3.5). Seventy-five percent of patients do not show any kind of symptoms at the time of diagnosis. Etiological factors (old age, male gender, positive family history, smoking, hypercholesteremia, hypertension, and existing vascular diseases) have varying influences on the rate of increase. In this connection, both male gender and smoking increase the risk of

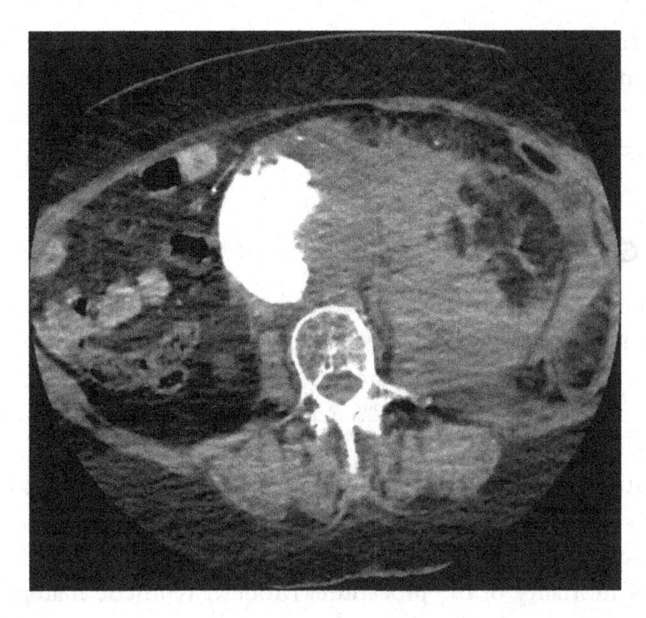

Fig. 3.5 Ruptured AAA.

AAA-rupture by a factor of 5 [12]. In the United States, ruptured AAA is the 10th most common cause of death in men over 55 years old. This number is quite undervalued, as 30–50% of all patients die before they reach a hospital, and a further 30–40% die after having reached a hospital, but without having been operated on. We have to expect a mortality rate of 40–50% in patients who underwent surgery, so we have to expect a total mortality rate of more than 80% in the case of rupture. Due to the considerable progress of surgical techniques and perioperative management, the mortality of elective surgical interventions could be reduced to less than 5%.

The effectiveness of elective AAA surgery implies that most cases of death caused by an AAA would theoretically be preventable.

2.3.3 Therapy

The essential strategy of AAA management is the prevention of rupture. The best medical treatment consists of hypertension control, cholesterol reduction, abstinence from smoking, and anticoagulation.

For an operative treatment, three individual aspects have to be addressed: the risk of rupture, the risk of an elective treatment, and the life expectancy of the patient.

The randomized studies show that regarding these aspects, it is justifiable to wait until the dilatation of the aneurysm has reached a diameter of 5 cm, when it is guaranteed that the patient will be observed in a special surveillance program [12]. The average diameter with women is normally about 0.5 cm smaller than with men. Further indications are a rapid increase in the size of the aneurysm, back pain or peripheral embolizations. In every individual case, the general guidelines have to be evaluated and reflected with regard to a risk-benefit analysis for each patient.

3. OPEN SURGICAL REPAIR

Most arterial aneurysms that are treated by surgery are localized at the infrarenal aorta. This results in approximately 60% of all operated aneurysms concerning the infrarenal aorta; iliac aneurysms, suprarenal, and thoracoabdominal together represent about 19%, while femoral are 6.4% and popliteal 2.2%. Multilevel aneurysms affect about 15% of all operated-on patients.

After the initial description of the successful resection and replacement of the terminal aorta by Dubost more than 60 years ago, numerous steps and preoperative assessment, surgical techniques, and anesthesiologic management have dramatically reduced the early mortality of the procedure. Modern synthetic materials have helped to achieve longer patency rates and thus have significantly reduced the incidence of so-called graft-related late complications.

But without a direct preparation and clamping of the aorta, an aortic reconstruction was not possible until the era of the catheterized femoral EVAR.

Now, 20 years after the introduction of this new technique, several late complications are known, which necessitate lifelong follow-up and which include material-referred complications of the stent, fracture, device migration, and the more frequent problem of endoleak. Open procedures still have a clear indication for patients who primarily do not indicate any substantial surgical or anesthesiological risk for the open procedure.

3.1 Indication

The disease progression measured based on increased aneurysmal diameter and volume still defines the fundamental risk of rupture of an AAA. Ischemic complications of the aneurysm with a complete thrombosis or distal embolism that, in case of an aortic occlusion, often lead to a chronic ischemia of the lower extremity are significantly much less common. Therefore the aim of the open treatment is the prevention of rupture, meaning the maintenance of an adequate arterial perfusion of the pelvis and the lower extremity [13].

While emergency care is justified for each kind of rupture, that is, clearly symptomatic AAAs, independent of its size, the indication for an elective surgical treatment is recommended at a size of 4.5–5.5 cm. This assessment is based on the maximum diameter of the aneurysm, corresponding to the guidelines of the European and the US-American societies of vascular surgery.

The historic data presented in this chapter are based on the data analysis of the ruptured AAAs that have undergone surgery. About 20% of these cases showed a diameter of 5 cm, 40% showed 6 cm, and a further 40% of these patients had an infrarenal AAA of at least 7 cm. The annual risk of rupture is 4% with a diameter of 5 cm, 7% with a diameter of 6 cm, and 20% with a diameter of 7 cm. Another factor is the rate of expansion of the aneurysm that exorbitantly raises the risk of a rupture of an AAA.

3.2 Preoperative Planning and Intraoperative Management

CTA is the essential preoperative diagnostic investigation that makes it possible to precisely illustrate and measure the aneurysm morphology, the anatomy of the patient, including anatomical anomalies of the infrarenal aorta and surrounding vital structures, as well as the vena cava, renal veins, renal arteries, the inferior mesenteric artery, and the iliac arteries and epigastric arteries. The evaluation of the wall structure, presence of the thrombus material, and calcification and inflammatory tissue on and around the aortic wall are also important information for case planning.

Intraoperative management is a dynamic, complex, and occasionally challenging process. Hemodynamic and metabolic changes that occur within the laparotomy, the cross-clamping and the declamping of the aorta, have to be considered with their corresponding consequences for organic ischemia during clamping and aortic replacement. Blood loss, with all its consequences relating to hypothermia and anesthesiologic management, is an inevitable component of the intervention.

The main alteration of the patients' physiology is initiated by aortic clamping, which causes the complex hormonal response of the renin-angiotensin system and of the sympathetic nervous system, and leads to an acidosis with an increase of the concentration of free oxygen radicals and a shift in the prostaglandin production of the neutrophiles and the complement activation, and generally to myocardial factors, caused by the aortic clamping, whereupon only the clamping plays a central role. The normalization of hemodynamics by declamping the aorta also occurs, along with considerable dynamic alterations. The consecutive hypotension should be kept within limits by an appropriate anesthesiologic management. In addition, the reperfusion of the lower extremity with the corresponding systemic circulation of metabolites and electrolytes has to be managed adequately at this stage.

3.3 Outcome

Most analyses and studies show an obvious relationship of the mortality and the number of comorbidities of the aged patient, female gendered, but also in particular the absence of an appropriate vascular surgical training and an inverse proportional relationship of hospital and surgeon volume [14, 15].

The greatest influence factors on the outcome of the open aortic therapy are comorbidities as cardiac ischemia, cardiac insufficiency, COPD that has with female gender an exceptionally negative influence. Nevertheless, the developments in surgical techniques, in the material and particularly in the anesthesiologic management, have realized even in large studies an operative mortality of less than 5% through all the patient groups.

3.4 Summary

Sixty years after the first description of this procedure, open aortic surgery represents a signature intervention of vascular surgery that can achieve excellent and permanently good results that are adjusted to the numerous characteristics of the particular patient in the sense of a personalized management. Even if more and more patients are treated endovascularly today, by using the new generation EVAR devices and the progress in technology of catheter-based interventions, the open procedure remains the procedure of choice for young patients with a high life expectancy and a low operative anesthesiologic risk.

4. ENDOVASCULAR ANEURYSM REPAIR

Since the middle of the last century, there has been a trend toward considering surgery as a special field of medicine with procedures that can considerably improve the quality of life of patients instead of being something that only exists to treat emergencies. Essentially, the trend to reach surgical goals with less invasive methods is influenced by the available techniques.

In the 1990s, many technical advances were made in surgical medicine. In the last 20 years, vascular surgery has integrated this technology into clinical daily routine.

It would be a mistake to frame endovascular surgery as an invention of the 1990s. Instead, it has been a continuous development based on the special characteristics of vascular medicine to reflect the technical possibilities in the clinical daily routine. Key figures such as Dos Santos [16] and Vollmar [17, 18] were able to demonstrate much earlier that procedures starting from the femoral artery could be effectively performed on other vessel segments by using the technique of thrombo-endarteriectomy with the help of ring strippers. In the same year (1964), Dotter and Judkins published the results of their so-called angioplasty of the femoropopliteal artery through the inguinal artery with the so-called coaxial-catheter [19], which was replaced a few years later by the dilatation balloon, until in the early 1990s, Parodi et al. [20] covered an AAA with a first endograft that was balloon expandable.

A laparotomy is no longer necessary for the treatment of an AAA because of the availability of endovascular therapy. Convalescence is reduced from 3 months to a few days. The inguinal access requires a significantly less burdening anesthesia. Aortic clamping is not necessary at any time. Blood loss is substantially lower as in open procedures.

Therefore, this procedure makes it possible to offer a therapy for patients with serious comorbidities who would not be candidates for an open procedure because of these comorbidities.

Despite all these technical innovations of the last 20 years, anatomical configuration remains the essential factor to determine planning, performance, and early and late results of the endovascular therapy.

In the following section, we present the main anatomical and technical aspects that have to be addressed in the planning of an endovascular intervention for AAA.

4.1 Blood Supply of the Abdominal Organs (Visceral and Renal)

The supply of the visceral organs by the celiac trunk, the superior mesenteric artery, and the inferior mesenteric artery should be clearly analyzed before a reconstruction, particularly because the inferior mesenteric artery is routinely covered by EVAR. An open superior mesenteric artery and an open celiac trunk, renal arteries, respectively, and an internal iliac artery are all requirements for the vital functioning of the abdominal organs. Compared to other abdominal organs, the kidneys can frequently be supplied by accessory arteries that do not anastomose intrarenally and present terminal arteries. As such, the presence of accessory renal arteries should therefore be identified preoperatively, and the presumable malfunction of renal parenchyma should be evaluated and considered for the operative planning.

4.2 Diameter of the Proximal Neck

By definition, an aorta with a diameter of more than 3 cm should be considered as aneurysmatic. As previously mentioned, pathophysiological alterations of the aorta wall can be expected regarding degeneration. In this way, an aneurysm neck of more than 3 cm can be a clear indicator for the occurrence of late complications in terms of a late degeneration process of the aortic wall, even if it can be technically treated with the available advanced techniques. Normally, the proximal neck with a size of less than 30 mm can be treated with a standard technique with an oversizing of 10–20%. However, an exact measurement of this distance is a basic prerequisite.

4.3 Length of the Proximal Neck

As there is no surgical anastomosis with this technique, the central sealing of the prosthesis and the aortic wall is essential for therapeutic success. Experiences of the last 20 years demonstrate that a minimum length of 10–15 mm is necessary to achieve a central sealing.

Statistical studies have shown that shorter necks have a considerable and significantly higher endoleak rate with any prosthesis.

Due to the availability of fenestrated and side-branch prostheses, a short neck is no longer a contraindication for endovascular treatment, even if the treatment by fenestration or side-branches becomes increasingly complex. The procedure is possible nowadays. Attempts to treat shorter necks with suprarenal fixation have regularly failed during recent years. A short neck remains short; the location of the fixation is only secondary.

4.4 Angulation of the Proximal Neck

Angulations of the proximal neck are common and they are more pronounced in larger aneurysms. It is one of the fundamental determinants for the primary outcome of EVAR. Minor angulations (<40 degree) can be treated very well by the common prostheses. More pronounced angulations imply a significantly higher risk for complications, in particular for development of type I endoleak [21].

4.5 Conical Configuration of the Proximal Neck

In statistical studies, an alteration of the contour of the proximal neck of more than 10% (beginning at the lower edge of the renal arteries to the beginning of the infrarenal aneurysm) can involve postoperative complications, particularly as it is more complex to achieve a sealing in these cases than at a neck that does not show any tapering [21]. In these cases, there is an evident limit for the oversizing with 10–20% that cannot be increased.

4.6 Calcifications and Mural Thrombus at the Proximal Neck

Calcifications and existing mural thrombus are common formations at the proximal neck. Their manifestation and configuration can considerably influence the sealing of the prosthesis and increase the risk of endoleaks, that is, considerably increase the risk of a stent graft migration if extended calcifications are present.

4.7 Anatomical Alterations of the Iliac Arteries

In general, the distal fixation of the endograft is the common iliac artery. One-third of all reinterventions after EVAR are still carried out because of alterations of the iliac level, so, analogical to the proximal neck, the diameter and the length are factors that influence the sealing of the prosthesis.

The Eurostar register data have shown that the application of conventional endografts in aortoiliac aneurysms can lead to an increased incidence of type Ib endoleaks, secondary interventions to late ruptures. Today, the problem of an adequate distal sealing is generally solvable by the application of iliac-side-branched prostheses.

The iliac arteries also play an essential role as access arteries to the terminal aorta and they are also a source of complications if their anatomy has not being carefully considered preoperatively. Kinkings and arteriosclerotic alterations of the iliac level are very common. The sheath systems that regularly have relatively large diameters (12–24 French) despite the low-profile prostheses can lead to extensive postoperative complications.

There are still no criteria for the decision regarding which anatomy can be overcome by which sheath system. The personal experience of the surgeon is still decisive regarding the devices that are used.

5. PREOPERATIVE PLANNING

5.1 CTA

While primarily the diameter and the localization of the aneurysm have been of interest in open surgical procedures, the surgical performance, and the postoperative outcome and the selection of the patients for endovascular treatment largely depend on the preoperative clinical diagnostics and its accuracy.

Sonography is certainly the basic examination method for the primary diagnosis of AAA. Spiral CTA is the imaging technique of choice for the planning of EVAR and an essential component of preoperative planning. It offers detailed information of the vascular anatomy that is important for patient selection as well as for the adequate selection of the stent graft [22]. The patients should not have taken any oral contrast agent, as this could definitely influence the 3D reconstruction negatively. There should be a scan from at least 15 cm above the separation of the coeliac trunk to far beyond the femoral bifurcation.

There are certainly 100 mL of contrast agent necessary for the common CTA-protocol. Layer thicknesses are normally 3–5 mm, for complex cases; however, 1 mm axial layers are required.

Potential misinterpretations can occur relatively often at the axial images, as the aorta can vary in its expansion in all dimensions and directions. An aneurysm can expand anterodorsally or laterally. The axial images do not always reflect this anatomy adequately, so a clear source of mistakes results from overestimating the diameter of the neck and underestimating the length of the neck.

The diagonal projections can be corrected at a workstation by using appropriate correction methods, and thus the real aortic shape can be illustrated. The multiplanar reconstructions from CTA data help in obtaining an adequate mapping (Fig. 3.6).

5.2 Arterial Angiography

Angiography can provide additional information that is superior to CTA, considering the demonstration of the degree of stenosis in particular of the side branches of

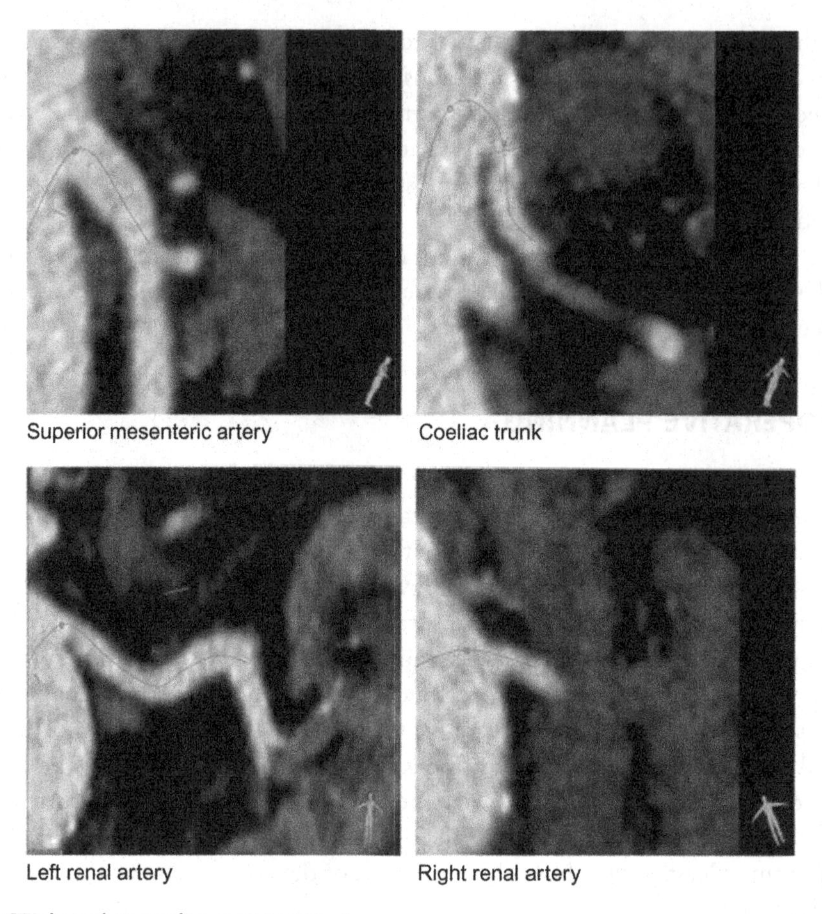

Fig. 3.6 CTA-based center line measurement.

the aorta. However, this technique is inferior for the evaluation of the aortic wall, particularly in the presence of a thrombus coat, and for the assessment of calcifications and their morphology. For many years, arterial angiography was necessary within the preoperative planning particularly for the adequate measurement of the length. From today's perspective, arterial angiography has become obsolete as a preoperative method because of the availability of CTA programs and modern endograft components.

5.3 Magnetic Resonance Angiography

Similar to CTA, magnetic resonance angiography (MRA) produces 2D and 3D images. The MRA remains inferior to CTA in terms of the resolution. As far as the technique is concerned, it is more complex and time-consuming, but the reduction of radiation is still an advantage of MR angiography.

5.4 Selection of Stent Grafts

The large potential market for endovascular products led to appropriate industrial innovations in this field. Intensive research of materials and techniques of stent grafts has made it possible to have four generations of stent grafts. The most frequently used covering material is polytetrafluorethylene (PTFE). Considering the abundance of technical characteristics and anatomical variations, the experiences of the last two decades have taught us that there will not always be stent grafts for all cases. The fundamental characteristics of the ideal stent graft can be listed as follows:

- stent graft-size range,
- durability,
- compatibility,
- low delivery-device size,
- delivery-device flexibility,
- sealing capacity,
- radio-capacity,
- low thrombogenicity,
- low delivery-device profile,
- delivery-device reliability, and
- excellent apposition of graft.

5.5 Components of the Endovascular Devices (Delivery System)

Currently, stent grafts are applied with the help of introducer sheaths, deployment capsules, and retractable covers. On the one hand, the profile of the introducer sheaths has to be small enough to enable passage through the iliac arteries without vascular injury. On the other hand, it has to be large enough to enable an easy passage of the stent graft without generating material damage. The sheaths have to be robust enough to overcome a kinking of the iliac arteries, but also flexible enough to participate in the angulations at the iliac level.

A hemostatic mechanism is available at modern delivery systems and this is an essential component for minimizing blood loss.

5.6 Graft Material

The graft material is the subject of similar strict requirements. It has to be strong enough to enable an appropriate unfolding of the covering material and to keep the flexion between the metallic part and the covering fraction. On the other hand, it has to be thin enough to be compacted into a delivery catheter.

Slim polyester, that is, PTFE, is now generally used as an alternative. Further materials such as polycarbon, polyurethane, and other polymers are being examined.

5.7 Vascular Stents

The vascular stents can be fabricated using different alloys (ELGILOY or nitinol, i.e., stainless steel). The fixation between the vascular wall and the stent graft is the fundamental mechanism of stability. They can be additionally fixated with screws, hooks and anchors, and barbs to generate a sealing inside the artery, that is, the aorta.

As a general comment on the deployment accessory, we should note that the fixation of a prosthesis does not necessarily imply a sealing. The location of the fixation, as well as the used accessories, has advantages and disadvantages, and it is hard to make general comparisons. In addition, the location of the fixation of the prosthesis requires certain individual planning. Another issue is the metallic structure of the endograft, which can be fixated inside or outside the covering material. The so-called exoskeleton can theoretically generate a better fixation for the reconstruction of the aortic wall. An endoskeleton would theoretically have advantages in terms of a better apposition to the wall and considerably less turbulence inside the vessel.

However, it remains to be proven whether infrarenal fixation, an exoskeleton or an endoskeleton, PTFE or Dacron, unibody, or modular design are connected with significant advantages or disadvantages. Vascular anatomy and the individual characteristics of the patient are still the deciding factors for stent graft selection.

5.8 Surgical Technique

EVARs are preferably performed in hybrid ORs. Operating theaters with appropriate equipment to catheter laboratories can potentially be used as locations. Stringent hygiene and sterility standards have to be maintained in this environment. The procedure can be carried out under general, regional, or local anesthesia.

5.9 Imaging

The minimum of radiological requirements is a mobile C-arm with the possibility of cineloop-angiography, DSA, roadmap, and frame-by-frame recline.

Compared to portable systems, ceiling- or floor-installed angiography-facilities, as built in hybrid ORs, have remarkable advantages, particularly the higher resolution, the limited use of a contrast agent, and the shorter radiation exposure for the patient and the staff. The first step is to reproduce the preoperatively analyzed CTA-reconstructions in the intraoperative angiography. The infrarenal aorta typically has an anterior angulation of 10–15 degree, which results from the anatomical lordosis of the spinal column in this segment. AAA patients have a significantly higher angulation in particular because the aneurysm detaches from the spinal column as it grows in all directions.

Therefore, the intraoperative fluoroscopy in a strict anterior-posterior position inevitably leads to projection mistakes that may result in a less optimal or even inadequate positioning of the stent graft in the infrarenal aortic neck. In general, further projection corrections in terms of a rotation of the C-arm are additionally required for an adequate illustration of the infrarenal aneurysm neck.

5.10 Graft Insertion

The common femoral artery is normally exposed with a surgical cut-down on both sides. The current method of using low-profile devices and sealing systems allow an entirely percutaneous approach. In this context, normally two explosure devices in the size of the introducer sheaths that are preloaded have to be used on each side.

A flexible guide wire is positioned suprarenal and an appropriate pigtail catheter, which is normally marked, and is primarily positioned at the level of the renal arteries. After illustrating the renal arteries, the anatomy of the infrarenal aorta is worked out by an appropriate positioning and correction of the C-arm. Then the pigtail catheter is used for the insertion of a stiff wire. It is again placed contralateral over a flexible wire above the renal arteries as a control catheter. Then the correspondingly big introducer sheaths are placed infrarenally over the stiff wire and afterward the endoprosthesis is placed at what should be the correct level of the renal arteries over the stiff wire.

Generally, at this step, the aortic anatomical changes, so that another angiography over the contralateral common femoral artery becomes necessary to evaluate the anatomy which has now been changed by the stiff wire and the prosthesis material, and if necessary to replace the prosthesis.

The so-called power injectors help to reduce the amount of contrast agent, as well as the required radiation, and to increase locally the precision of the imaging so that the usage of injectors is today an integral part of EVAR procedure. The main body and the ipsilateral leg can be delivered to the newly assessed position. This is normally performed under radiographic control. The short contralateral leg should now be cannulate. Many different techniques exist for this purpose; generally an angulated guiding catheter and a flexible wire with a flexible tip are required. If the primary retrograde cannulation is not successful, it is possible to cannulate the contralateral leg with a so-called crossover catheter over the delivered ipsilateral leg and a snare catheter. This is a technique that is significantly more time-consuming and that is connected with a higher dosage of radiation.

Another alternative solution for a difficult primary cannulation of the contralateral leg is the transbranchial access. However, this access is normally only used at revisions that require a stable position of the introducer sheaths particularly because of their complexity.

After cannulation of the legs, analogously to the ipsilateral side, the flexible wire is replaced with a stiff wire, suitable introducer sheaths are placed, and the contralateral legs are positioned with an appropriate overlapping. The position of the internal iliac artery is important; for most prostheses, a complete covering of the common iliac artery is recommended. After the release of the contralateral leg, the overlapping zone as well as the infrarenal sealing zone can be modulated with a balloon. With a pigtail catheter that is located at the level of the renal arteries, a final DSA should be performed postintervention. The discussion about the different techniques of releasing different stent grafts that are used is not within the scope of this chapter.

5.11 Perioperative Complications

Complications have been standardized: access- and wound-related, deployment- and implantation-related, and systemic complications. According to this, deployment complications can be categorized as follows:
- missing release with or without the possibility of conversion,
- aortic dissection of the iliac arteries, and
- corresponding mural hematoma bleeding.

Implantation-related complications can be categorized as follows:
- ruptured AAAs,
- stent graft-migration during implantation,
- device erosion caused by the vascular wall,
- intraoperative stent graft occlusion,
- early postoperative branch occlusion, and
- peripheral microembolization—which causes a postoperative renal malfunction by influencing the renal arteries as well as a corresponding gluteal ischemia by influencing the internal iliac arteries.

Furthermore, distal embolizations are caused by a displaced mural thrombus from the aneurysmal sack in peripheral vessels much more often. These range from small necrotic alterations up to a complete ischemia.

Surgeons should be aware of the fact that a device can potentially occlude an injury as long as it remains in the vessel, which primarily leads to bleeding after the removal of the device material. It is therefore important to verify the integrity of the access vessels before removing the wires and the sheaths. The removal should be performed exclusively with radiological control.

The systemic complications are generally codetermined by the significant comorbidities of the older patient population, and they can concern diverse vital functions, including:
- temporary or permanent renal insufficiency,
- cerebrovascular complications,
- deep vein thrombosis,
- pulmonary embolization,
- coagulopathy,
- spinal ischemia, and
- ischemia of pelvic organs.

The fact remains, however, that the incidence of perioperative complications within EVAR is significantly lower compared to the open procedure. It is an accepted fact that performing EVAR can also significantly reduce postoperative mortality compared to the open procedure. That being said, the complications of EVAR are often different from the complications of open procedure and they require appropriate analysis and correction procedures.

5.12 Specific EVAR-Complications

In this section, we will analyze endoleaks, endotension, device-failure, dilatation of the proximal aneurysm neck, and late rupture.

5.12.1 Endoleaks

By definition, an endoleak is a condition after implantation of an aortic stent with persistent blood-flow outside of the stent graft-lumen, but still inside the aneurysmal sack, that is, inside the over-stented vessel. The existence of an endoleak is an incomplete exclusion of the aneurysm from blood circulation.

According to current knowledge, an endoleak can cease spontaneously, persist furthermore without an increase of the aneurysm, or lead to an increase in size of the aneurysmal sack up to late rupture. In this connection, the systemic pressure inside the disconnected aneurysmal sack plays a role, which depends on the type and the mechanism of the endoleak [23].

Considering the morphology and the clinical aspect, endoleaks are classified into four groups with corresponding subcategories. Type I is the case when there is a persistent flow beside the endograft that is caused by an inadequate sealing either proximal (Ia) or distal (Ib) of the stent graft in the area of the so-called attachments zone (Fig. 3.7).

The much more frequent type II endoleak describes the retrograde flow from the inferior mesenteric artery (IIa) or lumbar arteries (IIb). Much less frequently, another collateral circulation is the cause for an endoleak IIb. However, it should be noted that there is no connection between the retrograde flow and the distal proximal attachment zone, which would lead to a type I endoleak classification.

A type III endoleak is caused by disconnection (IIIa) of the stent modules by manufacturing faults of the endoprosthesis with a corresponding leakage.

A type IV endoleak is caused by porosity of the endograft. In this case, there is a blood plasma passage into the aneurysmal sack despite an intact structure of the stent graft. Newer generations of endografts and materials have led to a decreasing occurrence of this type of endoleak. According to the relevant literature, the incidence of endoleaks is 10–44%, in more than 50% of the cases the bloodflow stops spontaneously; and a permanent endoleak remains in about 20%. The diameter of aneurysms normally

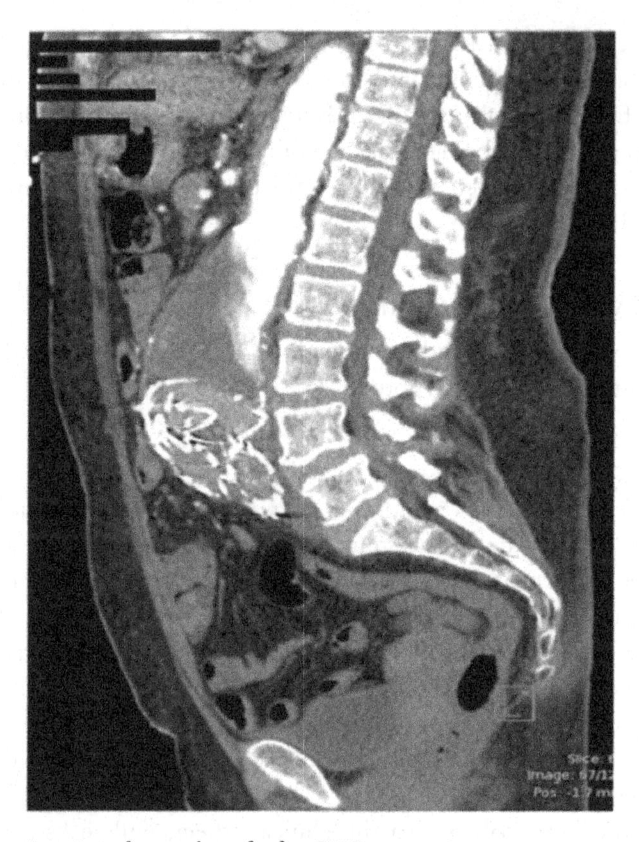

Fig. 3.7 Complete migration of an endograft after EVAR.

decreases after the cessation of the endoleak and reduction of the intraaneurysmatical pressure. However, secondarily appearing aneurysms can show a recurrence of an increasing growth. This is a fact that emphasizes the importance of permanent follow-up of these patients.

Type II endoleaks spontaneously disappear significantly more frequently than type I and type III endoleaks. A rupture based on a type II endoleak is also very rare. The risk is higher than zero, so these endoleaks should definitely be observed, although the risk of a rupture increase in size of the aneurysm is considerably lower than that of a persistent type I or type III endoleak. A persistent type I and type III endoleak is the fundamental combination that can lead to a late rupture, and so a definitive solution should be found.

The primary solution is, in many cases, the balloon modulation of the attachment zone with implantation of a large noncovered stent (Palmaz-Stent). This frequently leads to appropriate sealing, but if there is a migration, that is, a complete detachment, further interventions are necessary with the application of anchor devices that attach

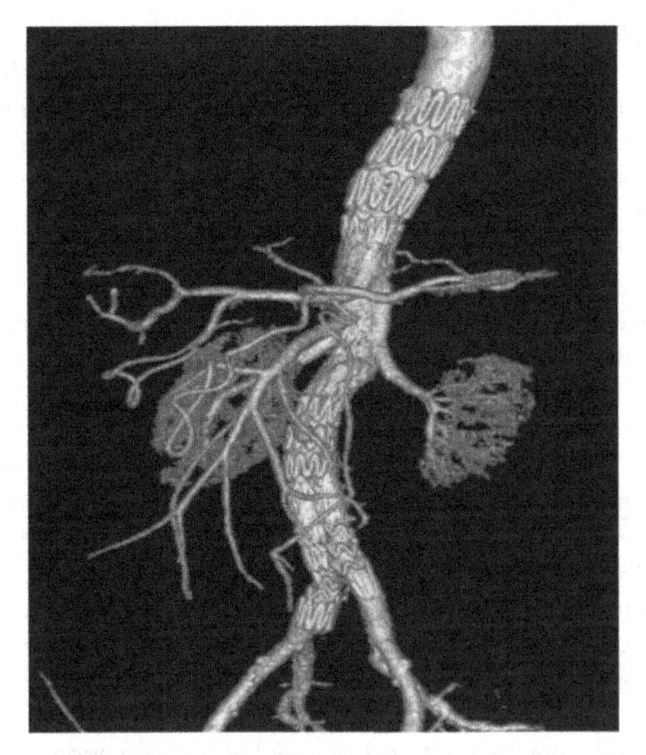

Fig. 3.8 Complex EVAR with stenting of both renal arteries, SMA and coeliac trunk.

the aneurysm to the endograft. In addition, further graft can be used for prolongation if there is a suitable anatomy. A change of the anatomical level with implantation of suprarenal endografts with corresponding fenestration to the renal arteries, that is, the superior mesenteric artery, is the next step. These procedures are technically much more complex, but in most cases they are the definite solution (Fig. 3.8).

If the endovascular procedures fail or if they do not seem to be promising in advance, surgical interventions with a complete resection and conversion have to be performed with the associated surgical risk for this high-risk patient group.

Generally, type II endoleaks can be observed conservatively. If an actual increase in size of the aneurysm occurs, embolization methods are available: by using super-selective catheter techniques, the circulation can be interrupted via the superior gluteal artery, that is, the superior mesenteric artery, or via translumbar puncture. Retroperitoneal endoscopic ligatures of the inferior mesenteric artery, as well as the lumbar arteries, have been described and they should be discussed as an alternative to the complex interventional technique.

In addition to the very common complication of the endoleaks, we are confronted with another phenomenon: so-called endotension. This describes the situation when the aneurysm still increases in size despite the absence of detectable endoleaks. To a certain extent, this condition resembles the so-called histologically well-known hygroma that is described as a transmission of blood components over the barrier of the stent graft, as well as a perigraft-reaction with an excessive accumulation of fluid around the stent graft. Moreover, patients with this condition probably show an increased activity of metalloproteases in the aortic wall, so that the pressure increase in the aneurysmal sack plays only a secondary role for the aneurysmatic progression. Endotension is certainly not a benign alteration and has to be treated. The primary step is relining with an endograft that is made of different material than the first one. If the expansion does not come to a standstill, surgical resection must subsequently be performed [23].

Another more common complication based on a similar mechanism is the dilatation of the proximal aneurysmal leakage. Morphological characteristics, as well as the mechanism of graft attachment but also pathophysiological alterations of the aorta, certainly play a role. Normally a 10–20% oversizing is used in the aneurysmatic neck. This generally leads to a sufficient sealing, but requires an intact aortic segment. The fact that the aneurysmal neck, in which an oversizing has taken place, has aneurysmatically considerably changed is a possible explanation for the neck dilatation. This leads to a continuous progression and dilatation that results in a dislocation of the endograft [24].

The main EVAR complications, such as endoleaks, endotension, and dilatations of the proximal neck, continue to be unpredictable. There is still no preventative solution available. EVAR remains an imperfect therapy regarding long-term results. The current limitations of the proximal aneurysm neck are significantly less relevant due to the existence of fenestrated and side-branch prostheses. The maintenance of the blood flow into the internal iliac artery is also easy to solve using side-branch prostheses. However, these solutions are only possible by passing the anatomical limitations of an infrarenal aneurysm. Related to standard EVAR, the technique is significantly more complex and there are still no long-term results available.

While it is remarkable how EVAR has revolutionized the morbidity of aneurysm-treatment, the dilemma remains that the solutions of today can become the problems of tomorrow.

6. IMAGING FOR PLANNING THE INTRAOPERATIVE PROCEDURE AND POSTOPERATIVE FOLLOW-UP

On examination, the experiences with endovascular therapy of aortic aneurysms of the last 20 years have constantly improved the modalities of these procedures. The challenge remains for EVAR to overcome its shortcomings as a procedure with a very high incidence of reinterventions. In contrast to the open procedure, imaging is an inseparable support process of this procedure that has a fundamental influence on preoperative planning, on intraoperative management, and in postoperative follow-up.

CTA has become a reliable imaging procedure for the anatomical evaluation of the aortic aneurysms and for the planning of endovascular treatment. Meanwhile, 3D reconstructions based on the centerline-flow have become an established procedure to reduce the influence of kinkings and curves of the vessel on the measurement while measuring the length and the diameter (Fig. 3.6). Also related to DSA angiography, 3D CTA is clearly advantageous. The programs that have been created for this during the last few years (Aquarius-Netz, Osirix, and Preview) also show a similar accuracy of the measurement in a head-to-head comparison. The basis of all three procedures is the measurement of the external diameter, as well as the centerline length measurement. Together, they finally permit a graft-diameter selection and, in the end, they are determinant for the graft selection. There is a steady flow of new software tools released for clinical use to improve the precision of the measurements. In addition to the diameter measurement and the centerline-based measurement that have been used until today, segmentation software adds another dimension of planning, namely the volume measurement [25]. New studies show that within the framework of preoperative indication, volume measurement provides an overview of the actual growth trend of the aneurysms that is significantly more accurate than the sole measurement of the diameter progression with the help of semiautomatic segmentation. This new dimension can represent a new main pillar of imaging in the preoperative indication as well as in the postoperative follow-up. This method offers a new dimension of measurement for the planning of the procedure, especially since the surface that will finally be covered by the prosthesis in the ideal case is illustrated and can be calculated in this case. In this way, the procedure has at least theoretical advantages over the centerline measurement, but still has to be evaluated and validated in practice.

Another new development, which can definitely lead to the assessment of the alterations of the aorta caused by cardiac and respiratory motion, is the 4D imaging with time-resolved 3D imaging, which can provide dynamical information about the structure and the function of the different aortic segments. The CTA technique that has been used until today does not permit a correlation of the images to the particular periods of the cardiac/respiratory cycle that can lead to an inadequate stent graft-sizing with a consecutively bad fixation and migration or an endoleak. With the help of the electrocardiac-gated dynamic CT, the elasticity of the aorta can be more accurately calculated, and its influence on the aortic diameter during the different hypotonic and hypertonic phases can be better demonstrated [26].

In sum, vascular surgery has made a considerable improvement in recent years for the treatment of aneurysmal aortic disease. Imaging plays a crucial role in the workflow with a high impact on all aspects of diagnosis, risk stratification, and selection of appropriate therapy modality in individual patients.

6.1 Final Remarks

Several well-known people have died from a ruptured aortic aneurysm. Albert Einstein was one of the most prominent individuals who had an identified aortic aneurysm. It was wrapped with cellophane in 1949, 3 years before the first successful resection and replacement. The aneurysm ruptured 6 years later, on April 13, 1955. At that time, 3 years of experience of surgical treatment has been collected, so he was recommended to undergo a surgical resection. Einstein refused the surgical treatment: "[. . .] because it is distasteful to extend life artificially. The time to go has come and I want to do it elegantly." On April 18, Albert Einstein died in Princeton at the age of 76.

Aortic surgery has a fascinating history. Two thousand years after its initial description by Antyllus, technical developments in the last 50 years have allowed for the abandonment of fatalism in favor of the preventative treatment of AAA.

REFERENCES

[1] Dubost C, Allary M, Oeconomos N. Resection of an aneurysm of the abdominal aorta. Arch Surg 1952;64:405–8.
[2] Clifton MA. Familial abdominal aortic aneurysm. Br J Surg 1977;64(11):765–6.
[3] Majumder PP. On the inheritance of abdominal aortic aneurysm. Am J Hum Genet 1991;48(1): 164–170.
[4] Wassef M. Pathogenesis of abdominal aortic aneurysms: a multidisciplinary research program supported by the national heart, lung and blood institute. J Vasc Surg 2001;34:730–8.
[5] Longo GM. Matrix metalloproteinases 2 & 9 work in concert to produce aortic aneurysms. J Clin Invest 2002;110:625–32.
[6] Beckman EN. Plasma cell infiltrates in atherosclerotic abdominal aortic aneurysms. Am J Clin Pathol 1986;85(1):21–4.
[7] Juvonen J. Demonstration of *Chlamydia pneumoniae* in the walls of abdominal aortic aneurysms. J Vasc Surg 1997;25(3):499–505.
[8] Henderson EL. Death of smooth muscle cells and expression of mediators of apoptosis by T lymphocytes in human abdominal aortic aneurysms. Circulation 1999;99(1):96–104.
[9] Peterson L. Mechanical prosperities of arteries in vivo. Circ Res 1960;8:622–33.
[10] The UK Small Aneurysm Trial Participants. Mortality results for randomised controlled trial of early elective surgery or ultrasound surveillance for small abdominal aortic aneurysms. Lancet 1998;352(9141):1649–55.
[11] Fillinger MF. In vivo analysis of mechanical wall stress and abdominal aortic aneurysm rupture risk. J Vasc Surg 2002;36(3):589–97.
[12] Lederle FA, Johnson GR, Wilson SE, Chute EP, Littooy FN, Bandyk D, et al. Prevalence and associations of abdominal aortic aneurysm detected through screening. Aneurysm Detection and Management (ADAM) Veterans Affairs Cooperative Study Group. Ann Intern Med 1997;126: 441–9.
[13] Bown MJ, Sutton AJ, Bell PR, Sayers RD. A meta-analysis of 50 years of ruptured abdominal aortic aneurysm repair. Br J Surg 2002;89:714–30.
[14] Johnston KW. Multicenter prospective study of nonruptured abdominal aortic aneurysm. Part II. Variables predicting morbidity and mortality. J Vasc Surg 1989;9:437–47.
[15] Young EL, Holt PJE, Poloniecki JD, Loftus IM, Thompson MM. Meta-analysis and systematic review of the relationship between surgeon annual caseload and mortality for elective open abdominal aortic aneurysm repairs. J Vasc Surg 2007;46:1287–9.

[16] Dos Santos JC. Leriche memorial lecture: from embolectomy to endarterectomy or the fall of a myth. J Cardiovasc Surg 1976;17(2):113–28.
[17] Vollmar J. Recent dissertations and theses in the field of surgery. Zentralbl Chir 1964;89:1939–44.
[18] Vollmar J, Laubach K, Gruss JD. Die chirurgische Behandlung des akuten Arterienverschlusses. Dtsch Med Wochenschr 1969;45:2315.
[19] Dotter CT, Judkins MP. Transluminal treatment of arteriosclerotic obstruction: description of a new technique and a preliminary report of its application. Circulation 1964;30:654–70.
[20] Parodi JC, Palmaz JC, Barone HD. Transfemoral intraluminal graft implantation for abdominal aortic aneurysms. Ann Vasc Surg 1991;5:491–9.
[21] Sternberg WC, Carter G, York JW, Yoselevitz M, Money SR. Aortic neck angulation predicts adverse outcome with endovascular abdominal aortic aneurysm repair. J Vasc Surg 2002;35:482–6.
[22] Broeders IA, Blankensteijn JD, Olree M, Mali W, Eikelboom BC. Preoperative sizing of grafts for transfemoral endovascular aneurysm management: a prospective comparative study of spiral CT angiography, arterial angiography and conventional CT imaging. J Endovasc Surg 1997;4:252–61.
[23] Veith FJ, Baum RA, Ohki T, Amor M, Adiseshiah M, Blankensteijn JD, et al. Nature and significance of endoleaks and endotension: summary of opinions expressed at an international conference. J Vasc Surg 2002;35:1029–35.
[24] Lee JT, Lee J, Aziz I, Donayre CE, Walot I, Kopchok GE, et al. Stent-graft migration following endovascular repair of aneurysms with large proximal necks: anatomical risk factors and long-term sequelae. J Endovasc Ther 2002;9:652–64.
[25] Singh-Rangar R, McArthur T, Corte MD, Lees W, Adiseshiah M. The abdominal aortic aneurysm sac after endoluminal exclusion: a medium term morphologic follow-up based on volumetric technology. J Vasc Surg 2000;31:490–500.
[26] Tsang W, Lang RM, Kronzon I. Role of real-time three dimensional echocardiography in cardiovascular interventions. Heart 2011;97:850–7.

CHAPTER 4

Overview of Different Medical Imaging Techniques for the Identification of Coronary Atherosclerotic Plaques

A. Taki[*], A. Kermani[†], S.M. Ranjbarnavazi[‡], A. Pourmodheji[§]

[*]Technical University of Munich (TUM), Munich, Germany
[†]Iran University of Science and Technology (IUST), Tehran, Iran
[‡]Tehran Azad University of Medical Science, Tehran, Iran
[§]Azad University, Dubai, UAE

Chapter Outline

1. INTRODUCTION

Atherosclerotic plaques are the most common cause of coronary artery diseases. Given that the distribution and morphology of components of these plaques define the severity of the lesion, analyzing atherosclerotic plaque composition is a helpful procedure in the diagnosis of such diseases. This chapter aims to address the set of imaging modalities used in the identification of atherosclerotic plaques. Although these plaques occur in both the coronary and the carotid arteries, the focus of the chapter is on the coronary arteries and the corresponding coronary diseases. The chapter starts with a clinical overview, followed by description of vulnerable plaques and the coronary artery diseases associated with them. The third section provides the clinical and technical descriptions of existing noninvasive and invasive imaging techniques to detect and characterize coronary atherosclerotic plaque compositions. Finally, the last section discusses the advantages and limitations of each imaging technique.

Computing and Visualization for Intravascular Imaging and Computer-Assisted Stenting
http://dx.doi.org/10.1016/B978-0-12-811018-8.00004-7

2. CLINICAL OVERVIEW

Despite significant advances in diagnosis and treatment, cardiovascular diseases remain to be a major cause of death in developed countries [1]. Cardiovascular diseases claim more lives than all forms of cancer combined [2].

2.1 Coronary Arteries

The two main coronary arteries originate separately from the base of the aorta (the main artery of the human body that arises from the left ventricle) just after it exits the left ventricle. These two main arteries are called the right coronary artery (RCA) and the left coronary artery (LCA). Each main coronary artery is 2–4 mm wide. Left and right coronary arteries travel over the surface of the heart and divide into smaller branches that penetrate through the muscular walls of the heart. These smaller arteries progressively divide and give rise to a network of vessels that provides oxygenated blood to every cell of the heart (Fig. 4.1).

- The LCA is a short vessel, as it immediately divides into two main branches:
  - **(a)** The left anterior descending (LAD) artery travels down the groove between the two ventricles. The LAD and its branches supply two-thirds of the interventricular septum and most parts of the left ventricle.
  - **(b)** The left circumflex (LCX) artery which encircles the heart, travels to the left, along the groove between the left atrium and ventricle. The circumflex and its branches supply blood to the left atrium, the lateral and posterior parts of the left ventricle and the back of the heart.

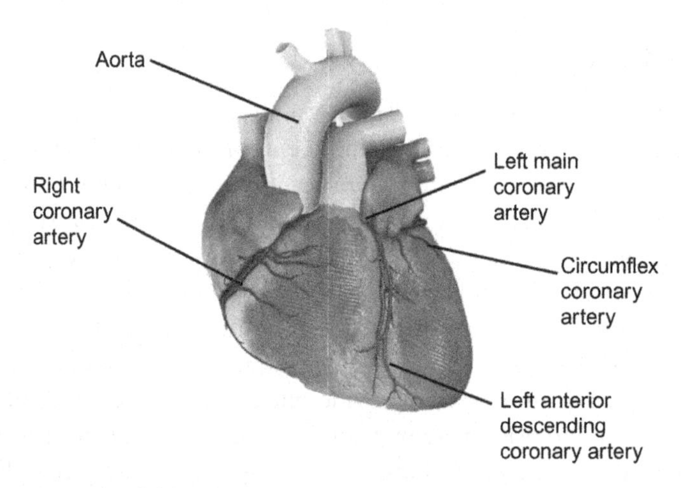

Fig. 4.1 Coronary arteries and the aorta.

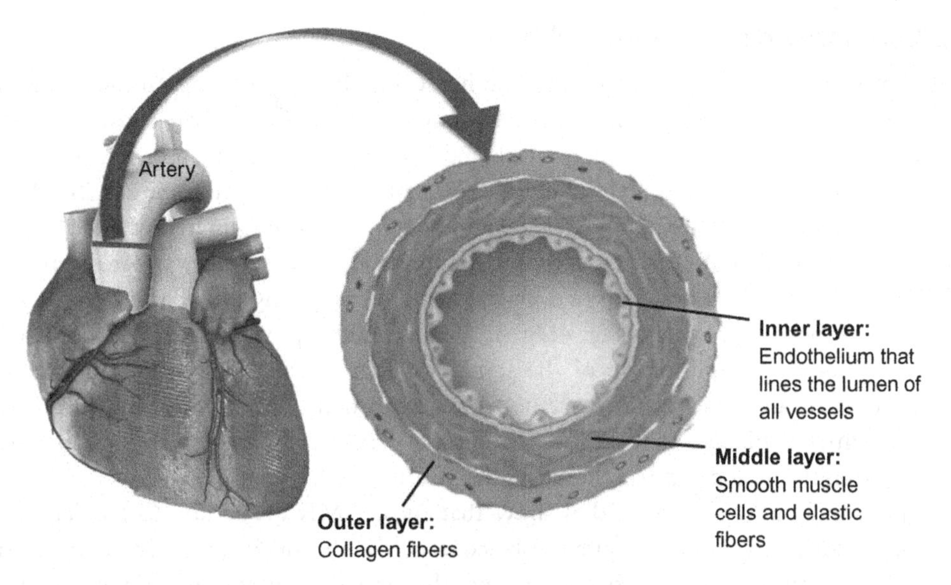

Fig. 4.2 Cross-section of an artery showing inner, middle, and outer layers.

- The RCA immediately gives off numerous branches, its main branch being the posterior descending artery, which travels through the groove between the two ventricles at the back (parallel to the LAD). The RCA and its branches supply blood to the right ventricle, the right atrium, and the sinoatrial (SA) and atrioventricular (AV) nodes.

2.1.1 Structure of Coronary Arteries

The coronary artery wall consists of three layers. The inner layer is called intima, the middle layer is called media, and the outer one is denoted the adventitia. These three layers are observable in a cross-sectional view of the artery, as shown graphically in Fig. 4.2.

The intima is a layer of endothelial cells and connective tissue which covers the luminal surface of the arteries. The media consists of connective tissue (tissue full of collagen, elastin, and other elastic fibers) and smooth muscle cells. Smooth muscle cells are able to contract and relax. The thickness of a normal media layer is between 125 and 350 μm (average 200 μm). However, the media in an atherosclerotic site is thinner and ranges between 16 and 190 μm (average 80 μm) [3]. The adventitia is made up of fibrous tissue, which is mostly elastin and collagen fibers, and fibroblasts. The intima has a variable thickness, which is expressed as a ratio of media thickness. Ratios of 0.1–1 are considered normal [4].

2.2 Coronary Artery Disease (CAD)

Since coronary arteries deliver blood to the heart muscle, any coronary artery disorder or disease can cause serious complications by reducing the flow of oxygen and nutrients to the heart. This can lead to a heart attack and possibly death. Sudden blockage in the blood supply of the heart muscle is defined as acute coronary syndrome (ACS). Two conditions, namely unstable angina (UA) and acute myocardial infarction (AMI), fall under this category. ACS is the most frequent complication of coronary atherosclerosis.

Atherosclerosis is the build-up of a plaque in the inner lining of an artery, causing it to narrow or become blocked (Fig. 4.3). As plaque builds up on the wall of a coronary artery, the result is compromise of the lumen of that artery. This condition can be diagnosed with an angiography system (to be explained later), which measures the luminal diameter. Plaques that narrow the coronary arteries by more than 60–70% are clinically significant and put patients at higher risk for ACS.

Many serial angiographic studies show that most AMIs occur due to blockage of coronary arteries that did not previously contain significant stenosis. Thus, in most cases, the severity of plaque formation and the patient's clinical outcomes are not related. In fact, a patient's outcome is poorly predicted on clinical and angiographic grounds. In the past decade, it has become clear that most plaques causing an AMI are associated with less than 70% stenosis on angiography [5]. These observations have led to the introduction of the concept of vulnerable plaque, a plaque categorized to be prone to rupture. Vulnerable plaque is a plaque that can cause AMI (culprit plaque).

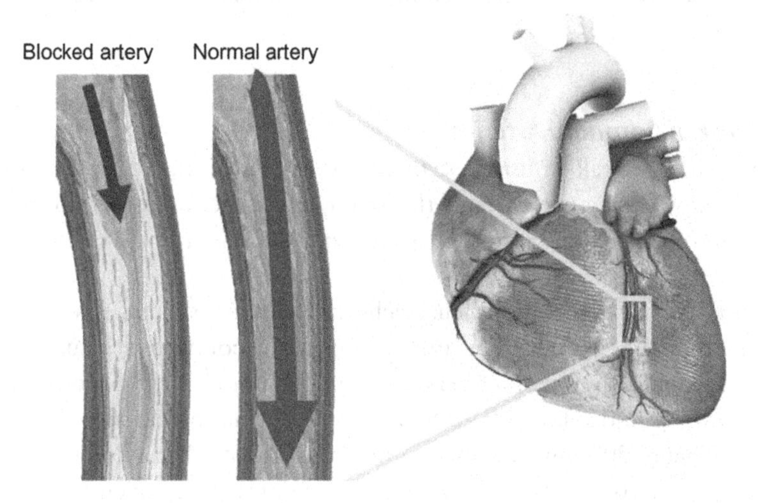

Fig. 4.3 Normal versus blocked artery.

2.3 Vulnerable Plaque

Lipids (fat droplets) in the form of low-density lipoproteins and cholesterol are absorbed by the arterial wall, which leads to release of cytokines from the arterial cells. Cytokines are low-molecular-weight proteins. All cells involved in atherosclerosis are capable of producing and responding to cytokines. Cytokines induce the adhesive molecules on the vascular endothelium, and these molecules, in turn, induce adherence and promote the emigration of one kind of blood cells called monocytes from the circulation to the site of inflammation on the arterial wall. Once monocytes are in the intima, they are activated by locally generated cytokines and transformed to macrophages. The massive uptake of the lipids changes macrophages to foam cells. In fact, foam cells are lipid-loaded macrophages.

A layer of fibrous connective tissue made up of macrophages, bundles of smooth muscle cells, lymphocytes, collagen, and elastin is formed on the site of inflammation. This layer is called the fibrous cap, which is prone to ulceration, rupture, and therefore, thrombosis.

Vulnerable plaques are defined as plaques at high risk for rupture, or for having the surface of their fibrous cap denuded; in either case, this leads to thrombus formation. In comparison to stable plaques, vulnerable plaques have these three histologic hallmarks: a larger lipid core, a thinner fibrous cap, and many inflammatory cells [5].

2.3.1 Different Types of Vulnerable Plaques

As shown in Fig. 4.4, there are seven types of vulnerable plaques, which have been categorized I–VII, and each plaque is described as follows [5]:

I. Rupture-Prone Vulnerable Plaque: a plaque that has a large necrotic lipid core covered by a thin fibrous cap infiltrated by macrophages and T cells. Plaque I is also called a thin-cap fibroatheroma (TCFA). TCFA is a nonthrombosed lesion composed of a necrotic core and a thin cap that covers it. The necrotic core is full of foam cells (macrophages loaded with lipid). The thin cap (which is defined as 65 μm or less) contains numerous macrophages and rare smooth muscle cells. TCFA is the most prone to rupture and is responsible for 70% of acute myocardial infarctions that cause sudden death [6].

This nonthrombosed lesion is most associated with acute death from acute myocardial infarction, and least commonly associated with plaque erosion. Although other plaques are also considered vulnerable, plaque I or TCFA is the classic definition of vulnerable plaques; in fact, TCFA is the most characteristic lesion in AMI.

II. Rupture/Healing Vulnerable Plaque: a plaque that has ruptured and is undergoing the healing process.

III. Erosion-Prone Vulnerable Plaque: a plaque that is rich in smooth muscle cells, and has a proteoglycan matrix. Plaque III is prone to erosion. There is no evidence of rupture at the site of erosion. In fact endothelium is absent at the erosion site and the inflammation is minimal.

IV. Eroded Vulnerable Plaque: a plaque that has been gradually destroyed and has a thrombus.

V. Vulnerable Plaque With Intra-Plaque Hemorrhage: a plaque with hemorrhage inside, which is due to leakage from the vasa vasorum (a network of small blood vessels that supply larger vessels).

VI. Vulnerable Plaque With Calcified Nodule: a plaque in which a calcified nodule is present. A calcified nodule refers to a calcified lesion rich in fibrin, with little or no necrotic core, and a thrombus without obvious rupture of the lesion [6, 7].

VII. Critically Stenotic Vulnerable Plaque: an old plaque that has chronically become stenotic. It has severe calcifications, old thrombus, and an irregular lumen.

Plaque rupture (plaque I or TCFA) is responsible for 70% of culprit lesions. The other 30% of culprit lesions are a result of erosion of the plaque (plaque III) or a plaque with a calcified nodule (plaque VI) [5].

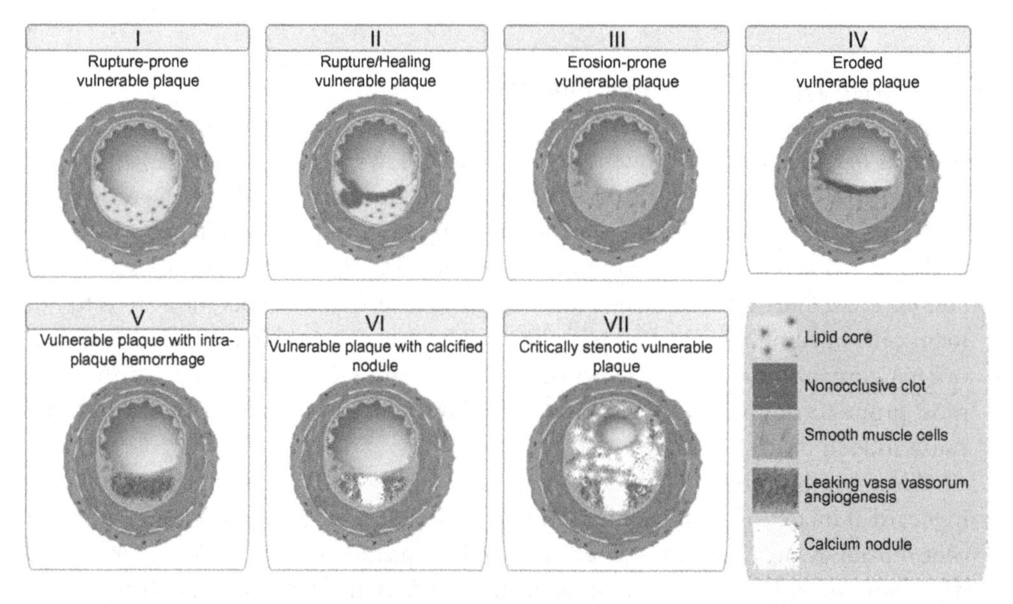

Fig. 4.4 Different types of vulnerable plaque.

3. IMAGING TECHNIQUES

Over many years, there have been huge efforts to develop imaging techniques for identifying plaques and other lesions in coronary arteries. In the last decade, science and technology have grown at an unbelievable rate and have taken physicians to a new frontier in medical technology. Trying to develop techniques that are more reliable for the physicians and patients has led to the development of numerous innovative techniques, due to the close collaboration of medical engineers and physicians throughout history.

This section provides clinical and technical descriptions of existing noninvasive and invasive imaging techniques to detect, and in some, characterize coronary atherosclerotic plaque compositions. According to the results of previous studies, invasive techniques, particularly intravascular ultrasound (IVUS) and optical coherence tomography (OCT), are more precisely in the early detection of atherosclerosis [8]. Thus, these two methods are explained in more detail compared to other modalities. In Table 4.1, the noninvasive and invasive imaging techniques are listed.

3.1 Noninvasive Imaging Techniques

Today, the general tendency is toward minimally invasive or noninvasive therapies and diagnostic procedures. These are associated with significant benefits, such as less downtime, shorter hospitalization, minimal trauma, and no incisions, which, in turn, prevent complications like excessive bleeding, reactions to anesthesia, and infections. These benefits can explain the tendency toward noninvasive cardiovascular imaging techniques, which are listed and compared in Table 4.2.

3.1.1 Computed Tomography (CT)

CT is a diagnostic medical imaging technique used to create representations of the inner structure of the human body by combining the principle of X-ray imaging with the power of computer processing. The typical CT scanner consists of an X-ray generating tube and beam detectors rotating in the opposite direction to create multiple slices in different angles (see Fig. 4.5).

Table 4.1 Noninvasive and invasive imaging techniques

Noninvasive techniques	Invasive techniques
3.1.1 Computed Tomography (CT)	3.2.1 Intravascular Angiography
3.1.2 Magnetic Resonance Imaging (MRI)	3.2.2 Intravascular Angioscopy
3.1.3 Nuclear Imaging	3.2.3 Thermography
	3.2.4 Raman Spectroscopy
	3.2.5 Intravascular Ultrasound (IVUS)
	3.2.6 Optical Coherence Tomography (OCT)

Table 4.2 Comparison of noninvasive imaging techniques

Imaging technique	MRI	CT	Nuclear
Spatial resolution (μ)	80–300	400–800	Poor
Vessel wall penetration (mm)	Good	Good	NA
Comment	Measurements of cap thickness and characterization of atherosclerotic lesion	More useful for detection of calcified plaques	Based on specific binding of radioactive labeled molecules to the target tissue

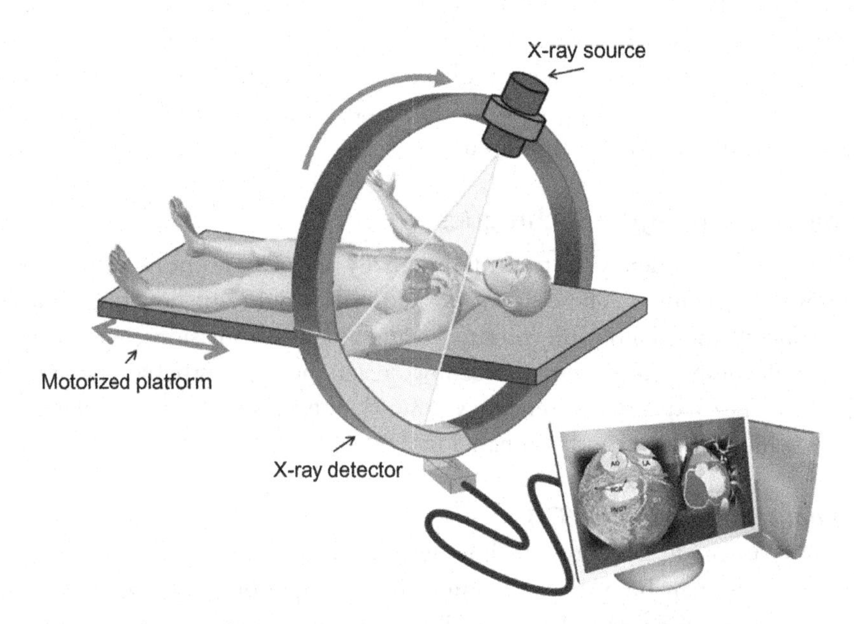

Fig. 4.5 Schematic of a typical CT scan system.

In regular radiography, a two-dimensional (2D) image is generated by subjecting the patient to an X-ray beam and calculating the amount of attenuation by the detectors placed on the same axis as the beam generator. The electromagnetic properties of X-ray radiation enable the beam to be absorbed by different physical structures at different rates. Simply said, rigid and dense structures absorb more radiation than softer, less dense structures. By using the gray-scale color coding, the system can represent structures like bone, which absorbs most of the radiation, to be the brightest; fat, less bright; and areas with cavities to be the darkest in the generated image [9].

Other than the harmful radiation effects on the body, radiography is limited in the amount of detail represented on the image. Because of the axial placement of the X-ray generator and the detectors, the produced images are a one-dimensional view of a three-dimensional (3D) structure, which lacks spatial details. This limitation was overcome by CT scanners, which were a revolution in the medical imaging field. This development not only helped produce more reliable images, but also computerized the process by which images are generated. It also enabled physicians to use the power of computers to manipulate acquired data and extract more information in order to enhance the quality of the images.

In cardiovascular imaging, the early generations of CT scanners were not able to produce a reliable image that could help the physician study the coronary arteries accurately. This was mainly due to hardware limitation and the natural movement of the heart, causing the generated image to be blurry in detail due to the slow scanning speed. Today, this limitation has been overcome by faster CT scanners, which have the ability to scan the heart by synchronizing the scan with the heart rhythm.

Two reliable CT methods for evaluating the coronary arteries are reviewed below.

Electron Beam Computed Tomography (EBCT)

EBCT, also known as ultrafast computed tomography (UCT), helical CT or spiral CT, is a CT technique that is able to quantify the amount of calcium in the coronary arteries at very early stages, and thus has been primarily investigated as a tool for predicting the risk of CAD (Fig. 4.6).

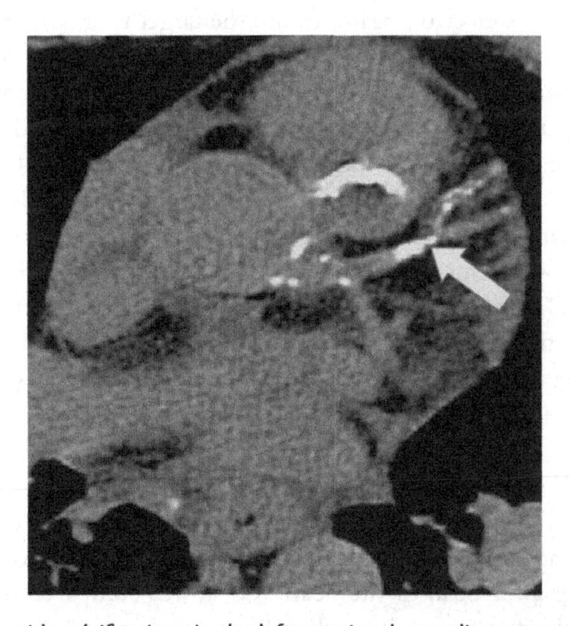

Fig. 4.6 EBCT sample with calcifications in the left anterior descending coronary artery (*arrowhead*) [10].

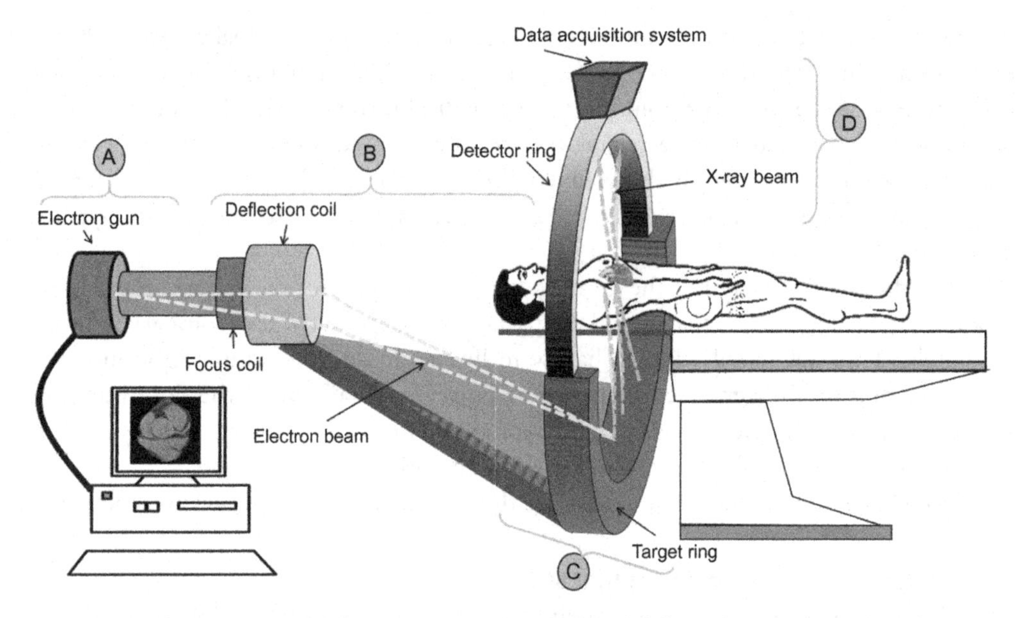

Fig. 4.7 Typical EBCT system: (A) electrons are produced, (B) and deflected to target rings. (C) The X-rays produced by the target rings, (D) they pass through the patient onto detectors.

What differentiates EBCT from the regular CT is the lack of any rotational tube. It functions by focusing the electron beam toward the target ring, which includes an array of anodes covering the bottom half of the gantry. The focused beam, in turn, excites the anodes to emit fan-shaped beams of X-ray that are collimated and detected by the detector ring in the upper half of the gantry. In order for the EBCT scanner to obtain the required data for image reconstruction, a beam current of 640 mA, for a period of 50–100 ms, combined with ECG-synchronized exposure, is needed. In addition, the anode and detector ring should be placed in a way to cover an angle of 270° surrounding the patient [11]. Fig. 4.7 demonstrates a typical EBCT system.

Some of the EBCT limitations are the inability to detect noncalcified atherosclerotic plaques, limited spatial resolution, and limited reproducibility of coronary calcium quantification. These factors, together with the limited functionality of EBCT in general medical imaging have restricted the use of the technology [11].

CT Angiography (CTA)

The evolution of CT scanners has introduced vast improvements, ranging from faster scan time to more accurate imaging. As mentioned earlier in this section, the time required for creating a full scan of the heart was slower compared to the heartbeat in the early CT models, due to mechanical and processing power limitations. Today, by the introduction of CT scanners with two rotating X-ray tubes, multiple detectors, and more powerful computers, comes the ability of scanning the heart with speeds up to 458 mm/s and a temporal resolution of 75 ms, reducing the average examination time to only 0.49 s, without the shortcomings of the older generation machines for patients suffering from irregular heart rhythms or atrial fibrillation [12].

In order to obtain a high resolution cardiac image with the least X-ray exposure, the CTA or cardiac CT procedure starts by synchronizing the CT scanner with the heart rhythm using ECG electrodes and injecting a contrast dye into the blood stream. Today, scanners are able to obtain 320 slices per revolution. Such rate, combined with digital image processing, enables the CT system to create three-dimensional, high-quality images of the heart, providing the ability to find the exact location of the plaques, in order to assist in the diagnosis [13].

The limitations of CTA are the inability to categorize the arterial strain and plaque composition in a way that can replace intravascular angiography.

Clinical Application of EBCT and CTA CT is one of the fastest imaging techniques that has been proven to be useful and reliable in many medical diagnostic applications. For the evaluation of the coronary arteries, each of the aforementioned CT techniques can be used with a specific diagnostic benefit. In the case of EBCT, the technique can help visualize and quantify arterial calcification in the early stages to predict the risk of CAD noninvasively. Meanwhile, CTA is able to assist in the visualization of the coronary arteries and to find the exact location of any arterial narrowing due to plaque build-up. CTA is mostly indicated for evaluating patients with symptoms of CAD who have a low or intermediate risk of having the condition. In patients with less than 20% risk of CAD who have had a positive stress echocardiography, CTA is used for follow-up to avoid catheterization of false-positive patients. Choosing each technique over the other is a task determined by the physician based on the diagnosis needed [11, 13].

3.1.2 Magnetic Resonance Imaging (MRI)

MRI is one of the fast growing noninvasive imaging techniques that enables clinicians to visualize the inner anatomy of the human body without any radiation effect, an advantage compared to other radiographic techniques. The general form of MRI machines comes in two variants, open and closed. Each machine consists of a superconductive magnet either surrounding the tube where the patient will be placed (closed MRI) or located at opposite sides (open MRI) to enable easier access for specific imaging needs. The way each machine obtains data and creates the images is generally the same. The physics behind MRI is based on the small magnetic field (magnetic moment) of the hydrogen nucleus. When the nuclei in the hydrogen atom is placed in the large and uniform magnetic field that is generated by the MRI strong main magnet coil (field strength range from 0.3 to 3 Tesla), the magnetic moment of the nucleus will shift to be aligned with the strong magnetic field.

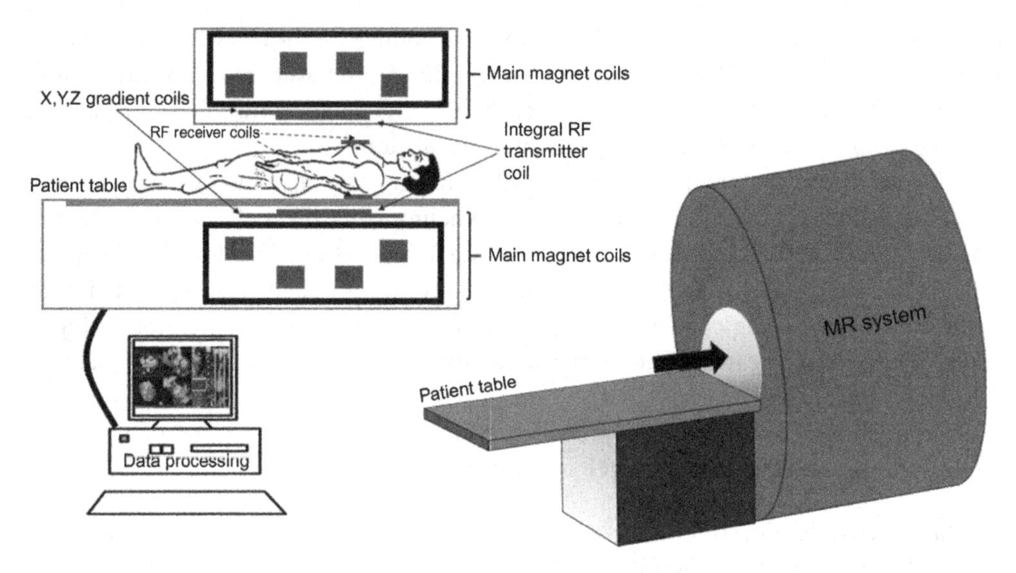

Fig. 4.8 Components of a typical MR system: main magnet coil, three gradient coils, and integral RF transmitter coil.

Application of a specific amount of energy in the form of radiofrequency (RF) pulses by the integral RF transmitter coil and addition of a small magnetic gradient in the x, y, and z axes by the gradient coil to the main static field causes the proton in the hydrogen nuclei to release faint amounts of radio waves proportional to the field strength and the RF that it is subjected to. The phenomenon in which the nucleus is emitting faint amounts of radio waves is called nuclear magnetic resonance (NMR). To generate an MR image, the variation of NMR is recorded in the form of signals by the RF receiver placed around the patient. The acquired signals will be processed with the help of a computer and by applying the Fourier transform to the signal obtained, a gray-scale image representing the human anatomy is generated. For better diagnosis, digital image processing can now be applied to the gray-scale image to show specific borders or highlight different parts of the image in color, making it easier to distinguish plaques from the surrounding tissues [14]. In Fig. 4.8, the components of a typical MR system are illustrated.

Clinical Application of MRI

MRI can identify plaque components on the basis of biophysical and biochemical parameters, such as chemical composition and concentration, water content, physical state, molecular motion, or diffusion. Hence, it has emerged as the potential, leading, noninvasive, in vivo imaging modality for atherosclerotic plaque characterization.

Compared to CT, MRI can be used without or with less allergenic contrast material. It has no radiation, better soft tissue resolution and can represent a functional animation of the heart; however, it is also more time consuming for each scanning session.

3.1.3 Nuclear Imaging

Medical nuclear imaging is based on the functionality and the metabolism of the body and the specific organ that is being examined [15]. By injecting radioactive tracers that emit radiation, physicians are able to trace the movement of the tracers in the desired location by the help of a gamma camera that records the radiation emitted from deep inside the body. The radioactive materials used have varying life spans because of the nuclear decay principle, in which the nucleus of an atom loses energy by emitting radiation to an extent that it has no more radioactive capabilities.

With regard to the coronary arteries, there are two nuclear imaging techniques that are same in principle, but differ in functions and tracers used.

Nuclear Scintigraphy

Nuclear scintigraphy is a two-dimensional imaging technique in which radioactive labeled molecules become bound to the atherosclerotic rupture-prone lesion. This process provides useful information about the severity of the plaques. A method that uses the nuclear scintigraphy principle is called myocardial scintigraphy. In this technique, the blood flow to the heart muscle is measured at rest and under pressure, in order to determine the narrowing of the coronary arteries. The radiotracers may be specific for the lipid core, macrophages or thrombus, and are able to predict clinically significant events.

Positron Emission Tomography (PET)/Single-Photon Emission Computed Tomography (SPECT)

PET and SPECT are three-dimensional nuclear imaging techniques for capturing the gamma rays emitted from the radioactive tracers injected (Fig. 4.9). Because of the spatial resolution created in the images, it helps the physicians to thoroughly examine the area in question and to make more accurate assessments. What sets PET and SPECT apart is the radiopharmaceutical (radioactive tracers) used in each method.

In nuclear cardiovascular imaging, PET is recommended over SPECT because of its improved image quality and viability determination [16], but SPECT remains the dominant technique in most instances, mainly because of the availability of the system. However, PET appears to be superior in patients with large body habitus, as well as in defining multivessel CAD [16].

Clinical Application of Nuclear Imaging Nuclear imaging is a costly modality when compared to other noninvasive techniques. The need for a specific radiopharmaceutical agent, the effect of the radioactive substance on the patient and the limited

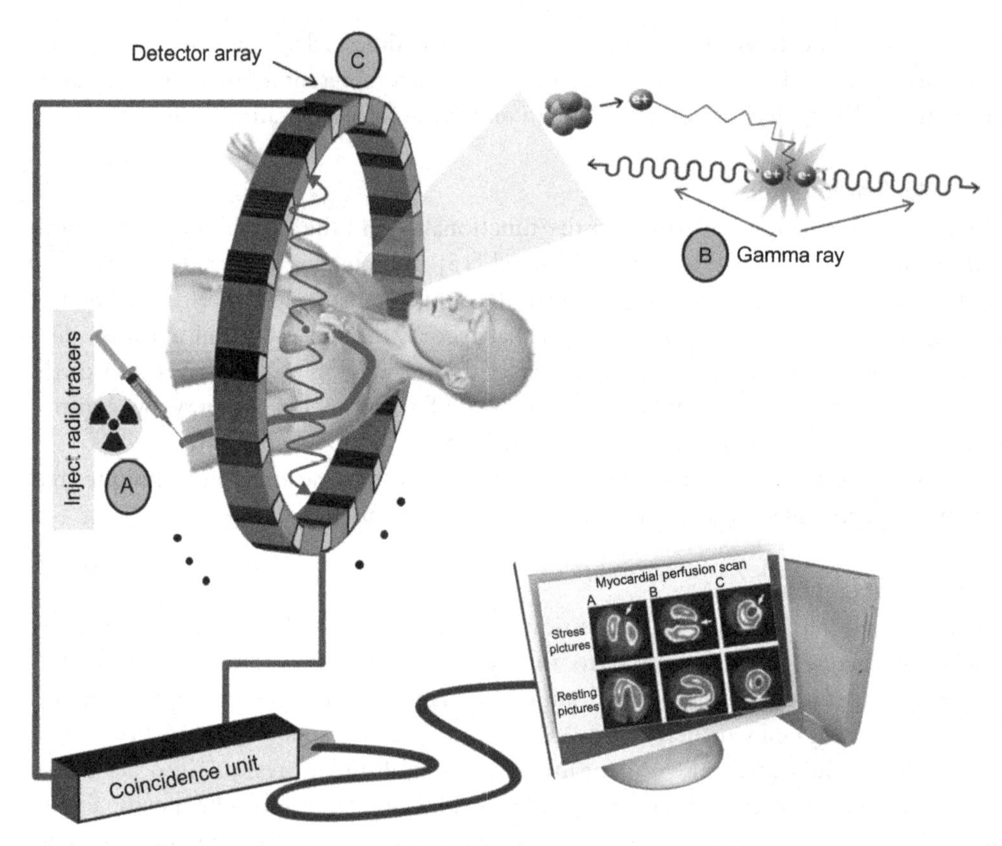

Fig. 4.9 PET imaging system: (A) radio tracer is injected, (B) positron of radionuclide interacts with normal matter and is converted to energy, (C) gamma rays is detected by detector array.

availability of the imaging systems compared to CT and MRI are some of the factors that limit the use of nuclear technology in the cardiac field. Still, PET and SPECT are more widely used than nuclear scintigraphy to assess the risks of CAD and to check the cardiac muscle. PET and SPECT are indicated for patients with stable chest pain. Coronary artery occlusion and aneurysms (an excessive localized swelling of the wall of an artery) can also be identified [17]. In addition, the fusion of MR and CT images with PET/SPECT has also improved the diagnostic process.

3.2 Invasive Techniques

With all the benefits of the noninvasive methods, they do have their limitations in providing accurate diagnosis and characterizing the atherosclerotic plaque components. This section provides clinical and technical descriptions of invasive techniques that enable physicians to visualize or study the core elements of a plaque. See Table 4.3 for a comparison of invasive imaging techniques.

Table 4.3 Comparison table of invasive imaging techniques

Imaging technique	OCT	IVUS	Angiography	Thermography	Angioscopy	Raman spectroscopy
Spatial resolution (μ)	10–20	80–120	100–200	500	Visual*	NA
Vessel wall penetration (mm)	1–2	10	NA	Poor	NA	1–1.5
Comment	Provides cross-sectional images of vessel wall and quantifies fibrous cap thickness and extent of lipid collections	Characterizes vessel wall and morphology, good for calcified plaque, poor for lipids	Reference standard for stenotic lesions	Images temperature heterogeneity due to rise in temperature in macrophage-rich areas in plaque	Direct visualization of lumen surface*. Using color and surface appearance to identify vulnerable plaques	Analyzing the chemical composition of plaques

3.2.1 Intravascular Angiography

Angiography is the gold standard for identifying coronary artery lesions [18]. It provides the practitioners with information about the severity of luminal narrowing and hence, enables the diagnosis of atherosclerotic disease. Angiography may show severe lesions, plaque disruption, luminal thrombosis, and calcification. It also serves as a decision making tool to direct therapy such as percutaneous coronary interventions (PCI) or coronary artery bypass surgery.

The invasive procedure starts by making a small incision and inserting a flexible guidewire into a large vein through a catheter toward the location of plaque. The exact location of the catheter will be monitored either by dynamic pressure reading from the tip of the catheter or by the help of fluoroscopy. Fluoroscopy is a medical imaging technique using a constant source of X-ray to show a live presentation of the inner structures of the human body. Fluoroscopy is similar to radiology in the sense that it uses X-ray technology and the detectors, but differs in the X-ray exposure time, and the representation of the obtained image. See Fig. 4.10 for a schematic illustration of an angiography system.

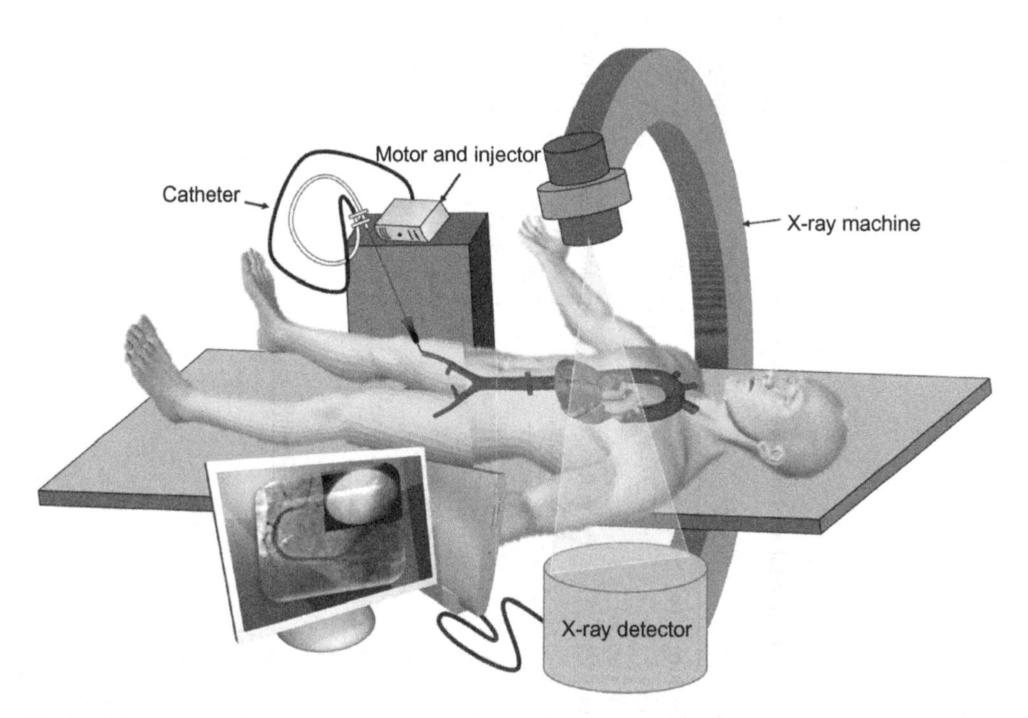

Fig. 4.10 Schematic of an angiography system: the catheter is passed through an artery and carefully moved into the heart. Then, a contrast material is injected into the catheter. X-ray images are taken to see how the material moves through the coronary.

Although the lumen boundaries can be assessed by using angiography, no information is given on plaque burden, its delineation, and components. Besides, it does not provide useful information regarding the vessel wall or atherosclerotic plaque characteristics, such as the vulnerable lipid-rich plaques or other histopathological features [1]. However, it is able to detect complex lesions.

The drawback is that angiography is unable to detect the majority of ulcerated plaques, as they are not big enough to be detected. In addition, while atherosclerotic disease may narrow the entire lumen of the artery, angiography may underestimate the degree of local stenosis. About 70% of acute coronary occlusions are in the areas where the angiography results seem to be normal [19]. Although angiography has a low-discriminatory power to identify the vulnerable plaque, a disrupted ulcerated plaque in angiography is highly suggestive of additional ruptured plaques. Fig. 4.11 shows some examples of angiography images.

Clinical Application of Intravascular Angiography

Application of angiography includes showing severe lesions, plaque disruption, luminal thrombosis, calcification, and diagnosis of atherosclerotic disease. In addition, it can serve as a decision-making tool to direct therapy such as PCI or angioplasty.

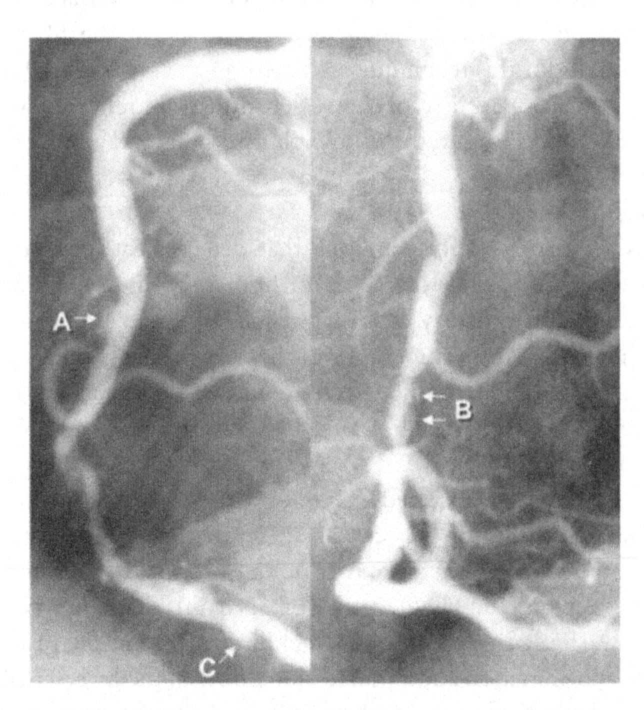

Fig. 4.11 Angiography shows a right coronary artery with three complex lesions containing ulcerations (A, B, and C) [20].

3.2.2 Intravascular Angioscopy

Angioscopy is the first intravascular imaging device that is based on fiber-optic transmission of visible light. It facilitates direct visualization of the plaque surface, color of the luminal surface, presence of thrombus, and macroscopic features of the arterial wall. The procedure starts by inserting a flexible fiber-optic guidewire into a large vein through a catheter.

Angioscopy is able to detect plaque features, such as ruptured caps and red discoloration (intra-plaque hemorrhage). Using this technique, the old thrombus or the normal appearance of the vessel surface is reflected as white. A red surface may indicate a fibrin or erythrocyte-rich thrombus and a lipid-rich core, and thin fibrous caps are yellow in color. These yellow plaques are not detectable by angiography. However, this technique has several drawbacks, one being the limited access to the arterial tree due to the size of the device. Another limitation of angioscopy is the inability to examine the different layers within the arterial wall. In addition, to make the intravascular visualization clear, the vessel has to be occluded and the remaining blood should be flushed away with saline solution (sodium chloride solution) [21].

Clinical Application of Intravascular Angioscopy

Angioscopy is mostly used to examine the arterial wall surface and the plaque characteristics based on the visible light transmitted through the fiber-optic catheter. Most of the assessments for this technique are based on the color seen. For example, old thrombus or the normal appearance of the vessel surface are reflected as white, while a red surface can show a fibrin or erythrocyte-rich thrombus and a lipid-rich core [21].

3.2.3 Thermography

Thermography is a catheter-based technique that detects heat released by the cells of atherosclerotic plaques. Atherosclerosis is an inflammatory disease and any temperature difference in activated inflammatory cells may reflect and predict plaque disruption and thrombosis. The idea behind thermography is the fact that vulnerable plaque is a very active metabolic area, and higher temperatures could be found, as opposed to that seen in a healthy vessel, due to the heat released by activated macrophages either on the plaque surface or under a thin cap.

Although researchers have revealed that most atherosclerotic plaques have higher temperatures compared to healthy vessel walls [22], the independent role of thermography remains limited because the structural definition obtained from high resolution imaging techniques is required. The necessity of a proximal balloon to provide a blood-free field is one of the limitations of this new technique. In addition, plaque temperature may be affected by inflammation and variable blood flow in the lumen making the results unreliable. Another limitation to this technique is being unable to provide information on eroded but noninflamed lesions [23].

Clinical Application of Thermography

This technique is currently under research and development to further improve the quality and reliability of the diagnosis [22].

3.2.4 Raman Spectroscopy

Spectroscopy is an optical technique that uses the reflected light from the plaques to determine the chemical composition of the tissue. Different chemical compositions scatter different wavelengths (and energies); thus, each tissue, due to its chemical composition (lipid, collagen, calcium, etc.), has a unique pattern of light absorbance. To apply this property and detect the chemical composition, many approaches are under development.

Raman spectroscopy collects light scattered by tissue illuminated with high-energy laser (see Fig. 4.12). While most of these scattered lights are at the same wavelength as the incident light, some are at different wavelengths. The amount of the wavelength shift, called Raman shift, depends on the characteristics of the molecule. Using this wavelength difference, Raman spectroscopy has a high-molecular sensitivity, but its tissue penetration is low [24].

Clinical Application of Raman Spectroscopy

Clinical application of Raman spectroscopy has been limited mainly due to the low-depth resolution and the inability to quantify the plaque or the location. This technique

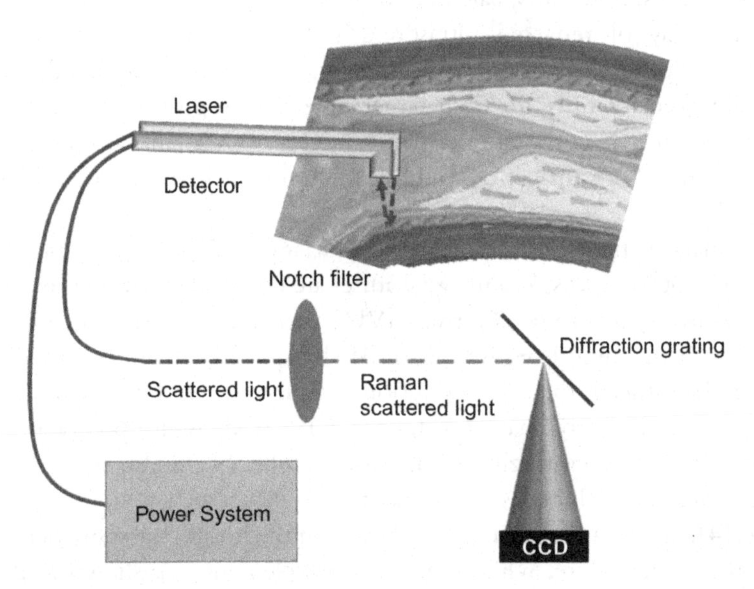

Fig. 4.12 Block diagram of Raman spectroscopy.

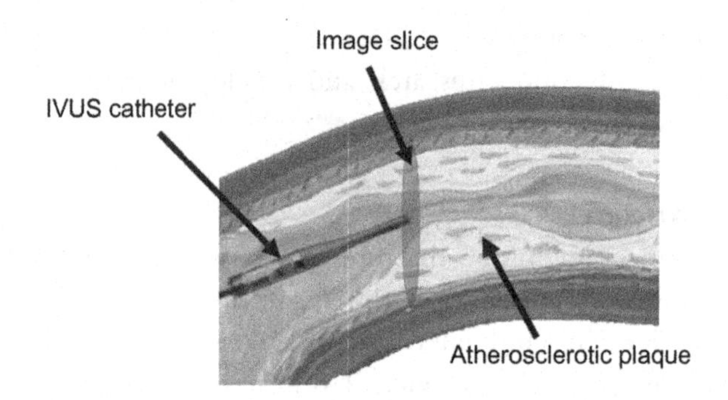

Fig. 4.13 Schematic of an IVUS catheter in a coronary artery.

can only identify the presence of the lipid core plaque in the artery. Further research regarding this technique is ongoing.

3.2.5 Intravascular Ultrasound (IVUS)

Over the past few years, IVUS has become very useful for studying atherosclerotic disease. IVUS is a catheter-based imaging technique providing real-time high-resolution images of the vessel wall and lumen (Fig. 4.13).

IVUS is useful in the evaluation of coronary disease due to its characteristics. Sound waves bounce back with varying intensities and different time delays according to the density of the plaques it encounters, and hence, allows the identification of plaque composition: fibrous, fibro-fatty, calcified, or mixed.

The time delay of reflected ultrasound waves is translated into spatial image information, while the intensity is converted to an intensity map encoded by a gray-scale. The displayed intensity for each target object is proportional to the amount of sound energy returned. For example, dense targets such as the calcified plaques are bright white in the IVUS image. In contrast, the soft tissues such as medial layer appear black.

The 2D image is then obtained from the variation of the angular position of the imaged line over 360 degrees, and image features are detected based on the echogenicity and the thickness of the vessel. Current IVUS catheters have a center frequency of 20–40 MHz with theoretical resolutions of 31–19 μm respectively. Depending on the distance from the catheter, the axial resolution is approximately 150 μm and the lateral is 300 μm. Frame rate is approximately 30 frames/s. IVUS allows for precise measurement of the plaque burden by visualizing plaque topography. IVUS also shows the shape of the lumen, as well as the thickness of layers of the wall.

In Fig. 4.14A, a typical IVUS image is shown. In the image, the more dense elements are shown as brighter white, while the less dense elements are shown as darker. It is important to note that most plaques are eccentric, which means there is more plaque on

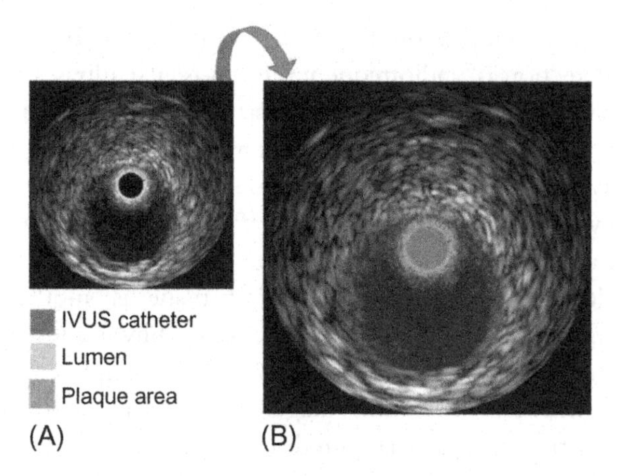

Fig. 4.14 (A) Typical IVUS image (20 MHz, Volcano System), (B) Characterization of an IVUS image.

one side of the vessel than the other. At the center of the IVUS image, the catheter can be seen as a bright white circle. In the following IVUS image, the intima and media can be seen, as well as the adventitia, which surrounds the vessel wall (Fig. 4.14B).

Plaque Components: All plaques with lower density than calcified plaques are called soft plaque. However, the more specific terms "fibrous" and "fibro-fatty" are better terms. Lipid-laden lesions appear hypoechoic, fibromuscular lesions generate low-intensity or soft echoes, and fibrous or calcified tissues are relatively echogenic.

IVUS-Based Plaque Characterization Techniques: Comparing IVUS and histology shows that the plaque calcification can be detected with a sensitivity of 86–97% [25]. Microcalcification sensitivity is about 60% [26]. Lipid pools are detected with a sensitivity of 78–95%. However, because of misclassification of echolucent areas, necrotic tissues could be detected with low sensitivity (<40%) [25]. Moreover, fatty plaques could be detected from the clinical samples with a sensitivity of 81–86% [26]. The limitation is that all the plaques analyzed must have a thickness >0.5 mm, and any lipid core must have a thickness of >0.3 mm. Therefore, it is not known whether it is possible to analyze thinner plaques or to identify very thin lipid cores with this method. Although IVUS characterization of plaques has been very promising, no one has yet produced a technique with sufficient spatial and parametric resolution to identify a lipid pool with a thin cap.

Despite its usefulness in diagnosis, gray-scale IVUS imaging is somehow limited with regard to analysis of plaque composition. Studies have so far demonstrated the potential to identify calcified plaques using IVUS images [27]; however, the identification of lipid pools is hampered by relatively low sensitivity and specificity. Besides, calcified and dense fibrotic tissues are not easily differentiable in IVUS images due to the echo reflections. To overcome these limitations, three main commercial RF-based techniques have been developed on IVUS imaging: Virtual Histology (VH-IVUS), spectral similarity detection of radiofrequency intravascular ultrasound signals (iMAP-IVUS), and integrated backscatter IVUS analysis (IB-IVUS). RF-based methods are sophisticated attempts for detailed tissue characterization and thus, the recognition of lipid-rich, atheromatous cores in potentially vulnerable plaques [27]. These techniques can distinguish between areas with low-echo reflections, which can be beneficial in addition to IVUS imaging.

Evaluation of local mechanical properties of tissue is another way of plaque characterization. Thus, intravascular ultrasound elastography is a technique that assesses the local strain in the artery wall and plaque [28]. The principle of IVUS elastography is illustrated in Fig. 4.15. An ultrasound image of a vessel-phantom with a hard vessel wall and a soft eccentric plaque is acquired at a low pressure. In this case, there is no difference in echogenicity between the vessel wall and the plaque, resulting in a homogeneous IVUS echogram. A second acquisition at a higher intraluminal pressure (pressure differential is approximately 5 mmHg) is performed. The elastogram (image of the radial strain) is plotted as a complimentary image to the IVUS echogram. The elastogram reveals the presence of an eccentric region with increased strain values, thus identifying the soft eccentric plaque [28].

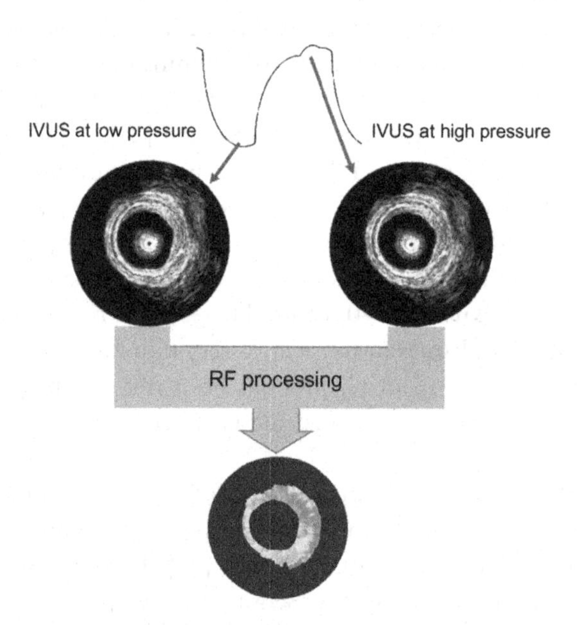

Fig. 4.15 Obtaining elastograph: two ultrasound RF images are acquired in two different pressures (one in low pressure and another in high pressure). Final elastograph image is obtained by RF processing.

Clinical Application of IVUS

IVUS should be considered supplemental, and not as an alternative, to angiography, and should be used as an adjunct. The best clinical use of IVUS by the physician is when the angiography results seem to be opaque and equivocal. IVUS allows measurements of cross-sectional area (CSA) and minimum lumen diameter (MLD), which can only be approximated with angiography. The most likely applications of IVUS are [29, 30]:

1. angiographically normal coronary vessels;
2. angiographically indeterminate lesions;
3. unstable plaque and thrombi;
4. left main coronary disease;
5. transplant coronary artery disease;
6. stenting of smaller vessels.

3.2.6 *Optical Coherence Tomography (OCT)*

The intravascular OCT technique measures the intensity of the back-reflected light. An optic wave with 1300 nm wavelength is used to reduce the energy absorption of vessel wall components. It is a useful technique, which can provide images with ultra-high resolution. It measures the intensity of reflected light and compares it with a reference. The reference is obtained by a mirror reflection on an arm. The mirror is dynamically translated in order to achieve cross-correlation at incremental penetration depths in the tissue. The measured intensity represents backscattering at a corresponding depth [31]. High-resolution images ranging from 4 to 20 μm can be achieved with a penetration depth of up to 2 mm [31]. The frame rate is about 15 frames/s. Lipid pools generate decreased signal intensity compared to fibrous regions. Compared to IVUS, OCT demonstrates superior delineation of the thin caps or tissue proliferation [31]. In Fig. 4.16, an OCT system is shown.

Compared with histology, the OCT image can be interpreted as follows: a lipid pool generates decreased signal areas with poorly delineated borders; a fibro-calcific plaque shows a sharply delineated region with a signal-poor interior; and a fibrous plaque produces a homogenous signal-rich lesion. It also provides useful information about structural details like thin caps or tissue proliferation [31].

One of the first investigations to demonstrate the feasibility of plaque characterization with OCT in vivo was performed by Jang et al. [31]. In addition, Kubo et al. compared the assessment of culprit plaque morphology on OCT to gray-scale IVUS and coronary angioscopy [31]. The authors concluded that OCT was superior in identifying the TCFA and thrombus, and that OCT was the only modality that could distinguish the thickness of the fibrous cap [32].

Postmortem studies demonstrated the accuracy of OCT in comparison with histology. The intravascular application of OCT has proven to be feasible in animal models. These studies showed that OCT can detect both normal and pathologic artery

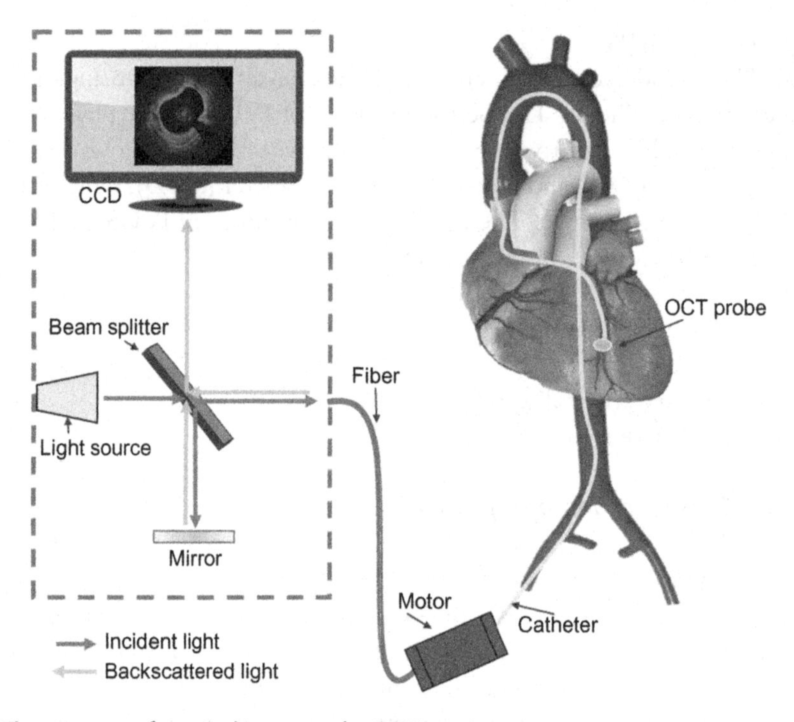

Fig. 4.16 The structure of a typical intravascular OCT imaging system.

structures. Recent experimental data suggest the possibility of detection of macrophages in atherosclerotic plaques. Detection of macrophage accumulation with OCT is based on the hypothesis that plaques containing macrophages have a high heterogeneity of optical refraction indices that exhibit strong optical scattering. Optical scattering results in a relatively high variance of the OCT signal intensity that can be expressed as normalized standard deviation (NSD) of the OCT signal. The analysis of NSD of OCT raw data reveals a high, positive correlation between OCT and fibrous cap macrophage density ($r = 0.84$, $P < .0001$) in vitro (Fig. 4.17).

Most coronary structures that were detected by IVUS could also be visualized with OCT. Intimal hyperplasia and echolucent regions, which may correspond to lipid pools, were identified more frequently by OCT than by IVUS [31].

Clinical Applications of OCT

There are several potential applications of OCT. With its high resolution and unique characteristics, it is a powerful modality for detection of vulnerable coronary plaque. The most frequent variant of a vulnerable plaque is characterized by three factors: a lipid pool, a thin fibrous cap, and increased macrophage infiltration. All of them can be detected by OCT. Another potential application of OCT is as an adjunct to PCI.

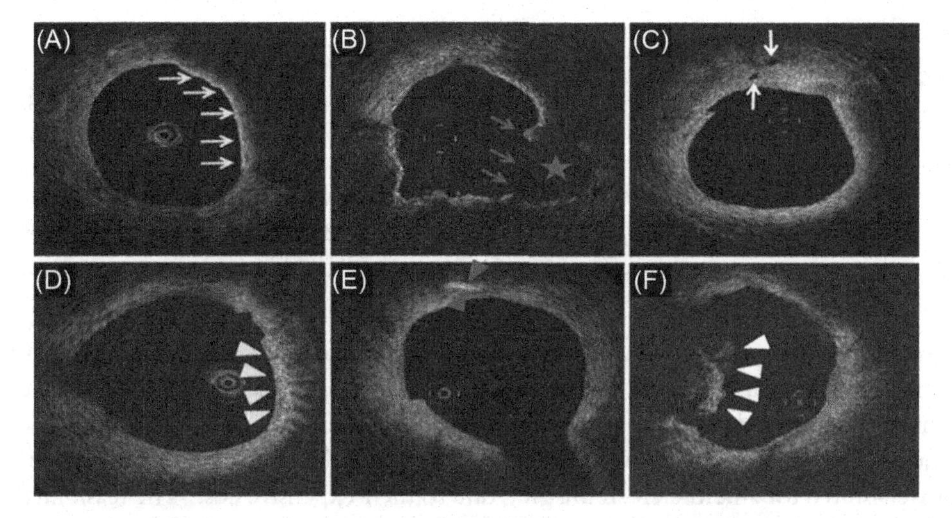

Fig. 4.17 OCT cross-section frames with: (A) TCFA (*arrows*), (B) Ruptured plaque (*arrows*) and cavity formation (*asterisk*) inside the plaque, (C) Microchannels (*arrows*), (D) Macrophage accumulations, shown as bright spots with high-signal variances (*arrowheads*), (E) Cholesterol crystals (*arrowheads*), (F) Thrombus (*arrowheads*) [33].

Detailed structural information before and after coronary intervention can be evaluated with greater accuracy compared with intravascular ultrasound.

OCT is now being introduced for in vivo human imaging at a resolution higher than any current imaging technology, which allows for the identification of TCFA. OCT was shown to identify structural features, such as lipid collections, thin intimal caps, and fissures characteristic of plaque vulnerability. OCT has also been directly compared with high-frequency intravascular ultrasound, the current clinical technology with the highest resolution. The superior resolution of OCT has been confirmed both quantitatively and qualitatively. OCT has several limitations: the low-penetration depth, which may hinder the study of large vessels; and the light absorbance by blood, which currently needs to be overcome by saline infusion or balloon occlusion [31].

4. DISCUSSION

New methods enable physicians to identify additional characteristics of atherosclerotic plaques to plan diverse treatments. Although multifocal diseases require systemic therapies, detecting vulnerable plaques can still help prevent AMI, stroke, and reduce the effort and cost of managing a systemic disease. Limitations, requirements of imaging vulnerable plaques, image resolution of different imaging modalities, and the best suited modality for identification of each plaque type, are listed in Table 4.4.

Table 4.4 Comparison table of main imaging techniques

Imaging technique	OCT	IVUS	Angiography	CT	MRI
Spatial resolution (μ)	5–20	80–120	100–200	400–800	80–300
Probe size (μm)	140	700	N/A	N/A	N/A
Thin cap	Yes	No	No	No	No
Best suited For	Thin caps of atheroma	Fibroatheroma	Inflammation and characterization	Calcium scoring	Lumen variation

In order to move the focus from the arterial lumen onto the wall, noninvasive methods are critical to reveal underlying atherosclerosis. MRI can provide diagnosis without ionizing radiation on plaque volume. On the other hand, inflammation can be quantified by PET/CT. However, the mentioned noninvasive methods are currently not well-suited for atherosclerosis imaging due to their expensive cost. The feasibility of combined noninvasive scanners (i.e., PET-MRI) due to hardware development, when taken together, could enhance the advantages of each modality. In contrast, catheter-based techniques can provide structural information with higher resolution than noninvasive methods. For instance, preliminary angiography techniques can characterize the presence of basic vulnerable plaques with a glistening surface, which can correlate with acute events. IVUS is also widely used in interventional cardiology. Moreover, intravascular OCT appears to be feasible and safe. OCT is able to recognize most of the architectural features detected by IVUS and may provide additional detailed structural information. As a result, the invasive methods have the highest potential to detect vulnerable plaques. Particularly, the development of IVUS and OCT data processing have improved the applicability of these techniques considerably.

4.1 Summary

Due to the high-mortality rate of coronary artery diseases, especially in developed countries, the importance of identification of the atherosclerotic plaques in today's medicine is undeniable. Many methods have been introduced within the past few decades. Noninvasive methods, such as CT, nuclear imaging (PET, SPECT), and MRI, have not yet been able to diagnose vulnerable plaques effectively. Invasive methods, such as angiography, angioscopy, IVUS, and OCT, are the main methods used for simultaneous diagnosis and treatment. While angiography remains to be the gold standard for identifying arterial lesions, using IVUS as an adjunct to angiography helps the clinicians to characterize the plaque by its components, due to different ultrasound reflections. Lower echo reflections, such as those in lipid pools and atheromatous cores, can be further distinguished by IVUS-based plaque characterization methods (e.g., VH and iMap). By adding elastography to IVUS, physicians would be able to discriminate between fatty, fibro-fatty, and fibrous materials, by measuring the local strain of the tissue in the artery. OCT, meanwhile, measures the reflected light in the artery. Although most structures detected by OCT are also visualized by IVUS, lipid pools that cause intimal hyperplasia and echolucent lesions are more frequently identified with OCT. It is important to note that none of the imaging techniques is perfect. From a clinical perspective, a combination of many of these imaging modalities may be required to identify a vulnerable patient (i.e., IVUS-OCT-PET). Extensive research has emerged on the wide variety of imaging techniques that assess the vulnerable plaques. In addition, prospective clinical trials are still needed to demonstrate the advantages and disadvantages of imaging approaches in clinical applications.

REFERENCES

[1] Topol EJ, Califf RM, et al. Textbook of cardiovascular medicine. 3rd ed. Lippincott Williams and Wilkins; 2007.
[2] Heart and Stroke Association Statistics, American Heart Association. Statistics At-a-Glance; 2016.
[3] Waller BF, Orr CM, Slack JD, Pinkerton CA, Van Tassel J, Peters T. Anatomy, histology, and pathology of coronary arteries: a review relevant to new interventional and imaging techniques. Clin Cardiol 1992;15:451–45.
[4] www.americanheart.org.
[5] Lau J, Kent D, Tatsioni A, Sun Y, Wang C, Chew P, et al. Vulnerable plaques: a brief review of the concept and proposed approaches to diagnosis and treatment. Rockville, MD: Agency for Healthcare Research and Quality; 2004.
[6] Virmani R, Burke AP, Farb A, Kolodgie FD. Pathology of the vulnerable plaque. J Am Coll Cardiol 2006;47(8):c13–8.
[7] Virmani R, Burke AP, Farb A, Kolodgie FD. Pathology of the unstable plaque. Prog Cardiovasc Dis 2002;44(5):349–56.
[8] Fleg JL, Stone GW, Fayad ZA, Granada JF, Hatsukami TS, Kolodgie FD, et al. Detection of high-risk atherosclerotic plaque: report of the NHLBI working group on current status and future directions. JACC Cardiovasc Imag 2012;5(9):941–55.
[9] Budoff MJ, Shinbane JS, editors. Cardiac CT imaging: diagnosis of cardiovascular disease. 3rd ed. Cham: Springer International Publishing; 2016. http://dx.doi.org/10.1007/978-3-319-28219-0.
[10] Becker CR, Ohnesorge BM, Schoepf UJ, Reiser MF. Current development of cardiac imaging with multidetector-row CT. Eur J Radiol 2000;36(2):97–103.
[11] Ohnesorge BM, Flohr TG, Becker CR, Knez A, Reiser MF. Multi-slice and dual-source CT in cardiac imaging: principles–protocols—indications—outlook. 2nd ed. Berlin, Heidelberg: Springer-Verlag GmbH; 2007. http://dx.doi.org/10.1007/978-3-540-49546-8.
[12] Siemens. Somatom definition flash. Technical specification.
[13] Halpern EJ. Clinical cardiac CT: anatomy and function. 2nd ed. New York: Thieme Medical Publishers, Inc.; 2011.
[14] Lee VS. Cardiovascular MRI: physical principles to practical protocols. Philadelphia: Lippincott Williams & Wilkins; 2006. p. 488.
[15] Knapp FF (Russ), Dash A. Radiopharmaceuticals for therapy. 1st ed. India: Springer; 2016.
[16] Heller GV, Hendel RC. Handbook of nuclear cardiology. 1st ed. London: Springer-Verlag; 2012.
[17] Cardona R, Gunabushanam G. Myocardial perfusion SPECT. Medscape 2016; http://emedicine.medscape.com/article/2114292.
[18] Ragosta M. Cardiac catheterization: an atlas and DVD. 1st ed. Philadelphia, PA: Saunders Elsevier; 2010.
[19] Agostoni P, Schaar JA, Serruys PW. The challenge of vulnerable plaque detection in the cardiac catheterization laboratory. Cardiovasc Med 2004;7:349–58.

[20] Maehara A, Mintz GS, Bui AB, Walter OR, Castagna MT, Canos D, et al. Morphologic and angiographic features of coronary plaque rupture detected by intravascular ultrasound. J Am Coll Cardiol 2002;40(5):904–10.
[21] Mizuno K, Takano M, editors. Coronary angioscopy. 1st ed. Japan: Springer; 2015.
[22] Tsakanikas VD, Fotiadis DI, Michalis LK, Naka KK, Bourantas CV. Intravascular imaging: current applications and research developments. Hershey, PA: Medical Information Science Reference; 2012. p. 165.
[23] Willerson JT, Cohn JN, Wellens HJJ, Holmes DR. Cardiovascular medicine. 3rd ed. London: Springer-Verlag; 2007.
[24] Falk E, Shah P, de Feyter P. Ischemic heart disease. London: Manson; 2007. p. 117.
[25] Taki A, Najafi Z, Roodaki A, Setarehdan SK, Zoroofi RA, Konig A, et al. Automatic segmentation of calcified plaques and vessel borders in IVUS images. Int J Comp Assist Radiol and Surg 2008;3(3–4):347–54.
[26] Taki A, Hetterich H, Roodaki A, Setarehdan SK, Unal G, Rieber J, et al. A new approach for improving coronary plaque component analysis based on intravascular ultrasound images. Ultrasound Med Biol 2010;36(8):1245–58.
[27] Konig A, Margolis MP, Virmani R, Holmes D, Klauss V. Technology insight: in vivo coronary plaque classification by intravascular ultrasonography radiofrequency analysis. Cardiovasc Med 2008;5(4):219–29.
[28] De Korte CL, Van der Steen AF. Intravascular ultrasound elastography: an overview. Ultrasonics 2002;40(1):859–65.
[29] Nissen SE, Yock P. Intravascular ultrasound, novel pathophysiological insights and current clinical applications. Circulation 2001;103(4):604–16.
[30] Boston Scientific Corporation. Scimed Division. The ABCs of IVUS; 1998.
[31] Yabushita H, Bouma BE, Houser SL, Aretz HT, Jang IK, Schlendorf KH, et al. Characterization of human atherosclerosis by optical coherence tomography. Circulation 2002;106:16405.
[32] Kermani A, Ayatollahi A, Taki A. Full-automated 3D analysis of coronary plaque using hybrid intravascular ultrasound (IVUS) and optical coherence (OCT). In: The second conference on novel approaches of biomedical engineering in cardiovascular diseases. 2015.
[33] Tian J, Ren X, Vergallo R, et al. Distinct morphological features of ruptured culprit plaque for acute coronary events compared to those with silent rupture and thin-cap fibroatheroma: a combined optical coherence tomography and intravascular ultrasound study. J Am Coll Cardiol 2014;63(21):2209–16.

SECTION II

Vascular and Intravascular Analysis of Plaque

CHAPTER 5

Implications of the Kinematic Activity of the Atherosclerotic Plaque: Analysis Using a Comprehensive Framework for B-Mode Ultrasound of the Carotid Artery

A. Gastounioti*, S. Golemati†, P. Mermigkas*, M. Prevenios*, K.S. Nikita*
*National Technical University of Athens, Athens, Greece
†National Kapodistrian University of Athens, Athens, Greece

Chapter Outline

1. INTRODUCTION

Cardiovascular disease represents a major cause of death and morbidity globally and is projected to remain the leading cause of death, particularly in high- and middle-income countries [1]. Carotid atherosclerosis, a degenerative disease causing lesions (i.e., atherosclerotic plaques) which narrow the carotid arteries, is characterized by arterial wall stiffening and thickening, and is the underlying cause of the majority of stroke events. The presence of atherosclerotic plaques, causing luminal narrowing of more than 50% in internal carotid arteries, has been positively associated with

Computing and Visualization for Intravascular Imaging and Computer-Assisted Stenting
http://dx.doi.org/10.1016/B978-0-12-811018-8.00005-9

a higher incidence of strokes in the elderly [2]. Various imaging modalities have been proposed to identify and evaluate the extent and vulnerability of atherosclerotic lesions in the carotid artery [3]. However, the unique features of ultrasound imaging, for example, relatively low-cost, noninvasiveness, widespread availability, and lack of radiation exposure, have established it as the cornerstone in the diagnosis and monitoring of carotid atherosclerosis [4, 5]. For example, the main criterion to evaluate plaque vulnerability is the ultrasonographically measured degree of stenosis [5], that is, the percentage of lumen area occupied by atheromatous material.

In addition to its valuable role in the diagnosis of carotid atherosclerosis, ultrasound has also demonstrated a substantial potential in providing novel, automated, risk markers, valuable in the diagnosis, evaluation, and treatment selection for the disease [4]. Among these ultrasound-image-derived risk markers, the gray-scale median (GSM), a measure of plaque echogenicity, has been repeatedly associated with plaque vulnerability, with symptomatic plaques corresponding to significantly lower GSM values than the asymptomatic ones [6, 7]. Another example of well-established, ultrasound-based risk markers is the intima-media thickness (IMT), which has an important prognostic and diagnostic role for the disease. Studies have consistently shown that the increase in IMT is able to predict the development of cardiovascular disease, to assess the progression or regression of atherosclerosis, and to evaluate plaque response to drug therapy [8, 9]. Texture and morphology features of the arterial wall were the first to be studied in computer-aided diagnosis for carotid atherosclerosis; however, the characterization of the arterial wall dynamics in ultrasound is also gaining increasing attention as an effective method for risk stratification [4].

Therefore, substantial research effort has been devoted to measure accurately the kinematic activity of the carotid artery wall from ultrasound image sequences [10–15]. Dynamic B-mode ultrasound imaging of longitudinal sections of the arterial wall allows the estimation of tissue motion in two dimensions, namely longitudinal or axial, that is, along the vessel axis, and radial, that is, along the vessel radius and perpendicular to the longitudinal one. The computational analysis of the carotid artery motion has produced important physiologically relevant findings related to carotid atherosclerosis [16, 17]. Radial and longitudinal motion patterns of the plaque itself, as well as of the normal, that is, nonatherosclerotic, arterial wall adjacent to the plaque have been described and associated with the risk for cerebrovascular complications, such as strokes or transient ischemic attacks [12, 18, 19]. Using motion analysis, studies have also shown a strong relationship of the severity of carotid stenosis with the arterial stiffness and the axial stress [20, 21]. Further, wall displacements of carotid artery areas have been measured in subjects with diabetes [11] and periodontal disease [22], where, compared to healthy volunteers, arterial tissue movements were found to be impaired in the presence of the disease.

This chapter is an attempt to leverage ultrasound-based kinematic analysis of the arterial wall in studying the mechanical phenomena taking place in atherosclerotic carotid arteries and in advancing imaging phenotyping of the plaque vulnerability. To this end, we present a comprehensive framework for quantifying the kinematic activity of the arterial wall from B-mode ultrasound images of the carotid artery. By applying this framework to retrospectively collected data from patients with carotid atherosclerosis, we then investigate potential implications of the arterial wall motion for the disease, focusing on bilateral asymmetry in kinematic features and associations with plaque vulnerability.

2. STUDY POPULATION AND ULTRASOUND IMAGE DATA

We retrospectively analyzed data from 96 patients (aged 50–86 years) with established carotid atherosclerosis (diagnosed carotid stenosis >30%), who were referred to the Attikon General University Hospital of Greece for carotid artery ultrasound scanning. Among those patients, 24 had experienced an ischemic cerebrovascular event (i.e., stroke or transient ischemic attack) associated with the carotid stenosis, and they form the "symptomatic" group of our study. The other 72 patients had no neurological symptoms within a 6-month follow-up time period from the time of examination, and they form the "asymptomatic" group of this study. In all cases, the presence or absence of symptoms had been validated with computed tomography (CT) or magnetic resonance imaging (MRI) scans of the brain. No statistically significant difference was found in the degrees of stenosis (Wilcoxon rank-sum test, $p = 0.10$), nor in the ages between the two groups (Wilcoxon rank-sum test, $p = 0.45$). The local institutional review board approved ultrasound image examinations and all subjects gave their informed consent to the scientific use of the data.

For each patient, the carotid artery was scanned in the longitudinal direction according to a standardized protocol (dynamic range, 60 dB; persistence, low) and a B-mode ultrasound image sequence was recorded at a rate higher than 25 frames/s for at least 3 s (2–3 consecutive cardiac cycles). Moreover, an experienced vascular physician manually traced four regions of interest (ROIs) at the first frame of each ultrasound image sequence (Fig. 5.1). These ROIs corresponded to the top surfaces (PTS) and bottom surfaces (PBS) of the atherosclerotic plaques and to the posterior wall-lumen (PWL) and anterior wall-lumen (AWL) interfaces. The PWL and AWL are healthy parts of the arterial wall adjacent to the plaques, the morphology and the kinematic activity of which have been associated with plaque vulnerability [12, 23].

To ensure comparable measurements between ultrasound image recordings, we applied an established preprocessing step for carotid ultrasound images, where, for each image sequence, image intensities ([0: black, 255: white]) are linearly adjusted so that the median gray level value of the blood was equal to 0, and the median gray level value of the adventitia was equal to 190 [24]. Additionally, for direct comparisons with the

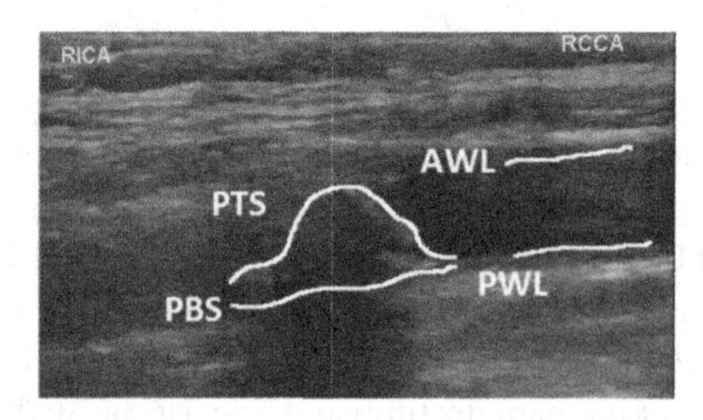

Fig. 5.1 Example of a B-mode ultrasound image of a carotid artery with atherosclerosis. The boundaries of the selected regions of interest are marked in *white*. *AWL*, anterior wall-lumen interface; *PBS*, plaque bottom surface; *PTS*, plaque top surface; *PWL*, posterior wall-lumen interface.

performance of well-established textural features in the analyses of this chapter, plaque texture was also evaluated using first (i.e., gray-level histogram features) and second-order (i.e., co-occurrence features) statistical properties [24], as well as multiresolution wavelet features [25] of the image intensities of the plaque region. Wavelet-based image decomposition is a state-of-the-art methodology in the field of texture analysis, which has also demonstrated a promising performance in characterizing the atherosclerotic tissue in B-mode ultrasound [25, 26]. For each image study, 286 texture features were estimated for specific instants of the cardiac cycle, namely systole and diastole, resulting in a total number of 572 features representing the texture of the plaque [27].

3. A COMPREHENSIVE FRAMEWORK FOR QUANTIFYING THE ARTERIAL WALL MOTION

In this section, we describe a comprehensive framework developed by our group for extracting quantitative measures of the kinematic activity of the arterial wall [28, 29]. In the first step, all pixels composing the four ROIs, as well as the pixels within the entire plaque region (i.e., the region contoured by PTS and PBS), are selected as motion targets. A motion tracking algorithm is then applied to estimate the radial and longitudinal positions of the targets across time. In this study, for this step of motion tracking, we used ABM_{KF-K2}, an adaptive block matching methodology with enhanced accuracy in the particular application, previously optimized and evaluated in both artifacts-free and artifacts-corrupted image recordings [12, 13]. Subsequently, the target-wise radial and longitudinal motion waveforms, which are produced in the previous step, are used to generate two different sets of kinematic measurements. The first set (spatiotemporal patterns) consists of waveforms representing patterns of the kinematic activity during the cardiac cycle [29], while the second one (spatiotemporal features) includes indices summarizing these patterns over time [28].

To generate the first set, the motion waveforms produced from motion tracking are processed according to the schematic representation of Fig. 5.2, thereby generating 120 kinematic (K1–K120) and 26 strain (S1–S26) waveforms. Specifically, a total of 24 kinematic waveforms are produced for each ROI by estimating target-wise velocity and displacement waveforms and then computing the mean and median waveforms over space (Fig. 5.2A). Based on similar steps, now applied to motion waveforms of pixel pairs, 26 strain waveforms are produced, expressing local deformations of each ROI over time and relative movements between (a) PWL and AWL, (b) PBS and PTS, (c) PBS and PWL or AWL, and (d) PTS and PWL or AWL, if the plaque was located at the posterior or the anterior wall, respectively. Specifically, we apply previously published mathematical formulas expressing strain and shear strain in the radial and longitudinal directions [14] to pixel pairs covering the corresponding ROI(s) [30], and then we estimate the mean and median waveforms over space (Fig. 5.2B).

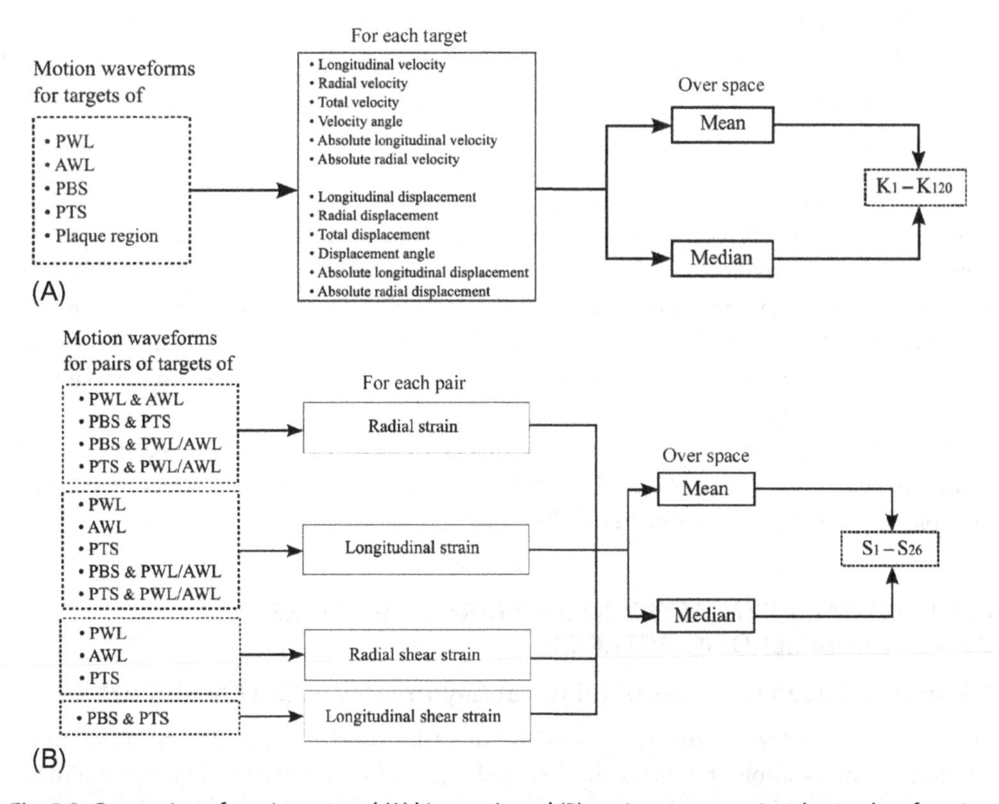

Fig. 5.2 Generation of spatiotemporal (A) kinematic and (B) strain patterns, using the results of motion analysis for an ROI or a pair of ROIs [29].

Table 5.1 Spatiotemporal features estimated using the produced motion waveforms of the targets of an ROI or a pair of ROIS

ROI(s)	Index encoding of Fig. 5.3	Spatiotemporal features
Kinematic indices		
PWL	I1–I154	M1–M154
AWL	I1–I154	M155–M308
PTS	I1–I154	M309–M462
PBS	I1–I154	M463–M616
Plaque region	I1–I528	M617–M1144
Strain indices		
PWL	I529–I535 I550–I556	M1145–M1158
AWL	I529–I535 I549–I556	M1159–M1172
PWL and AWL	I536–I542	M1173–M1179
PTS	I529–I535 I550–I556	M1180–M1193
PTS, PBS	I536–I549	M1194–M1207
PTS, PWL, or AWL	I529–I542	M1208–M1221
PBS, PWL, or AWL	I529–I542	M1222–M1235

Notes: *AWL*, anterior wall-lumen; *PBS*, plaque bottom surface; *PTS*, plaque top surface; *PWL*, posterior wall-lumen.

In the case of the second set consisting of spatiotemporal features, the produced motion waveforms are combined to generate a total of 1144 target-wise, kinematic indices and 91 pair-wise, strain indices, all corresponding to descriptive statistical measures of motion over both time and space. The kinematic indices represent (a) median and standard deviation in velocities during the cardiac cycle, (b) motion amplitudes, defined as the absolute difference between the corresponding maximum and minimum target positions, and (c) diastole-to-systole displacements of the selected ROIs (Fig. 5.3). Strain indices express again relative movements between ROIs or local deformations within single ROIs. The definitions of these kinematic indices are given in Table 5.1, following the index encoding of Fig. 5.3.

4. BILATERAL ASYMMETRY IN KINEMATIC FEATURES OF ATHEROSCLEROTIC ARTERIES

4.1 Image-Based Features of Bilateral Asymmetry in the Carotid Artery

"Bilateral asymmetry" is the term used to describe differences in paired anatomical territories, for example, between the left and right carotid arteries. These differences may be considered in terms of anatomical, geometrical, morphological, and functional features, occurring in healthy as well as in diseased conditions. In the latter case,

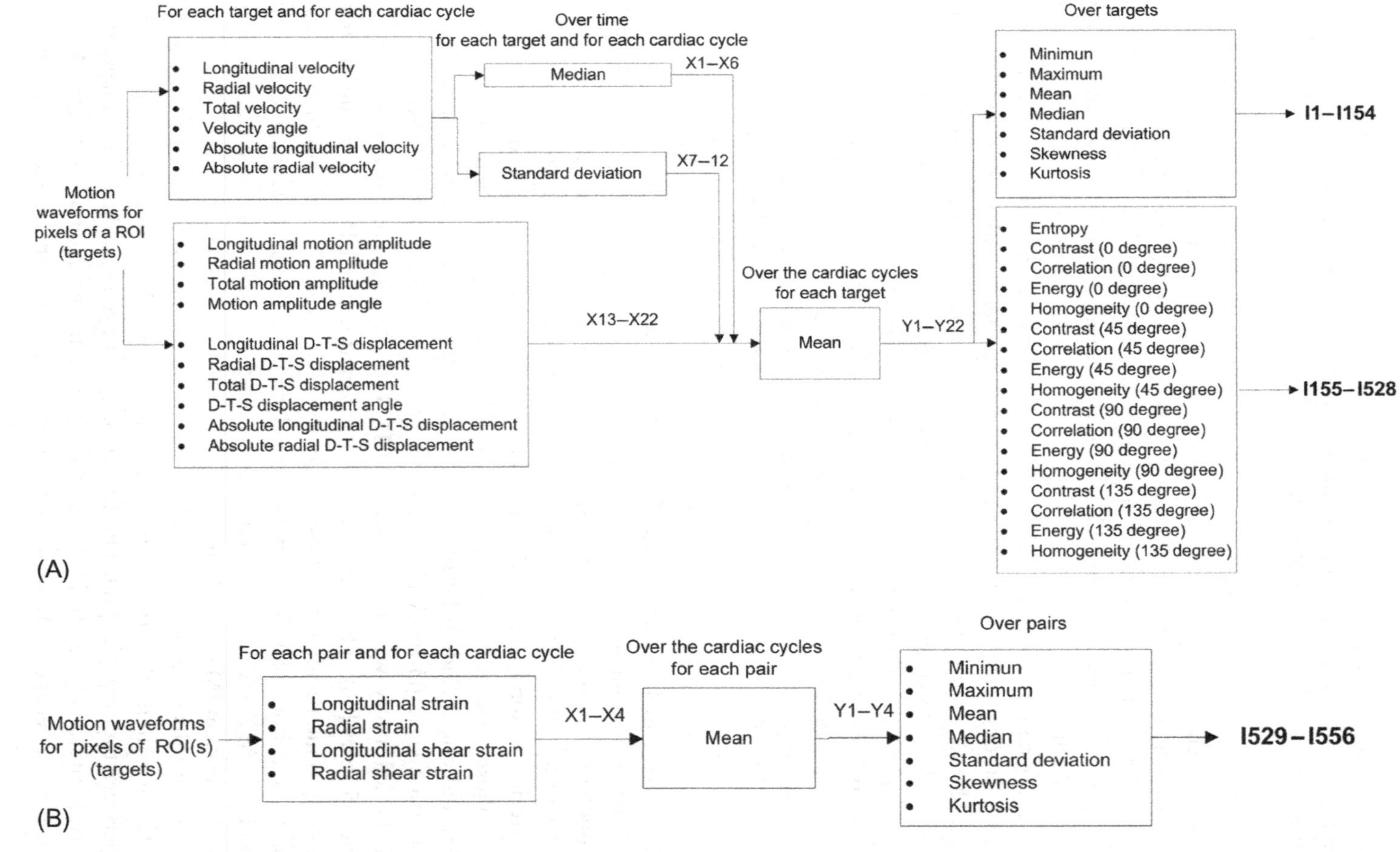

Fig. 5.3 Schematic presentation of the steps followed to estimate statistical measures of (A) pixel-wise kinematic and (B) pair-wise strain indices, using the results of motion analysis for an ROI or a pair of ROIs. *D-T-S*, diastole-to-systole [28].

indices of asymmetry may hold valuable information about disease pathophysiology, and therefore be useful in the diagnosis and required decision making. Cerebrovascular events, the major symptom of carotid atherosclerosis, have been shown to be diagnosed more frequently in the left, compared to the right, hemisphere [31]. This may be due either to easier event identification because of left hemisphere dominance in most people, or to increased vulnerability of the left carotid artery. Indeed, it was recently reported that left-sided plaques are more prevalent and more vulnerable than right-sided ones, with predominance of intraplaque hemorrhage and less calcification, both of which suggest reduced stability [32]. These observations were made from MRI acquisitions. Another study, however, reported no plaque prevalence in either side based on echographic features [33]. In addition to this, asymmetries in the normal (nonatherosclerotic) arterial walls are also important, because they may explain the preference of disease occurrence at specific sites. A relatively limited number of studies have addressed bilateral asymmetry in normal and diseased carotid arteries using a number of features derived from different imaging modalities.

In terms of anatomical features, it was demonstrated that the left common carotid artery, which arises directly from the aortic arch, was significantly longer (averaging 10 cm) than the right one (averaging 8 cm), which arises from the brachiocephalic artery off the arch [34]. The diameter of the right common carotid artery was shown to be larger, by about 6%, than the left, and the curvature of its thoracic part was also higher. These findings were derived from 28 participants (63 ± 12 years) of the VALIDATE (Vascular Aging—the Link that Bridges Age to Atherosclerosis) study using 3D contrast-enhanced magnetic resonance angiograms.

Morphological features, including the IMT, wall volume, and plaque calcification and echogenicity, have been evaluated in the context of bilateral asymmetry of the carotid arteries. Carotid artery IMT has been shown to be bilaterally similar in two different cohorts: one including apparently healthy individuals with a mean age of 46 years [33] and one including healthy subjects and patients with cardiovascular disease with a mean age of 63 years [35]. Another study, however, demonstrated that the IMT of the left artery was thicker than the IMT of the right one between the ages of 35 and 65 years [36]. The same study also reported that left IMT correlated better with hemodynamic parameters, whereas right IMT correlated better with biochemical markers. In a cohort of subjects with carotid atherosclerosis, a significant bilateral variation in the IMT was demonstrated, and was also found to be associated with vascular alteration better than IMT itself [23].

By imaging pairs of carotid arteries from cadaveric donors using MRI and electron-beam CT, it was found that total wall volume and plaque calcification were substantially similar between the two sides [37]. This finding suggests that individuals with atherosclerosis in one carotid artery are likely to have disease in the contralateral side. Along the same lines, a recent study demonstrated that ultrasound-image-based estimations of plaque echogenicity and texture were bilaterally similar when the ipsilateral plaque causes high-degree stenosis [38].

Image-based kinematic properties of the arterial wall have also been interrogated in the context of bilateral asymmetry. Ultrasound-based arterial strain measurements were found significantly higher at the right side compared to the left, in healthy and diabetic subjects with a mean age of 57 years [39]. The stiffness of bilateral carotid plaques was investigated by estimating the distensibility from MRI arterial cross-sections and tonometry-derived pulse pressures [40]. Bilateral asymptomatic plaques were less stiff than asymptomatic ones with contralateral symptomatic plaques [40]. Symptomatic plaques were found to be stiffer than their contralateral asymptomatic sides, despite a comparable plaque burden. On the other hand, Gnasso et al. [41] showed that wall shear stress was lower in arteries with plaque than in plaque-free arteries, suggesting that plaque presence affects wall mechanical properties. In that study, stress measurements for atheromatous cases were performed at plaque-free locations, upstream from the plaque.

4.2 Bilateral Asymmetry in a Cohort of Asymptomatic Carotid Atherosclerosis Subjects

Motivated by the above findings, the bilateral asymmetry of spatiotemporal features was investigated in a subgroup of asymptomatic patients of our dataset, diagnosed with bilateral carotid atherosclerosis [42]. This group of subjects included eight elderly subjects with carotid atherosclerosis in both the left and right carotid sides. There were no statistically significant differences in the degrees of stenosis between the two sides. As presented in the previous sections, features of interrogated arterial segments included kinematic and strain indices of the plaques and the normal wall adjacent to plaques. Differences between left and right sides were assessed with a Wilcoxon rank-sum test assuming a p-value of 0.05 or less to indicate statistical significance.

Eleven features were found statistically different between the left and right sides (Table 5.2). As we can see, in the case of the entire plaque, the energy of the total motion amplitude was higher in the left side; this is true for all four interrogated directions. According to its definition, the energy is an index of local image homogeneity. This implies that the left side had a more homogeneous distribution of motion amplitude than the right side. For a given side, the energy of total motion amplitude was similar for all directions, with slightly higher values in the horizontal direction (0 degree). Energy represents the opposite of entropy. Related to this, the entropy of radial velocity was found to be lower in the left side.

When looking at the distributions of feature values at the plaque top and bottom surfaces, we see that, out of the six significant features, five represent skewness and one kurtosis, both histogram-based metrics. When radial or total motion features were involved, skewness was higher at the left side, indicating right-tailed distributions, which

Table 5.2 Average ± STD of kinematic features which were found significantly different between left and right carotid sides

Kinematic feature	Left carotid	Right carotid
Entire plaque		
Entropy (over targets) of median (over time) radial velocity	0.75 ± 0.74	1.93 ± 1.28
Energy (0 degree) of total motion amplitude	0.68 ± 0.10	0.55 ± 0.12
Energy (45 degree) of total motion amplitude	0.65 ± 0.11	0.51 ± 0.12
Energy (90 degree) of total motion amplitude	0.65 ± 0.11	0.52 ± 0.12
Energy (135 degree) of total motion amplitude	0.64 ± 0.11	0.51 ± 0.12
PBS/PTS		
Skewness of LSI in PTS	1.33 ± 0.85	2.57 ± 1.28
Skewness of RSI between PBS/PTS	2.55 ± 0.97	1.26 ± 0.81
Skewness of longitudinal diastole-to-systole displacement in PTS	-0.07 ± 0.26	0.58 ± 0.53
Skewness of radial diastole-to-systole displacement in PTS	3.14 ± 0.72	2.40 ± 0.26
Skewness of total diastole-to-systole displacement in PTS	3.43 ± 0.93	2.35 ± 0.22
Kurtosis of total motion amplitude in PTS	2.68 ± 0.43	2.25 ± 0.35

Notes: *LSI*, longitudinal strain index; *PBS*, plaque bottom surface; *PTS*, plaque top surface; *RSI*, radial strain index.

are characterized by the presence of isolated high values well above the distribution mean. On the other hand, in the case of longitudinal direction, skewness was lower at the left side; in one case, skewness was negative, which corresponds to a left-tailed distribution, characterized by outliers below the mean. Kurtosis of total motion amplitude in the plaque's top surface was higher in the left side.

Our findings indicate differences in inhomogeneity of motion distribution within the plaque and along its top and bottom surfaces. They did not show differences in motion amplitudes between the left and right sides, neither in the plaque nor in the wall adjacent to it. Yang et al., however, reported differences of arterial strain between the left and right carotids of healthy and diabetic subjects, when considering normal (nonatherosclerotic) arterial segments [39]. The different observations of the two studies may be due to a number of factors, including the low number of subjects interrogated. A crucial issue for future investigation is whether bilateral mechanical asymmetries are related to asymmetries in plaque vulnerability [32].

The group of subjects interrogated here was also interrogated in terms of texture features of the atheromatous plaque [42]. However, only one texture feature was shown to be significantly different between the two sides. Specifically, a multiresolution-based feature was lower in the left side, indicating relatively higher homogeneity of gray levels in that side. The fact that the number of bilaterally different motion features was higher than textural ones suggests that motion-based features describe bilateral asymmetry better than textural ones.

In conclusion, our results show that bilateral asymmetries in the carotid artery may be quantified through motion analysis from ultrasound imaging. Such asymmetries may arise by differences in mechanical (blood pressures, tissue forces) and anatomical characteristics specific in the same individual between the right and the left carotid arteries, and merit further investigation.

5. RISK STRATIFICATION DRIVEN BY THE KINEMATIC ACTIVITY OF THE ARTERIAL WALL

5.1 Discriminating Symptomatic from Asymptomatic Plaques Using Motion and Strain Indices

The spatiotemporal features representing the kinematic activity of the arterial wall, extracted using the framework of Section 3, have been previously evaluated in terms of their ability to discriminate symptomatic from asymptomatic plaques on a smaller dataset, where they yielded higher than 88% classification accuracy [28]. In the same study, motion features were compared to texture features, which demonstrated a lower discriminatory capacity (80%). In an attempt to validate further the classification performance of motion versus texture, in this section, we comparatively evaluate the two types of features on a new, larger, dataset.

Our classification tool is a combination of a step-wise feed-forward feature selection step, in which an optimal subset of the whole feature representation is selected, with a classifier. The classifier used in this study is an implementation of support vector machines (SVMs), which, compared to other classification methods, is less affected by the so-called curse of dimensionality, and is therefore suitable for large sets of features [43]. SVMs are learning machines based on intuitive geometric principles, aiming to the definition of an optimal hyper plane which separates the training data so that a minimum classification error is achieved [44]. The training method is based on a nonlinear mapping of the dataset, using kernels that have to satisfy Mercer's theorem. In this study, we used a nonlinear kernel, namely a Gaussian radial basis function kernel with a scaling factor $s = 3.9^{27}$. The classification performance was measured in terms of accuracy and the area under the receiver operating characteristic curve (AUC), using leave-one-out cross-validation [45], where, in each round, a single observation (patient) is used as the testing sample, and the remaining observations compose the training dataset.

Table 5.3 lists the kinematic and the textural features which were deemed significant in our classification tool. The SVM model for motion incorporates 12 kinematic indices, which represent motion properties of AWL, PTS, and the whole plaque region, as well as strain indices, which express local deformations in PWL and relative kinematic activity between PTS and the healthy part of the wall adjacent to the plaque. This motion model achieved a classification accuracy of 0.85 and an AUC of 0.87. Six features were selected in the SVM model for texture, yielding a classification accuracy of 0.78 and a statistically lower (DeLong's test p-value < 0.05) AUC of 0.81 (Fig. 5.4).

Table 5.3 List of features selected in the classification models for motion and texture characterization

SVM model for kinematic features
Skewness of LSI in PWL
Kurtosis of LSI in PWL
Maximum (over targets) of median (over time) displacement angle in AWL
Kurtosis of median longitudinal velocity in the entire plaque
Kurtosis of median longitudinal displacement in the entire plaque
Correlation (90 degree of radial D-T-S displacement in the entire plaque
Skewness of absolute radial D-T-S displacement in PTS
Kurtosis of radial motion amplitude in PTS
Kurtosis of LSI in PWL
Mean of LSI between PTS/PWL
Skewness of LSI between PBS/PWL
Kurtosis of LSI between PBS/PWL
SVM model for textural features
Systolic images: cooccurrence energy, $\theta = 0$ degree
Systolic images: cooccurrence energy, $\theta = 90$ degree
Systolic images: mean of detail subimage $Dh_3Dh_2Dh_1$ [25]
Diastolic images: cooccurrence energy, $\theta = 0$ degree
Diastolic images: cooccurrence energy, $\theta = 90$ degree
Diastolic images: mean of detail subimage $Dh_3A_2A_1$ [25]

Notes: *AWL*, anterior wall-lumen; *D-T-S*, diastole-to-systole; *LSI*, longitudinal strain index; *PBS*, plaque bottom surface; *PWL*, posterior wall-lumen; *PTS*, plaque top surface.

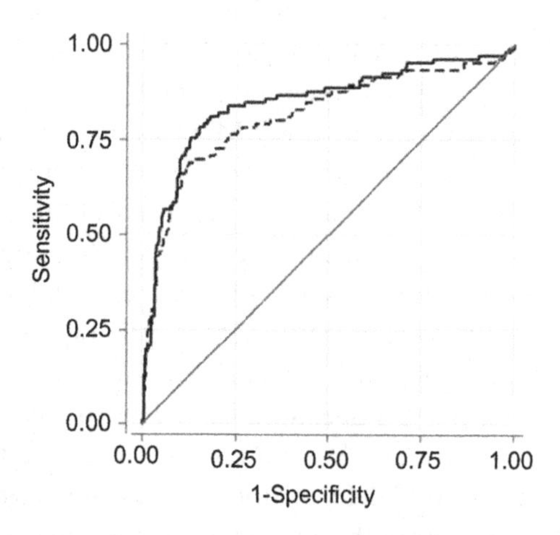

Fig. 5.4 Discriminatory capacity of motion (*solid line*) and texture (*dashed line*) features using SVM models.

5.2 Design of a Voice Recognition Analog for Motion and Strain Patterns

In an attempt to investigate the discriminatory capacity of spatiotemporal patterns produced from motion analysis, we designed a voice-recognition analog which is based on hidden Markov models (HMMs) and is guided by waveforms representing kinematic and strain activity in the arterial wall [29]. Our hypothesis is that, in correspondence with a voice-recognition system, the arterial wall dynamics which account for stable or vulnerable atherosclerotic lesions may vary among patients in the same way as identical words can be pronounced in different ways by humans with different voices. In this design, the spatiotemporal patterns correspond to the words (sets of phonemes) and a lexicon attributes the label "symptomatic" or "asymptomatic" to each word.

This idea is implemented with two majority voting schemes, each of which is fed with a subset of n spatiotemporal patterns (with $n \leq 146$) feeding into n classification models, one for each spatiotemporal pattern (Fig. 5.5). Each classification model is an implementation of an HMM, a stochastic state automaton, which, if properly trained, can decode an observation sequence (word) and hence recognize its underlying patterns [46]. Due to the periodic nature of arterial wall motion, the spatiotemporal patterns are periodically reproduced. Therefore, a left-to-right HMM, consisting of five states, was considered a suitable choice [47]. HMMs were implemented using the HTK Speech Recognition Toolkit, in which input signals are first sampled and converted to Mel-frequency cepstral coefficients; training is achieved through the Baum-Welch method, which has been employed successfully in cardiovascular applications [47, 48].

The first voting scheme generates the probability of the patient to belong in the "symptomatic" group (V_1), while the second one estimates the probability to be in the "asymptomatic" group (V_2). The vote of each scheme (V_j, with $j \in \{1, 2\}$) is estimated using the classification outputs, $p \in \{0\text{: false, } 1\text{: true}\}$, and some weights, $w \in [0, 1]$, of the classification models (Eq. 5.1). The final classification response, CAD_r, is produced using Eq. (5.2), with "−1," "0," and "1" representing "symptomatic," "not sure," and

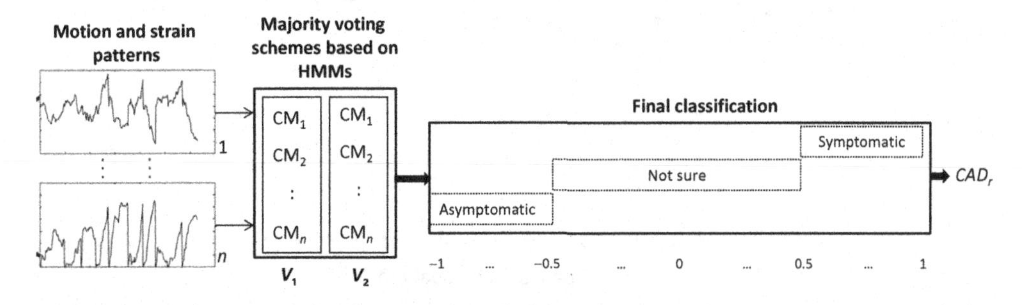

Fig. 5.5 Workflow for generating a computer-aided diagnosis response (CAD_r) using ultrasound-based spatiotemporal patterns of the arterial wall. *CM*, classification model.

"asymptomatic," respectively. The values of the parameters n and w were defined based on optimization and evaluation experiments on our study dataset, as explained below.

$$V_1 = \left[\frac{\sum_{i=1}^{n} p_i w_i}{n}\right] \in [0,1] \quad V_2 = -\left[\frac{\sum_{i=1}^{n} p_i w_i}{n}\right] \in [-1,0] \tag{5.1}$$

$$CAD_r = roundToInteger(V_1 + V_2) \in \{-1,0,1\} \tag{5.2}$$

Each HMM was parameterized in terms of (a) the implementation with monophones or triphones, where each word consists of three or nine phonemes, respectively, and (b) the preprocessing stage. The latter parameter involved two scenarios, in which the spatiotemporal patterns were (1) scaled and (2) not scaled in time to the maximum video duration among all patients. The optimization of each HMM lied in the maximization of the classification accuracy (i.e., percentage of correctly classified cases) for the corresponding spatiotemporal pattern, which was measured using leave-one-out cross validation.

Fig. 5.6 is a graphical presentation of the maximum classification accuracy, which was achieved for each spatiotemporal pattern by the corresponding optimized HMM. The classification performance ranged between 57% and 81%, and the average performance was 70%. Among all spatiotemporal patterns, we identified those with the strongest discrimination power (Fig. 5.7), that is, those which yielded a high (>75%) average value of specificity (i.e., correctly classified "asymptomatic" cases) and sensitivity (i.e.,

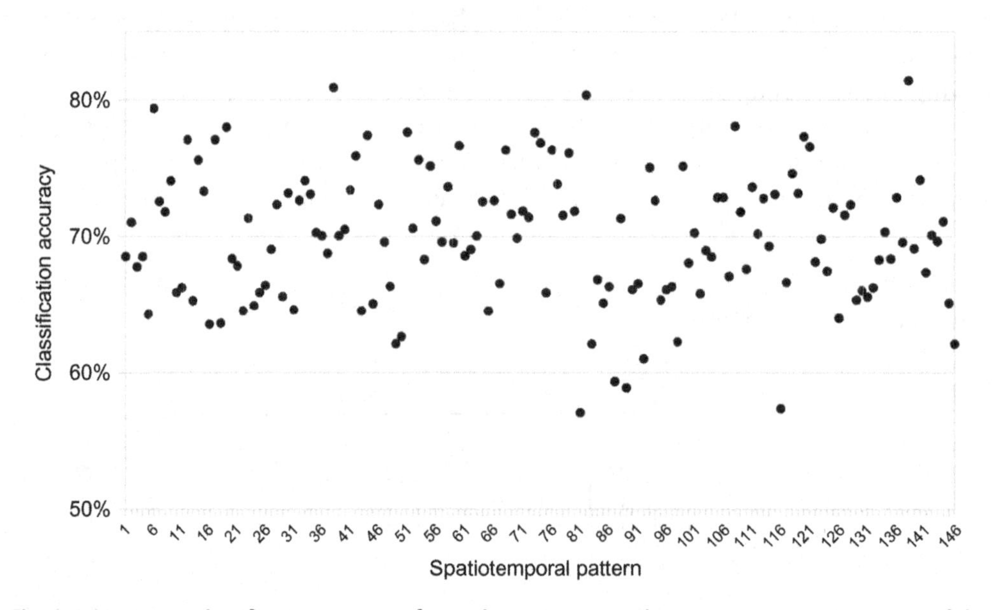

Fig. 5.6 Maximum classification accuracy for each spatiotemporal pattern, upon optimization of the corresponding HMM [29].

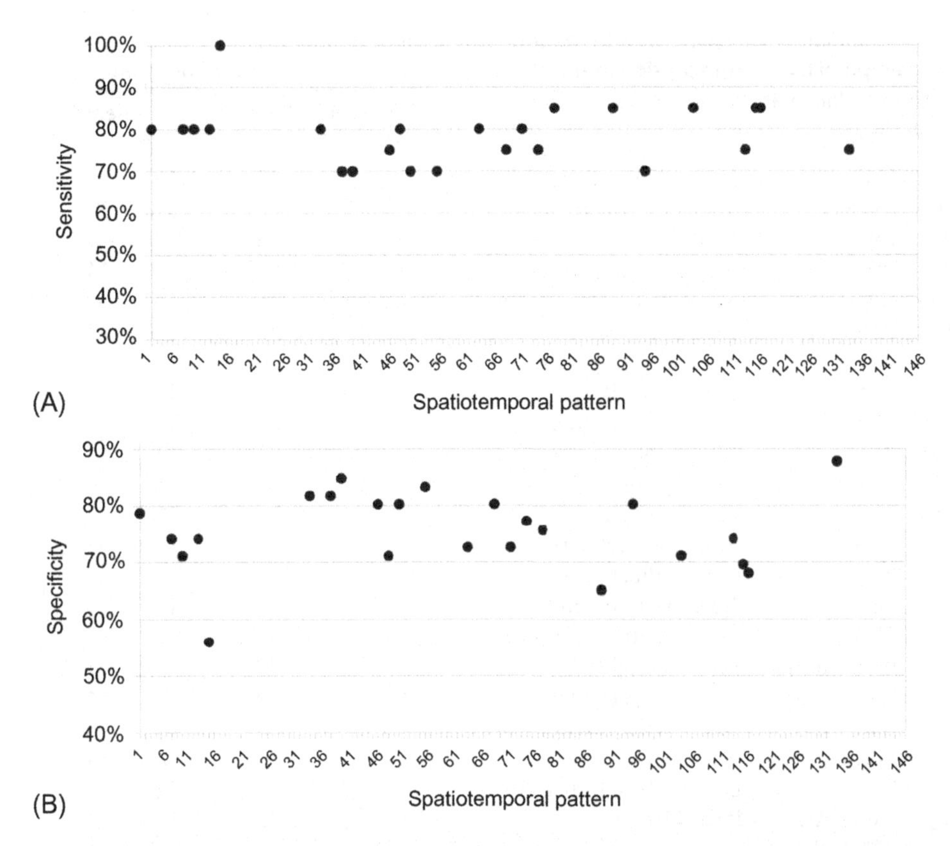

Fig. 5.7 (A) Sensitivity and (B) specificity values for the spatiotemporal patterns with the strongest discrimination power [29]. These values also correspond to the contributions (weights) of these patterns in the classification output (Eqs. 5.1, 5.2).

correctly classified "symptomatic" cases). For those $n = 24$ spatiotemporal patterns, Table 5.3 includes a short description, the most suitable HMM parameterization according to the optimization procedures, and the corresponding sensitivity and specificity results. Based on these results, the majority voting schemes of the final classification scheme are fed with the spatiotemporal patterns of Table 5.4, they consist of the corresponding optimized HMMs, and the weights w in V_1 and V_2 (Eq. 5.1) equal the corresponding sensitivity and specificity values, respectively.

The proposed classification scheme is able to assist treatment selection with accuracy between 76% and 79%. Given the results presented by related studies in the field [12] and the classification performance of the existing clinical practice on the same dataset [27], the aforementioned results are very encouraging for the potential of arterial-wall-motion patterns in assisting risk stratification for carotid atherosclerosis. The final classification scheme relies on 22 kinematic and 2 strain patterns which are related with the kinematic

Table 5.4 Spatiotemporal patterns with the strongest discrimination power

Spatiotemporal pattern description (region of interest – kinematic measure)			HMM optimization			
			Spec. (%)	**Sens. (%)**	**Parameterization**	
K1	PWL	Longitudinal velocity	80	70	NTS	M
K18	PWL	Abs. radial velocity	71	85	TS	M
K20	PWL	Radial displacement	74	75	TS	M
K21	PWL	Total displacement	68	85	TS	M
K24	PWL	Abs. radial displacement	70	85	TS	M
K36	AWL	Abs. radial displacement	88	75	TS	M
K57	PBS	Total displacement	80	75	TS	M
K60	PBS	Abs. radial displacement	83	70	TS	M
K64	PBS	Velocity angle	73	80	TS	M
K68	PBS	Radial displacement	73	80	TS	M
K69	PBS	Total displacement	77	75	TS	M
K76	PTS	Velocity angle	82	80	TS	M
K88	PTS	Velocity angle	85	70	TS	M
K90	PTS	Abs. radial velocity	82	70	TS	M
K91	PTS	Longitudinal displacement	80	75	TS	M
K94	PTS	Total displacement	80	70	TS	M
K95	PTS	Abs. longitudinal displacement	71	80	TS	M
K100	Entire plaque	Velocity angle	79	80	NTS	M
K104	Entire plaque	Radial displacement	71	80	TS	M
K105	Entire plaque	Total displacement	74	80	TS	M
K112	Entire plaque	Velocity angle	74	80	NTS	M
K115	Entire plaque	Longitudinal displacement	56	100	TS	M
S1	PWL/AWL	Radial strain	65	85	TS	M
S18	PTS/PWL	Longitudinal strain	76	85	NTS	M
Average			76	79		

Notes: For each pattern, this table presents the encoding of the pattern according to Fig. 5.2 and the specificity and sensitivity values which were achieved, upon optimization, by the corresponding HMM (together with the corresponding parameterization).
M, monophones; *NTS*, no time scaling; *TS*, time scaling.

activity of all the selected ROIs. This conclusion further reinforces the argument that the motion activity of the atherosclerotic lesion itself and healthy parts of the wall close to the lesion are equally important in risk stratification in the disease [12, 27].

A significant contribution of this study with respect to the related literature is that it suggested that the imaging phenotypes of symptomatic and asymptomatic carotid atherosclerosis differ in terms of not only mobility indices describing motion properties, but also in motion trajectories and strain patterns. This conclusion remains to be investigated further in future studies on larger datasets, which will reveal the full potential of the presented approach. In the same line of work, the effect of input variability (e.g., frequency and frame rate in ultrasound image recordings) on HMM performance will be examined, as well. Both the design principles and the results of this study are expected also to motivate the incorporation of motion analysis and spatiotemporal patterns in future related studies designing computer-aided diagnostic tools for carotid atherosclerosis.

6. DATA MINING OF ASSOCIATION-BASED PHENOTYPIC NETWORKS

In this section, we go one step further to examine potential imaging phenotypes of carotid atherosclerosis and their associations with two key biomarkers of the disease, namely the size (area) of the atherosclerotic lesions, A, and the degree of carotid stenosis caused by the lesions, DS. This work was inspired by related studies where networks of biomarkers were successfully used to investigate overlapping patterns in chronic renal failure symptoms [49] and phenotypes of the coronary heart disease [50]. In our study, we build a mutual-information-based phenotypic network of carotid atherosclerosis (Fig. 5.8) and we aim to identify groups of markers, or marker patterns that bear close

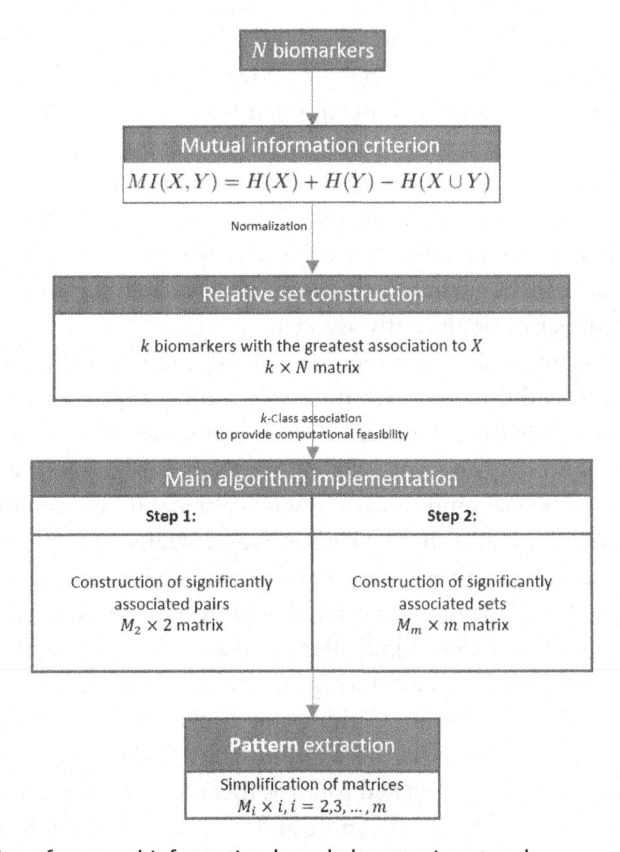

Fig. 5.8 Construction of a mutual-information-based phenotypic network.

association with each other, while also containing those two key biomarkers. The final results, exhibiting the most significant associations, are herein presented and visualized.

The mutual information (*MI*) between two random variables, X and Y, measures the amount of shared information and expresses the degree of association between them [51]. *MI* is defined as $MI(X, Y) = H(X) + H(Y) - H(X \cap Y)$, where H denotes the entropy, while $H(X \cap Y)$ denotes the joint entropy between the two variables. The entropy definitions differ slightly in the cases of discrete and continuous variables; however, the adequate sample size of our dataset allows the transformation of any continuous marker to its discrete equivalent through histogram construction. The metric *MI* is calculated for each possible pair of markers and is normalized in the interval (0, 1). Once *MI* values for all pairs are available, the relative sets $R(X)$ for each marker are constructed, which consist of k markers with the greatest association to X (excluding itself).

The next step involves the computation of all significantly associated pairs $\{X, Y\}$, which satisfy both of the conditions $X \in R(Y)$ and $Y \in R(X)$. By generalizing the same concept, tuples containing three or more significantly associated markers are constructed up to a maximum tuple size of k. However, the amount of markers involved is rather large and poses restrictions on the computational feasibility of the significantly associated sets. This limitation is overcome by ensuring that two conditions apply when deriving a $(p + 1)$-size tuple from two p-size tuples: (a) the two p-size tuples are ordered and identical in their first $p - 1$ members and (b) the pth members of the two p-size tuples form together a significantly associated pair. Then, the $(p + 1)$-size tuple composed of the common first $p - 1$ members plus the two pth members is also a significantly associated tuple and may be added to the p-size set. Patterns are finally defined as sets with maximum number of significantly associated variables.

The choice of parameter k, which represents the size of relative sets, has an important impact on the results and the total execution time, since it affects the amount of pairs, tuples and patterns which are generated. A very small value hinders the construction of higher-size tuples and completes fast, while a very large value has the opposite effect. Thus, a value of $k = 100$ was empirically chosen as a tradeoff between these two factors.

To visualize the outcome of this algorithm, we generated a graph where each node corresponds to a marker. Moreover, the size of each node is proportional to the sum of the associations (p) of the corresponding variable with these biomarkers: $p_A + p_{DS}$. The graph was created using ModelGUI [52] after importing the appropriate Pajek Network files and the additional text files for coloring and sizing of vertices.

The application of this algorithm to the set of kinematic and textural indices of our study is shown in Fig. 5.9. Based on these results, a few interesting conclusions may be derived. First, no significantly associated patterns included the biomarker *DS*. However, the biomarker A was found to be significantly associated with kinematic indices, suggesting that the size of the atherosclerotic lesion affects the kinematic activity of the

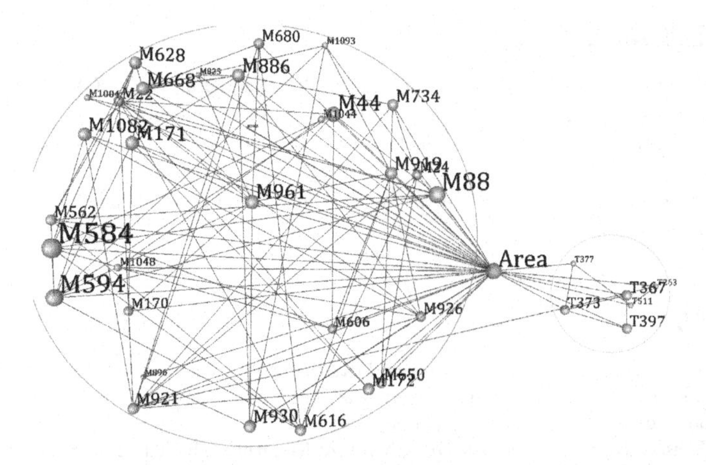

Fig. 5.9 3D graph visualizing mutual information patterns among motion features, textural indices, and the area of the atherosclerotic lesions.

arterial wall. Moreover, four kinematic indices (M44, M88, M584, and M594) raised the highest associations. These variables represent the diastole-to-systole total displacement of the entire plaque, and the longitudinal velocity and motion amplitude of the plaque top surface. Their key role in the phenotypic network of carotid atherosclerosis implies that they may have an important role in the characterization of the disease, particularly in relation with the extent of the atherosclerotic lesions.

7. CONCLUSION

By elucidating associations of motion features of the arterial wall with physiologically relevant characteristics of carotid atherosclerosis, this chapter leverages the role of kinematic analysis from B-mode ultrasound in evaluating atherosclerotic carotid arteries. Our results show that motion analysis from ultrasound imaging is able to effectively (a) quantify bilateral asymmetries in the carotid artery which may arise by differences in mechanical and anatomical characteristics, (b) discriminate symptomatic from asymptomatic patients with carotid atherosclerosis, valuable in assessing plaque vulnerability, and (c) reveal potentially different imaging phenotypes of the disease associated with the extent (size) of the atherosclerotic plaque. The extension of the presented methodologies and analyses to 3D ultrasound image data would be a valuable future perspective of our work toward further validating our conclusions and investigating the effect of out-of-plane motion which could not be captured in 2D B-mode ultrasound images.

ACKNOWLEDGMENTS

The authors would like to thank Dr. C. D. Liapis, Dr. N. P. E. Kadoglou, Dr. C. Gkekas, Dr. J. D. Kakisis, and Dr. A. Koulia from the Department of Vascular Surgery, Attikon General University Hospital of Greece for their invaluable contributions in data collection. The help of Dr. N. Tsiaparas on texture analysis is acknowledged.

The work of P. Mermigkas was supported in part by "IKY fellowships of excellence for postgraduate studies in Greece—Siemens program."

REFERENCES

[1] Mozaffarian D, Benjamin EJ, Go AS, Arnett DK, Blaha MJ, Cushman M, et al. Heart disease and stroke statistics—2016 update a report from the American Heart Association. Circulation 2015. http://dx.doi.org/10.1161/CIR.0000000000000350.

[2] Inzitari D, Eliasziw M, Gates P, Sharpe BL, Chan RK, Meldrum HE, et al. The causes and risk of stroke in patients with asymptomatic internal-carotid-artery stenosis. N Engl J Med 2000;342(23):1693–701.

[3] Nikita KS. Atherosclerosis: the evolving role of vascular image analysis. Comput Med Imaging Graph 2013;37(1):1–3.

[4] Golemati S, Gastounioti A, Nikita KS. Toward novel noninvasive and low-cost markers for predicting strokes in asymptomatic carotid atherosclerosis: the role of ultrasound image analysis. IEEE Trans Biomed Eng 2013;60(3):652–8.

[5] Gaitini D, Soudack M. Diagnosing carotid stenosis by Doppler sonography state of the art. J Ultrasound Med 2005;24(8):1127–36.

[6] Naylor A, Sillesen H, Schroeder T. Clinical and imaging features associated with an increased risk of early and late stroke in patients with symptomatic carotid disease. Eur J Vasc Endovasc Surg 2015;49(5):513–23.

[7] Naylor A, Schroeder T, Sillesen H. Clinical and imaging features associated with an increased risk of late stroke in patients with asymptomatic carotid disease. Eur J Vasc Endovasc Surg 2014;48(6):633–40.

[8] Molinari F, Zeng G, Suri JS. A state of the art review on intima-media thickness (IMT) measurement and wall segmentation techniques for carotid ultrasound. Comput Methods Prog Biomed 2010;100(3):201–21.

[9] Naqvi TZ, Lee M-S. Carotid intima-media thickness and plaque in cardiovascular risk assessment. J Am Coll Cardiol Img 2014;7(10):1025–38.

[10] Cinthio M, Ahlgren ÅR, Jansson T, Eriksson A, Persson HW, Lindström K. Evaluation of an ultrasonic echo-tracking method for measurements of arterial wall movements in two dimensions. IEEE Trans Ultrason Ferroelectr Freq Control 2005;52(8):1300–11.

[11] Zahnd G, Boussel L, Marion A, Durand M, Moulin P, Sérusclat A, et al. Measurement of two-dimensional movement parameters of the carotid artery wall for early detection of arteriosclerosis: a preliminary clinical study. Ultrasound Med Biol 2011;37(9):1421–9.

[12] Gastounioti A, Golemati S, Stoitsis J, Nikita K. Adaptive block matching methods for carotid artery wall motion estimation from B-mode ultrasound: in silico evaluation & in vivo application. Phys Med Biol 2013;58(24):8647–61.

[13] Gastounioti A, Golemati S, Stoitsis J, Nikita K. Comparison of Kalman-filter-based approaches for block matching in arterial wall motion analysis from B-mode ultrasound. Meas Sci Technol 2011;22(11):114008.

[14] Golemati S, Stoitsis JS, Gastounioti A, Dimopoulos AC, Koropouli V, Nikita KS. Comparison of block matching and differential methods for motion analysis of the carotid artery wall from ultrasound images. IEEE Trans Inf Technol Biomed 2012;16(5):852–8.

[15] Zahnd G, Salles S, Liebgott H, Vray D, Sérusclat A, Moulin P. Real-time ultrasound-tagging to track the 2D motion of the common carotid artery wall in vivo. Med Phys 2015;42(2):820–30.

[16] Golemati S, Gastounioti A, Nikita KS. Ultrasound-image-based cardiovascular tissue motion estimation. IEEE Rev Biomed Eng 2016; 9:208–18.

[17] De Korte CL, Fekkes S, Nederveen AJ, Manniesing R, Hansen HH. Review: mechanical characterization of carotid arteries and atherosclerotic plaques. Evid Based Complement Alternat Med 2016; 63(10):1613–23.
[18] Svedlund S, Gan LM. Longitudinal wall motion of the common carotid artery can be assessed by velocity vector imaging. Clin Physiol Funct Imaging 2011;31(1):32–8.
[19] Zahnd G, Orkisz M, Sérusclat A, Moulin P, Vray D. Evaluation of a Kalman-based block matching method to assess the bi-dimensional motion of the carotid artery wall in B-mode ultrasound sequences. Med Image Anal 2013;17(5):573–85.
[20] Soleimani E, Mokhtari-Dizaji M, Saberi H. A novel non-invasive ultrasonic method to assess total axial stress of the common carotid artery wall in healthy and atherosclerotic men. J Biomech 2015;48(10):1860–7.
[21] Mokhtari-Dizaji M, Montazeri M, Saberi H. Differentiation of mild and severe stenosis with motion estimation in ultrasound images. Ultrasound Med Biol 2006;32(10):1493–8.
[22] Zahnd G, Vray D, Sérusclat A, Alibay D, Bartold M, Brown A, et al. Longitudinal displacement of the carotid wall and cardiovascular risk factors: associations with aging, adiposity, blood pressure and periodontal disease independent of cross-sectional distensibility and intima-media thickness. Ultrasound Med Biol 2012;38(10):1705–15.
[23] Graf IM, Schreuder FH, Hameleers JM, Mess WH, Reneman RS, Hoeks AP. Wall irregularity rather than intima-media thickness is associated with nearby atherosclerosis. Ultrasound Med Biol 2009;35(6):955–61.
[24] Gastounioti A, Golemati S, Nikita KS, editors. Computerized analysis of ultrasound images: potential associations between texture and motion properties of the diseased arterial wall. In: 2012 IEEE International Ultrasonics Symposium (IUS). IEEE; 2012.
[25] Tsiaparas NN, Golemati S, Andreadis I, Stoitsis JS, Valavanis I, Nikita KS. Comparison of multiresolution features for texture classification of carotid atherosclerosis from B-mode ultrasound. IEEE Trans Inf Technol Biomed 2011;15(1):130–7.
[26] Acharya UR, Faust O, Alvin A, Krishnamurthi G, Seabra JC, Sanches J, et al. Understanding symptomatology of atherosclerotic plaque by image-based tissue characterization. Comput Methods Prog Biomed 2013;110(1):66–75.
[27] Gastounioti A, Kolias V, Golemati S, Tsiaparas NN, Matsakou A, Stoitsis JS, et al. CAROTID—a web-based platform for optimal personalized management of atherosclerotic patients. Comput Methods Prog Biomed 2014;114(2):183–93.
[28] Gastounioti A, Makrodimitris S, Golemati S, Kadoglou NP, Liapis CD, Nikita KS. A novel computerized tool to stratify risk in carotid atherosclerosis using kinematic features of the arterial wall. IEEE J Biomed Health Inform 2015;19(3):1137–45.
[29] Gastounioti A, Prevenios M, Nikita KS, editors. Using spatiotemporal patterns of the arterial wall to assist treatment selection for carotid atherosclerosis. In: Joint MICCAI-workshops on computing and visualization for intravascular imaging and computer assisted stenting, Cambridge, MA, USA; 2014.
[30] Gastounioti A, Golemati S, Stoitsis J, Nikita K. Carotid artery wall motion analysis from B-mode ultrasound using adaptive block matching: in silico evaluation and in vivo application. Phys Med Biol 2013;58(24):8647–61.
[31] Hedna VS, Bodhit AN, Ansari S, Falchook AD, Stead L, Heilman KM, et al. Hemispheric differences in ischemic stroke: is left-hemisphere stroke more common? J Clin Neurol 2013;9(2):97–102.
[32] Selwaness M, van den Bouwhuijsen Q, van Onkelen RS, Hofman A, Franco OH, van der Lugt A, et al. Atherosclerotic plaque in the left carotid artery is more vulnerable than in the right. Stroke 2014;45(11):3226–30.
[33] Bossuyt J, Van Bortel LM, De Backer TL, Van De Velde S, Azermai M, Segers P, et al. Asymmetry in prevalence of femoral but not carotid atherosclerosis. J Hypertens 2014;32(7):1429–34.
[34] Manbachi A, Hoi Y, Wasserman BA, Lakatta EG, Steinman DA. On the shape of the common carotid artery with implications for blood velocity profiles. Physiol Meas 2011;32(12):1885–97.
[35] Loizou CP, Nicolaides A, Kyriacou E, Georghiou N, Griffin M, Pattichis CS. A comparison of ultrasound intima-media thickness measurements of the left and right common carotid artery. IEEE J Transl Eng Health Med 2015;3:1–10.

[36] Luo X, Yang Y, Cao T, Li Z. Differences in left and right carotid intima-media thickness and the associated risk factors. Clin Radiol 2011;66(5):393–8.
[37] Adams GJ, Simoni DM, Bordelon CB, Vick GW, Kimball KT, Insull W, et al. Bilateral symmetry of human carotid artery atherosclerosis. Stroke 2002;33(11):2575–80.
[38] Doonan R, Dawson A, Kyriacou E, Nicolaides A, Corriveau M, Steinmetz O, et al. Association of ultrasonic texture and echodensity features between sides in patients with bilateral carotid atherosclerosis. Eur J Vasc Endovasc Surg 2013;46(3):299–305.
[39] Yang EY, Dokainish H, Virani SS, Misra A, Pritchett AM, Lakkis N, et al. Segmental analysis of carotid arterial strain using speckle-tracking. J Am Soc Echocardiogr 2011;24(11):1276–84.
[40] Sadat U, Usman A, Howarth SP, Tang TY, Alam F, Graves MJ, et al. Carotid artery stiffness in patients with symptomatic carotid artery disease with contralateral asymptomatic carotid artery disease and in patients with bilateral asymptomatic carotid artery disease: a cine phase-contrast carotid MR study. J Stroke Cerebrovasc Dis 2014;23(4):743–8.
[41] Gnasso A, Irace C, Carallo C, De Franceschi MS, Motti C, Mattioli PL, et al. In vivo association between low wall shear stress and plaque in subjects with asymmetrical carotid atherosclerosis. Stroke 1997;28(5):993–8.
[42] Golemati S, Gastounioti A, Tsiaparas NN, Nikita KS, editors. Bilateral asymmetry in ultrasound-image-based mechanical and textural features in subjects with asymptomatic carotid artery disease. In: 2014 IEEE-EMBS international conference on Biomedical and Health Informatics (BHI). IEEE; 2014.
[43] Goszczyński J. Texture classification using support vector machine. 2006;13(88):119–26.
[44] Burges CJ. A tutorial on support vector machines for pattern recognition. Data Min Knowl Disc 1998;2(2):121–67.
[45] Refaeilzadeh P, Tang L, Liu H. Cross validation, Encyclopedia of Database Systems (EDBS), vol. 6. Arizona State University, Springer; 2009.
[46] Duda RO, Hart PE. Pattern classification and scene analysis. New York: Wiley; 1973.
[47] Andreão RV, Dorizzi B, Boudy J. ECG signal analysis through hidden Markov models. IEEE Trans Biomed Eng 2006;53(8):1541–9.
[48] Baier V, Baumert M, Caminal P, Vallverdú M, Faber R, Voss A. Hidden Markov models based on symbolic dynamics for statistical modeling of cardiovascular control in hypertensive pregnancy disorders. IEEE Trans Biomed Eng 2006;53(1):140–3.
[49] Chen J, Xi G. An unsupervised partition method based on association delineated revised mutual information. BMC Bioinformatics 2009;10(Suppl. 1):S63.
[50] Chen J, Lu P, Zuo X, Shi Q, Zhao H, Luo L, et al. Clinical data mining of phenotypic network in angina pectoris of coronary heart disease. Evid Based Complement Alternat Med 2012;2012:546230.
[51] Cover TM, Thomas JA. Elements of information theory. New York: John Wiley & Sons; 2012.
[52] ModelGUI software website. https://launchpad.net/modelgui [accessed 23.06.16].

CHAPTER 6

Right Generalized Cylinder Model for Vascular Segmentation

L. Flórez-Valencia[*], M. Orkisz[†]
[*]Pontifical Xavierian University, Bogotá, Colombia
[†]Univ Lyon, CNRS UMR5220, Inserm U1206, INSA-Lyon, Université Lyon 1, CREATIS, F-69621, Lyon, France

Chapter Outline

In this chapter, the model was used to formulate a vessel tracking algorithm. This algorithm computes the model of a vessel at the same time as the image is segmented.

Chapter Points

- First, the direct right generalized cylinder (RGC) model is presented in Section 2.
- Second, an inversion method is presented in Section 3.
- Finally, this inversion method is used to infer RGC parameters from actual data (Section 4), such as contours extracted from segmented images of cylindrical shapes.

Computing and Visualization for Intravascular Imaging and Computer-Assisted Stenting
http://dx.doi.org/10.1016/B978-0-12-811018-8.00006-0

1. MOTIVATION

Vascular image segmentation aims at the extraction of vessels (arteries, veins, or even bronchi) from medical images to *quantify*, *simulate*, or *characterize* their morphology.

Vessel *quantification* is useful when a medical expert is interested in diagnosing the patient's degree of arterial illness. For example, in the case of stenoses (a pathological reduction of the vessel lumen), an important diagnose measure is the healthy-to-ill ratio, known as the *stenosis quantification*: while high values indicate a high health risk, low values indicate a low (or no) health risk. Such a measure is computed as the ratio between a healthy section of the vessel and the ill section of the vessel.

Vessel *simulations* are used to predict events or estimate values that cannot be measured directly by any imaging technique. Patient-specific simulations require the extraction of information such as vessel morphology from images based on measures taken from images, like flow simulation to predict thrombus formation [1], estimate physical properties of vessels, or prepare surgical interventions [2].

Vessel *characterization* aims at the topological description of vessels, that is, computing and presenting information that describes the shape of vessels, as in the study of the distribution of coronary arteries on the heart or describing the circle of Willis inside the skull [3].

The description of all the works that propose or define vascular segmentation techniques is out of the scope of this chapter; interested readers could refer to the reviews proposed in Refs. [4–6]. Unfortunately, in such works there is no unified taxonomy of vascular segmentation algorithms; in other words, each review proposes its particular taxonomy to classify algorithms. However, every algorithm results in objects that have the same goal: separate (segment) vascular shapes from their surroundings (background). Quantifying, simulating, and characterizing vessels from binary images could be a tedious and imprecise task since numerical results depend strongly on the sampling properties of the image, namely the voxel size.

Some works take into account these issues and propose cylindrical models to improve the segmentation procedure, such as [1, 7–11]. Although every model is different, they all have the same goal: to represent *continuous* cylinders (and trees) to simplify quantification [7, 8], simulation [1], or characterization [7], and even speed up the whole segmentation process [1, 8].

In this chapter, the RGC model is presented. This model has two purposes: a direct use (presented in Section 2) allowing a very flexible creation of cylindrical shapes that can be used in visualization, education, expert training, or digital phantom creation; and an inverse use (Section 3) providing a model for vessel segmentations by introducing an algorithm that allows us to infer RGC parameters from measures taken from, for example, a segmented image. Furthermore, this inversion algorithm can be used to guide a vessel tracking method (i.e., a segmentation method that follows a vessel along its elongated dimension) if used together with, for example, a Kalman filter (Section 4).

2. DIRECT MODEL

As first proposed in Ref. [12], the right generalized cylinder state model (RGC-sm) is inspired by the taxonomy proposed by Binford [13], which classifies cylinders in many categories, one of them being *right generalized cylinders*. Any cylinder belonging to such a category is composed by a generating curve and a stack of contours. It should be noted that the term "right" means the existence of a generating curve, rather than meaning that the cylinder is *straight*. Even though the model was originally presented as a *state model* (i.e., in terms of a system with different states and transitions between them), it can be also defined as a *geometrical model*, which is the main approach of this chapter.

More formally, Binford's taxonomy defines a right generalized cylinder (RGC from now on, without the "sm" suffix since, let us recall, the model is presented in this chapter from a *geometrical* point of view) $\mathcal{RGC}$ as an elongated geometrical object[1] that is composed of two simple objects: a generating curve $\mathcal{H}$ and a continuous stacking of two-dimensional contours $\mathcal{S}_\mathcal{C}$, attached to $\mathcal{H}$. Attachment between these two objects is realized by the following constraint: any two-dimensional contour $\mathcal{C}$, belonging to $\mathcal{S}_\mathcal{C}$, is locally perpendicular to $\mathcal{H}$. An example of this constraint is shown in Fig. 6.1.

In this figure, any contour $\mathcal{C}$ lies over its corresponding plane $\mathcal{P}$, which has an orientation represented by the orthonormal basis $\mathbf{B} \in \mathbb{R}^{3\times3}$ and a center defined by a point $\mathbf{h} \in \mathbb{R}^3$.

Hence, by simply assembling a linear algebra equation to locate and orient two-dimensional contours in 3D space, the model for an object $\mathcal{RGC}$[2] should be

$$\mathbf{rgc}(s,\omega) = \mathbf{B}(s)\,\mathbf{c}(s,\omega) + \mathbf{h}(s) \tag{6.1}$$

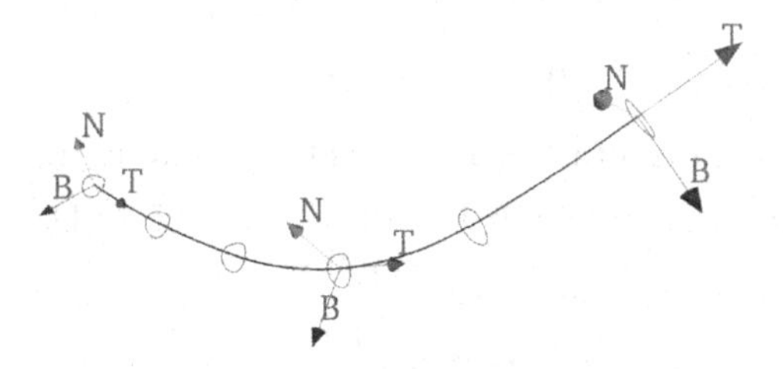

Fig. 6.1 Schematic representation of a right generalized cylinder: a generating curve $\mathcal{H}$ (*bold line*) localizes contours along it (*thin closed contours*). Each contour is located on a plane defined by three orthonormal vectors: **t** (the curve's tangent), **n** (the curve's normal), and **b** (the curve's binormal).

[1] Without any loss of generality, in this chapter, an *object* implies a *geometrical* object.

[2] Please be aware that while calligraphic expressions (like $\mathcal{RGC}$) represent an object in topological terms, bold symbols (like **rgc**) define a geometrical equation.

where

- $s \in [0, \Delta]$ and $\omega \in [0, 2\pi[$ are the curve's arc-length and contour's azimuthal parameters, respectively,
- $\Delta \in \mathbb{R}^{+}$ is the length of $\mathcal{H}$,
- $\mathbf{h}(s)$ describes three-dimensional points where perpendicular planes $\mathcal{P}$ are located,
- $\mathbf{B}(s)$ represents three-dimensional orthonormals basis to orient planes $\mathcal{P}$, and
- $\mathbf{c}(s, \omega)$ describe the continuous stacking of contours $\mathcal{S}_{\mathcal{C}}$.

The geometrical $\mathcal{RGC}$ model is derived as a rewriting process of Eq. (6.1); this process is guided by some theorems and lemmas (mainly taken from classical differential geometry) and by some assumptions on real-life cylindrical objects such as *continuity*, *manifoldness*, and *local complexity*. The rest of this section is organized as follows: Section 2.1 presents the general RGC model, then a particular piecewise constant model is shown in Section 2.2, followed by a particular contour model (Section 2.3). The contour model is then transformed into a surface model in Section 2.4.

2.1 Generating Curve

Informally, in nature, cylindrical shapes are *smooth* and *coherent*: this means that cylinders such as blood vessels, bronchi, trachea, intestines, or plant stalks have a smooth appearance and they do not self-intersect. In a more formal way, any $\mathcal{RGC}$ object that represents a real-life structure should (as well as its composing objects $\mathcal{H}$ and $\mathcal{S}_{\mathcal{C}}$) be *continuous* and *2-manifold*. In this section, the model for generating curves $\mathcal{H}$ is written in terms of its related equations for $\mathbf{h}(s)$ and $\mathbf{B}(s)$, taking into account the physical nature of such curves: they should be *continuous* (formally, at least C^2 differentiable).

We start by defining the curve's points equation:

$$\mathbf{h}(s) : [0, \Delta] \rightarrow \mathbb{R}^3 \tag{6.2}$$

which is a parametric equation that produces 3D points along the arc-length parameter $s \in [0, \Delta]$. From the classical differential geometry of curves [14], if such a curve is to be continuous (recall: C^2, at least), it should follow that

$$\mathbf{t}(s) = \frac{d\mathbf{h}(s)}{ds} \tag{6.3}$$

and

$$\mathbf{n}(s) = \frac{1}{\kappa(s)} \frac{d\mathbf{t}(s)}{ds} \tag{6.4}$$

and

$$\mathbf{b}(s) = \mathbf{t}(s) \times \mathbf{n}(s) \tag{6.5}$$

where $\mathbf{t}(s)$, $\mathbf{n}(s)$, and $\mathbf{b}(s)$ are, respectively, the *normalized* tangent, normal, and binormal vectors attached to $\mathcal{H}^3$ (where $\kappa(s)$ is the curve's curvature). These three vectors can be represented compactly using a matrix:

$$\mathbf{F}(s) = \begin{bmatrix} t_{\hat{\mathbf{x}}}(s) & n_{\hat{\mathbf{x}}}(s) & b_{\hat{\mathbf{x}}}(s) \\ t_{\hat{\mathbf{y}}}(s) & n_{\hat{\mathbf{y}}}(s) & b_{\hat{\mathbf{y}}}(s) \\ t_{\hat{\mathbf{z}}}(s) & n_{\hat{\mathbf{z}}}(s) & b_{\hat{\mathbf{z}}}(s) \end{bmatrix} \tag{6.6}$$

This is known as $\mathcal{H}$'s *Frenet-Serret* orthonormal basis [14]. This particular basis follows the *Frenet-Serret* formulas [14]:

$$\mathbf{F}^{\top}(s)\frac{d\mathbf{F}(s)}{ds} = \begin{bmatrix} 0 & -\kappa(s) & 0 \\ \kappa(s) & 0 & -\tau(s) \\ 0 & \tau(s) & 0 \end{bmatrix} = \mathbf{K}(s) \tag{6.7}$$

which relates any *Frenet-Serret* basis to its derivative in terms of the curve's curvature $\kappa(s) : [0, \Delta] \rightarrow \mathbb{R}$ and torsion $\tau(s) : [0, \Delta] \rightarrow \mathbb{R}$.

Although we can imagine that $\mathbf{F}(s)$ should be enough to orient planes in space, let us remember that planes (when defined as implicit functions) need a normal vector $\mathbf{t}(s)$ and an origin point $\mathbf{h}(s)$ in order to be defined. In natural shapes, nevertheless, these two objects are not enough: since 2D contours will be placed on such planes, an orientation should also be given to guarantee a good position/orientation

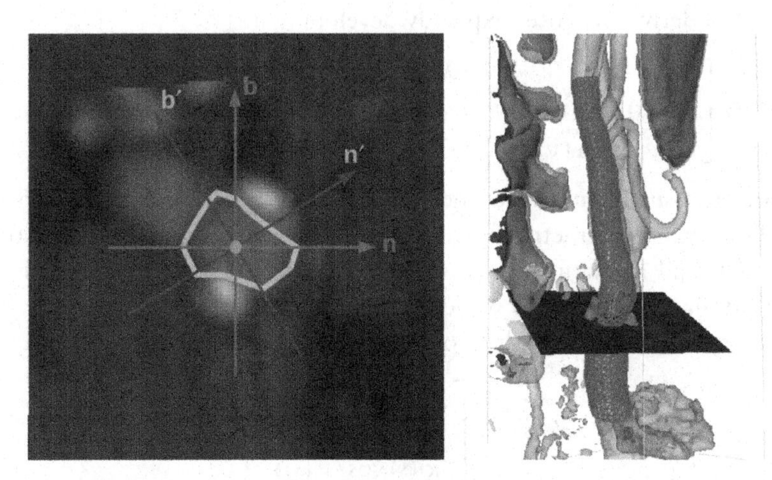

Fig. 6.2 Example of a contour segmented from a CT vascular image. Contour localization and orientation are shown on the *right*. On the *left*, in *light-colored arrows*, is shown the plane defined by the Frenet frame. As the contour "starts" at vector $\mathbf{n}'$, the localization is guaranteed by rotating the Frenet frame by ν around its first vector.

[3]These vectors are guaranteed to be normal since s is an arc-length parameter.

(see Fig. 6.2). Such a position/orientation will let any contour $\mathbf{c}(s, \omega)$ be more complicated/convoluted than circles.

As shown in Fig. 6.2, imagine a blood vessel such as a carotid or a coronary artery: its elongated shape could be tortuous enough for it to lead to important changes in its contour shape, thus creating highly convoluted contours. It is important to "know" where the contour starts and what its orientation is in 3D space.

In order to guarantee this "start point," RGC's orthonormal basis $\mathbf{B}(s)$ is thus defined in terms of the *Frenet-Serret* basis:

$$\mathbf{B}(s) = \mathbf{F}(s)\,\mathbf{R}\left(\hat{\mathbf{x}}, \nu(s)\right) \tag{6.8}$$

where $\mathbf{R}\left(\hat{\mathbf{x}}, \nu(s)\right)$ is the $\hat{\mathbf{x}}$-axis rotation matrix of $\nu(s)$ degrees that allows the contour's phase (i.e., first point $\mathbf{c}(s, 0)$) to be attached to a particular location in the plane, thus preserving its real appearance (see Fig. 6.2).

As *Frenet-Serret* formulas guarantee, at least, C^2 differentiability on $\mathcal{H}$ (in other words, it guarantees smooth curvilinear shapes), $\mathbf{B}(s)$ and $\mathbf{h}(s)$ can be written taking into account the *Frenet-Serret* formulas.

First, an operator related to the *Frenet-Serret* operator $\mathbf{K}(s)$ is written as

$$\boldsymbol{\Psi}(s) = \mathbf{B}^{\top}(s)\,\frac{d\mathbf{B}(s)}{ds} \tag{6.9}$$

which expresses the same idea as the classical *Frenet-Serret* formulas: it ties an orthonormal basis with its derivative. After explicitly developing it:

$$\boldsymbol{\Psi}(s) = \begin{bmatrix} 0 & -\kappa(s)\cos(\nu(s)) & \kappa(s)\sin(\nu(s)) \\ \kappa(s)\cos(\nu(s)) & 0 & -\tau(s) - \frac{d\nu(s)}{ds} \\ -\kappa(s)\sin(\nu(s)) & \tau(s) + \frac{d\nu(s)}{ds} & 0 \end{bmatrix} \tag{6.10}$$

which is, not surprisingly, similar to operator $\mathbf{K}(s)$ (see Eq. 6.7). Actually, this operator integrates all necessary parametric functions to describe $\mathcal{H}$ in space: its curvature $\kappa(s)$, its torsion $\tau(s)$, and a rotation function $\nu(s)$ that, let us remember, tightly attaches contours to orthogonal planes. To write this operator in a more succinct way, note that $\boldsymbol{\Psi}(s)$ is an antisymmetric matrix that could be written as the vector:

$$\boldsymbol{\psi}(s) = \begin{bmatrix} \tau(s) + \frac{d\nu(s)}{ds} \\ \kappa(s)\sin(\nu(s)) \\ \kappa(s)\cos(\nu(s)) \end{bmatrix} \tag{6.11}$$

where $\boldsymbol{\Psi}(s)$ is written in such terms: $\boldsymbol{\Psi}(s) \equiv [\boldsymbol{\psi}(s)]_{\times}$.[4] Now, $\boldsymbol{\psi}(s)$ is used to explicitly write an expression for $\mathbf{B}(s)$, starting from Eq. (6.9):

$$\mathbf{B}(s) = \exp\left(\int_0^s [\boldsymbol{\psi}(u)]_{\times}\,du\right)\mathbf{B}(0) \tag{6.12}$$

[4]The operator $[\cdot]_{\times}$ defines an antisymmetric matrix in terms of its unique elements ordered in a column vector.

which results in an exponential map of the input configuration vector $\psi(s)$, composed (in the sense of a transformation composition in linear algebra) with an initial orthogonal basis $\mathbf{B}(0)$. Although $\mathbf{B}(0)$ is a result of the integration process, this initial basis is actually useful to orient the cylinder's first contour in space.

The problem now is to write an expression for $\mathbf{h}(s)$. Fortunately, this equation is actually easy, since the relationship between $\mathbf{h}(s)$ and $\mathbf{B}(s)$ is given by Eq. (6.3)[5]:

$$\mathbf{h}(s) = \int_0^s \mathbf{B}(u)\,\hat{\mathbf{x}}du + \mathbf{h}(0) \tag{6.13}$$

It should be noted that, like $\mathbf{B}(0)$, $\mathbf{h}(0)$ is a result of the integration process: it is useful as the initial point where the first contour is located in space.

2.2 Piecewise Model

Note that Eqs. (6.12), (6.13) could be as difficult to solve as a cylinder designer wants, since they depend on the form of the curvature, $\kappa(s)$, torsion, $\tau(s)$, and rotation, $\nu(s)$, functions. However, simple functions (actually, constant functions) could be used to cope with a wide spectrum of complex cylindrical shapes. This could be done by letting the complete model be *piecewise constant*. In other words, the RGC curve operator $\psi(s)$ is defined by a set of n simpler functions (pieces), described by constant curvatures κ_i, torsions τ_i, and rotations ν_i $(1 \leq i \leq n)$:

$$\psi(s) = \begin{cases} \psi_1(s) & ; \quad 0 \leq s \leq s_1 \\ \quad\cdots & \\ \psi_i(s - s_{i-1}) & ; \quad s_{i-1} \leq s \leq s_i \\ \quad\cdots & \\ \psi_n(s - s_{n-1}) & ; \quad s_{n-1} \leq t \leq s_n \end{cases} \tag{6.14}$$

Note the overlapping limits on the arc-length parameter s: it guarantees that connecting basis and points are the same, hence guaranteeing up to C^2 continuity. Now, within this framework, Eq. (6.12) becomes

$$\mathbf{B}_i(s) = \mathbf{\Phi}_i(s)\,\mathbf{B}_{i-1}(s_{i-1}) \tag{6.15}$$

where $\mathbf{\Phi}_i(s) = \exp((s - s_{i-1})[\psi_i]_\times)$ is the exponential map of ψ_i at a parameter s, and, given the piecewise basic equation (6.14), $\mathbf{B}_{i-1}(s_{i-1})$ orients the first contour of the ith piece where the previous one ends. Also, note the recursive nature of Eq. (6.15), which can be expanded to

$$\mathbf{B}_i(s) = \mathbf{\Phi}_i(s) \underbrace{\prod_{k=i-1}^{1} \exp\left(\Delta_k [\psi_k]_\times\right) \mathbf{B}_0}_{\mathbf{\Gamma}_{i-1}} \tag{6.16}$$

[5]Recall that the first column vector of $\mathbf{B}(s)$ is the tangent vector $\mathbf{t}(s)$.

The missing part of this equation is the explicit formulation for the exponential map $\mathbf{\Phi}_i(s)$, which is expanded using the Taylor series. Now, the curve points are defined using Eq. (6.13):

$$\mathbf{h}_i(s) = \mathbf{P}_i(s)\,\mathbf{\Gamma}_{i-1}\mathbf{B}_0\hat{\mathbf{x}} + \mathbf{h}_{i-1}(s_{i-1}) \tag{6.17}$$

where (the explicit computation of $\mathbf{P}_i(s)$ is left as an exercise for the reader)

$$\mathbf{P}_i(s) = \int_{s_{i-1}}^{s} \mathbf{\Phi}_i(u)\,du \tag{6.18}$$

and, like Eq. (6.15), also has its own recursive definition:

$$\mathbf{h}_i(s) = \left[\mathbf{P}_i(s)\,\mathbf{\Gamma}_{i-1} + \sum_{k=1}^{i-1}\mathbf{P}_k(\Delta_k)\,\mathbf{\Gamma}_{k-1}\right]\mathbf{B}_0\hat{\mathbf{x}} + \mathbf{h}_0 \tag{6.19}$$

2.3 Contours Model

To represent the closed contours in an 2D plane, there exist a number of *direct domain parametric* representations (such as *splines*, *B-splines*, or *Bézier curves*) or transformed domain representations (such as *Fourier series* decomposition). While, in the case of splines, precision depends on the number and spatial location of control points [15], *Fourier series* (FS) decomposition depends on the number $q \in \mathbb{N}$ of *harmonics* taken into account. In Ref. [16], the authors argue that this leads to a lower number of parameters and that this number is more meaningful than the actual number of control points.

Before describing the connection between the curve model $\mathcal{H}$, described in previous sections, and the surface model $\mathcal{S_C}$, a brief summary on FS decomposition is presented.

2.3.1 Fourier Series of Closed Contours

A closed contour represented by Fourier series [16] is written as

$$c(\omega) = \sum_{l=-q}^{+q} z_l e^{jl\omega} \tag{6.20}$$

where

- $\omega \in [0, 2\pi[$ is an azimuthal parameter,
- j is the imaginary unit,
- $q \in \mathbb{N}$ is the contour's number of harmonics, and
- each complex number $z_l \in \mathbb{C}$ is the contour's lth Fourier coefficient.

For the sake of compactness, Fourier coefficients are contained in an ordered sequence $Z = \langle z_l \in \mathbb{C}; -q \leq l \leq q \rangle$. These coefficients have physical interpretations:

- z_0 represents the contour's center, and
- the tuple $\langle z_{-l}, z_l \rangle$ represents a *general* ellipse (i.e., an ellipse with two radii, an orientation, and a phase).

It should be noted that every tuple $\langle z_l, z_{-l} \rangle$, defining a general ellipse, has a phase depending on its order l. Thus, outer ellipses are "rotating" faster, by a factor of l, than their predecessors. Visually, we can interpret it as a planetary system where each planet rotates around its predecessor and we are interested in the total trajectory traced by the outermost ellipse (i.e., the qth ellipse). Fig. 6.3 shows an example of a complicated contour defined by just $q = 4$ ellipses.

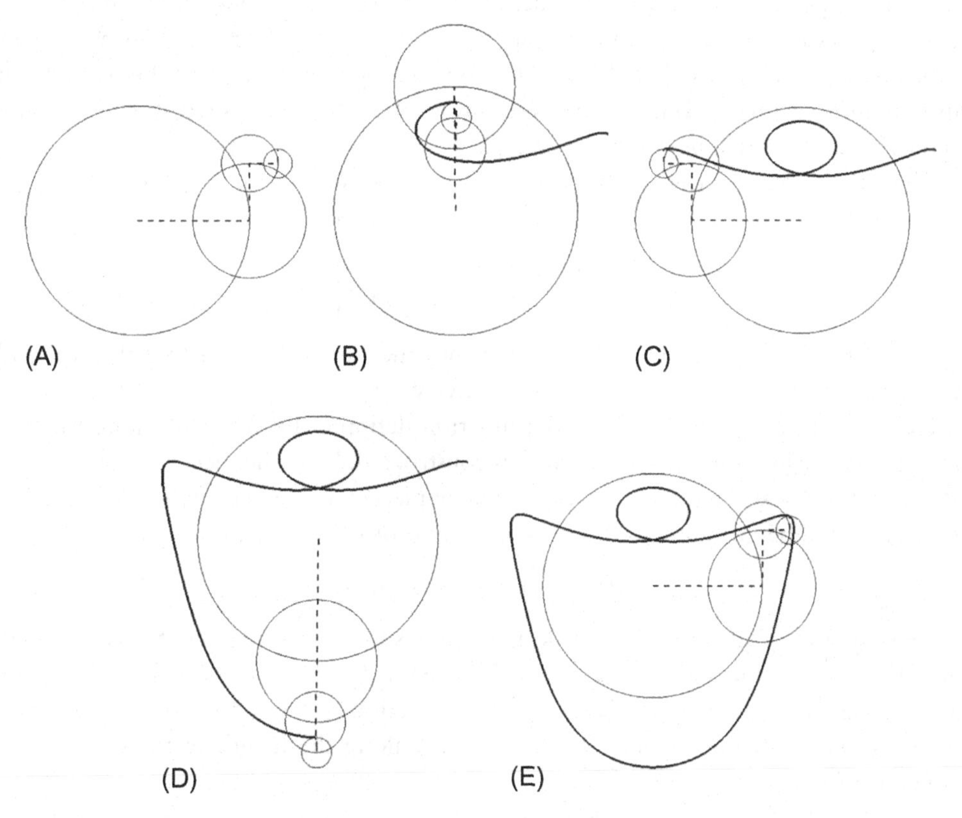

Fig. 6.3 Example of a contour (*bold line*) generated by a Fourier series of size $q = 4$. Each figure illustrates the localization of the ellipses (*circles* represented with thin lines) at five different values of the azimuthal parameter ω. The *dashed line* shows the connection between the ellipses. The coefficients are: $\mathbf{Z} = [z_l \in \mathbb{C}; -q \leq l \leq q] = [0.125j, 0.25, 0, 0, 0, 1, 0.5j, 0, 0]$. (A) $\omega = 0$. (B) $\omega = \frac{\pi}{2}$. (C) $\omega = \pi$. (D) $\omega = \frac{3}{2}\pi$. (E) $\omega = 2\pi$.

2.4 Surface Model

With Eqs. (6.16), (6.19) already written, the cylinder surface expression comes from rewriting Eq. (6.1) as its piecewise version:

$$\mathbf{rgc}_i(s,\omega) = \mathbf{B}_i(s)\,\mathbf{c}_i(s,\omega) + \mathbf{h}_i(s) \tag{6.21}$$

Regardless of the complexity of this expression (i.e., imagine it written explicitly with equations for $\mathbf{B}_i(s)$ and $\mathbf{h}_i(s)$), we can see that the contour part $(\mathbf{c}_i(s,\omega))$ should describe a closed contour in the $\hat{\mathbf{y}}\hat{\mathbf{z}}$ plane, since the plane tangent is perpendicular to the tangent vector $\mathbf{t}_i(s)$, related to the canonical vector $\hat{\mathbf{x}}$. Until this point, the contour part $\mathbf{c}_i(s,\omega)$ of this formula has been missing. It should be clear by now that the foundation model for this equation is the use of Fourier series.

First, note that a contour is located in a plane in $\mathbb{R}^3$, thus a correspondence between the complex plane (Fourier series are defined in the complex plane) and a 3D plane is defined as follows: the first column vector of $\mathbf{B}_i(s)$ (i.e., $\mathbf{t}_i(s)$) defines the normal vector to the contour plane; this plane has a 2D orthogonal system defined by the second and third column vectors of $\mathbf{B}(s)$. If this osculating plane is transformed to the absolute origin, the contour will be on the real $\hat{\mathbf{y}}\hat{\mathbf{z}}$ plane.

Thus, the localization of a complex contour $c_i(s,\omega)$ in the real $\hat{\mathbf{y}}\hat{\mathbf{z}}$ plane is

$$\mathbf{c}_i(s,\omega) \equiv \begin{bmatrix} 0 \\ \mathbb{R}\left[c_i(s,\omega)\right] \\ \mathbb{I}\left[c_i(s,\omega)\right] \end{bmatrix} \tag{6.22}$$

where $\mathbb{R}[\cdot] : \mathbb{C} \to \mathbb{R}$ and $\mathbb{I}[\cdot] : \mathbb{C} \to \mathbb{R}$ are the functions returning the real and imaginary parts of a complex number, respectively.

Now the stacking function (i.e., the function defining the contours along the axis according to the longitudinal parameter s) is constructed. First, the piecewise connection rules (see Eq. 6.14), saying that every cylinder piece starts where its predecessor has finished, should be guaranteed (this guarantees at least C^0 continuity):

$$\mathbf{c}_i(s_{i-1},\omega) = \mathbf{c}_{i-1}(s_{i-1},\omega)$$

The easiest way to guarantee this rule is to impose the first contour for each piece. The initial contour of the ith piece will be named $z_i(\omega)$ and it is defined by its FS $Z_i = \left[z_{i,l} \in \mathbb{C}; -q \le l \le +q\right]$. Then the stack function is defined in order to describe the evolution of the coefficients contained in Z_i, along the length of the curve piece defined by ψ_i:

$$\mathbf{c}_i(s,\omega) = (s - s_i)\,\boldsymbol{\lambda}_i(\omega) + \mathbf{c}_{i-1}(s_{i-1},\omega) \tag{6.23}$$

where each $\lambda_{i,l}$ is a factor acting linearly on each coefficient of the first piece's contour. These factors form an ordered complex sequence $\boldsymbol{\Lambda}_i = \left\langle \lambda_{i,l} \in \mathbb{C}; -q \le l \le +q \right\rangle$. Note that, like $\mathbf{B}(s)$ and $\mathbf{h}(s)$, $\mathbf{c}_i(s,\omega)$ also have a recursive definition:

$$\mathbf{c}_i(s,\omega) = \mathbf{z}_0(\omega) + \sum_{k=1}^{i-1} \Delta_k \boldsymbol{\lambda}_k(\omega) + (s - s_{i-1})\, \boldsymbol{\lambda}_i(\omega) \tag{6.24}$$

where $\mathbf{z}_0(\omega)$ is the first contour of the first RGC's piece. Now, Eqs. (6.16), (6.19), (6.24) can be blended together to complete Eq. (6.21).

3. PARAMETERS INVERSION

When constructing a cylinder (like the ones present in the results of a segmentation method performed on vascular images), usually a centerline is extracted, then a set of perpendicular contours is segmented, each one located in space by means of the initial extracted centerline. From these inputs, we can construct a RGC representation of the corresponding structure. In this section, an inversion algorithm that, from two 3D localized contours, computes the set of RGC parameters is described.

The *inversion* problem is stated as: from two discrete planar contours $C_0 = \langle \mathbf{p}_i \in \mathbb{R}^2; 0 \leq i < n \rangle$ and $C_1 = \langle \mathbf{q}_k \in \mathbb{R}^2; 0 \leq k < m \rangle$, each one localized in planes defined, respectively, by the tuples $\langle \mathbf{B}_0, \mathbf{h}_0 \rangle$ and $\langle \mathbf{B}_1, \mathbf{h}_1 \rangle$, compute the RGC axial parameters $\kappa \in \mathbb{R}$ (curvature), $\tau \in \mathbb{R}$ (torsion), $\nu \in \mathbb{R}$ (rotation), $\Delta \in \mathbb{R}^+$ (generating curve length), together with the linear evolution Fourier coefficients Λ of the best RGC model that passes through them. See Fig. 6.4 for a schematic of the situation.

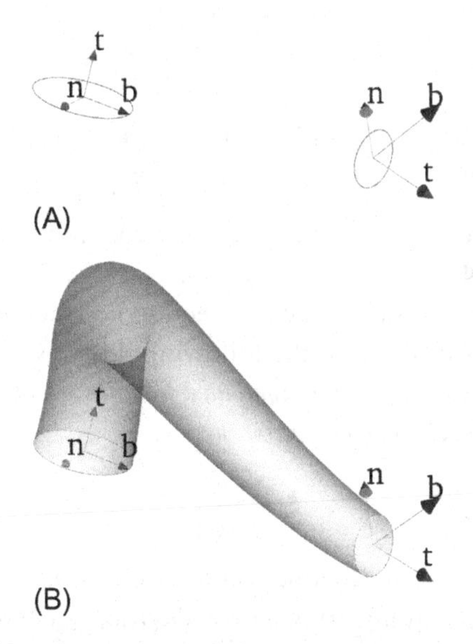

Fig. 6.4 Axis inversion schema. While inputs are shown in (A) (contours located in two planes in 3D space), the resulting cylinder is shown in (B).

Fig. 6.4 illustrates a possible configuration of the problem to solve:

1. ***Inputs***: Two 3D located contours $\mathcal{C}_\alpha = [\mathbf{B}_\alpha, \mathbf{h}_\alpha, C_\alpha]$ and $\mathcal{C}_\beta = \left[\mathbf{B}_\beta, \mathbf{h}_\beta, C_\beta\right]$ where C_α and C_β are *ordered* 2D point sequences, $\mathbf{B}_\alpha$ and $\mathbf{B}_\beta$ are 3D orthonormal basis representing the *orientation* of each contour and $\mathbf{h}_\alpha$ and $\mathbf{h}_\beta$ are the 3D points where contours are *located*.
2. ***Outputs***: An RGC model $\mathcal{RGC} = [\psi, \Delta, \mathbf{\Lambda}]$ that represents the best cylinder that passes through both input contours: ψ is the parameters vector, Δ is the positive cylinder's length, and $\mathbf{\Lambda}$ it the ordered set of evolution coefficients for the continuous contour stacking.

3.1 Axial Parameters Inversion

From the input data previously defined, the main challenge now is to build an RGC model in terms of its generating curve and contour stack. Since these inputs, when integrated into a RGC model, should follow the RGC's basis model, then $\mathcal{C}_\alpha$ and $\mathcal{C}_\beta$ would be the end contours of a RGC piece. As border contours, they should fully define the model operators:

$$\mathbf{P}(\Delta) = \mathbf{B}^\top(0)\left(\mathbf{h}(\Delta) - \mathbf{h}(0)\right) = \mathbf{B}_\alpha^\top\left(\mathbf{h}_\beta - \mathbf{h}_\alpha\right) = \mathbf{P}_\Delta \tag{6.25}$$

$$\mathbf{\Phi}(\Delta) = \mathbf{B}^\top(0)\,\mathbf{B}(\Delta) = \mathbf{B}_\alpha^\top \mathbf{B}_\beta = \mathbf{\Phi}_\Delta \tag{6.26}$$

Now, operator $\mathbf{\Phi}_\Delta$ can be represented in a vector-angle of rotation form $\langle \boldsymbol{\xi}, \theta \rangle$, since this operator is a rotation matrix that takes the first curve's basis $\mathbf{B}_\alpha$ and transforms it to the second basis $\mathbf{B}_\beta$. This conversion is interesting since two useful relationships exist:

$$\begin{cases} 2\cos(\theta) = \textsc{Trace}\left(\mathbf{\Phi}_\Delta\right) - 1 \\ 2\sin(\theta)\,[\boldsymbol{\xi}]_\times = \mathbf{\Phi}_\Delta - \mathbf{\Phi}_\Delta^\top \end{cases} \tag{6.27}$$

The first relation is the definition of the trace of a rotation matrix, which can be used to reinforce the fact that $\mathbf{\Phi}_\Delta$ is a rotation matrix.

The second relationship describes the connection between input basis and the actual parameters vector ψ. This is based on the following theorem.

Theorem 6.1. *The rotation vector $\boldsymbol{\xi}$, computed from the input rotation operator $\mathbf{\Phi}_\Delta$ is the normalized version of the RGC parameter vector ψ:*

$$\boldsymbol{\xi} = \frac{\psi}{|\psi|} \tag{6.28}$$

Proof. To show this theorem, an important observation arises since the vector ψ is (piecewise) constant: according to *Lancret's theorem*, a curve is a *helix* if and only if its curvature-to-torsion ratio $(\kappa(s)/\tau(s))$ is constant. In the case of the particular model presented here, this observation is (piecewise) true, since curvature and torsion values are (piecewise) constant. Then, a generating curve $\mathcal{H}$ can be represented using the classical helix equations (points and *Frenet frames*), where the parameter is angular $(\theta \in \mathbb{R})$[6]:

$$\mathbf{h}_\eta(\theta) = \begin{bmatrix} R\cos(\theta) \\ R\sin(\theta) \\ L\theta \end{bmatrix} \tag{6.29}$$

$$\mathbf{F}_\eta(\theta) = \begin{bmatrix} -\cos(\varphi)\sin(\theta) & -\cos(\theta) & \sin(\varphi)\sin(\theta) \\ \cos(\varphi)\cos(\theta) & -\sin(\theta) & -\sin(\varphi)\cos(\theta) \\ \sin(\varphi) & 0 & \cos(\varphi) \end{bmatrix} \tag{6.30}$$

where φ is the value that relates R and L: $\tan(\varphi) = L/R$. Before going into further detail and to explain the meaning of R, L, and φ, it should be noted that, geometrically, curvatures and torsions attached to any spatial curve are more difficult to interpret than their "inverses": in fact, helices are more easily described by its radius of curvature $(R \in \mathbb{R})$ and its loop separation coefficient $(L \in \mathbb{R})$. See Fig. 6.5 for an schema on these values.

The first problem to solve is the different parameter definition: while RGC curves are described by an arc-length parameter (s), classical helices use an angular

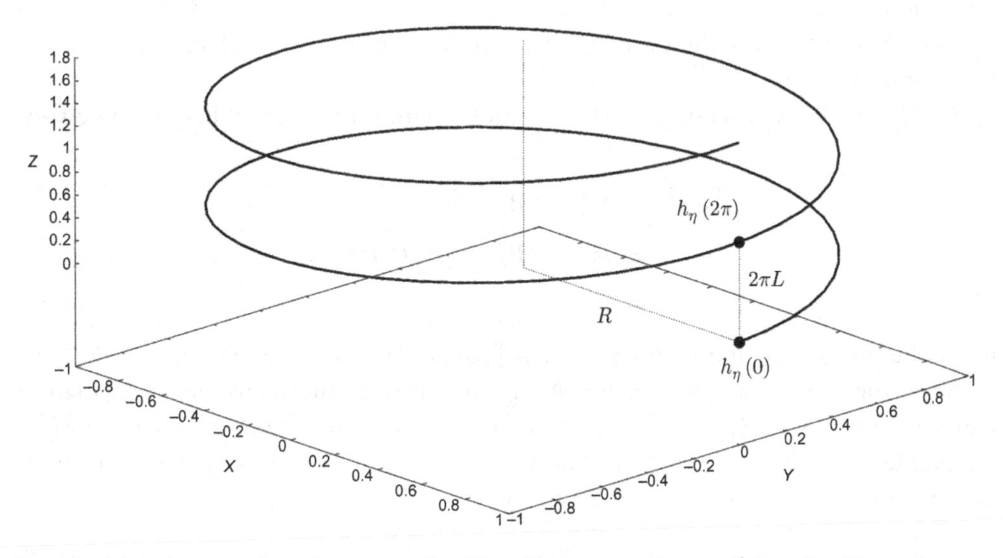

Fig. 6.5 Helix schema: radius of curvature ($R \in \mathbb{R}$) and loop separation coefficient ($L \in \mathbb{R}$) are shown as *thin dotted lines*.

[6]The η subindex allows us to differentiate between "classical" helices and RGC generating curves.

parameter (θ). However, converting between both parameters is easily calculated by deducing the helix length in terms of the angular parameter:

$$s(\theta) = \int_0^\theta \left| \mathbf{h}'_\eta(u) \right| du = \theta\sqrt{R^2 + L^2} \tag{6.31}$$

The second problem is the different global basis where each curve is located. In fact, RGC helices start with a contour located in the plane $\langle \mathbf{B}_\alpha, \mathbf{h}_\alpha \rangle$, and classical helices start at a plane located in $\left\langle \mathbf{F}_\eta(0), \mathbf{h}_\eta(0) \right\rangle$. This problem is solved when both representations are translated and rotated to the global orthonormal basis, hence allowing the rewriting of RGC operators $\mathbf{\Phi}_\eta(\theta)$ and $\mathbf{P}_\eta(\theta)$ in angular terms. Therefore, developing the second relation defined in Eq. (6.27) within the "classical helical" framework gives

$$\Xi = [\boldsymbol{\xi}]_\times = \begin{bmatrix} 0 & -\cos(\varphi)\cos(\nu) & \cos(\varphi)\sin(\nu) \\ \cos(\varphi)\cos(\nu) & 0 & -\sin(\varphi) \\ -\cos(\varphi)\sin(\nu) & \sin(\varphi) & 0 \end{bmatrix} \tag{6.32}$$

where

$$\boldsymbol{\xi} = \begin{bmatrix} \sin(\varphi) \\ \cos(\varphi)\sin(\nu) \\ \cos(\varphi)\cos(\nu) \end{bmatrix} \tag{6.33}$$

which is a constant unary vector that contains all the parameters that define a helix in terms of R and L (recall the meaning of φ) and introduces, to the classical framework, the rotation value ν.

Finally, the RGC parameter vector is written within the classical helical framework (see Eq. 6.11):

$$\begin{aligned} [\psi_\eta]_\times &= \mathbf{B}_\eta^\top(\theta)\,\mathbf{B}'_\eta(\theta) \\ &= \mathbf{R}_{\hat{\mathbf{x}}}^\top(\nu)\,\mathbf{F}_\eta^\top(\theta)\,\mathbf{F}'_\eta(\theta)\,\mathbf{R}_{\hat{\mathbf{x}}}(\nu) \\ &= [\boldsymbol{\xi}]_\times \end{aligned} \tag{6.34}$$

hence the unitary rotation vector $\boldsymbol{\xi}$ is parallel to the RGC characteristic vector ψ_i. □

Since the input rotation operator $\mathbf{\Phi}_\Delta$ is converted to the vector-angle of rotation representation tuple $\langle \boldsymbol{\xi}, \theta \rangle$ and due to the fact that Theorem 6.1 proves Eq. (6.28), the problem is reduced to find the $L2$-norm $|\psi|$ of the parameter vector ψ. In both frameworks (RGC and classical), this value is

$$|\psi| = \sqrt{\kappa^2 + \tau^2} = \frac{1}{\sqrt{R^2 + L^2}} \tag{6.35}$$

since the conversion between R and L to κ and τ is given by

$$\begin{bmatrix} \kappa \\ \tau \end{bmatrix} = \frac{1}{R^2 + L^2} \begin{bmatrix} R \\ L \end{bmatrix} \tag{6.36}$$

The base equation to find $|\psi|$ is the projection of the orthonormal tangent vector to the vector that connects the first input point with the second input point:

$$\begin{aligned}\hat{\mathbf{x}}^{\top}\mathbf{\Phi}_{\Delta}^{\top}\mathbf{P}_{\Delta} &= R\sin(\theta)\cos(\varphi) + L\theta\sin(\varphi)\\ &= \frac{\kappa}{|\psi|^2}\sin(\theta)\cos(\varphi) + \frac{\tau}{|\psi|^2}\theta\sin(\varphi)\\ &= \frac{\sin(\theta)\left(\left(\hat{\mathbf{y}}^{\top}\boldsymbol{\xi}\right)^2 + \left(\hat{\mathbf{z}}^{\top}\boldsymbol{\xi}\right)^2\right) + \theta\left(\hat{\mathbf{x}}^{\top}\boldsymbol{\xi}\right)^2}{|\psi|}\end{aligned}$$

Then, taking into account Eq. (6.33):

$$|\psi| = \frac{\sin(\theta)\left(\left(\hat{\mathbf{y}}^{\top}\boldsymbol{\xi}\right)^2 + \left(\hat{\mathbf{z}}^{\top}\boldsymbol{\xi}\right)^2\right) + \theta\left(\hat{\mathbf{x}}^{\top}\boldsymbol{\xi}\right)^2}{\hat{\mathbf{x}}^{\top}\mathbf{\Phi}^{\top}\mathbf{P}} \tag{6.37}$$

Finally, obtaining Δ (the total curve length) is a straightforward application of Eqs. (6.31), (6.35):

$$\Delta = \frac{\theta}{|\psi|} \tag{6.38}$$

3.2 Surface Parameters Calculation

Since the surface is defined by a linear evolution of Fourier coefficients, it is simply

$$\Lambda = \frac{Z_{\beta} - Z_{\alpha}}{\Delta} \tag{6.39}$$

However, the inputs are defined as ordered sequences C_{α} and C_{β}; hence, the remaining question is: how do we obtain the ordered sequence Z, given the set of n ordered points $C = \{\mathbf{p}_i \in \mathbb{R}^2; 0 \leq i < n\}$?

Note that Eq. (6.20) (contour points in terms of the coefficients) is the discrete inverse Fourier transform. Thus, its counterpart (to compute coefficients in terms of contour points) is performed using the *discrete Fourier transform* $Z = \mathcal{F}(C)$:

$$\mathcal{Z}_m = \frac{1}{n}\sum_{k=0}^{n-1} p_k e^{-\frac{2\pi jmk}{n}} \tag{6.40}$$

where $0 < m \leq n$ and $p_k = \mathbf{p}_{\hat{\mathbf{x}}} + j\mathbf{p}_{\hat{\mathbf{y}}}$ (recall Eq. 6.22, where the conversion between the complex plane and a real plane is presented). Note that the resulting coefficients sequence $\mathcal{F}(C) = [\mathcal{Z}_m \in \mathbb{C}; 0 \leq m < n]$ is different from the desired sequence $Z = [z_l \in \mathbb{C}; -q \leq l \leq +q]$. Actually, both sequences are equivalent, they just have different orders (i.e., they are *dephased*). Table 6.1 shows the correspondences.

Table 6.1 Discrete Fourier transform and Fourier series order correspondences

$\mathcal{F}(C)$	$\mathcal{Z}_0$	$\mathcal{Z}_1$	$\mathcal{Z}_2$	$\mathcal{Z}_3$	$\cdots$	$\mathcal{Z}_m$	$\mathcal{Z}_{m+1}$	$\cdots$	$\mathcal{Z}_{n-3}$	$\mathcal{Z}_{n-2}$	$\mathcal{Z}_{n-1}$
Z	z_0	z_1	z_2	z_3	$\cdots$	z_q	z_{-q}	$\cdots$	z_{-3}	z_{-2}	z_{-1}

This ordering is the result of the $\mathcal{F}(C)$ summation range going from 0 to $n-1$: the out-of-range coefficients are folded and come out as the negative coefficients. This is equivalent to the summation range $\left[-\frac{N}{2}, \frac{N}{2}\right]$. Nevertheless, the summation range $[0, N)$ is preferred since it handles odd and even numbers of points similarly, and since the first coefficient $\mathcal{Z}_0 = z_0$ is always the *center of gravity* of the given contour. This effect is due to the periodicity of the Fourier transformation [17].

3.3 Numerical Stability

Before presenting the final algorithm, we discuss some numerical problems that come with the use of this inversion method, mainly because the method is founded on rotation matrices. Fortunately, these stability problems have physical meaning in the inversion situation. In other words, each problem can be treated as a special case to generate special parameters.

3.3.1 Null Rotation

The first problem arises when both input orthonormal bases are the same ($\mathbf{B}_\alpha = \mathbf{B}_\beta$). In this case, the rotation operator becomes the identity matrix ($\mathbf{\Phi}_\Delta = \mathbf{I}$). The axis-angle representation of such matrix is $\langle \boldsymbol{\xi}, \theta \rangle = \langle \mathbf{0}, 0 \rangle$. This could introduce a potential numerical problem since Eqs. (6.37), (6.38) become null ($|\psi| = 0$) and undefined ($\Delta = \frac{0}{0}$), respectively. However, such a configuration means that the generating curve is a straight line ($\psi = \mathbf{0}$) with length $\Delta = \left|\mathbf{h}_\beta - \mathbf{h}_\alpha\right|$, since both contours are located in parallel planes, hence representing a straight cylinder.

3.3.2 Complementary Rotations

Every rotation matrix $\mathbf{\Phi}_\Delta$ has two possible vector-angle configurations, namely $\langle \boldsymbol{\xi}, \theta \rangle$ and $\langle -\boldsymbol{\xi}, 2\pi - \theta \rangle$ (which are complementary between them). Both rotations have the same effect on any point in space. Thus, two possible helices can be computed using the inversion method. The question is: which one to choose?

One can think of various decision heuristics, and the simplest one is to choose the configuration that minimizes

$$\min_{(\psi_+,\Delta_+),(\psi_-,\Delta_-)} \left|\mathbf{P}(\Delta) - \mathbf{P}_\beta\right| \tag{6.41}$$

where

$$\langle \psi_+, \Delta_+ \rangle = \left\langle |\psi|\,\boldsymbol{\xi}, \frac{\theta}{|\psi|} \right\rangle$$

$$\langle \psi_-, \Delta_- \rangle = \left\langle (|\psi| - \gamma)\, \boldsymbol{\xi}, \frac{\theta - 2\pi}{|\psi| - \gamma} \right\rangle$$

$$\gamma = \frac{2\pi \left(\hat{\mathbf{x}}^\top \boldsymbol{\xi}\right)^2}{\hat{\mathbf{x}}^\top \boldsymbol{\Phi}_\Delta^\top \mathbf{P}_\Delta}$$

This heuristic is based on the idea that the model's generating curve $\mathcal{H}$ passes as close as possible to the given contour origin center.

3.3.3 Negative Length

Even though Eq. (6.37) is supposed to compute the $L2$-norm $|\psi_i|$ (i.e., a positive real value), calculations could lead to negative values, hence leading to negative lengths ($\Delta_i < 0$, see Eq. 6.38). This case happens when the helical tangent-to-endpoint projection (Eq. 6.37's numerator) has a different sign from the RGC tangent-to-endpoint projection (Eq. 6.37's denominator).

This case is presented when inversion has no physical meaning: the best helix that passes between both planes goes from plane $\langle \mathbf{B}_\beta, \mathbf{h}_\beta \rangle$ to plane $\langle \mathbf{B}_\alpha, \mathbf{h}_\alpha \rangle$, not the contrary as desired by the user computing the inversion. Rather than a numerical error, this is a human-introduced error: both planes were incorrectly numbered (i.e., they are oriented backwards with respect to each other).

However, algorithmically, at this moment the precedent error (complementary rotations) should be taken into account, since one (or both) rotation configurations could lead to three situations:

1. Both solutions have positive lengths.
2. One solution has a positive length and the other one has a negative length.
3. Both solutions have negative lengths.

When one solution leads to a negative length, it could be safely ignored, since it represents a situation with no physical meaning. However, when both solutions lead to negative lengths, this really means that the used inputs were poorly given.

3.3.4 Generating Curve Centering

The angle θ, when a rotation matrix $\boldsymbol{\Phi}_\Delta$ is represented as the tuple axis-angle, is periodic. Then, any conversion strategy (using eigenanalysis or quaternions, for example) will produce a value $0 \leq \theta < 2\pi$. Hence, any input configuration will produce a generating curve that makes, at most, one loop. This could lead to cylinders where the generating curve does not pass through the contours' centers.

Two solutions are proposed for this problem: (1) an inversion-quality measure ϵ and (2) a distance minimization. While the first solution will help iterative algorithms to control the tracking speed (see Section 4 for more details), the second solution seeks to find the best helix that fits both input contours.

The first solution is very simple: the inversion algorithm is applied just once and the quality measure is the Euclidean distance between the given endpoint and the inverted one:

$$\epsilon = \left|\mathbf{P}(\Delta) - \mathbf{P}_\beta\right| \tag{6.42}$$

If $\epsilon \simeq 0$, that means that the inverted helix should pass exactly over the given endpoint; on the other hand, if $\epsilon > 0$ means that if the helix should pass exactly over the given endpoint, it should turn more than one loop. In this case, the tracking algorithm could backtrack to previous solution and slow down, since some information has been missed.

The second solution consists of adding loops in an optimization problem ($\theta_i = \theta_{i-1} + 2\pi$). Each iteration of this optimization makes both endpoints closer, hence finding the perfect helix that passes through both contours (Algorithm 1 shows this optimization).

Both solutions are valid but, in order to find representations of real cylinders, the first solution is considered to be preferable. Artificially adding loops to a cylinder will lead to folded surfaces, which will be unnatural.

3.4 Complete Inversion Algorithm

The complete pseudo-code for the inversion method, taking into account stability and input issues, is presented in Algorithm 1.

Algorithm 1 RGC Piece Inversion With Distance Minimization.

1: **procedure** INVERT_RGC($C_\alpha, C_\beta, \mathbf{B}_\alpha, \mathbf{B}_\beta, \mathbf{h}_\alpha, \mathbf{h}_\beta$)
2: $\quad \mathbf{\Phi}_\Delta \leftarrow \mathbf{B}_\alpha^\top \mathbf{B}_\beta$ $\qquad$ ▷ Transform inputs to global orthonormal basis.
3: $\quad \mathbf{P}_\Delta \leftarrow \mathbf{B}_\alpha^\top (\mathbf{h}_\beta - \mathbf{h}_\alpha)$
4: $\quad \langle \boldsymbol{\xi}, \theta \rangle \leftarrow$ MATRIX_TO_AXISANDANGLE ($\mathbf{\Phi}$)
5: $\quad$ **if** $\theta > 0 \wedge |\mathbf{P}_\Delta| > 0$ **then**
6: $\quad\quad a \leftarrow \sin(\theta) \left((\hat{\mathbf{y}}^\top \boldsymbol{\xi})^2 + (\hat{\mathbf{z}}^\top \boldsymbol{\xi})^2 \right) + \theta \, (\hat{\mathbf{x}}^\top \boldsymbol{\xi})^2$
7: $\quad\quad b \leftarrow \hat{\mathbf{x}}^\top \mathbf{\Phi}_\Delta^\top \mathbf{P}_\Delta$
8: $\quad\quad g \leftarrow 2\pi \, (\hat{\mathbf{x}}^\top \boldsymbol{\xi})^2$
9: $\quad\quad d_{opt} \leftarrow \infty$
10: $\quad\quad i \leftarrow 0$
11: $\quad\quad$ **while** $d_{opt} > 0$ **do**
12: $\quad\quad\quad \psi_+^i \leftarrow \boldsymbol{\xi} \, (a + gi) \, / b$
13: $\quad\quad\quad \psi_-^i \leftarrow \boldsymbol{\xi} \, (a - g\,(i+1)) \, / b$
14: $\quad\quad\quad \Delta_+^i \leftarrow ((\theta + 2\pi i) \, b) \, / \, (a + gi)$
15: $\quad\quad\quad \Delta_-^i \leftarrow ((\theta - 2\pi \, (i+1)) \, b) \, / \, (a - g\,(i+1))$
16: $\quad\quad\quad (\psi^i, \Delta^i) \leftarrow \left[\min\limits_{(\psi_+^i, \Delta_+^i), (\psi_-^i, \Delta_-^i)} |\mathbf{P}(\Delta) - \mathbf{P}_\Delta| \right]$

17: $d_{opt} \leftarrow \mathbf{P}(\Delta)$
18: $i \leftarrow i+1$
19: **end while**
20: $(\psi, \Delta) \leftarrow (\psi^{i-1}, \Delta^{i-1})$
21: **else**
22: $(\psi, \Delta) \leftarrow (0, |\mathbf{P}_\Delta|)$ ▷ Dealing with a straight line.
23: **end if**
24: **return** $\left[\psi, \Delta, \mathcal{F}(C_0), \frac{[\mathcal{F}(C_1)-\mathcal{F}(C_0)]}{\Delta}\right]$
25: **end procedure**

In line 4 we recommend using a matrix-to-quaternion conversion. Quaternions are known to avoid gimbal locks and to be numerically stable to represent any rotation, hence getting the axis-angle rotation representation is easy, since any rotation quaternion has the form:

$$r = \cos\left(\frac{\theta}{2}\right) + j\xi \sin\left(\frac{\theta}{2}\right)$$

Another key point in this algorithm is the condition presented in line 5. Note that not only the condition of having straight lines is tested: a second condition, when both planes are not just parallel, is also tested since they also have the same origin. In that case, a straight cylinder of zero length is generated.

4. MODEL-GUIDED IMAGE SEGMENTATION

In this chapter, a (theoretical) framework for extracting RGC models of vessels from medical images is presented. The idea behind such segmentation algorithm is based on a *tracking strategy*: the vessel is scanned along its *elongated shape* (axis) $\mathcal{P}_f$, keeping track of a certain number of *parameters* describing it; then the reconstructed RGC directly depends on the traces of these parameters.

The model parameters (see Section 4.2) are used to construct a *Kalman state estimator* [18, 19] that sequentially constructs the cylindrical model of the vessel. The Kalman strategy needs to be fed by a certain number of observations (measures). These observations are taken in slices locally perpendicular to the vessel using a two-dimensional segmentation algorithm. An overview of the algorithm is shown in Fig. 6.6.

An axis extraction algorithm from medical images is out of the scope of this chapter. Here, the algorithm proposed by Wink et al. [20] is used, since it is simple to implement (it is based on the well-known Dijkstra's minimum spanning tree extraction algorithm), easy to initialize (two end-points should be manually given), and fast to execute.

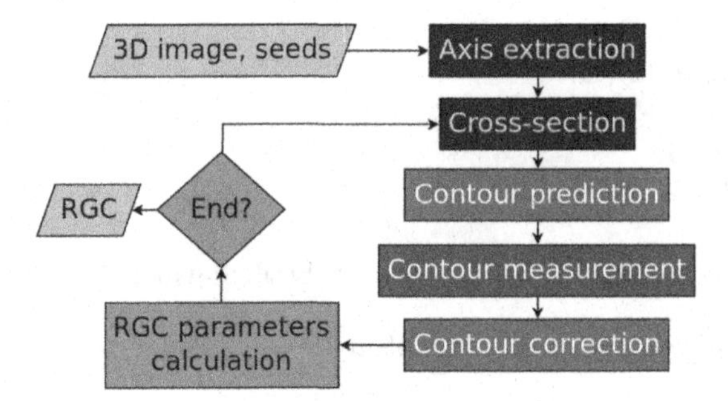

Fig. 6.6 Overview of the RGC-based segmentation. "Axis extraction" and "Cross-section" boxes represent algorithms that extract centerlines and segment 2D images, respectively. Remaining boxes (prediction, measurement, and correction) represent the main steps of a Kalman filter. "RGC parameters calculation" integrate the inversion algorithm presented in Section 3.

4.1 Kalman State Estimator

The Kalman state estimator (KSE) addresses the general problem of estimation of the x-dimensional state $\mathbf{x} \in \mathbb{R}^x$ of a *discrete-time* controlled process that is governed by the linear stochastic difference equation:

$$\mathbf{x}_i = \mathbf{A} \cdot \mathbf{x}_{i-1} + \mathbf{B} \cdot \mathbf{u}_i + \mathbf{w}_{i-1} \tag{6.43}$$

with an m-dimensional measurement $\mathbf{m} \in \mathbb{R}^m$:

$$\mathbf{m}_i = \mathbf{H} \cdot \mathbf{x}_i + \mathbf{v}_i \tag{6.44}$$

where the *random* variables $\mathbf{w}_i$ and $\mathbf{v}_i$ represent the *process* and *measurement* noises, respectively. They are assumed to be statistically independent between them, white and with normal probability distributions:

$$\begin{aligned} p(\mathbf{w}) &\sim N(0, \mathbf{Q}) \\ p(\mathbf{v}) &\sim N(0, \mathbf{R}) \end{aligned} \tag{6.45}$$

where $\mathbf{Q}$ and $\mathbf{R}$ are the respective covariance matrices.

The $x \times x$ matrix $\mathbf{A}$ in Eq. (6.43) relates the state at the previous step $i - 1$ to the state at the current step i, in the absence of either a driving function or process noise. The $x \times u$ matrix $\mathbf{B}$ relates the optional u-dimensional control input $\mathbf{u} \in \mathbb{R}^u$ to the state $\mathbf{x}$. The $m \times x$ matrix $\mathbf{H}$ in the measurement equation (6.44) relates the state to the measurement $\mathbf{m}$.

The KSE algorithm is basically a loop that updates the "time" i by projecting (*predicting*) the previous *estimated* state $\hat{\mathbf{x}}_{i-1}$ to $\hat{\mathbf{x}}_i^-$ and then *correcting* this *prediction* using the current *measurement* $\mathbf{m}_i$ to obtain the *newly estimated state* $\hat{\mathbf{x}}_i$. Furthermore, a measure of the estimation process' error is kept in the $x \times x$ matrices $\mathbf{P}_i$. The pseudo-code is written in Algorithm 2.

Algorithm 2 Kalman State Estimator Basic Algorithm as Presented in [19].

$\mathbf{Q}, \mathbf{R} \leftarrow \mathit{InitNoises}()$
$\hat{\mathbf{x}}_0 \leftarrow \mathit{InitState}()$
$\mathbf{P}_0 \leftarrow \mathit{InitProcessNoise}()$
while $\neg \mathit{StopCriteria}$:
  $\mathbf{u}_i \leftarrow \mathit{LoadInput}(i)$; Load current input
  $\hat{\mathbf{x}}_i^- = \mathbf{A} \cdot \widehat{nn}_{i-1} + \mathbf{B} \cdot \mathbf{u}_i$; Predict next state
  $\mathbf{P}_i^- = n \cdot \mathbf{P}_{i-1} \cdot \mathbf{A}^\top + \mathbf{Q}$; Predict the process noise
  $\mathbf{K}_i = \mathbf{P}_i^- \cdot \mathbf{H}^\top \cdot (\mathbf{H} \cdot \mathbf{P}_i^- \cdot n^\top + \mathbf{R})^{-1}$; Update the Kalman gain
  $\mathbf{m}_i \leftarrow \mathit{LoadMeasure}(i)$; Load current measure
  $\hat{\mathbf{x}}_i = \hat{\mathbf{x}}_i^- + n_i \cdot (\mathbf{m}_i - \mathbf{H} \cdot \hat{\mathbf{x}}_i^-)$; Estimate the next state
  $n_i = (\mathbf{I} - \mathbf{K}_i \cdot \mathbf{H}) \cdot \mathbf{P}_i^-$; Update process noise
elihw

The key to use a KSE is the definition of the matrices **A**, **B**, **H**, **Q**, **R**, the initialization vector $\hat{\mathbf{x}}_0$ and process noise $\mathbf{P}_0$, the time-depending input ($\mathbf{u}_i$) and measurement $\mathbf{m}_i$ vectors, and defining the *stop criteria* to terminate the estimation loop.

4.2 Kalman Equations for Vessel Tracking and RGC Construction

The matrices and vectors used in our KSE-based vessel tracking algorithm are as follows:

- The state vector $\mathbf{x}_i$ describes RGC pieces in terms of their axial parameters ψ_i, length Δ_i, and Fourier evolutions coefficients $\boldsymbol{\Lambda}_i$.
- The measurement vector $\mathbf{m}_i$ contains the same information as $\mathbf{x}_i$ ($\mathbf{m}_i \equiv \mathbf{x}_i$). In other words, it also describes RGC pieces (measured pieces).
- Following the same reasoning, the KSE algorithm is initialized as $\hat{\mathbf{x}}_0 = \mathbf{m}_0$; meaning that the initial estimated piece is the first measured piece.
- $\mathbf{u}_i = \mathbf{0} \wedge \mathbf{B} = \mathbf{0}$. In other words, there are no external inputs.
- $\mathbf{A} = \mathbf{I}$, i.e., predicted pieces are just projections of previously estimated pieces.
- $\mathbf{H} = \mathbf{I}$, i.e., measured pieces, in the absence of noise, should be considered as good estimations.
- Experimentally, noise covariance matrices to $\mathbf{Q} = \mathbf{I} \cdot 10^{-3}$ and $\mathbf{R} = \mathbf{I} \cdot 10^{-3}$ are experimentally fixed.
- Accordingly, the initial process noise matrix $\mathbf{P}_0 = n \cdot 10^{-3}$.

After each iteration, the estimated state $\hat{\mathbf{x}}_i$ is appended to the final RGC, thus constructing the final cylindrical representation of the chosen vessel.

4.3 Tracking the Vessel Along the Approximate Axis

At each tracking step i, a measurement $\mathbf{m}_i$ describing a cylinder piece is taken (see Section 3). This measurement is injected to the KSE to compute the estimation $\hat{\mathbf{x}}_i$ of the new RGC piece.

Tracking the vessel along the approximate axis $\mathcal{P}_f$ (Fig. 6.7 depicts some axes that are used to track vessels) grants access to planes defined by the tuple $[\mathbf{p}_i, \mathbf{t}_i]$, where $\mathbf{p}_i$ is the actual tracking point belonging to $\mathcal{P}_f$, and $\mathbf{t}_i$ is the tangent vector associated to it. In these planes a contour is segmented using the algorithm proposed in [22] (it goes without saying: one can change this 2D segmentation algorithm). Then the cylinder passing between successive contours is computed (see Section 3). In other words, it is important to have in mind that at each tracking step, a contour $\mathbf{c}_i(\Delta_i, \omega)$ is segmented on the plane $[\mathbf{p}_i, \mathbf{t}_i]$ in order to be connected with the last estimated one $\hat{\mathbf{c}}_{i-1}(\hat{\Delta}_i, \omega)$ to produce the cylinder measure $\mathbf{m}_i$.

Without taking into account the contour measurement method, which will be described later, the tracking process is composed by three methods:

1. initialization (Section 4.3.2),
2. computation of the tracking step (Section 4.3.1), and
3. definition of a stop criterion (Section 4.3.3).

In the next subsections, we present the tracking step computation and then the initialization method because it depends on the concepts defined to compute the tracking step. Finally the stop criterion is introduced.

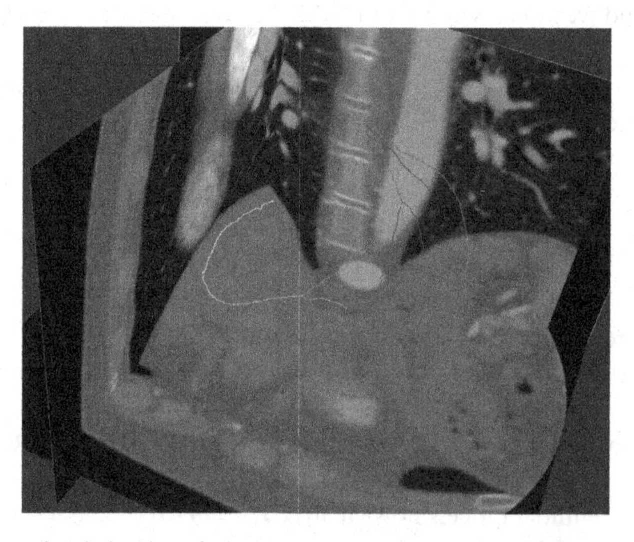

Fig. 6.7 Axes extracted with the algorithm proposed by Wink et al. [20]. The image is a CT image taken from the 2012 MICCAI coronary challenge [21] (http://coronary.bigr.nl/stenoses/).

4.3.1 Tracking Step

Suppose the algorithm has reached the $i-1$ iteration and it is ready to measure the next piece. As we are tracking the vessel along the approximate path $\mathcal{P}_f$ by segmenting contours, the thing to do (at each iteration) is to define the localization of the new contour segmentation process.

Let us start with an observation: RGC models should have a limited curvature in order to avoid self-intersecting surfaces (hence modeling real-world structures).

Suppose we take the last estimated contour of the cylinder, namely $\hat{\mathbf{c}}_{i-1}(\hat{\Delta}_{i-1}, \omega)$, then we compute its minimum diameter $\hat{d}_{i-1}$: a necessary (but not sufficient) condition to add a new piece, is that the last contour of this new piece should be located at a straight distance of, at most, $\hat{d}_{i-1}$ from its first contour, if the curvature and torsion for the newly added piece do not vary a lot with respect to the last estimated piece. Then, for the ith tracking step, the plane defined by the tuple $[\mathbf{p}_i, \mathbf{t}_i]$, which defines the location of the new contour's plane, is computed at a distance of $\hat{d}_{i-1}$ from the last tracking position.

Fig. 6.8 shows how the tracking system works: the minimum diameter of the last estimated contour defines a sphere, localized on the last estimated contour's center $\hat{\lambda}_{0,i-1}\hat{\Delta}_{i-1} + \hat{z}_{0,i-1}$, then we look for the intersection of $\mathcal{P}_f$ with this sphere; finally, this intersection is used to compute the tuple $[\mathbf{p}_i, \mathbf{t}_i]$.

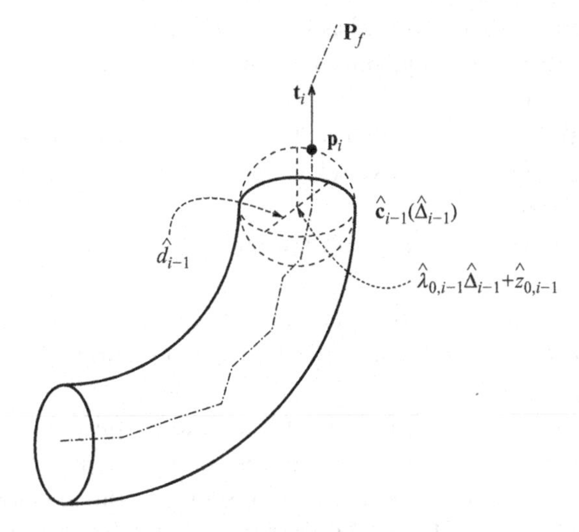

Fig. 6.8 Scheme for the tracking step computation. The sphere attached to the center of the last estimated contour $\hat{\mathbf{c}}_{i-1}(\hat{\Delta}_{i-1}, \omega)$ has a diameter equal to the minimum diameter $\hat{d}_{i-1}$ of this contour. The intersection between this sphere and the approximated axis $\mathcal{P}_f$ gives the tracking plane $[\mathbf{p}_i, n_i]$.

4.3.2 Initialization

The first estimated state is the first measured piece, as described in Section 4.2:

$$\hat{\mathbf{x}}_0 = \mathbf{m}_0$$

To perform this first measure, two contours have to be segmented to construct it, namely $\hat{\mathbf{c}}_{-1}(0,\omega)$ and $\mathbf{c}_0(d_{-1},\omega)$:

- The tuple defining the plane of the contour $\hat{\mathbf{c}}_{-1}(0,\omega)$ is the first point in the approximate axis $\mathcal{P}_f$ (which is $\mathbf{p}_{start}$) with its associated tangent.
- The tuple defining the second contour is defined exactly as described in the tracking step computation (previous section), taking $\hat{\mathbf{c}}_{-1}(0,\omega)$ as the previous estimated contour.

4.3.3 Stop Criterion

The KSE stop criterion is defined as

$$\mathbf{p}_i \geq \mathbf{p}_{end}$$

This simply means that the tracking process is stopped when the tracking point reaches (or overpasses) the second user given seed $\mathbf{p}_{end}$.

5. CONCLUSIONS

A cylindrical geometric model, the RGC model, was presented in this chapter. This model follows the definition given by Binford in his taxonomy defined in Ref. [13]. The model can define complex (as well as simple) cylinders with few parameters, and it could be used to construct digital phantoms, simulate vascular shapes, or even create 3D models to train physicians.

In this chapter, the model was used to formulate a vessel tracking algorithm. This algorithm computes the model of a vessel at the same time as the image is segmented. Its principal components are:

- a Kalman state estimator that controls the tracking of RGC parameters describing the vessel,
- a contour extraction strategy that can be user-given, and
- a method to find the cylinder parameters from two contours located in space (i.e., RGC inversion).

The constant piecewise version of this model is used to simplify the writing of the final equations used in vessel tracking. This leads to easy implementation and fast extraction.

However, the RGC model can be seen as a template: the axis formulation as well as the surface formulation can be modified to guarantee higher-order properties such as surface continuity, curvatures, and derivatives. This could lead to even more natural models.

REFERENCES

[1] Flórez-Valencia L, Dávila Serrano EE, Riveros Reyes JG, Bernard O, Latt J, Malaspinas O, et al. Virtual deployment of pipeline flow diverters in cerebral vessels with aneurysms to understand thrombosis. In: MICCAI-workshop on computer assisted stenting. Nice, France; 2012. p. 49–56. https://www.creatis.insa-lyon.fr/site/sites/default/files/thrombus.pdf.

[2] Marchenko Y, Volkau I, Nowinski W. Vascular editor: from angiographic images to 3D vascular models. J Digit Imaging 2010;23(4):386–98. http://dx.doi.org/10.1007/s10278-009-9194-8.

[3] DeVault K, Gremaud P, Novak V, Olufsen M, Vernières G, Zhao P. Blood flow in the Circle of Willis: modeling and calibration. Multiscale Model Simul 2008;7(2):888–909. http://dx.doi.org/10.1137/07070231X.

[4] Orkisz M, Flórez-Valencia L, Hoyos MH. Models, algorithms and applications in vascular image segmentation. MG&V 2008;17(1):5–33. http://dl.acm.org/citation.cfm?id=1534494.1534496.

[5] Lesage D, Angelini ED, Bloch I, Funka-Lea G. A review of 3D vessel lumen segmentation techniques: models, features and extraction schemes. MedIA 2009;13(6):819–45. http://dx.doi.org/10.1016/j.media.2009.07.011. Includes Special Section on Computational Biomechanics for Medicine, http://www.sciencedirect.com/science/article/pii/S136184150900067X.

[6] Singh N, Kaur L. A survey on blood vessel segmentation methods in retinal images. In: International conference on electronic design, computer networks automated verification (EDCAV); 2015. p. 23–8. http://dx.doi.org/10.1109/EDCAV.2015.7060532.

[7] Mille J, Cohen L. 3D CTA image segmentation with a generalized cylinder-based tree model. In: Proceedings of the 2010 IEEE international symposium on biomedical imaging: from nano to macro, Rotterdam, The Netherlands; 2010. p. 1045–48. http://dx.doi.org/10.1109/ISBI.2010.5490169.

[8] Flórez-Valencia L, Azencot J, Vincent F, Orkisz M, Magnin I. Segmentation and quantification of blood vessels in 3D images using a right generalized cylinder state model. In: International conference on image processing, Atlanta, GA, USA; 2006. p. 2441–44. http://dx.doi.org/10.1109/ICIP.2006.312770.

[9] Flórez-Valencia L, Azencot J, Orkisz M. Carotid arteries segmentation in CT images with use of a right generalized cylinder model. In: MICCAI Workshop—3D segmentation in the clinic: a grand challenge III. London, GB: Midas Journal; 2009. http://hdl.handle.net/10380/3106.

[10] Flórez-Valencia L, Azencot J, Orkisz M. Algorithm for blood-vessel segmentation in 3D images based on a right generalized cylinder model: application to carotid arteries. In: Bolc L, Kulikowski JL, Chmielewski LJ, Wojciechowski K, editors. International conference on computer vision and graphics, Warsaw, Poland. Lecture Notes in Computer Science, vol. 6374. Heidelberg: Springer; 2010. p. 27–34. http://dx.doi.org/10.1007/978-3-642-15910-74.

[11] Flórez-Valencia L, Orkisz M, Corredor Jerez RA, Torres González JS, Correa Agudelo EM, Mouton C, et al. Coronary artery segmentation and stenosis quantification in CT images with use of a right generalized cylinder model. In: MICCAI workshop 3D cardiovascular imaging segmentation challenge, Nice, France; 2012. https://www.creatis.insa-lyon.fr/site/sites/default/files/miccai2012_coronary_challenge_creandpuj.pdf.

[12] Azencot J, Orkisz M. Deterministic and stochastic state model of right generalized cylinder (RGC-sm): application in computer phantoms synthesis. Graph Models 2003;65(6):323–50. http://dx.doi.org/10.1016/S1524-0703(03)00073-0.

[13] Binford T. Visual perception by computer. In: Proceedings of the IEEE conference on systems and control, Miami, FL; 1971.

[14] do Carmo M. Differential geometry of curves and surfaces. Englewood Cliffs, NJ: Prentice-Hall, Inc.; 1976. p. 1–50. https://books.google.com.co/books?id=1v0YAQAAIAAJ.

[15] Bartels R, Beatty J, Barsky B. An introduction to splines for use in computer graphics and geometric modelling. In: Hermite and cubic spline interpolation. Los Altos, CA: Morgan Kaufmann Publishers, Inc.; 1998. p. 9–17. http://dl.acm.org/citation.cfm?id=35072.

[16] Staib L, Duncan J. Boundary finding with parametrically deformable models. IEEE Trans Pattern Anal Mach Intell 1992;14(11):1061–75. http://dx.doi.org/10.1109/34.166621.

[17] Briggs W, Henson V. The DFT: an owner's manual for the discrete Fourier transform. Philadelphia: Soc for Industrial & Applied Math; 1995. p. 434.

[18] Kalman R. A new approach to linear filtering and prediction problems. Trans ASME-J Basic Eng 1960;82(Series D):35–45.
[19] Welch G, Bishop G. An introduction to the Kalman filter. Chapel Hill, NC: University of North Carolina; 2001. http://www.cs.unc.edu/w̃elch/kalman/kalmanIntro.html.
[20] Wink O, Niessen W, Frangi A, Verdonck B, Viergever M. 3D MRA coronary axis determination using a minimum cost path approach. Magn Reson Med 2002;47(6):1169–75.
[21] Kirişli HA, Schaap M, Metz C, Dharampal AS, Meijboom WB, Papadopoulou SL, et al. Standardized evaluation framework for evaluating coronary artery stenoses detection, stenoses quantification and lumen segmentation algorithms in computed tomography angiography. MedIA 2013;17(8):859–76. http://dx.doi.org/10.1016/j.media.2013.05.007.
[22] Baltaxe Milwer M, Flórez-Valencia L, Hernández-Hoyos M, Magnin I, Orkisz M. Fast marching contours for the segmentation of vessel lumen in CTA cross-sections. In: Conference proceedings of IEEE engineering in medicine and biology society. Lyon, France: IEEE; 2007, p. 791–94. http://dx.doi.org/10.1109/IEMBS.2007.4352409.

CHAPTER 7

Domain Adapted Model for In Vivo Intravascular Ultrasound Tissue Characterization

S. Conjeti[*], A.G. Roy[*,†], D. Sheet[†], S. Carlier[‡], T. Syeda-Mahmood[§], N. Navab[*,¶], A. Katouzian[§]

[*]Technical University of Munich, Munich, Germany
[†]Indian Institute of Technology Kharagpur, Kharagpur, West Bengal, India
[‡]University of Mons, Mons, Belgium
[§]IBM Almaden Research Center, San Jose, CA, United States
[¶]Johns Hopkins University, Baltimore, MD, United States

Chapter Outline

Computing and Visualization for Intravascular Imaging and Computer-Assisted Stenting
http://dx.doi.org/10.1016/B978-0-12-811018-8.00007-2

1. INTRODUCTION

Coronary atherosclerosis occurs to accumulation of plaque within the arteries that often leads to instable or stable angina pectoris (chest pain or discomfort), myocardial infarction (heart attack), and sudden death due to arterial occlusion under plaque rupture. As the disease progresses, the occlusion of the arterial cross-section leads to deficient supply of blood, oxygen, and nutrients to the cardiac muscles. Naghavi et al. suggested using the terminology of "vulnerable plaque" as a standardized term to indicate all thrombosis-prone plaques and plaques that exhibit a high probability of undergoing rapid progression resulting in arterial occlusion and subsequent death [1]. Assessment of vulnerable atherosclerotic plaque and its rupture have led to identification of multiple clinical image based morphological and pathological markers fibrotic cap thickness, lip core size, percentage of stenosis (arterial narrowing), mechanical stability features (arterial wall stiffness and elasticity), calcification burden and patterns (nodular, scattered calcifications, superficial vs. deep, etc.), change in arterial flow patterns, and ventricular remodeling.

Over the past few decades, interventional imaging modalities have found increasing applicability during percutaneous coronary interventions as it provides the interventional cardiologist with information about the imperative characteristics of the plaques that can lead to better identification of vulnerability cues and potentially better clinical decisions on the subsequent intervention. Routinely, coronary angiography (both computed tomography based and magnetic resonance based) is performed to visualize narrowing in the arterial vessel tree and guide interventional procedures like balloon angiography or stent deployment. However, this imaging modality lacks in its ability to provide sufficient geometrical and pathological information about the plaque constituents and its morphological attributes. Toward this end, secondary imaging modalities like intravascular ultrasound (IVUS), optical coherence tomography (IV-OCT), near-infrared have been employed in conjunction with angiography.

IVUS is often deemed as the best complementary interventional imaging modality as it provides real-time cross-sectional images of the arterial wall with sufficient spatial resolution and penetration. Additionally, interaction between acoustic signals and pathological tissues leads to distinct signature features that can be leveraged for tissue characterization and vulnerability assessment. IVUS also offers precise spatial information and helps directly visualize clinically relevant attributes like fidelity of the fibrotic cap, extent of the lipid pool, calcification patterns and degree of stenosis. These aforementioned factors are used to guide the interventional procedure and can also be deployed in retrospective studies to evaluate the success of angioplasty or stenting. In this work, we focus on the task of IVUS TC that aims to probabilistically characterize heterogeneous vascular plaques in vivo to provide tissue-compositional information for better assessment of their vulnerability.

To date, the majority of the IVUS TC techniques were designed, developed, and validated upon in vitro collected data and directly deployed for in vivo application without taking into account the effect of dynamic factors like blood [2]. However, the underlying ultrasound physics behind image acquisition indicates that the signal transmission media between the source and acoustic scatterers significantly influences the nature of the recorded backscattered signals. This holds true in in vitro versus in vivo IVUS signal acquisition as the transmission media between the two settings differs (blood for in vivo and saline for in vitro). Empirical observations on ultrasound attenuation power law fit for whole blood components for frequency range of 0–70 MHz report an attenuation factor of $0.0546f^{1.58}$ dB/cm for whole blood with speed of sound of $1590(\pm2.8)$ m/s and that for deionized water (very similar to saline) as $0.00139f^2$ dB/cm with speed of sound of 1524 m/s, where f is frequency in MHz [3]. This changing physical properties between the two transmission media lead to significant variability in the observed back-scattering patterns. These aforementioned factors limit the direct deployment of in vitro trained IVUS TC models in in vivo settings. Alternatively, one could propose to train a new IVUS TC model exclusively for deployment under in vivo settings. This approach meets with the bottled necks of (1) lack of appropriate collectible samples, and (2) inconsistency among expert-defined labels (*due to nonavailability of histological validation*). In the in vivo setting, where conventional histology assessment cannot be performed on living patients, it is deemed effective to device techniques that can effectively leverage the exhaustive in vitro dataset and suitably modify the models trained on it to provide a realistic solution for in vivo TC. This acts as the premise for this work and we explore in detail the challenges involved within in vitro to in vivo translation and propose to modify the in vitro model suitably for reliable in vivo TC despite limited availability of learning examples. Fig. 7.1 provides a schematic of the overall framework proposed for supervised domain adaptation of decision forests for in vivo TC. Fig. 7.2 illustrates the effect of the domain shift by contrasting features signatures observed in vitro to in vivo domains.

The presented chapter is adapted from two of our prior published works [4, 5]. Sheet et al. introduced leveraging ultrasonic backscattering statistical physics and signal confidence based features for in vitro tissue characterization [4]. Conjeti et al. extended their work for effective characterization in in vivo domain through domain adaptation of decision forests [5]. This chapter summarizes their major contributions and is organized as: Section 2 covers a brief literature survey on the state of the art in vitro IVUS tissue characterization methods and briefly delves into the related work from domain adaptation literature, thus laying the premise for the proposed domain adaptation method. Following this, we present the mathematical model of the ultrasonic backscattering and signal propagation physics limitation in evaluating its classical form represented as mixture of multiple components spread across multiple scales, and our solution through a dual representation and random forests is detailed in Section 3.

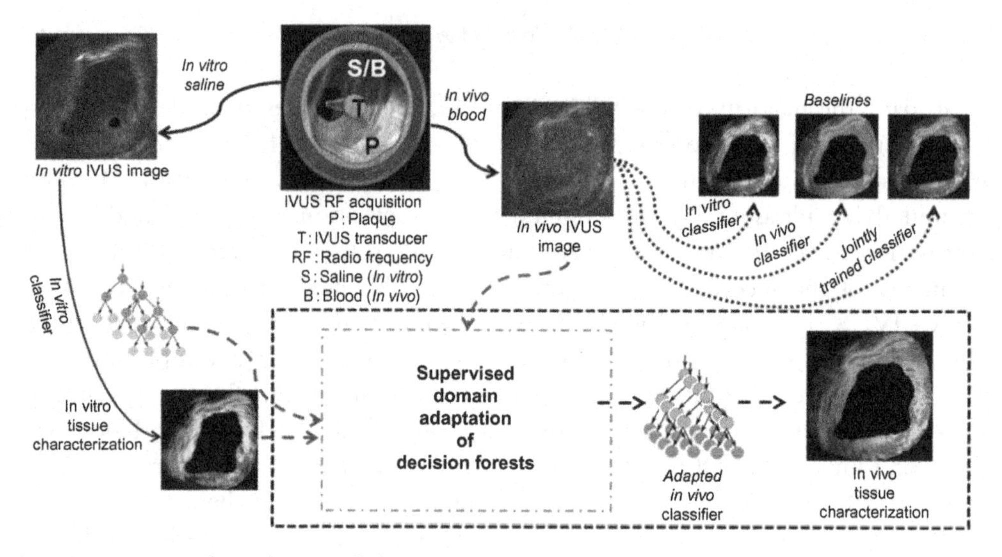

Fig. 7.1 Overview of our framework for in vivo atherosclerotic plaque characterization using IVUS employing decision forests. We adapt classifiers trained on in vitro acquired data (circulating saline in lumen) through supervised domain adaptation to clinically better-suited in vivo settings (blood circulating in the lumen).

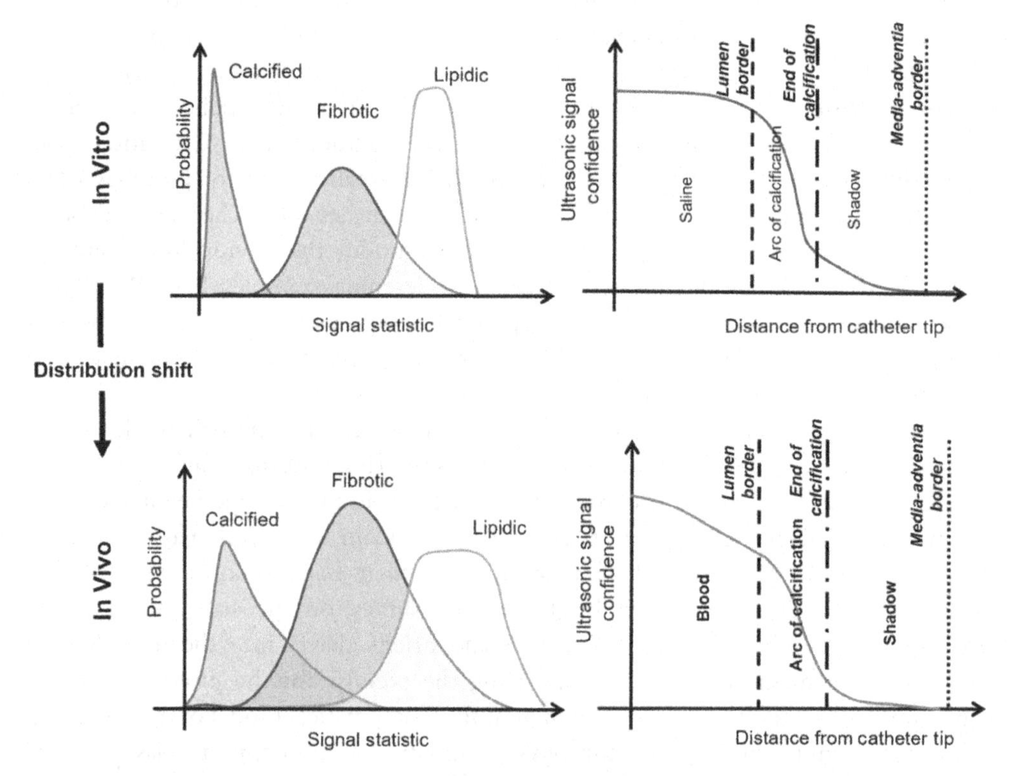

Fig. 7.2 Difference between in vitro and in vivo scenarios: We use tissue-specific back-scattering signal statistics and ultrasonic signal confidence for learning a TC model. Transitioning to in vivo, the presence of blood in the lumen and artery motion results in incoherently attenuated IVUS signals, resulting in a shift in the distribution of signal backscattering statistics and lowering of the ultrasonic signal confidence.

Next, we delve into the domain adaptation of in vitro trained models for translation to in vivo settings is discussed in Section 4. Finally, we conclude the chapter reflecting on the major chapter highlights and scope for its further clinical translatability in Section 6.

2. STATE OF THE ART

2.1 In Vitro IVUS Tissue Characterization

In comparison to other interventional imaging modalities, IVUS has been a primary choice for atherosclerotic tissue characterization and has been a topic of active research which has resulted in a number of popular approaches including radio-frequency based [6–8] and texture based [9–12] approaches. We briefly review these approaches in light of their pros and cons and how they lead up to the proposed hybrid approach.

Radio-Frequency-Based TC

Raw RF signals are deemed to be better representation of ultrasonic interaction with biological tissues as it avoids artifacts arising from pixel interpolation in the process of generating a log-Compressed B-Mode image. Further, raw RF signals a higher spatial resolution and small regions of plaque can be confidently distinguished [9]. In this context, Kawasaki et al. proposed to employ classical integrated backscattered coefficients (IBs) to perform TC on 40 MHz IVUS signals acquired with a single element transducer into three major tissue types viz. fibrotic, lipidic, and calcified [13]. In another related work, Nair et al. proposed RF features motivated by observations reported in Ref. [14] that are indicative of tissue scatterer size and density. These included the slope and intercept of linear regression fit to normalized windowed tissue spectra, spectral signatures within the functional bandwidth like midband-fit, minimum and maximum powers along with their corresponding frequencies. These features were employed in conjunction with a classification tree-based classifier to perform TC to identify fibrotic, fibrofatty, calcified, and necrotic tissues.

Texture-Based TC

In an earlier study by Katouzian et al., it has been reported that spectral features are subject to inconsistencies owing to challenges involved in global data calibration, selection of optimal signal bandwidth, etc. and particularly demonstrated the inconsistency among these features in the presence of highly heterogeneous atherosclerotic tissues. Consequently, they proposed the first texture driven TC framework based on overcomplete wavelet packet representations with a nonparametric K-means based classifier to characterize tissues into fibrotic, lipidic, and calcified tissues [2]. Thereafter, the algorithm was further improved in [11] by employing iterative self-organizing data analysis clustering for adaptive clustering over a weak K-means classifier and the proof-of-concept was presented for both 20 and 40 MHz IVUS signals and was demonstrated to be superior in TC performance over spectral features based TC proposed in Ref. [15]. Despite promising results, the method lacked for accounting for the co-located nature of heterogeneous atherosclerotic tissues, which becomes a major focus of the proposed TC framework. Further, they proposed using RF features in conjunction with texture-driven features into a superior machine learning framework that is able to account for tissue heterogeneity effectively.

2.2 Domain Adaptation

Learning mathematical models to predict tissue composition using IVUS data requires reliable labels for training and validating the learnt models. The underlying assumption is that the training data and the unseen testing data arise from the same statistical distribution (termed as a *domain*). In real world, this assumption often does not hold owing to uncontrolled factors such as dataset bias and differences in domains, termed as *domain shift*. Extending this argument forward to IVUS TC in an in vivo setting, due to domain differences arising as an effect of uncontrolled factors like blood flow and arterial dynamics, the nature of the backscattered signals would change, potentially degrading the performance of any TC model trained on in vitro data. In a related work by Ciompi et al., they concluded that the common hypothesis of assuming the difference between in vivo and in vitro IVUS signals as negligible to be incorrect. They sought to extend TC to in vivo domain by selectively fusing with in vitro data and training a TC classifier.

Methods to avoid classifier degradation owing to domain shift has been an active area of research within the machine learning and computer vision communities [16–18]. Popular approaches toward this include transfer learning or multitask learning [19], self-taught learning [20], semisupervised learning [21], concept drift [22], multiview learning [23], and domain adaptation [18]. Domain adaptation (DA) aims at leveraging knowledge (such as decision models and datasets) acquired from one or more *source* domains, and apply it to a closely related *target* domain, where there is no or few labeled data [18]. The availability of source and target domain data has led to four major categorization of DA methods viz. Supervised, Unsupervised, Multisource, and Heterogeneous. Supervised DA methods use large labeled source data and few labeled target data, while unsupervised approaches use large labeled source data and unlabeled target data. Multisource implies that adaptation happens from multiple sources domains onto a related target domain and heterogeneous DA refers to the scenario where the feature representations differ between the two domains. The current work falls under the category of supervised domain adaptation as we adapt learned models for in vivo TC by leveraging in vitro data and models (source) along with few labeled in vivo examples (target).

3. MATHEMATICAL MODELING OF ULTRASONIC BACKSCATTERING AND SIGNAL PROPAGATION PHYSICS IN HETEROGENEOUS TISSUES

The propagation of ultrasonic signals through biological tissue results in them either being backscattered, refracted, or absorbed depending on the nature of the transmission media. In ultrasound signal acquisition, we primarily record the backscattered acoustic signals as an echo reflected back from acoustic scatterers within the tissue. The task of TC employing ultrasonic backscattering based features derived from RF signals requires understanding the nature of the originating scatterers, which would directly translate into identifying the underlying tissues.

3.1 Multiscale Nakagami Distribution

As is the case with majority of biological tissues, atherosclerotic tissues are often co-localized with heterogeneous composition and the constituent structures of interest can potentially span multiple spatial scales. This implies that the tissue composition is not homogeneous (i.e., purely belonging to a particular tissue type) and consists of a mixture of constituents. Their ultrasonic signatures characterizing them can be observed at different resolutions due to variations in scatterer size, density, composition, and the characteristics of the incident acoustic wave. This heterogeneous nature of the underlying tissues is reflected directly in the statistical nature of the ultrasonic pulse echo (say R). Take a purely stochastic perspective on R, the probabilistic characterization of a tissue belonging to type y using a recorded pulse r can be formulated using the Bayesian paradigm as

$$p\left(y|r\right) = \frac{p\left(r|y\right) P\left(y\right)}{p\left(r\right)} \tag{7.1}$$

where $p\left(y|r\right)$ is the estimated posterior probability of occurrence of a particular tissue type $y \in \mathcal{Y}$, $p\left(r|y\right)$ is the learned tissue specific likelihood of generating a particular backscattered ultrasonic pulse r and $p\left(r\right)$ is the priori distribution for any observation r.

Box 7.1 Nakagami Distribution

The *Nakagami* distribution, initially proposed for describing Radar returned echoes, is a simpler universal model capable of describing all of the above conditions [24]. Under this model, probability density of the enveloped ultrasonic echo signal is given by

$$\mathcal{N}(r|m,\Omega) = \frac{2m^m r^{2m-1}}{\Gamma(m)\Omega^m} \exp\left(-\frac{m}{\Omega}r^2\right) U(r) \tag{7.2}$$

(Continued)

Box 7.1 Nakagami Distribution—cont'd

where m is the Nakagami parameter and Ω is the scaling parameter. These parameters are obtained from the moments of the envelope and are expressed as

$$m = \frac{\left(E\left[R^2\right]\right)^2}{E\left[R^2 - E\left[R^2\right]\right]^2} \tag{7.3}$$

$$\Omega = E\left[R^2\right] \tag{7.4}$$

where $E[\cdot]$ is the mathematical expectation operator.

The direct approach of learning this likelihood in limited sample population is through parametric estimation $p(r|y) \propto f(r|\ldots)$. The Nakagami distribution model $p(r|y) \propto \mathcal{N}(r|\Omega, r)$ has been contextually proven to be appropriate in perspective of ultrasonic statistical physics [24]. Box 3.1 briefly presents the mathematical exposition behind the Nakagami distribution. Though this method works outright in controlled experiments, yet in context of biological structures with co-localized tissue heterogeneity, likelihood of r would be manifested as a mixture of multiple of Nakagami distributions. Destrempes et al. have modeled this likelihood as a mixture of Nakagami distributions [25]

$$\begin{aligned} p(r|y) &= f(r|p_1, \ldots, m_1, \ldots, \Omega_1, \ldots; y) \\ &= \sum_{l=1}^{L} p_l \mathcal{N}(r|m_l, \Omega_l) \end{aligned} \tag{7.5}$$

where $\mathcal{N}(r|m_l, \Omega_l)$ represents the lth parametric Nakagami distribution constituting to the likelihood with prior probability p_l. Extending this formulation to multiple spatial scales for estimation of the distribution statistics, the likelihood can be re-stated as:

$$\begin{aligned} p(r|y) &= f(r|(p_1, \ldots, m_1, \ldots, \Omega_1, \ldots)_\tau, (\ldots), \ldots; y) \\ &= \sum_{\tau} \left(p_\tau \sum_{l=1}^{L} p_{l,\tau} \mathcal{N}(r|m_l, \Omega_l)_\tau \right) \\ &+ \sum_{\tau_1, \tau_2} p_{\tau_1,\tau_2} \left(\sum_{l=1}^{L} p_{l,\tau_1} \mathcal{N}(r|m_l, \Omega_l)_{\tau_1} \times \sum_{l=1}^{L} p_{l,\tau_2} \mathcal{N}(r|m_l, \Omega_l)_{\tau_2} \right) \\ &+ \sum_{\tau_1,\tau_2,\tau_3} p_{\tau_1,\tau_2,\tau_3} (\ldots) \\ &+ \cdots \end{aligned} \tag{7.6}$$

where $\mathcal{N}(r|m_l, \Omega_l)_\tau$ is the parametric form of Nakagami distribution component as estimated using observations at scale τ and $p_{i,\tau}$ is the corresponding prior probability of occurrence of the lth component. p_{τ_1,τ_2}, p_{τ_1,τ_2,τ_3}, etc. are higher-order joint probabilities between similar mixture components occurring across multiple scales. In total, 2^S are required for completing this representation for a total of S unique spatial scales. For mathematical tractability of the solution and to avoid expensive analytic solutions to Eq. (7.6), we propose to model the likelihood on the dual form using multiscale Nakagami parameter estimations as:

$$\begin{aligned} p(r|y) &= f\left(r|(p_1, \,, m_1, \ldots, \Omega_1, \ldots)_\tau, (\ldots), \ldots; y\right) \\ &= \frac{p\left((p_1, \ldots, m_1, \ldots, \Omega_1, \ldots)_\tau, (\ldots), \ldots |r; y\right)}{p\left((p_1, \ldots, m_1, \ldots, \Omega_1' \ldots)_\tau, (\ldots), \ldots; y\right)} p(r) \end{aligned} \tag{7.7}$$

where $p(r)$ is the prior probability of origin of r due to scattering from tissue of type y. The form for probabilistic characterization of tissue types as in Eq. (7.1) can now be written using Eq. (7.7) as

$$p(y|r) = \frac{p\left((p_1, \ldots, m_1, \ldots, \Omega_1, \ldots)_\tau, (\ldots), \ldots |r; y\right)}{p\left((p_1, \ldots, m_1, \ldots, \Omega_1, \ldots)_\tau, (\ldots), \ldots; y\right)} P(y) \tag{7.8}$$

As the likelihood of Nakagami parameters given r originating from a tissue type y is same as the likelihood of the Nakagami parameters causing an observed signal r, we can write $p\left(\ldots|r; y\right) = p\left(\ldots; r|y\right)$, which enables probabilistic TC purely in terms of the estimated Nakagami features as:

$$\begin{aligned} p\left(y|\mathbf{x}; r\right) &= \frac{p\left(\mathbf{x}; r|y\right) P\left(y\right)}{P\left(\mathbf{x}; r\right)} \\ &= \mathcal{H}(y|\mathbf{x}; r) \end{aligned} \tag{7.9}$$

where $p\left(\mathbf{x}; r|y\right)$ is the likelihood of appearance of the feature vector $\mathbf{x}$ which is generated from observation r given that the underlying tissue is of type $y \in \mathcal{Y}$. Here $p\left(\mathbf{x}; r|y\right)$ is the learned likelihood of features $\mathbf{x}$ comprising of $((m_1, \ldots, \Omega_1, \ldots)_\tau, (\ldots), \ldots; r)$ constituent of observed multiscale co-localized Nakagami parameter estimations. The joint form of the estimator can be represented using a posterior estimating learner $\mathcal{H}(y|\mathbf{x}; r)$ which can intrinsically jointly learn $p\left(\mathbf{x}; r|y\right)$, $P\left(y\right)$ and $P\left(\mathbf{x}; r\right)$ from available observations $\{\mathbf{x}\}$.

3.2 Ultrasonic Signal Confidence

Signal attenuation is an important factor to be taken into account during this learning. Severely attenuated ultrasonic signal sensed by the transducer is generally overloaded with spurious noisy components affecting susceptibility in learning of tissue specific

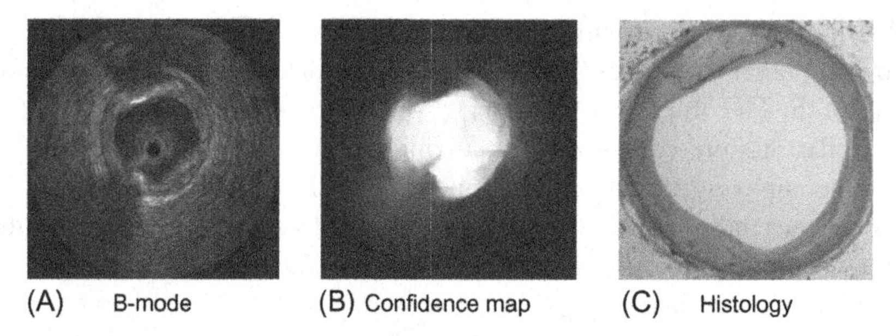

Fig. 7.3 An example case with (A) log-compressed B-mode image of IVUS of the vessel section under examination along with (B) corresponding confidence map, and (C) Hematoxylin and Eosin histology section of plaque. *(Adapted from Karamalis A, Katouzian A, Carlier SG, Navab N. Confidence estimation in IVUS radio-frequency data with random walks. In: Proceedings of ISBI; 2012. p. 1068–71.)*

signatures. In an effort to account for such anomalous readings, we also estimate confidence of the received ultrasonic signal, directly translating in minimizing erroneous characterizations. The envelope of ultrasonic echoes received after backscattering from randomly distributed scatterers in the media being imaged, is often treated as a random walk. Under the assumption that an ultrasonic pulse and the backscattered echo traveling along the same path through a heterogeneous media are subjected to the same attenuation, we estimate confidence of the received ultrasonic signal treating its propagation as a random walker along an ultrasonic scan-line [26]. The confidence of received signal is estimated as the probability of a random walker starting at a node on the scan-line and reaching the virtual transducer element placed at the origin of each scan-line. Under this framework, we employ the electrical network equivalent analogy of the random walks steady state solution, to solve for these probabilities analytically, in paradigm of discrete graph theory [27]. A detailed description of the random walks and confidence estimation framework would be outside the scope of this article and can be found in Ref. [26]. An example from our evaluation dataset of vessel cross-section as seen using IVUS along with the ultrasonic confidence map and the corresponding histology is shown in Fig. 7.3.

In view of achieving the above objective it is intuitionistic to use an ensemble learning engine independent of any constrained parametric form. Random forests introduced by Breiman [28] naturally appears as a potent solution in such a situation due to its inherent ability of learn metric spaces from sparse representations and immunity to overfitting.

3.3 In Vitro Tissue Characterization

As previously introduced, our first objective in this work is to characterize different lesions constituting vascular plaques imaged in IVUS acquired under in vitro settings. In view of limitation of existing framework of ultrasonic tissue characterization with

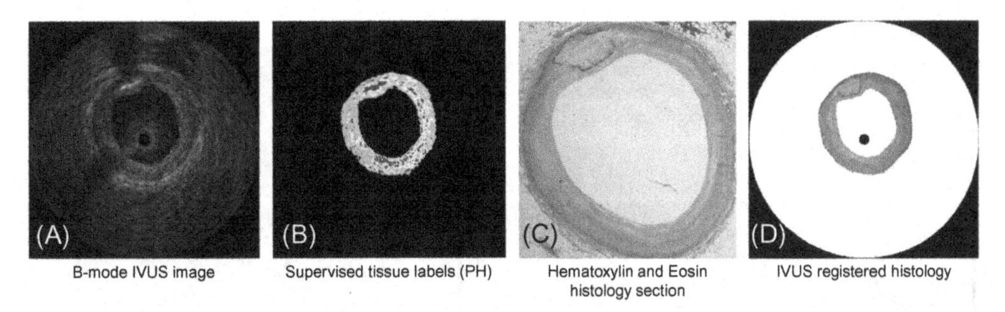

Fig. 7.4 An example case with (A) log-compressed B-mode image of IVUS of the vessel section under examination along with (B) supervised tissue labels obtained by PH. (C) Hematoxylin and Eosin histology section of plaque, and (D) histology deformable registered on the IVUS. The different tissue types are labeled as: *Black*, nonlesion areas; *Yellow*, fibrotic lesion; *Blue*, calcified lesions; and *Pink*, lipidic lesion. In learning phase, features based on speckle statistics and signal attenuation are extracted at each resolution cell. Together with the supervised labels, we train a random forest classifier to probabilistically estimate the composition of the underlying tissues. (For interpretation of the references to color in this figure legend, the reader is referred to the web version of this article.) *(Adapted from Sheet D, Karamalis A, Eslami A, Noël P, Chatterjee J, Ray AK, et al. Joint learning of ultrasonic backscattering statistical physics and signal confidence primal for characterizing atherosclerotic plaques using intravascular ultrasound. Med Image Anal 2014;18(1):103–17.)*

finite mixture models (Section 3) as used by Destrempes et al. [25], we employ its dual form of representation as in Eq. (7.9) to achieve a form of mathematically tractable solution. Analytically we say that $\mathcal{H}^{src}(y|\mathbf{x}; r)$ is the posterior probability predictor which has learned the marginal posterior probability space of $p(y|r)$ using some labeled input examples $\{(\mathbf{x}; r|y)\}$. As previously discussed (Section 3), this vector $\mathbf{x}$ constitutes of Nakagami parameters estimated at multiple scales and also signal confidence measures. For the purpose of achieving analytical solutions we estimate the multiscale Nakagami parameters for a finite number of scales (say S), thus resulting in $2S$ features. The estimated ultrasonic signal confidence adds as another constituent on the vector $\mathbf{x}$. The dimension of the feature vector is thus $\mathcal{D} = 2S + 1$. Supervised labels $\mathcal{Y}$ were generated using the prognosis histology (PH) images from our previously developed texture-based TC algorithm [11]. The reason was the fact that constructed PH images represented the underlying tissue composition well and could be used more reliably for labeling in contrast with manual tracing (illustrated in Fig. 7.4B).

Since the estimation of the tissue posterior probability is to be learned with sparse number of samples, which have been approximately labeled, we choose to employ the oblique random forests [28–30] in view of its affinity to such learning paradigms (discussed in Section 3.4).

3.4 Oblique Random Forests

Oblique random forests use oblique decision splits instead of traditional axis-aligned decision boundaries at split nodes. In addition to the earlier discussed advantages that

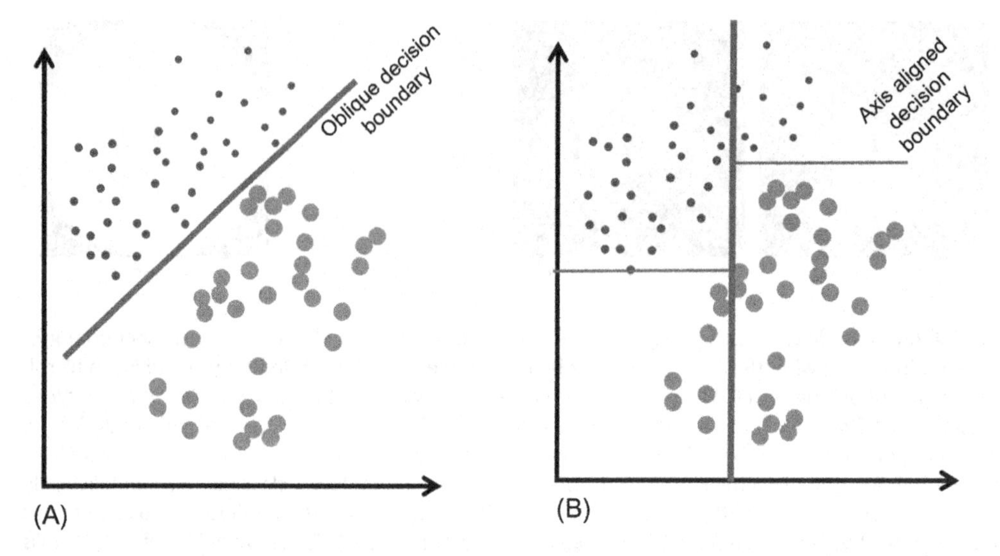

Fig. 7.5 Illustration of class separation using oblique splits (as shown in A) in contrast to nested axis aligned splits (shown in B).

decision forests entail, oblique forests are superior in the following aspects: (a) ability to separate distributions that lie between the coordinate axes with a single multivariate split, which might have required deep nested axis-aligned splits otherwise, and (b) less bias of the learnt decision trees to the geometrical constraints imposed by the base learner [30]. This property of oblique decision boundaries is illustrated in Fig. 7.5, where the data points are distributed between the coordinate axes. It can be seen that its easier the two classes easier with an oblique decision split in contrast to axis-aligned splits which require two levels of nesting to separate the classes.

The oblique random forest is an ensemble of T decorrelated decision trees where each decision tree (indexed as $t \in 1, \ldots, T$) recursively partitions the data domain ($\mathbf{x}$) based on a learnt binary oblique split function $\phi_n(\mathbf{x})$ at a split node n, until a leaf node l is reached. The split function ϕ_n is a multivariate decision rule and each leaf l contains a learned posterior class distribution $p_l(y|\mathbf{x}; r)$ (illustrated in Fig. 7.6). The final posterior probability is estimated by averaging over the posteriors estimated for the T trees as: $p_{\mathcal{H}}(y|\mathbf{x}; r) = \frac{1}{T}\sum_{t=1}^{T} p_{h_t}(y|\mathbf{x}; r)$.

3.5 In Vitro Learning and Prediction

For consistency with the exposition on domain adaptation, we term the in vitro setting as the *Source* domain and append the superscript $(\cdot)^{src}$ to the hypotheses and models associated with the source domain. During *learning*, given the training set in which each

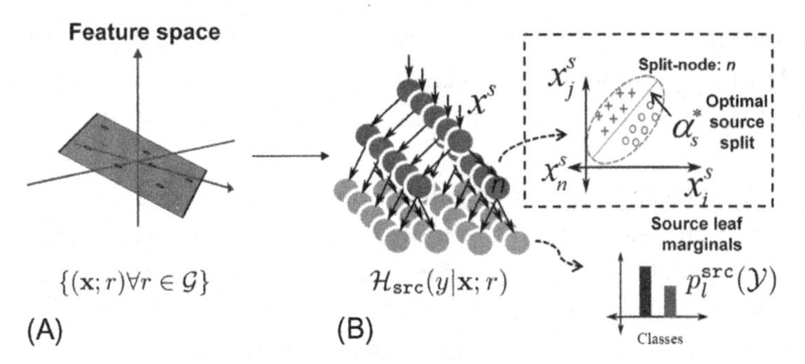

Fig. 7.6 Schematic illustrating the training of the in vitro source forests. We start with the feature vector $\{(\mathbf{x};\mathbf{r})\,\forall r \in \mathcal{G}\}$ extracted at spatial location r and its corresponding label y (as shown in A). Following this, we train an ensemble of decision trees with oblique decision boundaries at the split node using the methodology discussed in Section 3.5 (shown in B). Each of the trees end in a group of leaf nodes l that carry an estimate of the posterior probability of the tissue types ($p^l_{\mathtt{src}}(\mathcal{Y})$).

case contains ultrasonic signals and their corresponding tissue labels, the feature vector $\mathbf{x}^{\mathtt{src}}; r^{\mathtt{src}}$ consisting of multiscale estimated Nakagami parameters and ultrasonic signal confidence is computed and represented as $\{(\mathbf{x}^{\mathtt{src}}; r^{\mathtt{src}})\forall r \in \mathcal{G}\}$. Fig. 7.4 A presents the log-compressed and scan-converted B-mode IVUS image corresponding to $\mathcal{G}$. Given the tissue specific labels $y \in \mathcal{Y}$ corresponding to the grid points $r \in \mathcal{G}$, we train an oblique forest $\mathcal{H}^{\mathtt{src}}(y|\mathbf{x}^{\mathtt{src}}; r^{\mathtt{src}})$ using the features vector $\{(\mathbf{x}^{\mathtt{src}}; r^{\mathtt{src}})\forall r \in \mathcal{G}\}$. The different types of tissues considered are nonlesion, fibrotic lesion, calcified lesion, lipidic lesion, and necrotic lesion as illustrated for this example in the PH labels $\mathcal{L}$ in Fig. 7.4B. The hematoxylin and eosin histology of this case is in Fig. 7.4C and the same view is deformably registered with the IVUS image and shown in Fig. 7.4D using deformable registration process as detailed in Ref. [31].

Each tree is trained independently on random bootstrapped subsets of the training data $(\mathbf{x}^{\mathtt{src}}; r^{\mathtt{src}})$. From the root node, the split functions $\phi_n^{\mathtt{src}}(\mathbf{x}^{\mathtt{src}}; r^{\mathtt{src}})$ recursively split the data reaching the node into left and right subsets. We assume $\mathbf{x}_n^{\mathtt{src}}$ denote the source-domain input feature vector which reaches node n. Here, k features for the split are drawn from the input feature vector for node n of tree $t^{\mathtt{src}}$ (say $x_i^{\mathtt{src}}$ and $x_j^{\mathtt{src}}$ for $k = 2$). The oblique split function is constructed as follows:

$$\phi_n^{\mathtt{src}}(\mathbf{x}^{\mathtt{src}}; r^{\mathtt{src}}) = \left(\alpha_{i,n}^{\mathtt{src}}\left(\frac{x_i^{\mathtt{src}} - \mu_{i,n}^{\mathtt{src}}}{\sigma_{i,n}^{\mathtt{src}}}\right) + \alpha_{j,n}^{\mathtt{src}}\left(\frac{x_j^{\mathtt{src}} - \mu_{j,n}^{\mathtt{src}}}{\sigma_{j,n}^{\mathtt{src}}}\right) + \alpha_{\theta,n}^{\mathtt{src}}\right) \leq 0 \tag{7.10}$$

where $\alpha_{i,n}^{\mathtt{src}}, \alpha_{j,n}^{\mathtt{src}}$ and $\alpha_{\theta,n}^{\mathtt{src}}$ are coefficients generated from a normal distribution $\mathcal{N}(0,1)$. To account for differences in dynamic ranges of variables, we normalize using the mean $(\mu_{i,n}^{\mathtt{src}}, \mu_{j,n}^{\mathtt{src}})$ and standard deviations $(\sigma_{i,n}^{\mathtt{src}}, \sigma_{j,n}^{\mathtt{src}})$ estimated from $\mathbf{x}^{\mathtt{src}}; r_n^{\mathtt{src}}$. At each split node, several candidate split functions are generated randomly and the one that maximizes the expected information gain at the node is chosen (say $\phi_n^*(\mathbf{x}^{\mathtt{src}}; r^{\mathtt{src}})$ with parameters $\alpha_{i,n}^{\mathtt{src},*}, \alpha_{j,n}^{\mathtt{src},*}$ and $\alpha_{\theta,n}^{\mathtt{src},*}$) (see Fig. 7.6B). This recursive training continues till the termination criterion for leaf node is reached. Let $n_l^{\mathtt{src}}$ samples reach the leaf node $l^{\mathtt{src}}$ (denoted as $\mathbf{x}_l^{\mathtt{src}}$ with associated class labels $y_{l^{\mathtt{src}}}$), the leaf posteriors $p_{l^{\mathtt{src}}}(y|\mathbf{x}^{\mathtt{src}}; r^{\mathtt{src}})$ are estimated as normalized histogram of the class labels of these samples $\mathtt{hist}(y_{l^{\mathtt{src}}})$.

Each tree within the ensemble trained independently, introducing a certain level of randomness through bagging and random subspace selection to constitute the final hypothesis $\mathcal{H}^{\mathtt{src}}$. This is done so that the resultant trees are decorrelated, and thus improve the generalization of the forests. During *prediction*, given an previously unknown case with discretely sampled ultrasonic grid $\mathcal{G}'$, the feature vector is computed and represented as $\{(\mathbf{x}^{\mathtt{src}}; r^{\mathtt{src}})\forall r \in \mathcal{G}'\}$ and used for predicting $p^{\mathtt{src}}(y|\mathbf{x}^{\mathtt{src}}; r^{\mathtt{src}}) = \mathcal{H}^{\mathtt{src}}(y|\mathbf{x}^{\mathtt{src}}; r^{\mathtt{src}})\forall(\mathbf{x}; r) \in \mathcal{G}'$. The posterior probability estimated under in vitro setting for the different tissue types is represented as $p^{\mathtt{src}}(y|\mathbf{x}^{\mathtt{src}}; r^{\mathtt{src}})$.

4. DOMAIN ADAPTATION FOR IN VIVO TC

From the perspective of DA, we define the in vitro acquired data as the source domain and the limited in vivo dataset as the target domain. Extracting features through multiscale prediction of Nakagami parameters and signal confidence (as discussed in Section 3.5), we constitute a $2T + 1$ dimensional feature vector for each ultrasound resolution cell. Following which, the TC random forest classifier trained on in vitro data is adapted to be maximally discriminative on the target domain by correcting for domain shifts and errors in the decision boundaries using the adaptation mechanisms presented in Section 4.2. This adapted classifier is the final outcome of the proposed DA method. In the subsequent sections, we discuss in detail the individual steps proposed for the DA method.

4.1 Hypothesis

Consider registered IVUS frames $\mathcal{I}^{\mathtt{src}}$ and $\mathcal{I}^{\mathtt{tar}}$ under in vitro (Source) and in vivo (Target) settings respectively. For an arbitrary spatial location, say $\mathbf{p}$, let $\mathbf{x}^{\mathtt{src}}; r^{\mathtt{src}}$ and $\mathbf{x}^{\mathtt{tar}}; r^{\mathtt{tar}}$ represent the feature vector recorded under in vitro and in vivo settings respectively. The posterior probability on the source domain: $p^{\mathtt{src}}(y|\mathbf{x}^{\mathtt{src}}; r^{\mathtt{src}}; \mathbf{p})$ and target domain $p^{\mathtt{tar}}(y|\mathbf{x}^{\mathtt{tar}}; r^{\mathtt{tar}}; \mathbf{p})$, observed at the same arbitrary location $\mathbf{p}$, do not vary, as the tissue composition is domain-invariant.

The task of supervised domain adaptation is to adapt each reliably trained source hypothesis $h^{\mathtt{src}}$ to be maximally discriminative on a closely related target domain. We propose to domain adapt each hypothesis independently to generate the corresponding adapted hypothesis $h^{\mathtt{tar}}$. In the context of decision forests, this implies adapting the split nodes and the leaf nodes of each decision tree. It is important to preserve the de-correlated nature of the trees even after adaptation. For a properly adapted target hypothesis, testing on the matched in vivo location with the corresponding in vitro location, it is reasonable safe to expect that

$$p_{h^{\mathtt{src}}}(y|\mathbf{x}^{\mathtt{src}};r^{\mathtt{src}};\mathbf{p}) - p_{h^{\mathtt{tar}}}(y|\mathbf{x}^{\mathtt{tar}};r^{\mathtt{tar}};\mathbf{p}) \leq \epsilon \tag{7.11}$$

with very small $\epsilon \to 0$.

4.2 Domain Adaptation of Decision Forests

We take a random forest classifier trained on the source domain, which reliably estimates the marginal probabilities for each tissue type (as described in Section 3.5). Conceptually, if this forest is to be adapted, the split function $\phi_n^{\mathtt{tar}}(\mathbf{x})$ and the leaf-wise posterior class distribution $p_{l^{\mathtt{src}}}(y|\mathbf{x}^{\mathtt{tar}};r^{\mathtt{tar}})$ be adapted from the corresponding $\phi_n^{\mathtt{src}}(\mathbf{x})$ and $p_{l^{\mathtt{src}}}(y|\mathbf{x}^{\mathtt{src}};r^{\mathtt{src}})$ respectively. We perform DA tree-wise as the trees in the forest are trained independently.

We propose a supervised domain adaptation technique wherein a set of labeled target domain data $(\mathbf{x}^{\mathtt{tar}};r^{\mathtt{tar}},y)$ is available and modify the split function construction as well as leaf node posterior estimation. The amount of transfer relaxation between the source and target decision boundaries is controlled by parameter Γ. At node n, we consider the optimal split function $\phi_n^*(\mathbf{x}^{\mathtt{src}};r^{\mathtt{src}})$ of the source tree $t^{\mathtt{src}}$ as an anchor (retaining the same features used in the source split function as shown in Fig. 7.7A and B) about which the new random split functions $\phi_n^{\mathtt{tar}}(\mathbf{x}^{\mathtt{tar}};r^{\mathtt{tar}})$ are generated (see Fig. 7.7C). The target split is estimated as follows:

$$\phi_n^{\mathtt{tar}}(\mathbf{x}^{\mathtt{tar}};r^{\mathtt{tar}}) = \left(\alpha_{i,n}^{\mathtt{tar}}\left(\frac{x_i^{\mathtt{tar}} - \mu_{i,n}^{\mathtt{tar}}}{\sigma_{i,n}^{\mathtt{tar}}}\right) + \alpha_{j,n}^{\mathtt{tar}}\left(\frac{x_j^{\mathtt{tar}} - \mu_{j,n}^{\mathtt{tar}}}{\sigma_{j,n}^{\mathtt{tar}}}\right) + \alpha_{\theta,n}^{\mathtt{tar}}\right) \leq 0 \tag{7.12}$$

where $\alpha_{i,n}^{\mathtt{tar}} \sim \mathcal{N}(\alpha_{i,n}^{\mathtt{src},*},\Gamma)$, $\alpha_{j,n}^{\mathtt{tar}} \sim \mathcal{N}(\alpha_{j,n}^{\mathtt{src},*},\Gamma)$ and $\alpha_{\theta,n}^{\mathtt{tar}} \sim \mathcal{N}(\alpha_{\theta,n}^{\mathtt{src},*},\Gamma)$. Further, the data is locally normalized with the mean $(\mu_{i,n}^{\mathtt{tar}},\mu_{j,n}^{\mathtt{tar}})$ and standard deviations $(\sigma_{i,n}^{\mathtt{tar}},\sigma_{j,n}^{\mathtt{tar}})$ estimated from $\mathbf{x}_n^{\mathtt{tar}}$. From a number of randomly generated split functions, the one that produces maximal information gain at that node is chosen (see Fig. 7.7C–E). This is performed recursively till adapted tree terminates into a leaf node. Further, for nodes where no target data reaches, the source tree structure is retained without any adaptation. Given $n_l^{\mathtt{tar}}$ samples reaches leaf node $l^{\mathtt{tar}}$, its posterior $p_{l^{\mathtt{tar}}}(y|\mathbf{x}^{\mathtt{tar}};r^{\mathtt{tar}})$ is estimated as:

$$p_{l^{\mathtt{tar}}}(y|\mathbf{x}^{\mathtt{tar}};r^{\mathtt{tar}}) = \frac{1}{n_l^{\mathtt{tar}} + n_l^{\mathtt{src}}}\left(\left(n_l^{\mathtt{src}} * \mathtt{hist}\left(y_l^{\mathtt{src}}\right)\right) + \left(n_l^{\mathtt{tar}} * \mathtt{hist}\left(y_l^{\mathtt{tar}}\right)\right)\right) \tag{7.13}$$

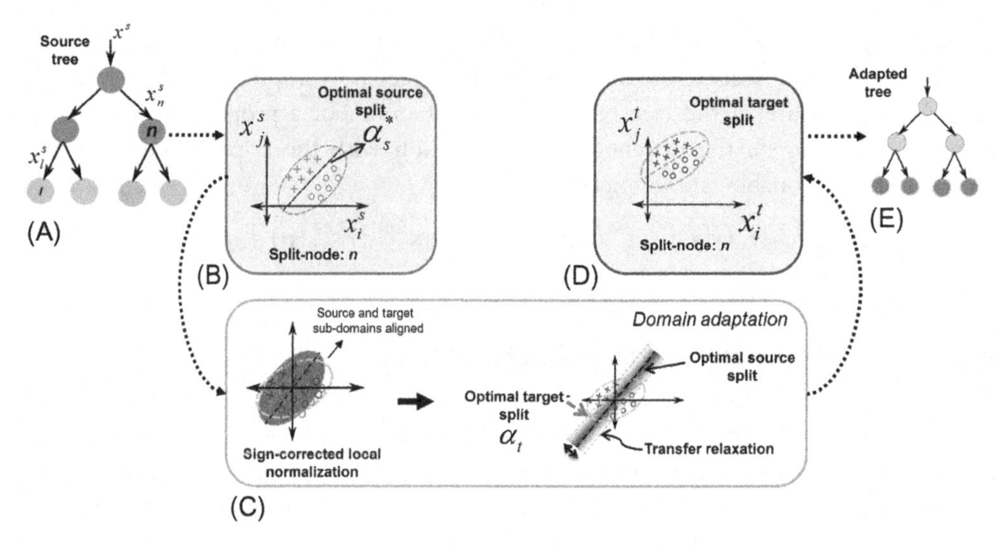

Fig. 7.7 Schematic illustrating local normalization and transfer relaxation to adapt the source boundary locally to target domain. Firstly, the target and source data reaching a particular split node *n* are locally aligned by normalizing them, following which the optimal target hypothesis is sought on a search space anchored at the source split (controlled by transfer relaxation parameter Γ). The grayscale gradient of the search space next to the optimal source split is indicative of the probability of generating a candidate hypothesis in that subspace during randomized node optimization. After optimization, the target split is reprojected back in the original feature space and the recursive adaptation continues.

where $\mathtt{hist}(y_l^{\mathtt{tar}})$ is the normalized histogram of class labels of target samples that reach node $l^{\mathtt{tar}}$. This adaption procedure is repeated on all the trees of $h^{\mathtt{src}}$ and the ensemble of the adapted trees constitutes adapted forest ($\mathcal{H}^{\mathtt{tar}}$). The computational complexity of the domain adapting individual trees of the source forest is $\Omega(T\tilde{N}\log^2\tilde{N})$ for T number of trees to be adapted and N number of samples in the adapting dataset ($\tilde{N} = 0.632N$), due to bootstrap sampling. It is lesser than training complexity ($\Omega(TK\tilde{N}\log^2\tilde{N})$, where K is the number of features randomly drawn at each node during training), as the forest is adapted node-wise retaining the same features as the source classifier.

Effect of Transfer Relaxation

In this chapter, we evaluate the performance of the proposed DA algorithm, varying the transfer relaxation parameter (Γ). In case of higher Γ, the search space for target split parameter is quite large, resulting in a reduced bias toward the parameters of the source split. In other words, the influence of the source split toward determining split in the target domain is low, which is quite similar to selecting coefficients through randomized node optimization solely based on target domain data. Reducing the Γ (from high toward low relaxation), leads to an increasing bias toward the optimal source split as a result of statistically constraining the parameter search space around the source split parameters.

4.3 Baselines

In addition to employing DA as a solution for in vivo TC, we explore other plausible approaches as incremental baselines for comparative analysis. A naïve solution would be to deploy the in vitro trained classifier directly on in vivo data. This approach is not suited as it does not consider the domain shift between the two domains and would lead to erroneous predictions. Alternatively, one could train a supervised classifier (say another oblique random forest) exclusively on the limited labeled in vivo data. One may expect that such a model would potentially overfit due to limited training data and hence would not generalize well to unseen test cases. Following on the lines similar to that proposed by Ciompi et al., one could train a domain-invariant classifier by fusing the in vitro and in vivo datasets [12]. We validate the performance of the proposed algorithm by constituting baselines based on the aforementioned plausible approaches to the problem. They are listed as follows:

- **Baseline 1 (BL1):** Forest trained on source data and tested on source data ($\mathcal{H}^{\mathtt{src}}$ on $\mathcal{I}^{\mathtt{src}}$). This baseline also doubles up as the *ground truth* for contrasting with our predictions on corresponding in vivo data;
- **Baseline 2 (BL2):** Forest trained on source data and tested on target data ($\mathcal{H}^{\mathtt{src}}$ on $\mathcal{I}^{\mathtt{tar}}$);
- **Baseline 3 (BL3):** Forest trained on target data and tested on target data ($\mathcal{H}^{\mathtt{tar\text{-}NA}}$ on $\mathcal{I}^{\mathtt{tar}}$) sans domain adaptation learnt from limited training data on target domain;
- **Baseline 4 (BL4):** Forest trained using combined source and target data and tested on target data ($\mathcal{H}^{\mathtt{combined}}$ on $\mathcal{I}^{\mathtt{tar}}$); and
- **Proposed (DA):** Adapted forest tested on target data ($\mathcal{H}^{\mathtt{tar}}$ on $\mathcal{I}^{\mathtt{tar}}$).

5. EXPERIMENTS AND DISCUSSION

5.1 Data Description and Configuration Settings

We validate our proposed method in characterizing heterogeneous atherosclerotic tissues on IVUS, where direct deployment of in vitro trained classifier for in vivo application is challenging due to the presence of blood which induces a domain distribution shift. The in vitro radio frequency ultrasound data used in these experiments is acquired from 53 coronary artery cross-sections from postmortem or transplantation of 13 human hearts [32], circulating saline in the lumen using commercially available[1] single-element 40 MHz catheter. The radio frequency data was acquired in the polar domain with each cross-section of 2048 × 256 at a sampling frequency of 400 MHz. Subsequently,

[1] Atlantis IVUS Catheter, Boston Scientific, Fremont, CA, USA.

conventional histology (Hematoxylin and Eosin and Movat Pentachrome) was obtained for validation. (For detailed description of the protocol, refer to Ref. [33].) The ground-truth labels for training the forest in the in vitro dataset is derived from prognostic histology from our previously developed texture-based TC algorithm [11].

The in vivo data is available for seven artery cross-sections from 3 human hearts. We obtained near-realistic in vivo setting by circulating human blood maintaining arterial pressure and pulsatile flow. This approach is observed to mimic closely IVUS acquisition for in vivo settings as corroborated by an expert interventional cardiologist. The cost and complexity involved and strict inclusion criteria in terms of the disease stage, availability of histology and human blood made in vivo data collection extensively challenging.

To remove any bias that could be potentially introduced in training/adaptation due to a possibly close by match between the in vitro and in vivo datasets from the 7 common arteries, we completely separate the 53 in vitro samples into the in vitro training dataset comprising of 46 samples and the matched dataset comprising of the 7 cross-sections. We match the in vitro data from these 7 artery cross-sections to their corresponding in vivo data and utilize it to contrast the predicted tissue posterior probabilities (shown in pseudo color in Fig. 7.8E).

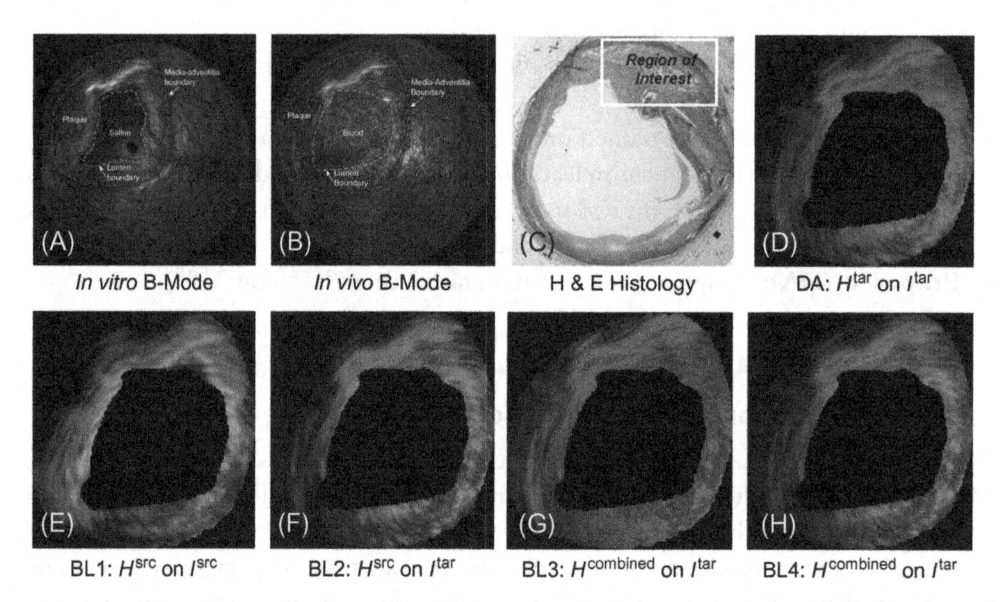

Fig. 7.8 Illustration of the source domain (in vitro) image $\mathcal{I}^{\mathtt{src}}$ (A) with corresponding target domain $\mathcal{I}^{\mathtt{tar}}$ (in vivo) image (B) and conventional histology (C). The results using the proposed method (D) and baselines (E)–(H) are rendered in pseudo-color (*blue*, calcified; *pink*, fibrotic; *yellow*, lipidic; and *red*, necrotic). (A) In vitro B-Mode. (B) In vivo B-Mode. (C) H & E Histology. (D) DA: $\mathcal{H}^{\mathtt{tar}}$ on $\mathcal{I}^{\mathtt{tar}}$. (E) BL1: $\mathcal{H}^{\mathtt{src}}$ on $\mathcal{I}^{\mathtt{src}}$. (F) BL2: $\mathcal{H}^{\mathtt{src}}$ on $\mathcal{I}^{\mathtt{tar}}$. (G) BL3: $\mathcal{H}^{\mathtt{combined}}$ on $\mathcal{I}^{\mathtt{tar}}$. (H) BL4: $\mathcal{H}^{\mathtt{combined}}$ on $\mathcal{I}^{\mathtt{tar}}$. (For interpretation of the references to color in this figure legend, the reader is referred to the web version of this article.)

The ground truth for the application is derived from matched in vitro TC results which are estimated using the in vitro trained decision forest discussed in Section 3.3. These labels were correlated with the corresponding histology slice and corrected by an expert interventional cardiology using ImageJ [34]. Histological validation revealed that plaques exhibit co-located heterogeneity. Manual tracings on IVUS images of tissues generally mark large regions to be belonging to one tissue type, which is contrary to what is observed on histology. As such, matched results from in vitro TC are more reliable representatives of tissue heterogeneity in contrast to manual labels. Additionally, since the datasets are matched, the tissue composition should be invariant of the transmission medium in the lumen (saline for in vitro and blood for in vivo), which further justifies use of matched in vitro TC observations as ground truth. The in vivo and the matched in vitro artery cross-sections are registered using control-point registration with thin plate splines deformation model. The inferred deformation field is used to transform probabilistic labeling derived from in vitro TC to align with the corresponding in vivo artery cross-section. The maximum a posteriori class of tissues is used as the discrete label for adaptation and evaluation. These labels were then transformed to the polar coordinate space of in vivo IVUS RF data for creating learning examples for this work.

Configuration Settings

In this application, we utilize bivariate splits at each split node. While choosing higher order splits ($\texttt{featSplit} > 2$), the number of free-parameters grows in proportion to the chosen order. In addition to higher training/adaptation computational time incurred, if the order is chosen very high, the forest might over fit during randomized node optimization while training and adaptation, due to limited number of target data available. The algorithms were implemented using MATLAB R2014b on workstation with 64GB RAM memory and 2.67GHz Intel Xeon (R) processor. It is important to note that, DA is controlled by one free parameter only, the transfer relaxation Γ_c, and the rest of the configuration is retained from the source forest. In an typical setting, we assume that the source forest is reliably trained where the choice of configuration parameters is optimal and well validated such that a source forest is readily available for adaptation to a target domain.

5.2 Evaluation Metrics

IVUS tissue characterization falls under the premise of multiclass probabilistic segmentation algorithms due to the presence of heterogeneous tissues which inherently prevent assignment of a *single* ground truth label that can be used for comparison. Further, as discussed in Section 4.1, the posterior probability estimated on in vivo data using the adapted hypothesis ($p_{\mathcal{H}^{\texttt{tar}}}(y|\mathbf{x}^{\texttt{tar}}; r^{\texttt{tar}}; \mathbf{p})$) should be ideally equal to the posterior probability estimated on matched in vitro data using the optimal source hypothesis

($p_{\mathcal{H}^{\text{src}}}(y|\mathbf{x}^{\text{src}}; r^{\text{src}}; \mathbf{p})$) at matched spatial location $\mathbf{p}$. Hence, we chose *logloss* metric over conventional metrics for evaluation as we are comparing posterior probabilities generated by the two hypothesis, which requires a probabilistic measure for both true and predicted factors in the metric (to validate Eq. 7.11). Secondly, as reported by Skalak et al. [35], the cross-entropy (or *logloss*) is the best of comparative metrics for cross-metric evaluation of classifier loss under uncertainty. To account for multiclass nature of tissue characterization, we use a multiclass extension of probabilistic log-loss, with $p_{h^{\text{tar}}}(y|\mathbf{x}^{\text{tar}}; r^{\text{tar}}; \mathbf{p})$ as predicted probability distribution as shown in Eq. (7.14).

$$logloss = -\frac{1}{N}\sum_{\forall \mathbf{p}}\sum_{\forall j} p_{\mathcal{H}^{\text{src}}}(y|\mathbf{x}^{\text{src}}; r^{\text{src}}; \mathbf{p}) \log(p_{\mathcal{H}^{\text{tar}}}(y|\mathbf{x}^{\text{tar}}; r^{\text{tar}}; \mathbf{p})) \quad (7.14)$$

where N is total number of pixel locations in the plaque and y_j is posterior probability estimated for the jth tissue of 5 tissue classes characterized from IVUS (viz. fibrotic, lipidic, calcified, necrosis, and background). It must be noted that *logloss* is a very sensitive metric and lower the value, the better the performance.

5.3 Transfer Relaxation (Γ) Versus Size of Target Database

The margin over which the optimal target split deviates from the source split is controlled by the transfer relaxation parameter (Γ). In Table 7.1, we understand the influence of Γ in relationship to the size of labeled target data available for training/adapting. We perform a quantitative evaluation of our dataset and report the results (multiclass *logloss*) with respect to the baselines discussed in Section 5. The configuration parameters of all the random forest classifiers are: 100 trees, 2 feature split-functions and infinite depth until minimum samples for split of 50 is reached. At each split node, 30 randomly generated splits were evaluated. In **DA**, Γ is varied in orders of 10 from 10^{-4} to 1. The percentage of labeled training samples available (*%adapt*) is also varied from 5% to 80% in steps of 15%. In Table 7.1, the best performance among the baselines and proposed method for a fixed *%adapt* is shown in **boldface** and best

Table 7.1 Experiment 1: Transfer relaxation Γ versus *%adapt*

					DA (with Γ =)				
%adapt	**BL1**	**BL2**	**BL3**	**BL4**	10^{-4}	10^{-3}	10^{-2}	10^{-1}	1
5 %			1.2576	1.1926	1.1629	1.1632	1.1644	**1.1455**	1.1832
20 %			1.2144	1.1525	1.1367	1.1353	1.1351	**1.1340**	1.1682
35 %	0.8121	1.2796	1.1867	1.1418	1.1241	**1.1234**	1.1246	1.1286	1.1552
50 %			1.1924	1.1275	1.1254	**1.1230**	1.1239	1.1306	1.1516
65 %			1.1979	1.1246	1.1238	**1.1232**	1.1245	1.1300	1.1573
80 %			1.2013	1.1288	**1.1276**	1.1277	1.1279	1.1348	1.1552

performance for each baseline with varying %*adapt* is underlined. Postanalysis, the best result among the comparative (*closest to the* **BL1**) is frameboxed.

Discussion

The comparative analysis of the proposed **DA** versus the baselines (shown in Table 7.1), we observe that for all tested %*adapt*, **DA** is closest to **GT**. This validates the superiority of adapting over retraining by combining source and target domains (**BL4**). This is further corroborated in Fig. 7.8, where the estimated tissue-specific marginals are rendered in a pseudo-color scheme. It must also be noted that **BL4** is more computationally expensive due to co-training of target and source data over the other methods. Visual assessment demonstrates that the performance of the proposed **DA** (Fig. 7.8D) classifier is visually closest to **GT** (Fig. 7.8E) and significantly improves over the best results of the comparative baselines **BL2** (Fig. 7.8F), **BL3** (Fig. 7.8G), and **BL4** (Fig. 7.8H).

We observe that optimal performance of adaptation requires choice of an appropriate transfer relaxation (Γ) depending on quantity and quality of target data available for adaptation. We also infer that for situations with smaller fractions of %*adapt*, **DA** requires a larger search margin ($\Gamma = 10^{-1}$) as local alignment of source and target subdomains through normalization may be weak. With increasing %*adapt*, the required search margin narrows ($\Gamma = 10^{-3}$) implying stronger alignment of subdomains. Table 7.1 further validates the hypothesis that *transfer learning* requires lesser training data in comparison to *self-training* for achieving the same performance. The proposed **DA** (with $\Gamma = 10^{-3}$) method reaches a better performance at 35% of training data in comparison to 65% required by the closest baseline **BL4**.

5.4 Transfer Relaxation Γ Versus Number of Trees (*M*)

During testing, marginals from individual trees are aggregated for the final result. As an evaluation of efficacy of the proposed **DA** classifier over the baselines, we compare their performances varying the number of trees. Here we fix the %*adapt* to 50% and retain other configuration parameters from Section 5.3. The number of trees (T) is varied from 20 trees to 100 in steps of 20 and the resultant multiclass *logloss* tabulated in Table 7.2. The best observed result (*closest to* **BL1**) is frameboxed and best results for each T value **boldfaced**.

Discussion

Table 7.2 shows the computational efficacy of **DA** over the baselines as well as the relationship between transfer relaxation (Γ) and number of trees (T). We observe that, **DA** reaches superior performance to co-training **BL4** with 40 trees in comparison to 100 trees for **BL4** (shown in underline in Table 7.2). Further, optimal Γ (*here* 10^{-3}) is invariant to T for sufficient number of trees and smaller T requires a larger relaxation

Table 7.2 Influence of Γ versus number of trees T

					DA(with Γ =)				
T	GT	BL2	BL3	BL4	10^{-4}	10^{-3}	10^{-2}	10^{-1}	1
20	0.8240	1.2929	1.1965	1.1485	1.1629	1.1369	**1.1346**	1.1492	1.1838
40	0.8173	1.2895	1.1947	1.1409	1.1467	**1.1257**	1.1278	1.1349	1.1695
60	0.8141	1.2863	1.1944	1.1365	1.1309	**1.1244**	1.1264	1.1336	1.1596
80	0.8133	1.2848	1.1937	1.1295	1.1295	**1.1239**	1.1246	1.1323	1.1574
100	0.8121	1.2796	1.1924	1.1275	1.1254	**1.1230**	1.1239	1.1306	1.1516

margin for optimal performance. The improvement in *logloss* of **DA** for $\Gamma = 10^{-3}$, with increasing M is much lesser than **BL4**, suggesting fast convergence of performance for **DA** upon proper choice of Γ.

6. CONCLUSIONS

Tissue characterization (TC) using ultrasonic back scattered signals aims at providing a probabilistic estimation of the tissue composition in the presence of highly heterogeneous tissues, beyond what is interpretable through a visual assessment ("eyeballing") of the log-compressed B-mode scans. In this work, we delved into statistical physics models behind tissue acoustic energy interaction and leveraged ensemble learning methods (decision forests) for performing TC for atherosclerotic tissues as seen through IVUS scans. Following this, we delved into greater detail on a novel approach for supervised domain adaptation of decision forests to facilitate the transition from models trained on in vitro data to realistic in vivo settings with minimalistic use of labeled data samples. We demonstrated that in the presence of a distribution shift, our approach of domain adapting is a superior alternative to methods like directly deploying the source classifier and training a domain-invariant classifier by combining both datasets. In the presence of few examples, the proposed approach is successful in factoring out the effect of domain shift at local and global levels by adapting decision boundaries in a regularized fashion to work on the target domain through the formulation of error-correcting hierarchical transfer relaxation and updating leaf posteriors thorough locally adaptive re-weighting.

Achieving aforementioned objectives has enabled us to better understand tissue-specific backscattering and attenuation concepts within heterogeneous media. They will eventually leverage translation of existing lab-scale practices associated with in vitro non real-time practice into clinical environment. From clinical perspective, development of a reliable TC algorithm will enable radiologists to perform tissue-specific US imaging and real-time imaging of tissue characteristics, which ultimately improve diagnostic sensitivity and specificity. Since these TC algorithms would employ the full spectrum of information conveyed by the backscattered signals, the outcome shall be as informative as conventional histology. In practical applications like assessment of vascular plaques, where conventional histology cannot be performed on living patients, this tool chain would provide a realistic solution.

Despite all technological advances in medical diagnosis, the community still lacks of intra-operative real-time in vivo histology so surgeons could make the most confident decisions and opt their treatment strategies. For example, one of the main challenges in existing algorithms is to detect reliably necrotic tissues [36] and resolve ambiguity of tissue characterization in hypoechoic areas particularly behind arc of calcified plaques [37]. This not only demands for development of TC algorithms upon features with ultrasonic physics backgrounds (something beyond classical signal processing signatures) but also leverage appropriately designed adaptation approaches to in vitro driven algorithms for in vivo tissue characterization, resulting in improving quality of health care delivery and reducing overall costs.

ACKNOWLEDGMENTS

This research was performed as a part of the research project "Computational Modeling of Ultrasonic Backscattering Statistical Physics for In Situ Tissue Characterization" which is financed by the Samsung Global Research Outreach (GRO) Program 2013.

REFERENCES

[1] Naghavi M, Libby P, et al. From vulnerable plaques to vulnerable patients: a call for new definitions and risk assessment strategies: Part I. Circulation 2003;108:1664–72.
[2] Katouzian A, Baseri B, Konofagou EE, Laine AF. Challenges in atherosclerotic plaque characterization with intravascular ultrasound (IVUS): from data collection to classification. IEEE Trans Inf Technol Biomed 2008;12:315–27.
[3] Treeby BE, Zhang EZ, Thomas AS, Cox BT. Measurement of the ultrasound attenuation and dispersion in whole human blood and its components from 0–70 MHz. Ultrasound Med Biol 2011;37(2):289–300.
[4] Sheet D, Karamalis A, Eslami A, Noël P, Chatterjee J, Ray AK, et al. Joint learning of ultrasonic backscattering statistical physics and signal confidence primal for characterizing atherosclerotic plaques using intravascular ultrasound. Med Image Anal 2014;18(1):103–17.
[5] Conjeti S, Katouzian A, Roy AG, Peter L, Sheet D, Carlier SG, et al. Supervised domain adaptation of decision forests: transfer of models trained in vitro for in vivo intravascular ultrasound tissue characterization. Med Image Anal 2016;32:1–17.
[6] Nair A, Kuban BD, Tuzcu EM, Schoenhagen P, Nissen SE, Vince DG. Coronary plaque classification with intravascular ultrasound radio frequency data analysis. Circulation 2002;106(17):2200–6.
[7] Ohota M, Kawasaki M, Ismail TF, Hattori K, Serruys PW, Ozaki Y. A histological and clinical comparison of new and conventional integrated backscatter intravascular ultrasound (IB-IVUS). Circulation 2012;76(7):1678–86.
[8] Sathyanarayana S, Carlier S, Li W, Thomas L. Characterisation of atherosclerotic plaque by spectral similarity of radiofrequency intravascular ultrasound signals. EuroIntervention 2009;5(1):133–9.
[9] Escalera S, Pujol O, Mauri J, Radeva P. Intravascular ultrasound tissue characterization with sub-class error-correcting output codes. J Sign Process Syst 2009;55(1–3):35–47.
[10] Taki A, Hetterich H, Roodaki A, Setarehdan SK, Unal G, Rieber J, et al. A new approach for improving coronary plaque component analysis based on intravascular ultrasound image. Ultrasound Med Biol 2010;38(8):1245–58.

[11] Katouzian A, Karamalis A, Sheet D, Konofagou E, Baseri B, Carlier SG, et al. Iterative self-organizing atherosclerotic tissue labeling in intravascular ultrasound images and comparison with virtual histology. IEEE Trans Biomed Eng 2012;59(11):3039–49.
[12] Ciompi F, Pujol O, Gatta C, Rodríguez-Leor O, Mauri-Ferré J, Radeva P. Fusing in-vitro and in-vivo intravascular ultrasound data for plaque characterization. Int J Cardiovasc Imaging 2010;26(7): 763–79.
[13] Kawasaki M, Takatsu T, Noda T, Sano K, Ito Y, Hayakawa K, et al. In vivo quantitative tissue characterization of human coronary arterial plaques by use of integrated backscatter intravascular ultrasound and comparison with angioscopic findings. Circulation 2002;105:2487–92.
[14] Lizzi FL, Greenebaum M, Feleppa EJ, Elbaum M. Theoretical framework for spectrum analysis in ultrasonic tissue characterization. J Acoust Soc Am 1983;73:1366–73.
[15] Nair A, Kuban BD, Obuchowski N, Vince DG. Assessing spectral algorithms to predict atherosclerotic plaque composition with normalized and raw intravascular ultrasound data. Ultrasound Med Biol 2001;27:1319–31.
[16] Shimodaira H. Improving predictive inference under covariate shift by weighting the log-likelihood function. J Stat Plan Inference 2000;90(2):227–44.
[17] Ben-David S, Blitzer J, Crammer K, Kulesza A, Pereira F, Vaughan JW. A theory of learning from different domains. Mach Learn 2010;79(1–2):151–75.
[18] Patel VM, Gopalan R, Li R, Chellappa R. Visual domain adaptation: a survey of recent advances. IEEE Signal Process Mag 2015;32(3):53–69.
[19] Caruana R. Multitask learning. Mach Learn 1997;28(1):41–75.
[20] Raina R, Battle A, Lee H, Packer B, Ng AY. Self-taught learning: transfer learning from unlabeled data. In: Proceedings of international conference on machine learning. 2007. p. 759–66.
[21] Chapelle O, Schlköpf B, Zien A. Semi-supervised learning. Cambridge, MA, USA: MIT Press; 2010.
[22] Widmer G, Kubat M. Learning in the presence of concept drift and hidden contexts. Mach Learn 2011;23(1):69–101.
[23] Sun S. A survey of multi-view machine learning. Neural Comput Appl 2013;23(7–8):2031–38.
[24] Shankar PM. A general statistical model for ultrasonic backscattering from tissues. IEEE Trans Ultrason Ferroelectr Freq Control 2000;47:727–36.
[25] Destrempes F, Meunier J, Giroux MF, Soulez G, Cloutier G. Segmentation in ultrasonic b-mode images of healthy carotid arteries using mixtures of nakagami distributions and stochastic optimization. IEEE Trans Med Imaging 2009;28:215–29.
[26] Karamalis A, Katouzian A, Carlier SG, Navab N. Confidence estimation in IVUS radio-frequency data with random walks. In: Proceedings of ISBI. 2012. p. 1068–71.
[27] Grady L. Random walks for image segmentation. IEEE Trans Pattern Anal Mach Intell 2006;28:1768–83.
[28] Breiman L. Random forests. Mach Learn 2001;45:5–32.
[29] Criminisi A, Shotton J, Konukoglu E. Decision forests: a unified framework for classification, regression, density estimation, manifold learning and semi-supervised learning. Found Trends Comput Graph Vis 2012;7:81–227.
[30] Menze BH, Kelm BM, Splitthoff DN, Koethe U, Hamprecht FA. On oblique random forests. In: Proceedings of machine learning and knowledge discovery and data. 2011. p. 453–69.
[31] Katouzian A, Karamalis A, Lisauskas J, Eslami A, Navab N. IVUS-histology image registration. In: Dawant BM, Christensen GE, Fitzpatrick JM, Rueckert D, editors. Biomedical image registration. Lecture notes in computer science, vol. 7359. 2012. p. 141–9.
[32] Katouzian A, Sathyanarayana S, Li W, Thomas T, Carlier SG. Challenges in tissue characterization from backscattered intravascular ultrasound signals. In: Proceedings of SPIE, vol. 6513. 2007. p. 65130O–65130O-11.
[33] Katouzian A, Laine AF. Methods in atherosclerotic plaque characterization using intravascular ultrasound (IVUS) images and backscattered signals. In: Suri JS, et al., editors. Atherosclerosis Disease Management. New York: Springer; 2010. p. 121–52. ISBN: 978-1-4419-7221-7.
[34] Schneider CA, Rasband WS, Eliceiri KW. NIH image to ImageJ: 25 years of image analysis. Nat Methods 2012;9(7):671–5.

[35] Skalak DB, Niculescu-Mizil A, Caruana R. Classifier loss under metric uncertainty. In: Proceedings of European conference on machine learning. 2007. p. 310–22.
[36] Sheet D, Karamalis A, Eslami A, Noel P, Navab N, Katouzian A. Hunting for necrosis in the shadows of intravascular ultrasound. Comp Med Imag Graph 2014;38(2):104–12.
[37] Tanaka K, Carlier SG, Katouzian A, Mintz G. Hunting for necrosis in the shadows of intravascular ultrasound. Am Coll Cardiol 2007;49(9):29B.

CHAPTER 8

Intracoronary Optical Coherence Tomography

G.J. Ughi[*], T. Adriaenssens[†,‡]
[*]Massachusetts General Hospital and Harvard Medical School, Boston, MA, United States
[†]University Hospitals Leuven, Leuven, Belgium
[‡]KU Leuven, Leuven, Belgium

Chapter Outline

1. INTRODUCTION

1.1 Physical Principles of Optical Coherence Tomography

The principles of optical coherence tomography (OCT) evolved from one-dimensional (1D) optical low-coherence reflectometry, that uses a Michelson interferometer and a broad-band laser light source [1, 2]. The term OCT was first used in 1991, when the addition of transverse scanning enabled two-dimensional (2D) imaging of human retina in vitro [3]. OCT rapidly expanded to numerous medical and biological applications

Computing and Visualization for Intravascular Imaging and Computer-Assisted Stenting
http://dx.doi.org/10.1016/B978-0-12-811018-8.00008-4

[4–8], including endoscopic imaging [9, 10]. Several OCT catheters and endoscopes have been developed for invasive and semi-invasive imaging of human organs. Systems for the in vivo imaging of human eye, skin, esophagus, and vasculature are commercially available and additional medical application (e.g., cancer detection) are currently under investigation [11–13].

State-of-the-art intracoronary OCT technology uses light sources with a central wavelength of 1300 nm, spanning over a range of approximately 80–100 nm (i.e., laser bandwidth) [14, 15]. The optimal choice of the wavelength is driven by tissue scattering and absorption properties. OCT provides a penetration depth in the coronary wall of 1–2 mm and an axial resolution in tissue of approximately 10–15 μm. OCT scan range is related to coherence properties of the light source and commercial intracoronary imaging systems are typically limited to approximately 5 mm. Compared to intravascular ultrasound (IVUS), OCT shows a limited imaging depth in atheromas (lipid-rich plaques), but can image through calcium where IVUS cannot and it provides one order of magnitude improved resolution (IVUS axial resolution >120 μm) [16]. As such, OCT provides significantly improved visualization of lumen morphology, coronary wall microstructure, and interaction between the vessel and implanted devices (e.g., intracoronary stents) during percutaneous coronary intervention (PCI).

An OCT image is formed by collecting backscattered light from the artery wall, "measuring" the amplitude and the time of flight for the emitted light to travel between the catheter optics (i.e., the focusing lens) and the tissue (Axial-scan line, or simply A-line). Multiple A-lines are acquired by rotating the imaging catheter to create a 2D image (B-scan) as shown in Fig. 8.1. Since the speed of light is much faster than sound (i.e., light propagates at a speed of 3E8 m/s in air), a direct measurement of light "echo time delay" is unfeasible and the use of interferometric techniques (i.e., Michelson interferometer) is necessary. OCT uses interferometry to obtain a measurement of the intensity of single scattered light (carrying information about tissue optical properties) and to reject multiple scattering (i.e., ballistic imaging modality) [17].

Earlier implementations of OCT technology required the use of a spatial-scanning reference mirror to match the length of light path in tissue and generate constructive interference at different depths (time-domain OCT or TD-OCT). However, TD-OCT imaging speed is significantly limited by the mechanical movement of such a reference reflector, as the acquisition of each A-line requires a full scan of the mirror. Fourier-domain OCT (FD-OCT) circumvents this limitation by using a rapid variable wavelength light source (i.e., swept laser)[1] that allows the detection of reflections from

[1] Two different implementations of FD-OCT exist: the one described here, swept-source OCT, and a second one named spectral domain OCT. They both use the same principles and they only differ by the method used to extract the data from the interferometer. The specific description of each modality is outside the scope of this chapter.

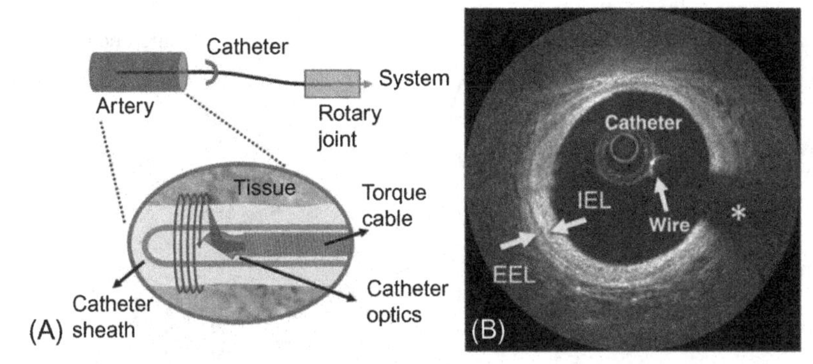

Fig. 8.1 Schematic of a generic OCT imaging catheter. (A) Cross-sectional images are obtained by rotating the catheter optical element inside the vessel lumen. An optical rotary joint is used to interface the spinning catheter to the OCT imaging system. In this schematic, the illustrated optical element is a ball-lens, polished at ~42 degrees for obtaining lateral focusing of the light (i.e., side viewing). (B) An example of OCT imaging of normal coronary wall. Both the internal elastic lamina (IEL) (i.e., the interface between the vessel intima and the media) and the external elastic lamina (EEL) (i.e., the interface between media and adventitia) are visible. Between 5 and 11 o'clock it is possible to observe the thickening of the neointimal layer. The *star* (*) indicates the shadow generated by the guide wire that used to guide the OCT catheter through the coronary arteries for in vivo imaging.

all the echo time delays simultaneously, enabling higher imaging speeds (e.g., 100,000 A-lines/s) without limiting the OCT signal detection sensitivity (>105 dB). As blood is a highly scattering medium in the near-infrared (NIR), arteries need to be cleared from blood in order to acquire OCT data. The high frame rate (>100 frames/s) of FD-OCT, scanning an entire artery volume in few seconds, enabled the widespread use of OCT for coronary applications [18–20].

Fig. 8.2 shows a generic schematic of an FD-OCT system. In the interferometer the light from the swept laser is split: a small portion is sent to the reference arm (e.g., 5–10%) and the remaining to vessel wall (e.g., 90–95%). For coronary applications, the intensity of the light delivered to the vessel wall (sample arm) is typically <20 mW. The light back-reflected by the tissue is merged with the one returning from the reference mirror by an optical combiner, generating an interference pattern. These interference fringes are detected by a fast, dual-balanced, photo detector, and a fast Fourier transform (FFT) is used to process the signal and display the magnitude of tissue optical reflections as function of depth in real-time (A-scan line). Multiple A-scan lines are acquired sequentially at high-speed (e.g., 100 kHz) and combined to generate a cross-sectional image.

OCT imaging catheters deliver and collect reflected light by the means of a single-mode optical fiber. Single-mode light is focused by a miniaturized optical element spliced to the tip of the optical fiber (Fig. 8.1). The catheter optics can be created

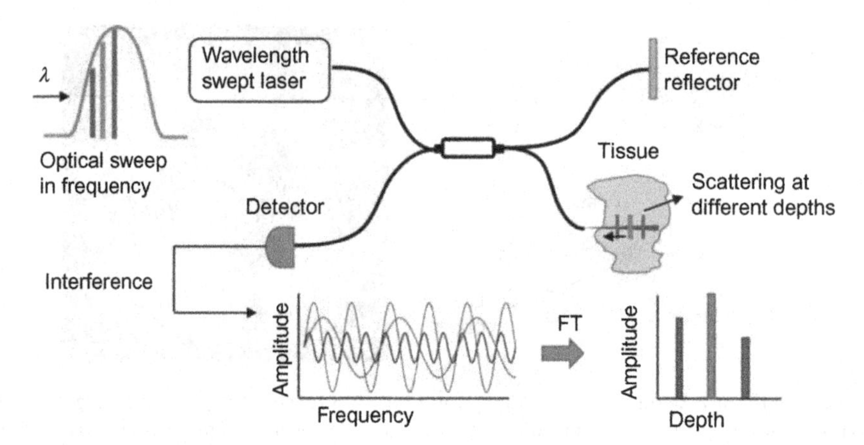

Fig. 8.2 Schematic of a generic, fiber-based, FD-OCT imaging system. Light from a swept laser is split using a fiber optics beam splitter: 95% of the light is directed to the tissue and the remaining 5% to a reference reflector. The illuminated tissue backscatters the light at different depths, and the returning light is combined with the one reflected by the reference mirror generating an interference pattern. This interferometric technique allows the rejection of multiple scattered light and retaining of single scattered light only. The interference patterns are detected by the means of a dual-balanced photo detector. The interference fringes are processed by the means of an FFT and sample reflectivity at different depths is reconstructed generate an intensity profile (A-line).

by using a GRIN (graded index) or a ball lens polished to <42 degrees for side viewing. Lenses are designed to achieve a working distance of approximately 0.5–4 mm and a lateral resolution of 30 μm (measured at catheter focal distance), optimized for coronary artery imaging. Commercial catheter size is 2.7F (i.e., outer diameter equal to approximately 1 mm) with an insertable length of approximately 1.5 m, and it is designed with a monorail tip to be navigated through the artery lumen over a coronary wire. The entire catheter optics (i.e., the optical fiber and the imaging element) is rotated at high speed (e.g., 160–180 frames/s [21]) by the means of a hollow torque cable containing the optical fiber and a proximal motor located in the so-called optical rotary joint, that serves as an interface between the spinning catheter core to the OCT system. A volumetric OCT dataset is acquired by pulling back the spinning optical core within a protective, optically transparent sheath, performing a helical scan of the vessel (Fig. 8.1).

1.2 Coronary OCT Image Acquisition Techniques

Over the past decade, intracoronary OCT has been increasingly adopted, both for clinically oriented research in the field of coronary artery disease and in the assessment and guiding of treatment of coronary lesions in the individual patient. As with many diagnostic imaging modalities, it is important to spend meticulous care during the process of acquisition of the images, to ensure safety of the patient, and to provide the best quality images possible.

The safety of the current generation of clinically applicable FD-OCT systems has been assessed in several reports and is reflected in an increasingly growing uptake of the technology by the interventional community [22, 23]. The choice between different imaging modalities, such as OCT or IVUS, will depend on the operator's experience with a specific technique and on economic aspects such as cost of the catheter and reimbursement issues which differ importantly between countries.

In contrast to IVUS, where there is no interference of the circulating red blood cells with the emitted sound waves, the acquisition of light-based OCT images necessitates a complete clearance of the lumen from blood in order to achieve a blood-free imaging field. This makes the success of the OCT procedure substantially more operator-effort dependent, as correct positioning of the imaging and guiding catheter and application of an adequate flush of contrast medium (either manually or via an automated infusion pump) are crucial steps for optimal image acquisition. Typically, coronary arteries are flushed with a 3–4 mL/s contrast injection for 3–5 s in total.

Although beyond the scope of this section, it is obvious that the operator should make him/herself familiar with all aspects of material and setup and with the different particularities of the procedure. An intravascular OCT system generally consists of a table-fixed or mobile unit containing the light source and the hardware for computation of the images, a docking station or patient interface unit (containing the optical rotary joint), and disposable intracoronary imaging catheters for sterile use in the coronary artery. The imaging catheter is introduced over a regular coronary wire and connected with the system via a patient interface or docking unit. Different systems have different designs of catheters as well. A clear understanding of the position of radio opaque markers on the catheters and ability to handle fluidly the different steps for starting a pullback run is indispensable for safe and correct image acquisition.

With respect of image interpretation, it is advisable in all circumstances, before engaging in a more in-depth study of the OCT cross sections, to assess general quality and interpretability of the pullback immediately after its acquisition. Three important aspects of the acquisition need to be assessed: (1) verification whether the segment of interest is completely captured in the pullback delivered, which can be assessed versus landmarks such as bifurcation points and the tip of the guiding catheter, and newer techniques using angiography-OCT coregistration can provide additional information as well; (2) assessment of the quality of clearance of the lumen from blood; and (3) assessment of catheter performance. Most importantly, it should be checked that no blood accumulating between the different layers of the OCT imaging catheter is interfering with image quality. In the case of a suboptimal acquisition, it should be decided immediately whether an additional pullback in the vessel is mandatory and which measures improving acquisition quality should be taken. Good communication between the person operating the OCT console and the interventional cardiologist is key in this regard.

While the passage of the OCT imaging catheter over the coronary wire usually is straightforward, this part of the procedure can be more challenging when confronted with tortuous vessels or severely calcified lesions. While the use of extra support guide wires or the application of guiding catheter extensions can be of help in these situations, it has to be accepted that some, more difficult, lesions cannot be crossed with the imaging catheter before adequate predilation. Another anatomical difficulty might be the case with acute angulation from the left main coronary artery (LMCA) toward the left circumflex artery. Several reports have illustrated the potential utility of child-in-a-mother or guiding extension catheters in such circumstances. Moreover, the assessment of the vessel through the extension catheter is still possible and consists another potential advantage of such a strategy [24].

1.2.1 Strategies to Limit Contrast Use

When using an OCT-guided PCI strategy, often multiple OCT acquisitions are performed during the same procedure. The operator should be aware of the risk of contrast-induced nephrotoxicity, especially in high-risk patients. An important measure is the coordination of cineangiography with OCT runs to accomplish both angiography and OCT imaging simultaneously [25].

Although the great majority of clinical OCT examinations is performed using radiographic contrast agents to achieve clearance of the vessel lumen from blood, some alternatives have been described. In several studies in humans, low molecular weight dextran provided comparable image quality and measurements (after correction for refractive index) compared with radiographic contrast [26, 27]. This currently remains, however, an off-label approach.

1.2.2 Coregistration of OCT and X-Ray Angiography

With respect to optimization of PCI procedures using OCT for the selection of appropriate size and length of stents and assistance in the actual deployment from healthy to healthy segments, a precise correspondence of OCT images and the angiogram is critical. Formerly, this was achieved by the operator identifying corresponding structures and segments using landmarks thereby mind-mapping the intracoronary scanning images to the angiogram. Recently, automated coregistration of intracoronary scanning modalities and X-ray angiography were introduced, with promising capacity for enhancing clinical applicability of OCT-guided PCI [28].

1.2.3 OCT in Assessment of the LMCA

Several studies reported on the feasibility of achieving high image quality in LMCA assessment using OCT [29]. The need to flush the lumen to achieve a blood-free field, makes the assessment of the ostium of the LMCA with OCT a challenging task, however. Only when precise positioning of the guiding catheter at the ostium of the left main can be reached, it is possible to achieve visualization of the ostium of the left main while ensuring adequate flush of the lumen during the pullback [30, 31]. But even then, the occurrence of some image artifacts at the level of the ostium of the left main is not unusual [29]. Not with standing these limitations and challenges, more and more operators feel more comfortable in the assessment of the LMCA with OCT, certainly triggered by the fact that OCT offers additional value with respect to stent optimization after implantation.

1.3 Commercial OCT Systems

The first commercially available intracoronary OCT solution was the Light Lab M2 time-domain (TD-OCT) imaging system and "image-wire" catheter (LightLab/St. Jude, Westford, MA). In 2008, the first commercial high-speed FD-OCT system has been released in Europe, South America, and Asia (C7-XR, LightLab) that was subsequently approved in the United States in 2010. This first high-speed FD-OCT catheter was named Dragonfly (LightLab/St. Jude), capable of acquiring images at 100 frames/s and a pullback speed of 20 mm/s, eliminating the need of balloon occlusion. The higher speed of the C7 system, eliminating the need of artery occlusion required by TD-OCT, significantly expanded the adoption of intravascular OCT worldwide. The size of the Dragonfly catheter is 2.7F, including a drive shaft and monorail tip. The C7-XR imaging system is claimed to provide an axial resolution <20 μm. A latest generation OCT system was released in 2014 (Ilumien Optis, St. Jude Medical, MN), featuring a faster imaging speed if compared to C7-XR system (i.e., 180 frames/s), integration with angiography and console compatibility with Fractional Flow Reserve catheters (St. Jude Medical). In 2013, Terumo released their first product in Europe and Asia, a high-speed FD-OCT imaging system (Lunawave, Terumo, JP) and catheter (Fastview, Terumo). Current Terumo generation features a 2.4F crossing profile, monorail imaging catheter capable to acquire data at a speed of 160 frames/s and resolution comparable to St. Jude Medical imaging products. It is estimated that approximately 500,000 OCT imaging cases have been performed worldwide since the release of the first intravascular system, that approximately 100,000 cases are currently performed every year and that intracoronary OCT market is growing yearly of approximately 20%.

2. CORONARY IMAGING

2.1 Assessment of Lumen Morphology

Intravascular OCT enables the detailed assessment of vessel lumen morphology. In particular, OCT has the capability to assess the lumen area with elevated accuracy [32], to visualize ostium of side-branches, vessel stenosis, and to determine vessel reference area. Additionally, OCT has the ability to visualize both small and large wall dissections and to determine the presence of intravascular thrombosis. OCT can also help to quantify thrombus characteristics (e.g., area and composition), in cases of plaque rupture and stent thrombosis.

2.2 Assessment of Coronary Atherosclerosis

Current understanding of plaque biology suggests that ~80% of plaque rupture events are associated with inflamed thin-cap (<65 μm) fibroatheromas [33]. The presence of calcium nodules (~10%) and plaque erosion has been also suggested as mechanisms of vessel thrombosis [34]. OCT visualizes coronary artery microstructure in high-resolution, providing detailed information about plaque composition and morphology. Moreover, OCT has the capability of visualizing plaque fibrotic cap in detail, allowing the assessment of cap thickness, rupture, and erosion.

It has been shown that OCT enables the characterization of atherosclerotic plaque components [5, 35–37], relying on the visualization of tissue optical properties such as attenuation and backscattering coefficients (μ_a and μ_s, respectively). Studies demonstrated the ability of OCT to distinguish between the three different plaque types (i.e., fibrotic, calcium, and lipid) [15, 38, 39]. It has been shown that fibrotic tissue shows low μ_a and elevated μ_b, resulting in homogeneous OCT signal (Fig. 8.3A). As such, OCT signal can penetrate deeper in fibrotic tissue, and the vessel external elastic membrane (i.e., the interface between the media layer and the adventitia—EES) is typically visible. Fibrocalcific plaques show both low μ_a and μ_b, resulting in a signal poor region with sharply delineated borders (Fig. 8.3B). Given the limited optical attenuation of calcium in the NIR, OCT is capable of visualizing the full extension of the lesion and plaque external borders are typically visible. Lipid-rich fibroatheromas and necrotic cores present both elevated backscattering and attenuation coefficients and the penetration depth of OCT is very limited in such plaques. This results in a fast OCT signal drop-off in those lesions, and only the plaque luminal surface is visible. Opposite to calcium, the interface between the fibrotic cap and the lipid shows poorly delineated luminal borders (Fig. 8.3C). A known limitation for the assessment of atheromas is that OCT alone does not have the ability to differentiate between lipid accumulations on the vessel wall and the presence of a necrotic core, as both conditions exhibit an elevate μ_a [15]. On the basis of such observations, quantitative methods have been proposed trying to achieve an automated characterization of atherosclerosis by intracoronary OCT [40, 41].

Although OCT cannot visualize the complete plaque burden of lipid and necrotic cores, it has been demonstrated that OCT can accurately identify the presence of thin-capped fibroatheromas (TCFA), allowing the precise quantification of the thickness

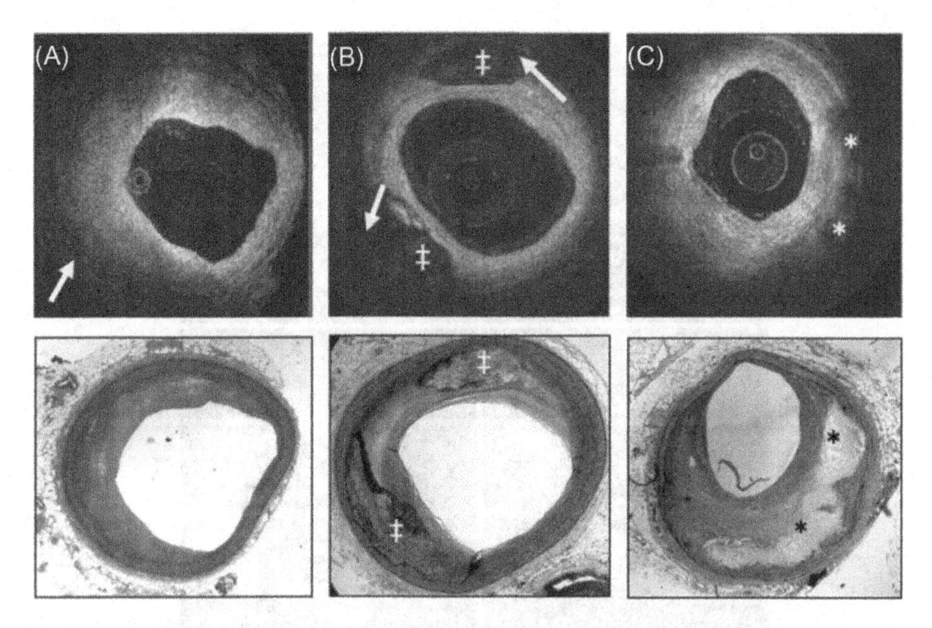

Fig. 8.3 Characterization of human atherosclerosis. (A) Fibrotic plaque showing low attenuation and elevated signal intensity (i.e., backscattering) and corresponding histological section. The plaque shows a homogeneous appearance and EEM is visible (*arrow*). (B) Example of fibrocalcific plaque. Calcium can be identified by OCT as it appears as a low backscattering, low attenuation, and sharply defined borders region. Plaques external boundaries are visible in the image. (C) A lipid pool characterized by elevated OCT signal attenuation. Fibrotic cap thickness can be evaluated; however, plaque external borders cannot be assessed due to the rapid OCT signal fall-off. *(The image is reproduced from Bezerra HG, Costa MA, Guagliumi G, Rollins AM, Simon DI. Intracoronary optical coherence tomography: a comprehensive review clinical and research applications. JACC Cardiovasc Interv 2009;2:1035–46, with permission from the publisher.)*

and morphology of the fibrous cap [42] (Fig. 8.4A and B). It has been shown that OCT visualizes in detail plaque cap rupture (Fig. 8.4C), erosion (Fig. 8.4D) [43] and luminal thrombus overlying thin, disrupted plaques [44]. Examples of clinical OCT plaque assessment is the use in clinical trials to quantify the effect of statins on plaque cap thickness [45] and to identify culprit lesions in acute coronary syndrome (ACS) patients [43, 46, 47]. OCT was also used to assess the role of small calcium nodules in plaque rupture and to identify plaque erosion [43, 44]. Furthermore, several studies demonstrated the capability of OCT to localize thrombus in the vessel lumen and over metallic intracoronary devices, and to characterize thrombus phenotype (i.e., white vs. red) on the basis of optical attenuation [48–50].

Additional data have been published about the ability of OCT to identify macrophages and plaque inflammation [51, 52]. It was shown that clusters of

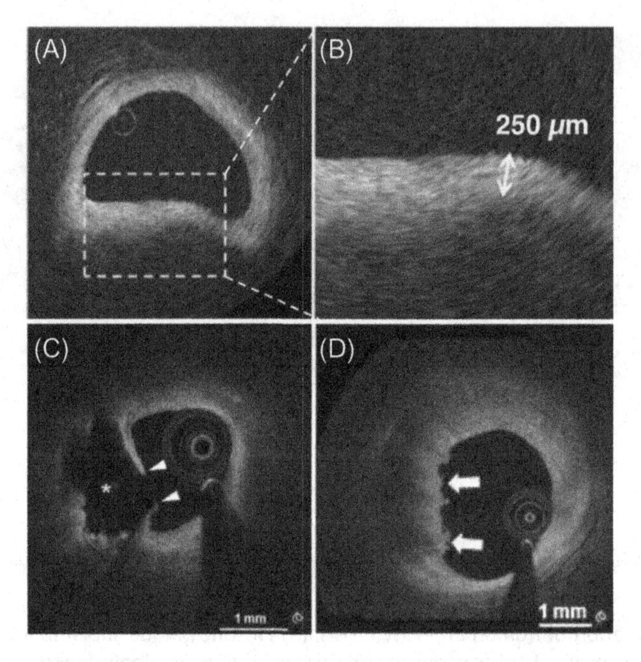

Fig. 8.4 (A and B) Example of fibroatheroma and fibrotic cap measurements. Cap minimal thickness is quantified to be approximately 250 μm. (C) Example of plaque rupture (*arrows*) and (D) plaque erosion and mural thrombi (*arrows*). *((A and B) Modified from Bezerra HG, Costa MA, Guagliumi G, Rollins AM, Simon DI. Intracoronary optical coherence tomography: a comprehensive review clinical and research applications. JACC Cardiovasc Interv 2009;2:1035–46 and (C and D) modified from Jia H, Abtahian F, Aguirre AD, Lee S, Chia S, Lowe H, et al. In vivo diagnosis of plaque erosion and calcified nodule in patients with acute coronary syndrome by intravascular optical coherence tomography. J Am Coll Cardiol 2013;62:1748–58. All images are reproduced with permission from the publisher.)*

macrophages yield strong optical signals, generating elevated OCT signal attenuation and backscattering. Studies suggested that OCT can determine the presence of macrophages in ACS patients [51]; however, other vessel wall components may generate similar features and it has been suggested that the presence of macrophages should only be analyzed in atheromas fibrotic caps [52]. Nevertheless, an extensive validation of algorithms for macrophage quantification using commercially available systems still remains to be conducted and the ability of OCT to reliably assess the presence of macrophages is currently subject of debate [52–54].

2.3 Assessment of Intracoronary Stent

With an axial resolution of ~15 μm, 10 times better than IVUS (resolution around 100 μm), OCT has rapidly become the gold standard for the assessment of intracoronary stents in the immediate phase after stent implantation, as well as during any phase of the healing process.

In clinical practice, it is important to perform a structured and comprehensive assessment of an OCT acquisition of a stented segment of a coronary artery. The analysis of the edge segments is of paramount importance for determination of reference areas and detection of possible edge dissections during the initial stent implantation phase. In a later phase after stent implantation, edge segments are important for detection of edge in-stent restenosis, plaque overgrowth and for the detection of vulnerable plaque rupture in close proximity to the stent edges as a potential cause of stent thrombosis.

With respect to the in-stent segment analysis, stent underexpansion is the most important parameter to be analyzed, with respect to potential PCI optimization measures. It has been proposed that stent areas provided by OCT measurements tend to deliver slightly smaller values compared to IVUS measurements at the same location in the vessel. The most plausible explanation for this is a better and more reliable delineation of lumen contour when using OCT, especially in difficult anatomical settings, such as an important burden of calcium in the vicinity of the stent contour [55].

At the level of the individual strut, apposition and coverage are the two most important characteristics. For detailed definitions, we refer to the manuscript of the Consensus Standards for Acquisition, Measurement, and Reporting of Intravascular Optical Coherence Tomography Studies [15]. Malapposition of stent struts is considered to be present when the axial distance between the strut's surface to the luminal surface is greater than the strut thickness (including polymer, if present). If the distance is less than the strut thickness, the strut is considered apposed. Struts are termed covered by OCT if tissue can be identified above the struts (Fig. 8.5). They are considered uncovered if no evidence of tissue can be visualized above the struts [15]. While the discriminatory capacity of OCT for detection of coverage above struts is to be considered excellent, the assessment might be more difficult in specific situations, such as in the case of struts with important blooming artifact. Moreover, some struts are covered by thin layers of thrombus or fibrin and it is not always possible to discriminate this tissue from a healthy neointimal layer [56] (Fig. 8.6).

The assessment of healing aspects of intracoronary stents attracted a lot of attention after an increased risk of very late stent thrombosis, related to delayed healing and long-term persistence of uncovered struts was observed with first generation drug-eluting stents [57–59]. The availability of OCT for application in human coronary arteries in clinical practice allowed for in vivo assessment of the underlying causes of stent thrombosis and boosted research projects investigating newer designs of DES [60, 61].

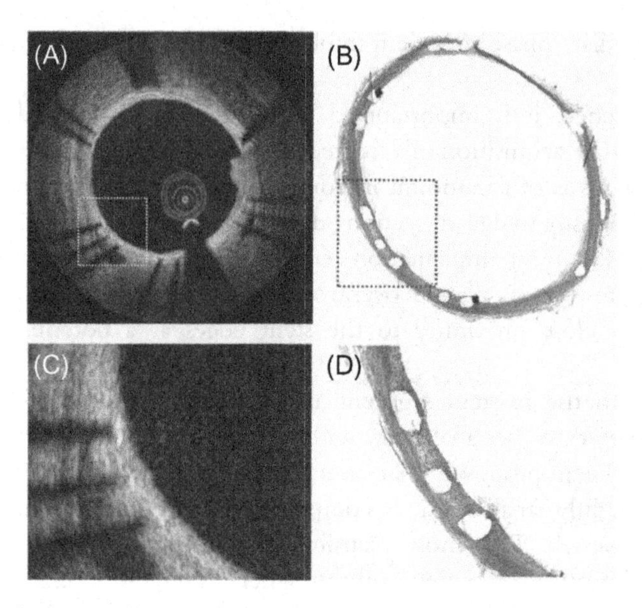

Fig. 8.5 Example of OCT visualization of stent coverage compared to histology. (A and B) OCT image and corresponding histological session. (C and D) Strut-level registration between the two modalities, illustrating the ability of OCT to accurately assess the thin layer of neointimal tissue covering the struts. *(The image is reproduced from Ughi GJ, Van Dyck CJ, Adriaenssens T, Hoymans VY, Sinnaeve P, Timmermans JP, et al. Automatic assessment of stent neointimal coverage by intravascular optical coherence tomography. Eur Heart J Cardiovasc Imag 2014;15:195–200, with permission from the publisher.)*

2.3.1 Bioresorbable Vascular Scaffolds

Inherent to their design and composition of poly-L-lactic acid (PLLA), bioresorbable vascular scaffolds (BVS) have an OCT appearance with is completely different from their metallic counterparts (Fig. 8.7E and F). Struts appear as boxes, with no retro optic shadow and gradually resorb over the course of 3 years, leaving no trace of the initial scaffold behind [62].

2.3.2 Assessment of Stent Failure

Until recently, description of mechanisms of stent failure leading to stent thrombosis (Fig. 8.8), based on OCT imaging, had been limited by small patient numbers. Over the course of the last year, results of larger patient number registries of patients presenting with ST assessed with OCT have been presented [63, 64]. These studies highlight the importance of stent underexpansion and malapposition as important mechanisms or contributors to the development of stent thrombosis in the early phase after initial PCI. When stent thrombosis occurs in a later phase after stent implantation, the most important mechanisms appear to be the persistence of uncovered struts and

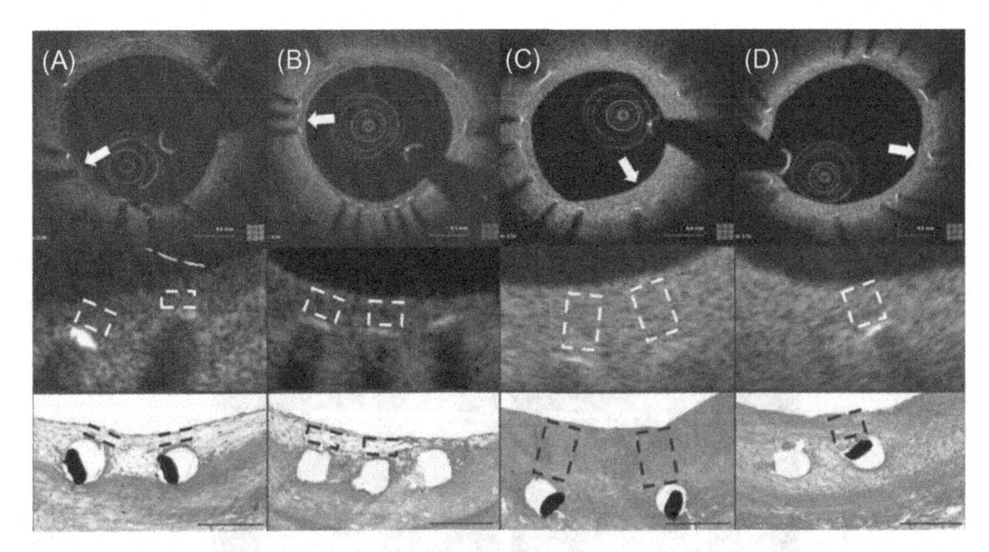

Fig. 8.6 OCT assessment of neointimal tissue maturity. (A and B) Examples of immature neointimal tissue (lack of smooth muscle cell and presence of fibrin and extracellular matrix) assessed by OCT and vessel histology. If compared to mature neointimal tissue (C and D), immature tissue shows a lower backscattering coefficient. This allows assess not only the presence of neointimal coverage, but also the composition of covering tissue providing information about the stent healing process. However, such an analysis may be challenging when struts are covered by a thin layer of tissue. *(This image is reproduced from Malle C, Tada T, Steigerwald K, Ughi GJ, Schuster T, Nakano M, et al. Tissue characterization after drug-eluting stent implantation using optical coherence tomography. Arterioscler Thromb Vasc Biol 2013;33:1376–83, with written permission of the publisher.)*

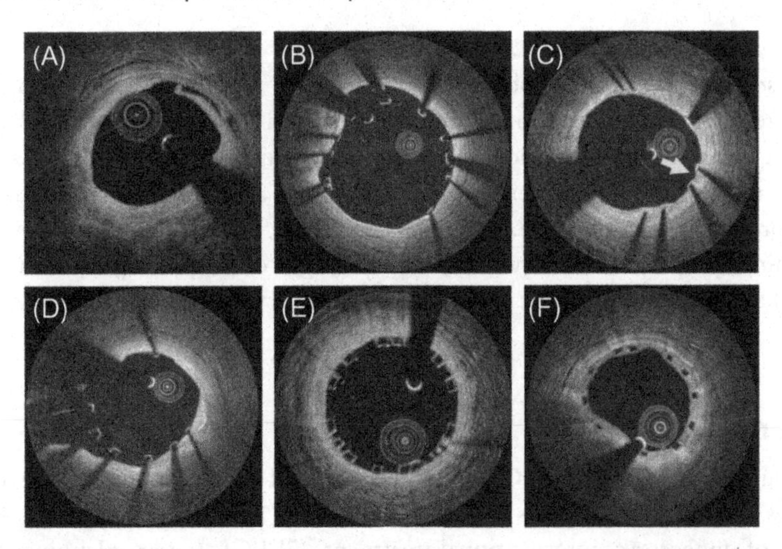

Fig. 8.7 Examples of stent assessment by OCT. (A) Visualization of stent proximal edge dissection. (B) Strut malapposition (11–1 o'clock) due to calcific plaque and lack of conformability. (C) Example of lack of neointimal coverage at follow-up examination (*arrow*) and (D) stent jailing a large side branch. (E) Visualization of BVS at baseline and (F) BVS stent healing at 3-month follow-up showing thin amount of neointimal coverage.

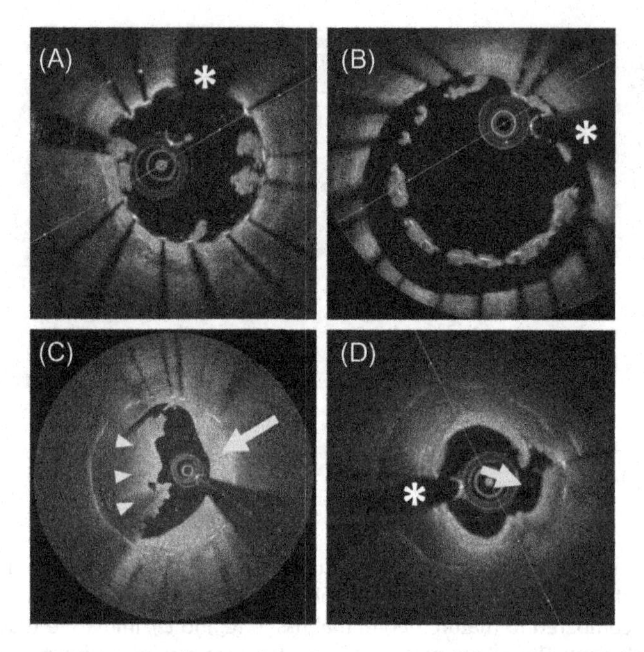

Fig. 8.8 Underlying mechanisms of stent thrombosis. (A) Accumulation of *white* thrombus above uncovered struts. (B) Severe malapposition and underexpansion with thrombus adherent to struts. (C) Stent thrombosis (*) and restenosis showing thick fibrotic tissue covering the stent (*arrow*). (D) Neoatherosclerosis with rupture (rupture site indicated by *arrow*). The *asterisk* is indicating the guide wire artifact.

neoatherosclerosis. This latter phenomenon was first recognized 5 years ago and is characterized histologically by an accumulation of lipid-laden foamy macrophages with or without necrotic core formation and/or calcification within the neointima [65, 66]. On OCT images, neoatherosclerosis is recognized as a region of lipid accumulation, with important signal attenuation, covered by a fibrous cap of variable thickness. In some cases, a thin cap fibroatheroma (with or without signs of plaque rupture and overlying thrombus) and/or clusters of macrophages can be observed (Fig. 8.8D).

2.4 OCT-Guided PCI

With its unsurpassed detail, OCT is especially well suited to guide intracoronary procedures. Three main aspects can be described: (1) the indication for PCI and avoidance of unnecessary interventional treatment, (2) PCI strategy planning including the delineation of the segment to be treated and the choice of adequate stent size and length, and (3) optimization of the result after stent placement.

Optimal stent implantation has been reported to be associated with a lower incidence of restenosis and stent thrombosis, and may affect favorably clinical outcomes [67, 68]. By intravascular imaging, the most important determinant of early stent thrombosis and restenosis after stent implantation is the minimum stent area (MSA) achieved [67]. In line with studies assessing the safety of OCT acquisition, the use of OCT in the guidance of PCI in studies like CLI-OPCI or ILUMIEN I has not been associated with an increase in procedural complications [69, 70]. It seems intuitive that an OCT-guided PCI, with further optimization measures performed based on OCT acquisitions, goes along with a small increase in procedural duration and fluoroscopy times, as documented in the ILUMIEN I study. With respect to the total amount of contrast used, there are some conflicting results between different studies assessing this issue [69, 70]. Generally speaking, the operator should spend maximal effort to minimize contrast exposure by acquiring poststent angiographic cine image during FD-OCT pullback with a single contrast injection.

The value of the guidance of PCI based on intracoronary imaging has been established earlier in the IVUS era [71]. Recently, the ILUMIEN II study showed that OCT and IVUS guidance result in a comparable degree of stent expansion. In a study by Kim et al., OCT guidance achieved comparable results to IVUS in mid-term clinical outcomes, suggesting that OCT can be an alternative tool for stent placement optimization [72]. From a practical point of view, for most operators, OCT provides more accurate and faster delineation of the lumen or stent border, thereby making it more suitable for online use in the cardiac catheterization laboratory [55]. Pending larger clinical trials establishing firm criteria for OCT guidance, different strategies are currently being used in clinical practice. With respect to stent expansion, we propose an MSA of 90% compared to the mean of proximal and distal reference areas to be achieved. In longer stented segments, especially in tapering vessels such as the LAD, it is advisable to analyze separately the proximal and the distal half of the stented segment, as was done in the currently ongoing ILUMIEN III trial. The question of what degree of acute incomplete stent apposition (ISA) that is clinically worth correcting is still largely undetermined. Gutiérrez-Chico et al. demonstrated that all ISA sites with an acute maximal ISA distance <270 μm were reapposed at follow-up [73]. In the OCTACS study, lesions with an acute maximal ISA distance of <335 μm were resolved at follow-up [74]. As a general rule, with respect to malapposition, a burden of cluster of malapposition, comprising multiple struts with >250–300 μm of malapposition distance in a frame and extending for several frames, is probably worth to be corrected by further balloon postdilation.

Beyond the obvious beneficial effects of PCI guidance (correction/reduction of underexpansion and malapposition, avoidance and treatment of residual stenosis and edge dissections), there is a potential advantage of reduction of periprocedural myocardial OCT-guided optimization of stent positioning, avoiding diseased landing zones [75].

2.5 Image Artifacts and Other Limitations

As any imaging modality, OCT can be affected by artifacts that may complicate interpretation. A typical artifact is due to incomplete blood displacement (residual blood) in the vessel lumen, resulting in signal attenuation (Fig. 8.9A and B). In some cases, when suboptimal vessel flushing occurs but blood is sufficiently diluted with injected contrast or saline solution, the artery wall is still visible and a correct image interpretation is possible. However, in some other cases, an excessive amount of blood in the lumen may significantly scatter the signal from the artery and a second acquisition is typically required to achieve an OCT datasets of adequate quality.

In the vast majority of cases, OCT imaging is performed having the catheter in an eccentric position with respect to the vessel wall. Care is recommended to assess plaque

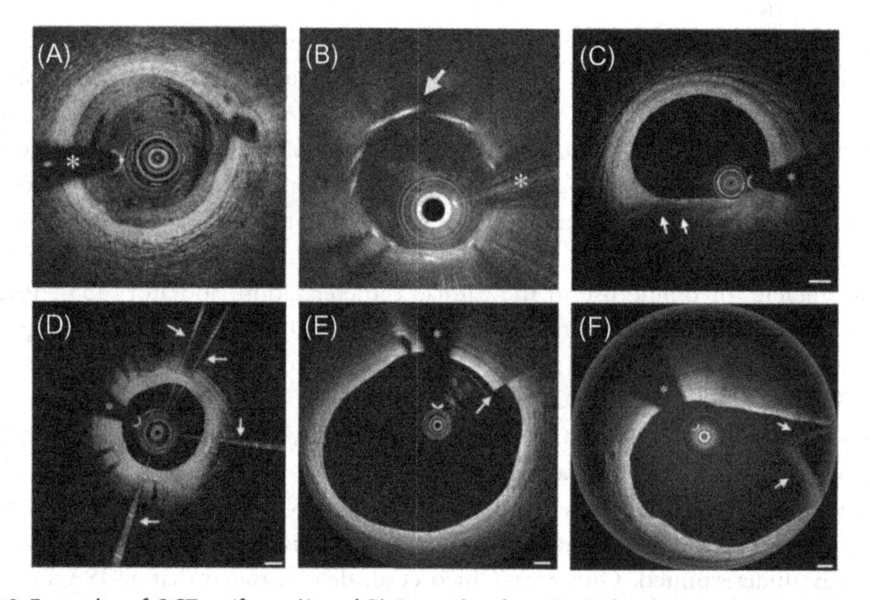

Fig. 8.9 Examples of OCT artifacts. (A and B) Example of residual blood in the lumen. Although in the first case blood is not shadowing the tissue behind, in the second case image quality dropped significantly (both signal drop and multiple scattering artifacts are visible) and a correct analysis of stent coverage is unfeasible. (C) Tangential signal dropout, creating a false TCFA image appearance. (D) Saturation artifact over stent strut luminal surface and (E) sew-up artifact due to catheter and vessel relative motion. (F) Part of the vessel wall is outside the catheter field of view, generating a fold over artifact (*arrows*). *(This image reproduced from Tearney GJ, Regar E, Akasaka T, Adriaenssens T, Barlis P, Bezerra HG, et al. Consensus standards for acquisition, measurement, and reporting of intravascular optical coherence tomography studies: a report from the international working group for intravascular optical coherence tomography standardization and validation. J Am Coll Cardiol 2012;59:1058–72, with permission from the publisher.)*

compositions in these situations as the optical beam may be directed nearly parallel to the tissue surface. When this happens, a signal poor region below the luminal surface may be seen and misinterpreted as a lipid-rich plaque or a TCFA (Fig. 8.9C). Similarly, when the catheter lies against the vessel wall being in an eccentric position, stent struts appear being oriented toward the catheter.

Similar to IVUS, OCT has no frame of reference for the catheter spatial orientation with respect to the vessel wall, making it difficult to register and compare to other noninvasive imaging modalities used in the clinical catheterization laboratory, such as X-ray and CT angiography. Additionally, when the catheter is not parallel to the longitudinal axis of the vessel wall (i.e., the optical beam is not at a 90 degree angle with respect to the artery wall), the image may undergo elliptical distortion (vessel curvature artifact), making OCT measurements less accurate. This effect has been often observed when measuring area and diameter of the ostium of the left-main artery, large branches and vessel diagonals.

When a strong reflector is encountered (e.g., metallic stent struts), the OCT signal can be reflected at very high intensity, causing detection artifacts called "saturation" effect [15, 39]. This phenomenon appears as a fully saturated A-line, sometimes including multiple harmonic reflections due to nonlinear effects of the photodiode (Fig. 8.9D). This effect needs to be taken into account when measuring stent apposition and thickness of neointimal tissue over struts.

The motion of the imaging catheter with respect to the artery can result in multiple artifacts and it is generally related to heart beating during OCT acquisition. Both axial and longitudinal motions may be observed in a single dataset and may significantly affect three-dimensional (3D) morphology assessment. Longitudinal motion can disrupt the continuity of the pullback resulting in the repeated appearances of the same stent portion and vessel wall features. Axial motion can cause a discontinuity between A-lines within a cross-sectional image ("seam line" or sew-up artifact) (Fig. 8.9E). Moreover, a larger lumen area can be observed during diastole as the vessel changes its dimensions during a cardiac cycle. If compared to older TD-OCT imaging system, the higher speed of FD-OCT reduces motion artifacts as the entire dataset can be acquired within a reduced number of cardiac cycles.

Nonuniform rotational distortion (NURD) is an artifact that can occur in tortuous anatomy or when the catheter is placed through a narrow stenosis. It is a consequence of mechanical friction along the catheter torque-cable that may cause rotation at variable speed within a single rotation, causing distortion in cross-sectional images. Another common artifact encountered by FD-OCT systems is called the "foldover" effect (Fig. 8.9F). Such an artifact happens when the vessel is larger than the system imaging range causing a portion of the vessel to appear to fold over in the image.

Finally, several additional effects need to be considered when assessing vessel wall pathology. The presence of a (thin-cap) fibroatheroma is determined observing signal attenuation, however, other components such as cluster of macrophages, proteoglycans, loose connective tissue, and cellular fibrous tissue may present a similar appearance [53, 76]. Furthermore, no study was able to show that OCT can distinguish between lipid accumulations in pathological intimal thickening conditions and necrotic cores [15]. Nevertheless, the accuracy of OCT to assess presence of lipid has been extensively validated showing elevated specificity and sensitivity [38]. However, care is required when analyzing complex and heterogeneous plaques, as other tissue components can mislead image interpretation presenting similar appearance of TCFA, lipid, and necrotic cores (Fig. 8.10).

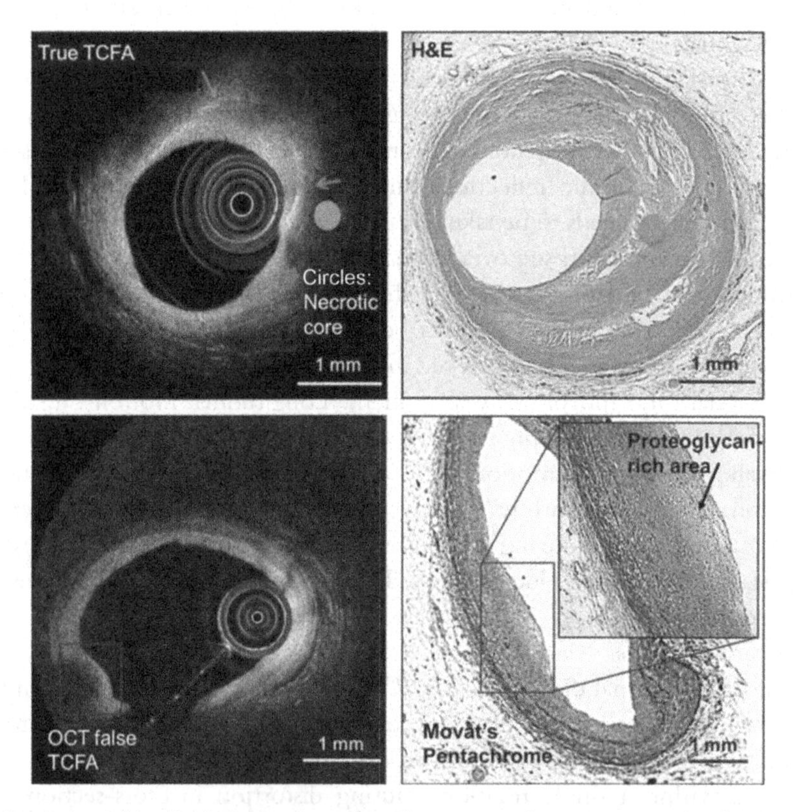

Fig. 8.10 Example of false lipid plaque detection. The *top row* shows an example of true TCFA (*circle* correspond to a necrotic core). The *second row* shows an example of false TCFA detection, due to a proteoglycan-rich area causing a rapid signal drop-off (elevated μ_a). *(This image reproduced from Imanaka T, Hao H, Fujii K, Shibuya M, Fukunaga M, Miki K, et al. Analysis of atherosclerosis plaques by measuring attenuation coefficients in optical coherence tomography: thin-cap fibroatheroma or foam cells accumulation without necrotic core? Eur Heart J 2013;34:P5482, with permission from the publisher.)*

3. OCT CLINICAL RESEARCH

3.1 Clinical Trials

Over the last decade, the adoption of OCT in clinically oriented research has been overwhelming. In a first phase, immediately following the important concerns on the increased risk of very late stent thrombosis with first-generation drug-eluting stents, there was important interest in trials comparing long-term healing characteristics of these devices compared to bare metal stents on one hand, and second-generation drug-eluting stents on the other hand. The latter generation of devices had been specifically designed to eliminate certain characteristics, associated with bad healing properties, of the earlier generation stents. In parallel, trials were run to evaluate differences in healing process according to specific lesion types (e.g., bifurcation, CTO, and in-stent restenotic lesions). As one of the potential responsible factors for delayed healing in DES has been the polymer coating on stent struts, several drug-eluting stents with biodegradable polymer coating have been designed, and subjected to comparative analysis with durable polymer DES using OCT as well [77].

The role of OCT has been even more fundamental in the scientific development and the introduction of fully BVS into clinical practice [62]. With a much more complex healing process, leading to full absorption of the device over the course of 3 years after implantation, OCT has been an integral part of almost any trial assessing BVS. As it is appreciated that careful lesion selection, preparation, and optimization after device implantation are even more crucial with BRS compared to their metallic counterparts, it is to be anticipated that trials focusing on OCT-guided BRS implantation will attract a lot of attention over the course of the next years.

The unsurpassed detail of the images and the ease of use and safety of the procedure, together with an explosion of scientific reports and use of OCT during live case transmissions at international congresses, have gradually convinced the great majority of interventional cardiologists of the important value of OCT in the catheterization laboratory. In daily cath laboratory work, OCT is now mainly used for several indications: (1) augmentation of diagnostic accuracy in the case of ambiguity on coronary angiography; (2) planning of PCI strategy and optimization of PCI result; and (3) assessment and guidance of treatment in the case of stent failure.

Especially in the field of optimization of PCI, several efforts aiming at providing robust evidence supporting the use of OCT, have been spent over the course of the last years. The CLI-OPCI study was a nonrandomized study where 335 patients undergoing OCT guided were matched with 335 patients undergoing angiography only guidance [69]. Angiography plus OCT guidance was associated with a significantly lower risk of cardiac death or MI even at multivariable analysis adjusting for baseline and procedural differences between the groups. In the CLI-OPCI II study, a further analysis of 1002 lesions in 832 patients, suboptimal stent deployment (such as MSA < 4.5 mm [2], dissection > 200 μm at the distal stent edge, reference lumen area < 4.5 mm [2] at either distal or proximal stent edges) was associated with an increased risk of MACE during follow-up [78].

In the OPINION (optical frequency domain imaging vs. IVUS in PCI) study, a prospective, multicenter randomized controlled trial, OCT guidance of PCI appeared to be noninferior compared to IVUS guidance PCI with respect to clinical and angiographic outcomes at 1 year [79]. Finally, the ILUMIEN series of studies aims at assessing the role of OCT in optimizing PCI outcomes. In the ILUMIEN I study, a nonrandomized observational study, physician decision making was affected by OCT imaging prior to PCI in 57% and post-PCI in 27% of all cases. Additional interventions included predominantly in-stent balloon postdilations and—to a lesser extent—additional stent implantation [70].

The ILUMIEN II study sought to determine whether OCT guidance results in a degree of stent expansion comparable to what is obtained with IVUS guidance. To that purpose, the relative degree of stent expansion after OCT-guided stenting in patients in the ILUMIEN I study and IVUS-guided stenting in patients in the ADAPT-DES study were compared. In the matched-pair analysis, as well as after adjustment for baseline differences, the degree of stent expansion was not different between the two imaging modalities [80]. In the currently ongoing ILUMIEN III study, OCT-guided PCI is randomly compared with an IVUS guided and an angiography alone guided strategy, with a post-PCI MSA as assessed by OCT as the primary endpoint.

4. OCT IMAGE PROCESSING

4.1 Segmentation

Multiple methods have been proposed for the automated analysis of OCT data. Multiple groups proposed solutions for the segmentation of the vessel lumen [81, 82, 84] and automated lumen tracing is now available on commercial systems. Image binarization techniques, morphological image processing, and vessel spatial continuity are common procedures used by the different methods. Those algorithms can automatically determine the lumen area through an entire dataset and locate vessel stenosis with respect to a reference frame, facilitating a rapid analysis during coronary interventions. Such methods can also be complemented by automated side branches detection and segmentation. Additionally, lumen morphology can be used in combination with 3D X-ray angiography for the quantification of endothelial shear stress (ESS) [85, 86].

Several methods for the automatic segmentation of stents have also been proposed [81, 82, 84, 87–89] with the aim to identify and segment the luminal surface of metallic struts. Various techniques have been described including individual A-line analysis and classification, and assessment of 2D image features, including the presence of a shadow and bright reflections. If used in combination with lumen contour, stent

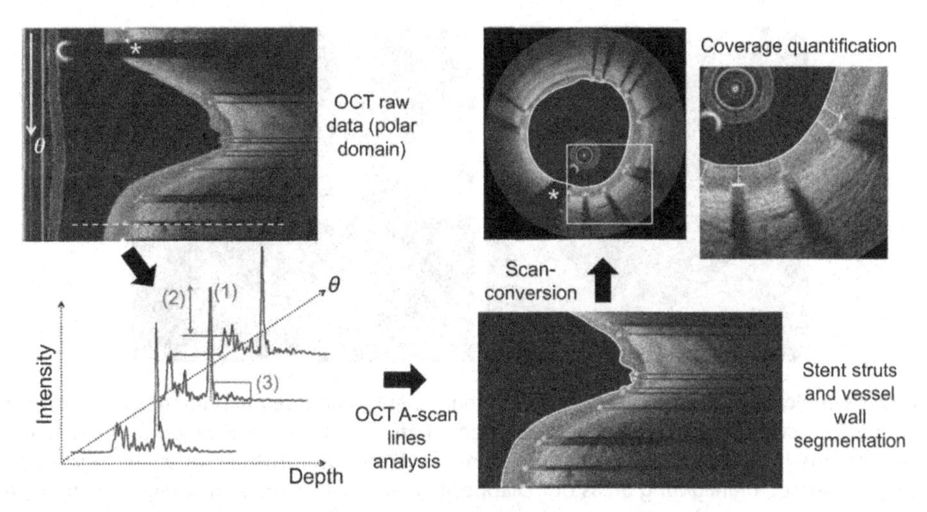

Fig. 8.11 Algorithm for automatic segmentation and quantification of stent coverage. Polar domain images (angle vs. depth) are analyzed by identifying A-scan lines containing struts. Classification is obtained using properties relative to the presence of a shadow and struts bright reflections. Spatial continuity is applied to generate a smooth vessel and stent contour and, following scan conversion, stent strut coverage can automatically be quantified. A similar approach can be used for automated assessment of stent malapposition. *(This image is reproduced from Ughi GJ, Van Dyck CJ, Adriaenssens T, Hoymans VY, Sinnaeve P, Timmermans JP, et al. Automatic assessment of stent neointimal coverage by intravascular optical coherence tomography. Eur Heart J Cardiovasc Imag 2014;15:195–200, with permission from the publisher.)*

segmentation enables the automatic quantification of strut apposition and neointimal coverage (Figs. 8.11 and 8.12). Additionally, strut detection is used to generate detailed 3D OCT reconstructions, highlighting stent apposition in an entire coronary segment [90] and to register coronary datasets acquired at different time points [91, 92]. Furthermore, this methodology has been used to visualize in 3D the struts covering the ostium of a side-branch, to identify the optimal stent cell for branch rewiring, and optimize bifurcation stenting techniques in the clinic [91, 93, 94]. Recently, the feasibility of automated OCT segmentation of BVS has also been demonstrated [95].

4.2 Automated Tissue Analysis

As discussed in the previous section, OCT can differentiate atherosclerotic tissue type on the basis of tissue optical properties in the NIR. It has been demonstrated that optical attenuation and backscattering can be used for the characterization of plaque components [35]. Automatic quantification of tissue optical attenuation (μ_a) was proposed to identify high-risk features in the vessel wall [36, 40]. Ex vivo validation showed the ability of attenuation analysis to detect lipid/necrotic cores,

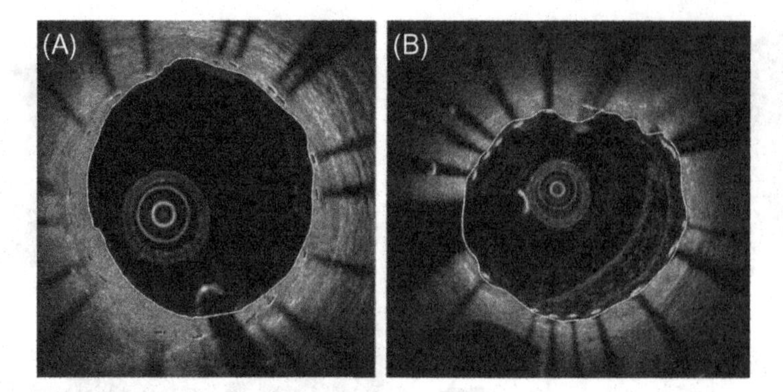

Fig. 8.12 Examples of automatic stent segmentation. (A) Automated segmentation of a stent covered by neointimal tissue. (B) Stent and vessel wall segmentation in presence of blood in the lumen and uncovered struts (i.e., freshly implanted stent). This method can perform a fully automatic segmentation of an entire dataset, highlighting areas of malapposition and uncovered struts, segment stent struts overlying a side-branch and to obtain detailed 3D reconstruction that may provide important information when performing bifurcation PCI. *(This image is reproduced from Ughi GJ, Adriaenssens T, Onsea K, Kayaert P, Dubois C, Sinnaeve P, et al. Automatic segmentation of in-vivo intra-coronary optical coherence tomography images to assess stent strut apposition and coverage. Int J Cardiovasc Imag 2012;28:229–41, with written permission from the publisher.)*

macrophage accumulations, and other components with elevated attenuation (Fig. 8.13) [40, 96]. The use of optical attenuation has been further expanded by following studies, combining optical attenuation with statistical image features (as an "indirect" measurement of tissue backscattering properties) to differentiate plaque components (Fig. 8.14). A supervised classification scheme has been successfully applied using optical attenuation, image statistics (e.g., cooccurrence matrix) and geometrical features (e.g., the distance between the imaging catheter and the vessel wall) as classification features, achieving an automated differentiation of fibrotic, calcified, and lipid tissue [41].

Additional methods for the semiautomatic segmentation of fibrocalcific plaques [97], fibrous cap thickness [98, 99], morphological image analysis (i.e., to assess automatically the thickness of vessel layers to differentiate between normal vs. abnormal vessel wall), and methods for macrophage detection (based on image statistical features) [51, 54] have been proposed. Nevertheless, automatic OCT tissue analysis still remains a difficult task and clinical use has rarely being reported. Due to such limitations, multimodality approaches (i.e., the combination of OCT with other spectroscopic imaging techniques) have been proposed to complement the ability of OCT to assess high-risk plaque components, adding biological and molecular information about the progression of atherosclerotic disease.

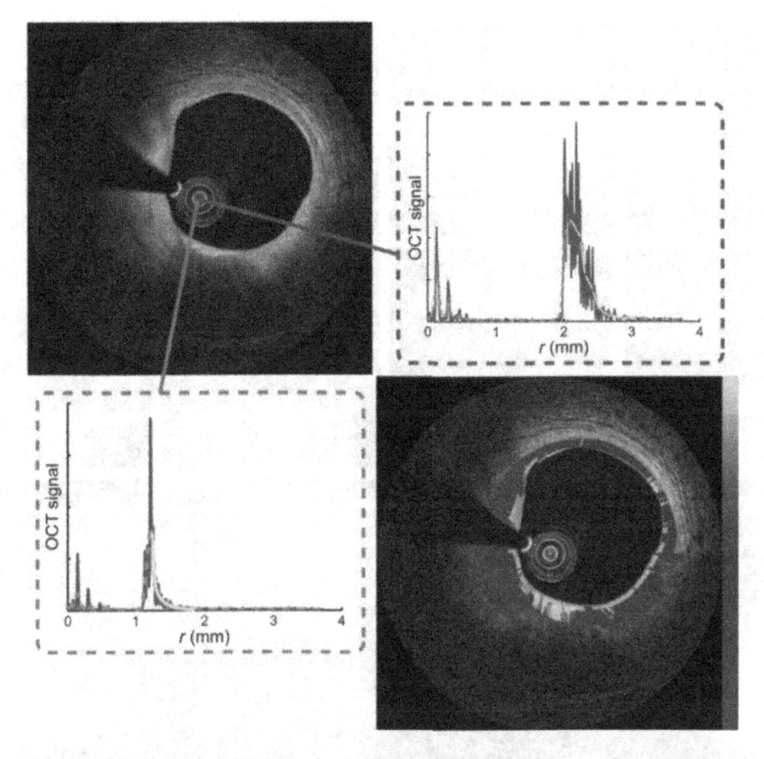

Fig. 8.13 Illustration of OCT attenuation analysis. OCT intensity profiles are first preprocessed to reduce speckle noise by means of a technique called spectral binning or by spatial filtering. Subsequently a multilayer exponential decay model is fitted through the data to estimate automatically the attenuation coefficient. *Bottom right image* shows an example of attenuation imaging results. An area of high attenuation is visible between 4 and 7 o'clock. *(This image is reproduced from White S, van Soest G, Johnson TW. Development of tissue characterization using optical coherence tomography for defining coronary plaque morphology and the vascular responses after coronary stent implantation. Curr Cardiovasc Imag Rep 2014;7:1–10, with written permission of the publisher.)*

4.3 Real-Time OCT Data Analysis

Automated algorithms can be applied: (1) for the online processing of data as a tool to guide PCI or (2) for offline analysis in core-lab settings, to analyze data that are part of research studies and clinical trials. For online applications, algorithm requirements are more stringent, as the data need to be analyzed efficiently with a reduced processing time (i.e., few seconds) to be of clinical utility. To date, automatic segmentation and quantification of vessel lumen area and diameter are available in commercial OCT system. Such tools can be used online, to visualize lumen morphology efficiently in

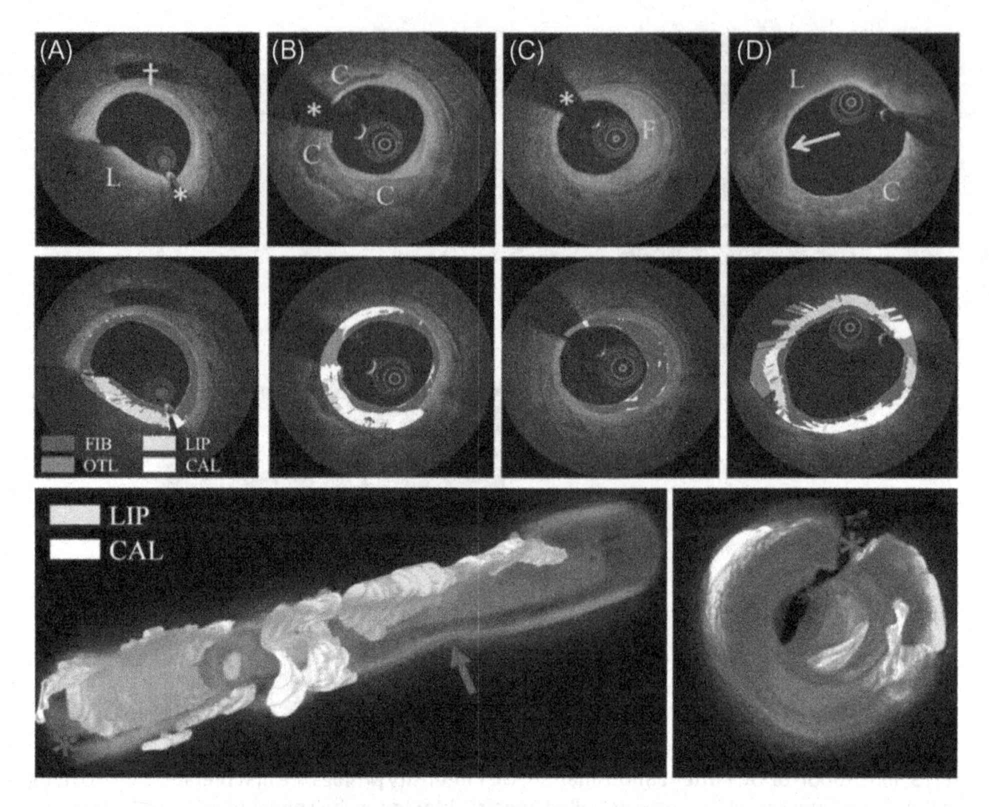

Fig. 8.14 Example of automatic tissue classification. (A) An example of lipid plaque characterization, (B) calcium, (C) fibrotic plaque, and (D) mixed plaque (OTL = outliers, corresponding to tissue that cannot reliably be classify). Three-dimensional OCT rendering illustrates how tissue analysis can provide volumetric information about plaque distribution in the vessel. *(This image is modified from Ughi GJ, Adriaenssens T, Sinnaeve P, Desmet W, D'hooge J. Automated tissue characterization of in vivo atherosclerotic plaques by intravascular optical coherence tomography images. Biomed Opt Express 2013;4:1014–30.)*

a comprehensive way, helping the user to identify vessel reference segment and stenosis. Recently, automatic segmentation and 3D visualization of vessel and stent, including stent strut malapposition, were made available in commercial systems to provide guidance during coronary intervention. Similarly, tools for the registration of OCT and X-ray angiography were made available and the registration between the two modalities can be rapidly obtained online through a semi-automated framework. Currently, tools for the quantification of stent coverage and plaque segmentation/classification are not available in commercial systems. Algorithm prototypes for stent coverage assessment have been proposed [87] and may became available in the catheterization laboratory in a near-future.

5. FUTURE OUTLOOK

5.1 High-Speed OCT

Commercial OCT systems acquire pullbacks within 2–4 cardiac cycles and recorded datasets may be affected by motion artifacts, limiting the accuracy of OCT to depict the true lumen morphology. "High-speed" FD-OCT has been proposed to perform a full acquisition in less than a cardiac cycle (e.g., 0.5 s), minimizing artifacts generated by cardiac motion. In a recent study, a 500 frames/s system and catheter was used to perform ECG triggered acquisitions in an animal model in vivo, performing imaging during a single late diastolic phase (i.e., avoiding acquisition during QRS and T-wave), resulting in "motion-free" OCT datasets [100] (Fig. 8.15). A second study proposed the use of a system with an order of magnitude improved imaging speed, acquiring OCT data at an A-line speed of 2.88 MHz. The design of such a system included the use of a Fourier-domain mode-locked (FDML) sweep laser in combination with an imaging catheter featuring a 1-mm outer diameter micro-motor, located distally to the optical imaging element [101]. In this study, ECG triggered datasets have been acquired in vivo in swine coronaries at a speed of 4000 frames/s at 100 mm/s pullback, demonstrating the capability of such a technique (named by the authors "Heartbeat OCT") to acquire motion-free OCT datasets [101, 102]. The use of such techniques ensures the acquisition of 3D volumes free of distortions in both longitudinal and axial directions, providing an anatomically corrected 3D visualization of human coronary arteries in vivo.

5.2 Polarization Sensitive OCT

Polarization sensitive OCT (PS-OCT) is an extension of OCT imaging technology, aiming to provide information about light polarization properties of a sample in addition to reflectivity [103, 104]. PS-OCT allows the quantification of the birefringence magnitude and orientation in a depth-resolved fashion. Frequency-multiplexing approaches have been proposed to acquire high-speed intravascular birefringence images through a fiber-optic catheter [105]. Frequency multiplexing is used to measure reflectance of two incident polarization states simultaneously and ignore temporal variations of catheter fiber birefringence. Additionally, approaches to limit the effect of polarization mode dispersion [106] below the axial resolution of the imaging system have been proposed, allowing for reproducible measurements of sample birefringence [107].

Preclinical ex vivo studies suggested that PS-OCT can be used to assess collagen content and smooth cell density in atherosclerotic plaques by means of local retardation maps [108] (Fig. 8.16). Additionally, immunohistochemistry validation of PS-OCT is ongoing investigating the dependence of PS-OCT with respect to plaque collagen quantity and fiber architecture [109]. First-in-man intravascular PS-OCT studies are in progress using clinically approved experimental imaging system and catheters [110, 111],

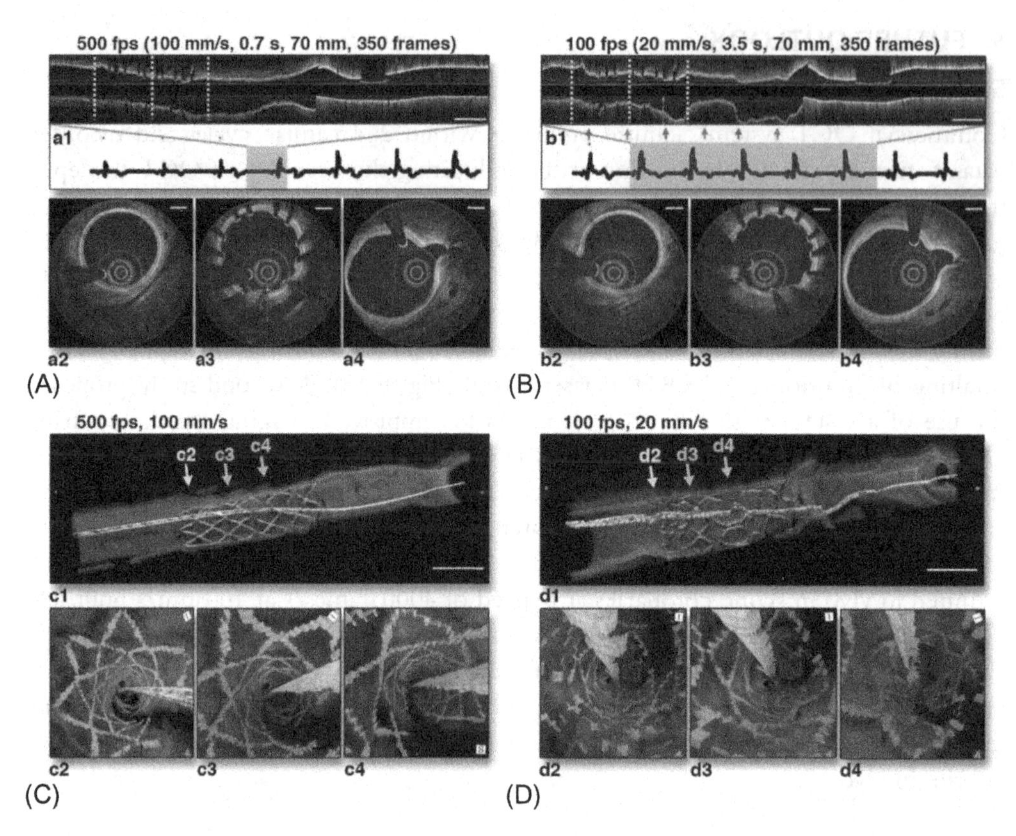

Fig. 8.15 High-speed, ECG triggered OCT acquisitions. Comparison of 0.7 s pullback during diastole (A) versus a standard 100 frames/s, 3.5 s pullback (B). OCT cross-sectional images (b2, b3, and b4) show artifacts due to heart beating during image acquisition in the case of the "slower" pullback. Moreover, from the OCT 3D renderings, it is possible to appreciate that a quick pullback during diastole can provide a much more accurate 3D visualization (C) if compared to (D) where multiple motion artifacts in both the axial and longitudinal directions are visible. *(The image is reproduced from Jang SJ, Park HS, Song JW, Kim TS, Cho HS, Kim S, et al. ECG-triggered, single cardiac cycle, high-speed, 3D, intracoronary OCT. JACC Cardiovasc Imag 2016;9:623–5, with written permission from the publisher.)*

to demonstrate feasibility and potential benefits of clinical use of tissue polarization measurements. Preliminary results suggested that PS-OCT may complement OCT with additional information about coronary plaques prone to rupture, assessing fibroatheroma cap birefringence and lack of collagen fibers (i.e., low birefringence) as potential indicator of plaque instability. Additionally, the use of PS-OCT has been suggested as a tool to investigate stent healing in vivo. Nevertheless, extensive ex vivo validation and future clinical studies including large number of patients are required to assess clinical utility of polarization measurement extension of intravascular OCT.

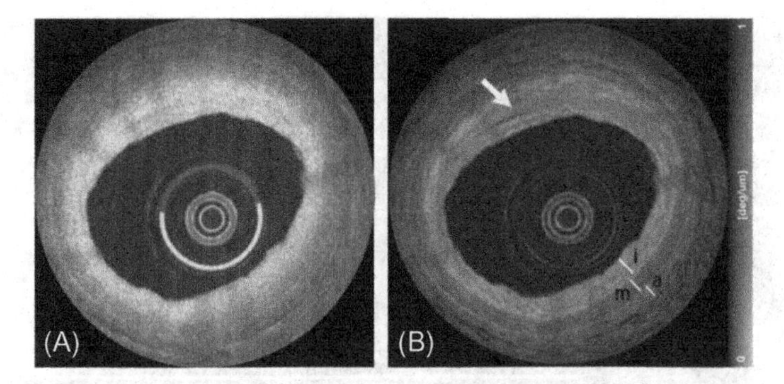

Fig. 8.16 Polarization sensitive OCT (PS-OCT). The *red* arc in image (A) indicates a fibrotic plaque. (B) Intensity OCT image overlaid with local retardation map. The *arrow* points to the region of increased light birefringence within the fibrotic plaque. Additionally, birefringence contrast helps in the visualization of the vessel layered architecture (i.e., intima, media, and adventitia). *(This figure is reproduced with permission from Villiger M, Zhang EZ, Nadkarni SK, Oh WY, Vakoc BJ, Bouma BE. Spectral binning for mitigation of polarization mode dispersion artifacts in catheter-based optical frequency domain imaging. Opt Express 2013;21:16353–69.)*

5.3 Multimodality OCT Imaging

As described in the previous sections, OCT is a high-resolution imaging technique for the visualization of vessel wall microstructure, lumen morphology, and implanted coronary devices (e.g., intracoronary stent) in vivo. OCT significantly contributed in the advancement of PCI techniques and is currently used to guide intervention and assess safety and efficacy of novel therapies, devices, and interventions. Although OCT visualizes plaques providing morphological information and some insights about composition, atherosclerosis includes complex interactions between structural, chemical, biomechanical, biological, and molecular processes happening in the vessel wall [112]. In this section, we will discuss how OCT can be complemented with other modalities to provide currently missing information about vessel wall atherosclerotic processes.

5.3.1 Endothelial Shear Stress

The use of intracoronary imaging techniques in combination with X-ray angiography allows for the comprehensive evaluation of vessel geometry, plaque distribution, and lumen morphology. In computational fluid-dynamic (CFD) applications, angiography can be used to model the vessel curvature and intravascular imaging techniques can provide high-resolution details about the morphology of the lumen. Three-dimensional models obtained combining both modalities have been used to study the association between vessel hemodynamic, local ESS, stent healing, and plaque progression (Fig. 8.17). In a recent clinical study [85], IVUS and angiography have been used to study the effect

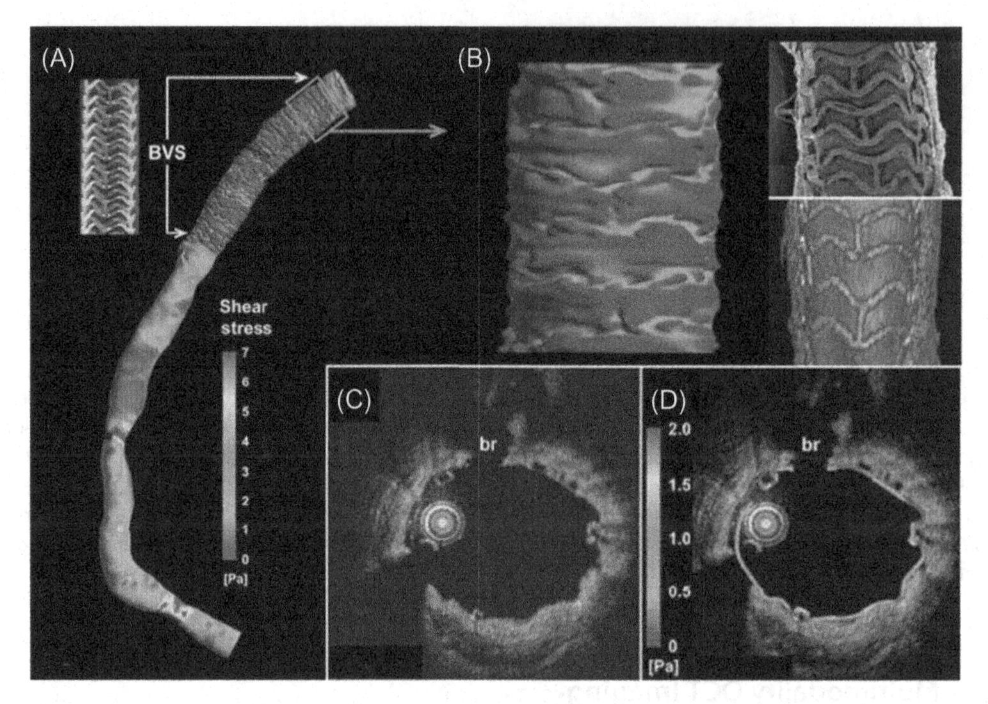

Fig. 8.17 OCT-ESS calculation. (A) 3D rendering of vessel geometry, lumen, and ESS distribution along the vessel obtained combining high-resolution vessel wall morphology extracted from OCT and true artery anatomy extracted from biplane X-ray angiography. (B) ESS evaluation around the BVS struts. (C and D) ESS results overlaid to OCT cross-sectional image. *(This material is reproduced from Papafaklis MI, Bourantas CV, Farooq V, Diletti R, Muramatsu T, Zhang Y, et al. In vivo assessment of the three-dimensional haemodynamic micro-environment following drug-eluting bioresorbable vascular scaffold implantation in a human coronary artery: fusion of frequency domain optical coherence tomography and angiography. EuroIntervention 2013;9:890, with written permission from the publisher.)*

of ESS on plaque progression in 374 patients at two different time points (i.e., 6 and 10 months after ACS), showing that local ESS is both an independent and addictive predictor of future coronary events. A similar study using OCT in combination with angiography demonstrated that areas of low ESS are associated with a higher prevalence of thin-cap fibroatheromas (TCFA) [86]. Furthermore, the use of OCT in combination with X-ray angiography has been suggested to assess local ESS around stent struts with the aim of optimize stent selection for a given vessel anatomy. Although those studies suggested clinical potential for ESS measurements, required processing and analysis are time consuming and labor intensive and, as a result, ESS measurements cannot be made available online, at the time of intervention. Further research is required to develop methods to compute ESS in real time and to evaluate the utility of such a measurement in the clinic.

5.3.2 Combination of OCT and IVUS

The main advantage of OCT with respect to IVUS is the improved resolution (i.e., ~10–15 vs. >100 μm), enabling a more detailed visualization of arterial wall morphology, TCFAs, plaque erosion and rupture, thrombi, calcium nodules, and stent neointimal tissue. However, the penetration depth of OCT in lipid-rich fibroatheromas is limited to a few millimeters. On the contrary, IVUS is capable of deeper imaging quantifying plaque burden and remodeling, but at a much lower spatial resolution than OCT. Furthermore, IVUS cannot visualize calcified plaque in depth; however, OCT can provide information about fibrocalcific plaques outer boundaries. As such, the combination of OCT and IVUS shows a unique advantage in plaque characterization (Fig. 8.18). The use of two separate catheters is difficult in the catheterization laboratory

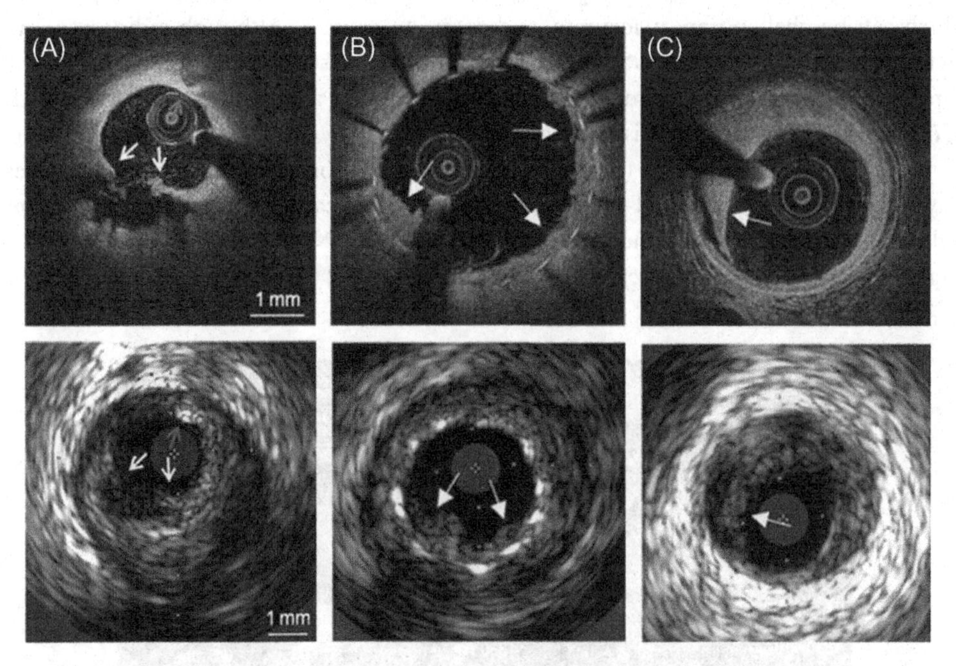

Fig. 8.18 Coregistered in vivo OCT and IVUS images. (A) A case of plaque rupture. OCT visualizes in detail the ruptured cap (arrows at 6 and 8 o'clock) and IVUS can visualize the entire plaque burden. The *arrow* at 1 o'clock points to a calcified nodule. (B) OCT-IVUS assessment of stent coverage. The higher resolution of OCT reveals the presence of thrombus (irregular tissue surface) overlying the stent struts (*arrows*). (C) OCT visualization of vessel wall dissection compared to limited resolution of IVUS. *((A) Modified from Higuma T, Soeda T, Abe N, Yamada M, Yokoyama H, Shibutani S, et al. A combined optical coherence tomography and intravascular ultrasound study on plaque rupture, plaque erosion, and calcified nodule in patients with ST-segment elevation myocardial infarction: incidence, morphologic characteristics, and outcomes after percutaneous coronary intervention. JACC Cardiovasc Interv 2015;8:1166–76 and (B and C) modified from Maehara A, Ben-Yehuda O, Ali Z, Wijns W, Bezerra HG, Shite J, et al. Comparison of stent expansion guided by optical coherence tomography versus intravascular ultrasound. JACC Cardiovasc Interv 2015;8:1704–14.)*

and image coregistration cannot be accomplished during the procedure and may be subject to artifacts and inaccuracies. The feasibility of hybrid OCT-IVUS imaging has been demonstrated integrating an IVUS transducer with OCT optics [113, 114]. In a recent study, in vivo imaging of swine coronary arteries was demonstrated using a 3.4F catheter at a 5 mm/s pullback speed [114, 115]. Despite such recent progress, several challenges still need to be addressed for clinical use, including increase the imaging rate of IVUS to match the higher speed of OCT, visualization, and catheter miniaturization.

5.3.3 Near-Infrared Spectroscopy

Since both OCT and IVUS are structural imaging modalities, the detection of plaque composition is based on spatial pattern recognition and image interpretation, which can be challenging, inaccurate, and prone to artifact. Although radio frequency backscattering analysis of IVUS and spectroscopic OCT signals have been proposed to improve the accuracy of plaque characterization, results are highly dependent on the algorithm used and are inconsistent when the signal is low [116, 117]. The combination of near-infrared spectroscopy (NIRS) with both IVUS or OCT may overcome these drawbacks. Both hybrid OCT-NIRS and IVUS-NIRS catheters have been developed and demonstrated, showing the feasibility of combining NIRS with structural imaging (Fig. 8.19). IVUS-NIRS is commercially available and validated for native arteries

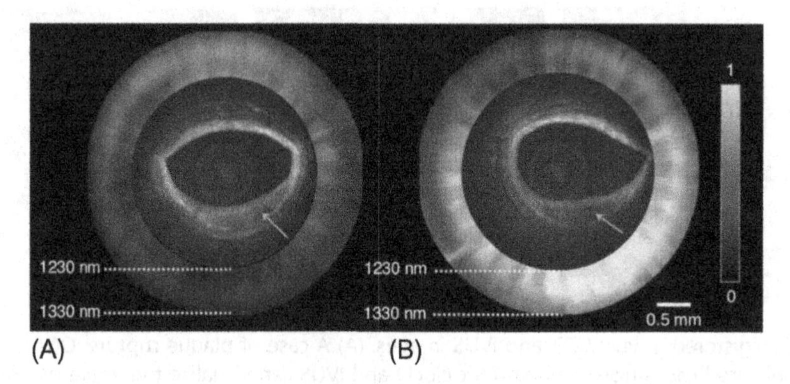

Fig. 8.19 OCT-NIRS ex vivo imaging. This example shows how NIRS signature can complement and simplify OCT image interpretation. (A) OCT-NIRS cross-sectional image of vessel intimal thickening (absence of lipid accumulation in the intima) and (B) OCT-NIRS image presenting a case of pathological intimal thickening with presence of lipid in the vessel wall. NIRS assesses the presence of lipid in the vessel wall by quantifying optical absorption coefficient in the NIR. The colormap range is between zero and one, with 1 being high attenuation in the NIR suggesting the presence of lipids. *(The images are reproduced with permission from the publisher from Fard AM, Vacas-Jacques P, Hamidi E, Wang H, Carruth RW, Gardecki JA, et al. Optical coherence tomography-near infrared spectroscopy system and catheter for intravascular imaging. Opt Express 2013;21:30849–58.)*

imaging. IVUS-NIRS has been used in clinical studies assessing the effect of lipid lowering treatments on coronary arteries [8] and to characterize culprit lesions in an STEMI population suggesting a predictive value of this technology [118]. Similar to IVUS-NIRS, an OCT-NIRS catheter has been recently proposed [119].

5.3.4 Near-Infrared Fluorescence

Although both IVUS/OCT and NIRS imaging provide morphological and chemical plaque information, these technologies cannot provide information about the "functional" state of coronary plaques (e.g., inflammation, macrophage activation, and other molecular processes). The combination of OCT with near-infrared fluorescence (NIRF) and autofluorescence (NIRAF) molecular imaging may fill this gap. An all-optical catheter and system for dual-modality OCT and NIRF has been recently developed and in vivo imaging in coronary-sized vessel demonstrated in an animal model of atherosclerosis [120]. More recent studies showed that the use of exogenous agents (i.e., indocyanine green and ProSense 750) allows NIRF to provide information about plaque inflammation, hemorrhage, and enzymatic activity [121–123]. Additionally, NIRF agents have been developed and demonstrated for detection of stent fibrin deposition in vivo [124].

NIRAF is an endogenous signal (i.e., tissue autofluorescence), excited in the red (630–650 nm) and detected in the NIR (700–900 nm). NIRAF is advantageous because it does not require the use of exogenous agents and therefore, its path to clinical use is relatively straightforward compared to NIRF. Studies on human aortic and coronary cadaver plaques have shown that NIRAF is elevated in necrotic core plaques, with respect to other lesion types [125]. Noticeably, a clinical OCT-NIRAF system has been developed and a first-in-human coronary study conducted acquiring high-resolution morphological OCT images synchronized with plaque autofluorescence [83] (Fig. 8.20). Results of this study showed elevated NIRAF signature in cases of plaque ruptures and TCFAs. Interestingly, elevated NIRAF signal presented a focal appearance over a more extended lipid-arc, suggesting potential ability of NIRAF to detect locally coronary plaque inflammation. Nevertheless, the molecular origin of such a red-excited NIRAF in coronary plaques is currently under investigation and considerable work is required to refine the acquisition and interpretation of different NIRAF signals and to elucidate their molecular sources from numerous potential candidates in plaque, such as oxidation modified lipids and proteins, hemorrhage-related porphyrins, and others [83, 126].

5.3.5 Combination of IVUS With Photoacoustics Imaging

Photoacoustics (PA) is an imaging modality capable of assessing light absorption properties of tissue in a depth-resolved fashion, providing information about its chemical

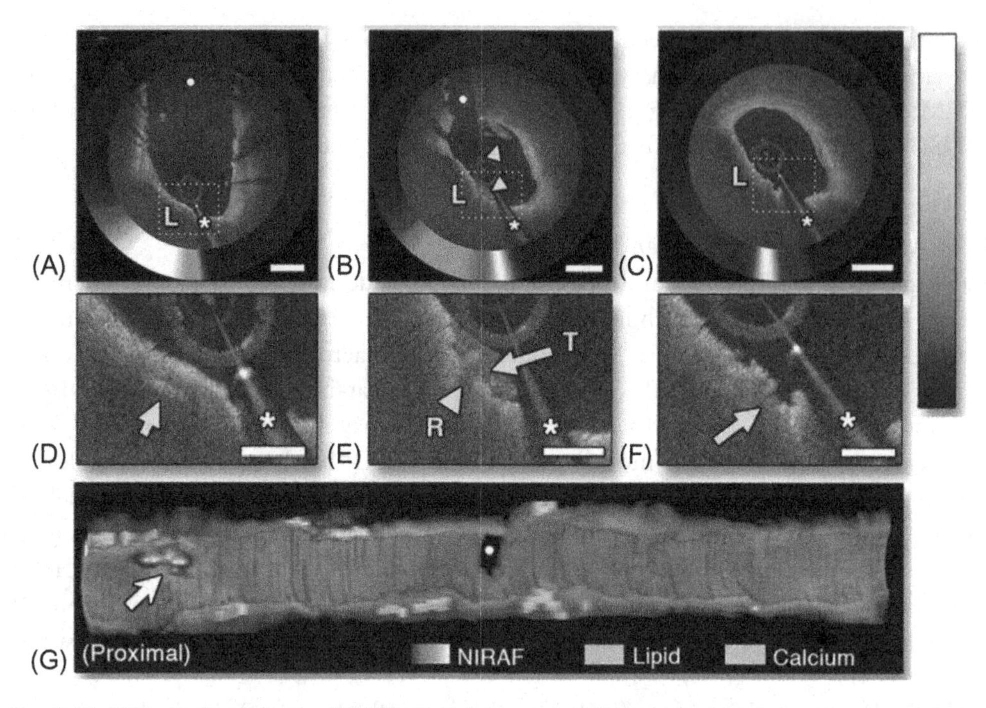

Fig. 8.20 OCT-autofluorescence (NIRAF) clinical imaging. NIRAF is displayed as a coregistered ring around the OCT image. The colormap ranges between low to high autofluorescence intensity. In this example, OCT visualizes an example of spontaneous TCFA rupture in vivo. (C and F) The rupture site is located within a large TCFA (from 5 to 11 o'clock) and high NIRAF is colocalized with the rupture site. (B and E) An organized thrombus sealing the rupture site in a frame adjacent to the rupture site and (A and D) elevated NIRAF signal colocalized with OCT appearance of a cholesterol crystal, which is a known marker of plaque inflammation [130]. (G) 3D OCT-NIRF reconstruction showing the focal appearance of elevated NIRAF signal within a large fibroatheroma, colocalized with the rupture site. *Scale bars* on (A–C) is equal to 1 mm; on (D–F) equal to 0.5 mm. *(The image is reproduced from Ughi GJ, Wang H, Gerbaud E, Gardecki JA, Fard AM, Hamidi E, et al. Clinical characterization of coronary atherosclerosis with dual-modality OCT and near-infrared autofluorescence imaging. JACC Cardiovasc Imag 2016. http://dx.doi.org/10.1016/j.jcmg.2015.11.020, with written permission from the publisher.)*

composition (e.g., lipid). A short laser pulse is sent to the tissue inducing thermal expansion, which generates an acoustic wave that is detected by the catheter. IVUS-PA catheters require the use of an ultrasound transducer and a light delivery system (i.e., fiber optics). Ex vivo studies described the potential of this imaging modality for intravascular application to provide information about atherosclerotic disease composition (i.e., lipid) and inflammation by the means of exogenous contrast agents (e.g., nanoparticles) [127, 128]. The feasibility of a hybrid system for dual-modality OCT-PA has been demonstrated [129]; however, not yet applied to intravascular imaging. Translation to the clinic of this imaging modality will require significant developments aiming to increase

PA imaging speed (currently limited to few frames/seconds for intravascular imaging applications [131]) and regulatory studies are required to demonstrate safety and efficacy of exogenous PA contrast agent for human use [132].

5.3.6 High Resolution, Micro-OCT

Commercial intravascular OCT systems have an axial resolution of approximately 10–15 μm and cannot image cellular and extracellular components in vivo. Micro-OCT (μ-OCT) technology is an improved resolution OCT imaging technique. It uses a light source with a very broad bandwidth (i.e., 650–950 nm) in a common-path FD-OCT implementation, providing images with an axial and lateral resolution of approximately 2 μm [133, 134]. Recently, a high NA, common-path, fiber optic probe has been proposed using an imaging element with a size of 500 μm potentially suitable for a 3F coronary catheter. This study showed that this type of probe can achieve a 2 μm axial resolution over a depth range of approximately 1 mm [135].

Ex vivo imaging of cadaver coronary arteries showed the capability of μ-OCT to visualize cells in situ (Fig. 8.21). Foam cells (macrophages) and cholesterol crystals were visualized with clarity within the thin-cap of fibroatheromas (Fig. 8.21D). Additionally, μ-OCT can also visualize other cellular features of the coronary wall and atherosclerosis such as smooth muscle cells, fibrin, platelet aggregation, and microcalcifications [133]. Furthermore, the unique capabilities of μ-OCT can be applied for the visualization

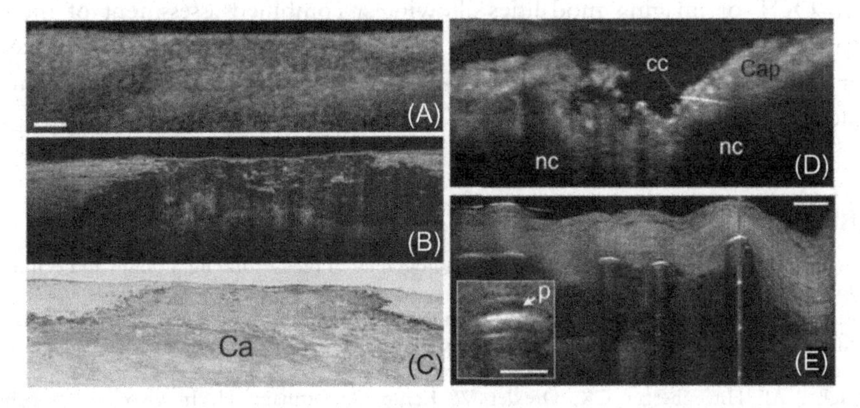

Fig. 8.21 μ-OCT ex vivo images. (A and B) Comparison of OCT versus μ-OCT and (C) corresponding histological session showing a calcified lesion. μ-OCT enables the visualization of cellular features that cannot be assessed by conventional OCT. (D) μ-OCT visualization of a necrotic core showing individual foam cells (macrophages) infiltrated in the cap and a single cholesterol crystal (cc) perforating the cap transversally. (E) μ-OCT visualization of drug-eluting stent struts. The *arrow* indicates the polymer overlying the metallic stent strut surface. *(This image is reproduced with publisher permission from Liu L, Gardecki JA, Nadkarni SK, Toussaint JD, Yagi Y, Bouma BE, et al. Imaging the subcellular structure of human coronary atherosclerosis using micro-optical coherence tomography. Nat Med 2011;17:1010–4.)*

of intracoronary stents, visualizing previously unseen microstructural details of stent endothelial coverage, fibrin accumulation, drug-eluting stent polymer degradation, and cell infiltration assessing local strut inflammation (Fig. 8.21E). Micro-OCT showed potential to improve our understanding of human CAD, enabling the identification of cellular features of high-risk coronary plaques and stent failure in vivo. However, μ-OCT technology research is currently ongoing and further development in several aspects of this imaging technology are required to make it available in the clinic.

5.4 Clinical Perspective

With respect to the future direction of the use of OCT in clinical practice, a more widespread use of OCT, according to insights in coronary artery disease pathophysiology and results from studies assessing the value of OCT in guidance of PCI, can be foreseen. It will probably become common practice to use OCT in cases of diagnostic uncertainty (e.g., haziness on angiography), to assess coronary artery segments with serial lesions, and to guide more complex coronary interventions (e.g., long and calcified lesions and bifurcation lesions), PCI for stent failure (stent thrombosis and in-stent restenosis) and interventions using BVS. With gradually more maturing insights into the development and natural history of different types of vulnerable plaques in patients presenting with ACS and improved OCT catheter designs, a differential interventional approach according to OCT defined type of underlying lesion of ACS is within reach. For the mid-term future, it is to be hoped and expected for that technological innovations, such as micro-OCT or imaging modalities allowing a combined assessment of molecular or inflammatory characteristics, together with OCT, would allow us to explore on a larger scale particularities of the coronary artery which remain hidden from us at this time.

REFERENCES

[1] Duguay MA, Mattick AT. Ultrahigh speed photography of picosecond light pulses and echoes. Appl Opt 1971;10:2162–70.
[2] Webb RH, Hughes GW, Pomerantzeff O. Flying spot TV ophthalmoscope. Appl Opt 1980;19:2991.
[3] Huang D, Swanson E, Lin C, Schuman J, Stinson W, Chang W, et al. Optical coherence tomography. Science 1991;254(80):1178–81.
[4] Fercher AF, Hitzenberger CK, Drexler W, Kamp G, Sattmann H. In vivo optical coherence tomography. Am J Ophthalmol 1993;116:113–4.
[5] Schmitt JM, Knüttel A, Yadlowsky M, Eckhaus MA. Optical-coherence tomography of a dense tissue: statistics of attenuation and backscattering. Phys Med Biol 1994;39:1705–20.
[6] Brezinski ME, Tearney GJ, Bouma BE, Izatt JA, Hee MR, Swanson EA, et al. Optical coherence tomography for optical biopsy. properties and demonstration of vascular pathology. Circulation 1996;93:1206–13.
[7] Tearney GJ, Brezinski ME, Southern JF, Bouma BE, Boppart SA, Fujimoto JG. Optical biopsy in human gastrointestinal tissue using optical coherence tomography. Am J Gastroenterol 1997;92: 1800–4.

[8] Boppart SA, Brezinski ME, Pitris C, Fujimoto JG. Optical coherence tomography for neurosurgical imaging of human intracortical melanoma. Neurosurgery 1998;43:834–41.
[9] Tearney GJ, Brezinski ME, Fujimoto JG, Weissman NJ, Boppart SA, Bouma BE, et al. Scanning single-mode fiber optic catheter-endoscope for optical coherence tomography. Opt Lett 1996; 21:543.
[10] Tearney GJ, Brezinski ME, Bouma BE, Boppart SA, Pitris C, Southern JF, et al. In vivo endoscopic optical biopsy with optical coherence tomography. Science 1997;276:2037–9.
[11] Vakoc BJ, Fukumura D, Jain RK, Bouma BE. Cancer imaging by optical coherence tomography: preclinical progress and clinical potential. Nat Rev Cancer 2012;12:363–8.
[12] Kut C, Chaichana KL, Xi J, Raza SM, Ye X, McVeigh ER, et al. Detection of human brain cancer infiltration ex vivo and in vivo using quantitative optical coherence tomography. Sci Transl Med 2015;7:292ra100.
[13] Zysk AM, Nguyen FT, Chaney EJ, Kotynek JG, Oliphant UJ, Bellafiore FJ, et al. Clinical feasibility of microscopically-guided breast needle biopsy using a fiber-optic probe with computer-aided detection. Technol Cancer Res Treat 2009;8:315–21.
[14] Bezerra HG, Costa MA, Guagliumi G, Rollins AM, Simon DI. Intracoronary optical coherence tomography: a comprehensive review clinical and research applications. JACC Cardiovasc Interv 2009;2:1035–46.
[15] Tearney GJ, Regar E, Akasaka T, Adriaenssens T, Barlis P, Bezerra HG, et al. Consensus standards for acquisition, measurement, and reporting of intravascular optical coherence tomography studies: a report from the international working group for intravascular optical coherence tomography standardization and validation. J Am Coll Cardiol 2012;59:1058–72.
[16] McDaniel MC, Eshtehardi P, Sawaya FJ, Douglas JS, Samady H. Contemporary clinical applications of coronary intravascular ultrasound. JACC Cardiovasc Interv 2011;4:1155–67.
[17] Wang LV, Wu H. Biomedical optics: principles and imaging. Hoboken, NJ: Wiley-Interscience; 2007.
[18] Yun SH, Tearney GJ, Vakoc BJ, Shishkov M, Oh WY, Desjardins AE, et al. Comprehensive volumetric optical microscopy in vivo. Nat Med 2007;12:1429–33.
[19] Adler DC, Chen Y, Huber R, Schmitt J, Connolly J, Fujimoto JG. Three-dimensional endomicroscopy using optical coherence tomography. Nat Photonics 2007;1:709–16.
[20] Tearney GJ, Waxman S, Shishkov M, Vakoc BJ, Suter MJ, Freilich MI, et al. Three-dimensional coronary artery microscopy by intracoronary optical frequency domain imaging. JACC Cardiovasc Imag 2008;1:752–61.
[21] Cho HS, Jang SJ, Kim K, Dan-Chin-Yu AV, Shishkov M, Bouma BE, et al. High frame-rate intravascular optical frequency-domain imaging in vivo. Biomed Opt Express 2014;5:223.
[22] Imola F, Mallus MT, Ramazzotti V, Manzoli A, Pappalardo A, Di Giorgio A, et al. Safety and feasibility of frequency domain optical coherence tomography to guide decision making in percutaneous coronary intervention. EuroIntervention 2010;6:575–81.
[23] Lehtinen T, Nammas W, Airaksinen JKE, Karjalainen PP. Feasibility and safety of frequency-domain optical coherence tomography for coronary artery evaluation: a single-center study. Int J Cardiovasc Imag 2013;29:997–1005.
[24] Mitomo S, Naganuma T, Nakamura S, Tahara S, Ishiguro H, Nakamura S. Potential advantages of the guideliner catheter: insights from optical coherence tomography. Cardiovasc Interv Ther 2015. http://dx.doi.org/10.1007/s12928-015-0365-x.
[25] Lopez JJ, Arain SA, Madder R, Parekh N, Shroff AR, Westerhausen D. Techniques and best practices for optical coherence tomography: a practical manual for interventional cardiologists. Catheter Cardiovasc Interv 2014;84:687–99.
[26] Frick K, Michael TT, Alomar M, Mohammed A, Rangan BV, Abdullah S, et al. Low molecular weight dextran provides similar optical coherence tomography coronary imaging compared to radiographic contrast media. Catheter Cardiovasc Interv 2014;84:727–31.
[27] Ozaki Y, Kitabata H, Tsujioka H, Hosokawa S, Kashiwagi M, Ishibashi K, et al. Comparison of contrast media and low-molecular-weight dextran for frequency-domain optical coherence tomography. Circ J 2012;76:922–7.

[28] Hebsgaard L, Nielsen TM, Tu S, Krusell LR, Maeng M, Veien KT, et al. Co-registration of optical coherence tomography and x-ray angiography in percutaneous coronary intervention the does optical coherence tomography optimize revascularization (DOCTOR) fusion study. Int J Cardiol 2015;182:272–8.
[29] Burzotta F, Dato I, Trani C, Pirozzolo G, De Maria GL, Porto I, et al. Frequency domain optical coherence tomography to assess non-ostial left main coronary artery. EuroIntervention 2015;10:e1–8.
[30] Bing R, Yong ASC, Lowe HC. Percutaneous transcatheter assessment of the left main coronary artery: current status and future directions. JACC Cardiovasc Interv 2015;8:1529–39.
[31] Fujino Y, Bezerra HG, Attizzani GF, Wang W, Yamamoto H, Chamié D, et al. Frequency-domain optical coherence tomography assessment of unprotected left main coronary artery disease—a comparison with intravascular ultrasound. Catheter Cardiovasc Interv 2013;82:E173–83.
[32] Gerbaud E, Weisz G, Tanaka A, Kashiwagi M, Shimizu T, Wang L, et al. Multi-laboratory inter-institute reproducibility study of IVOCT and IVUS assessments using published consensus document definitions. Eur Heart J Cardiovasc imaging 2015;19:207–29.
[33] Schaar JA, Muller JE, Falk E, Virmani R, Fuster V, Serruys PW, et al. Terminology for high-risk and vulnerable coronary artery plaques. Report of a meeting on the vulnerable plaque, June 17 and 18, 2003, Santorini, Greece. Eur Heart J 2004;25:1077–82.
[34] Farb A, Burke AP, Tang AL, Liang TY, Mannan P, Smialek J, et al. Coronary plaque erosion without rupture into a lipid core. A frequent cause of coronary thrombosis in sudden coronary death. Circulation 1996;93:1354–63.
[35] Xu C, Schmitt JM, Carlier SG, Virmani R. Characterization of atherosclerosis plaques by measuring both backscattering and attenuation coefficients in optical coherence tomography. J Biomed Opt 2008;13:034003.
[36] Faber DJ, van der Meer FJ, Aalders MCG, van Leeuwen TG. Quantitative measurement of attenuation coefficients of weakly scattering media using optical coherence tomography. Opt Express 2004;12:4353.
[37] Koskinas KKC, Ughi GJ, Windecker S, Tearney GJ, Räber L. Intracoronary imaging of coronary atherosclerosis: validation for diagnosis, prognosis and treatment. Eur Heart J 2016;37:524–35.
[38] Yabushita H, Bouma BE, Houser SL, Aretz HT, Jang IK, Schlendorf KH, et al. Characterization of human atherosclerosis by optical coherence tomography. http://dx.doi.org/10.1161/01.CIR.0000029927.92825.F6.
[39] Prati F, Regar E, Mintz GS, Arbustini E, Di Mario C, Jang IK, et al. Expert review document on methodology, terminology, and clinical applications of optical coherence tomography: physical principles, methodology of image acquisition, and clinical application for assessment of coronary arteries and atherosclerosis. Eur Heart J 2010;31:401–15.
[40] van Soest G, Goderie T, Regar E, Koljenović S, van Leenders GLJH, Gonzalo N, et al. Atherosclerotic tissue characterization in vivo by optical coherence tomography attenuation imaging. J Biomed Opt 2010;15:011105.
[41] Ughi GJ, Adriaenssens T, Sinnaeve P, Desmet W, D'hooge J. Automated tissue characterization of in vivo atherosclerotic plaques by intravascular optical coherence tomography images. Biomed Opt Express 2013;4:1014–30.
[42] Tanaka A, Imanishi T, Kitabata H, Kubo T, Takarada S, Kataiwa H, et al. Distribution and frequency of thin-capped fibroatheromas and ruptured plaques in the entire culprit coronary artery in patients with acute coronary syndrome as determined by optical coherence tomography. Am J Cardiol 2008;102:975–9.
[43] Jia H, Abtahian F, Aguirre AD, Lee S, Chia S, Lowe H, et al. In vivo diagnosis of plaque erosion and calcified nodule in patients with acute coronary syndrome by intravascular optical coherence tomography. J Am Coll Cardiol 2013;62:1748–58.
[44] Otsuka F, Joner M, Prati F, Virmani R, Narula J. Clinical classification of plaque morphology in coronary disease. Nat Rev Cardiol 2014;11:379–89.
[45] Chia S, Raffel OC, Takano M, Tearney GJ, Bouma B, Jang IK. Association of statin therapy with reduced coronary plaque rupture: an optical coherence tomography study. Coron Artery Dis 2008;19:237–42.

[46] Kubo T, Imanishi T, Takarada S, Kuroi A, Ueno S, Yamano T, et al. Assessment of culprit lesion morphology in acute myocardial infarction. J Am Coll Cardiol 2007;50:933–9.
[47] Ino Y, Kubo T, Tanaka A, Kuroi A, Tsujioka H, Ikejima H, et al. Difference of culprit lesion morphologies between ST-segment elevation myocardial infarction and non-ST-segment elevation acute coronary syndrome. JACC Cardiovasc Interv 2011;4:76–82.
[48] Kume T, Okura H, Kawamoto T, Akasaka T, Toyota E, Watanabe N, et al. Fibrin clot visualized by optical coherence tomography. Circulation 2008;118:426–7.
[49] Kume T, Akasaka T, Kawamoto T, Ogasawara Y, Watanabe N, Toyota E, et al. Assessment of coronary arterial thrombus by optical coherence tomography. Am J Cardiol 2006;97:1713–7.
[50] Kubo T, Xu C, Wang Z, van Ditzhuijzen NS, Bezerra HG. Plaque and thrombus evaluation by optical coherence tomography. Int J Cardiovasc Imag 2011;27:289–98.
[51] MacNeill BD, Jang IK, Bouma BE, Iftimia N, Takano M, Yabushita H, et al. Focal and multi-focal plaque macrophage distributions in patients with acute and stable presentations of coronary artery disease. J Am Coll Cardiol 2004;44:972–9.
[52] Tearney GJ. Imaging of macrophages. JACC Cardiovasc Imag 2015;8:73–5.
[53] Hoyt T, Phipps J, Vela D, Buja M, Milner T, Feldman M. Mechanisms for false thin-cap fibroatheromas identified by optical coherence tomography. J Am Coll Cardiol 2015;65: A1936.
[54] Phipps JE, Vela D, Hoyt T, Halaney DL, Mancuso JJ, Buja LM, et al. Macrophages and intravascular OCT bright spots: a quantitative study. JACC Cardiovasc Imag 2015;8:63–72.
[55] Bezerra HG, Attizzani GF, Sirbu V, Musumeci G, Lortkipanidze N, Fujino Y, et al. Optical coherence tomography versus intravascular ultrasound to evaluate coronary artery disease and percutaneous coronary intervention. JACC Cardiovasc Interv 2013;6:228–36.
[56] Nakano M, Vorpahl M, Otsuka F, Taniwaki M, Yazdani SK, Finn AV, et al. Ex vivo assessment of vascular response to coronary stents by optical frequency domain imaging. JACC Cardiovasc Imag 2012;5:71–82.
[57] Kastrati A, Mehilli J, Pache J, Kaiser C, Valgimigli M, Kelbaek H, et al. Analysis of 14 trials comparing sirolimus-eluting stents with bare-metal stents. N Engl J Med 2007;356:1030–9.
[58] Finn AV, Joner M, Nakazawa G, Kolodgie F, Newell J, John MC, et al. Pathological correlates of late drug-eluting stent thrombosis: strut coverage as a marker of endothelialization. Circulation 2007;115:2435–41.
[59] Finn AV, Nakazawa G, Joner M, Kolodgie FD, Mont EK, Gold HK, et al. Vascular responses to drug eluting stents: importance of delayed healing. Arterioscler Thromb Vasc Biol 2007;27:1500–10.
[60] Shite J, Matsumoto D, Yokoyama M. Sirolimus-eluting stent fracture with thrombus, visualization by optical coherence tomography. Eur Heart J 2006;27:1389.
[61] Guagliumi G, Sirbu V. Optical coherence tomography: high resolution intravascular imaging to evaluate vascular healing after coronary stenting. Catheter Cardiovasc Interv 2008;72: 237–47.
[62] Serruys PW, Ormiston JA, Onuma Y, Regar E, Gonzalo N, Garcia-Garcia HM, et al. A bioabsorbable everolimus-eluting coronary stent system (ABSORB): 2-year outcomes and results from multiple imaging methods. Lancet 2009;373:897–910.
[63] Souteyrand G, Amabile N, Mangin L, Chabin X, Meneveau N, Cayla G, et al. Mechanisms of stent thrombosis analysed by optical coherence tomography: insights from the National PESTO French Registry. Eur Heart J 2016;37:1208–16.
[64] Taniwaki M, Radu MD, Zaugg S, Amabile N, Garcia-Garcia HM, Yamaji K, et al. Mechanisms of very late drug-eluting stent thrombosis assessed by optical coherence tomography. Circulation 2016;133:650–60.
[65] Otsuka F, Byrne RA, Yahagi K, Mori H, Ladich E, Fowler DR, et al. Neoatherosclerosis: overview of histopathologic findings and implications for intravascular imaging assessment. Eur Heart J 2015;36:2147–59.
[66] Nakazawa G, Otsuka F, Nakano M, Vorpahl M, Yazdani SK, Ladich E, et al. The pathology of neoatherosclerosis in human coronary implants bare-metal and drug-eluting stents. J Am Coll Cardiol 2011;57:1314–22.

[67] Fujii K, Carlier SG, Mintz GS, Yang Y, Moussa I, Weisz G, et al. Stent underexpansion and residual reference segment stenosis are related to stent thrombosis after sirolimus-eluting stent implantation: an intravascular ultrasound study. J Am Coll Cardiol 2005;45:995–8.
[68] Cook S, Wenaweser P, Togni M, Billinger M, Morger C, Seiler C, et al. Incomplete stent apposition and very late stent thrombosis after drug-eluting stent implantation. Circulation 2007;115: 2426–34.
[69] Prati F, Di Vito L, Biondi-Zoccai G, Occhipinti M, La Manna A, Tamburino C, et al. Angiography alone versus angiography plus optical coherence tomography to guide decision-making during percutaneous coronary intervention: the centro per la lotta contro l'infarto-optimisation of percutaneous coronary intervention (CLI-OPCI) study. EuroIntervention 2012;8:823–9.
[70] Wijns W, Shite J, Jones MR, Lee SWL, Price MJ, Fabbiocchi F, et al. Optical coherence tomography imaging during percutaneous coronary intervention impacts physician decision-making: ILUMIEN I study. Eur Heart J 2015;36:3346–55.
[71] de Jaegere P, Mudra H, Figulla H, Almagor Y, Doucet S, Penn I, et al. Intravascular ultrasound-guided optimized stent deployment. Immediate and 6 months clinical and angiographic results from the multicenter ultrasound stenting in coronaries study (MUSIC Study). Eur Heart J 1998;19: 1214–23.
[72] Kim IC, Yoon HJ, Shin ES, Kim MS, Park J, Cho YK, et al. Usefulness of frequency domain optical coherence tomography compared with intravascular ultrasound as a guidance for percutaneous coronary intervention. J Interv Cardiol 2016;29:216–24.
[73] Gutiérrez-Chico JL, Wykrzykowska J, Nüesch E, van Geuns RJ, Koch KT, Koolen JJ, et al. Vascular tissue reaction to acute malapposition in human coronary arteries: sequential assessment with optical coherence tomography. Circ Cardiovasc Interv 2012;5:20–9. S1–8.
[74] Antonsen L, Thayssen P, Maehara A, Hansen HS, Junker A, Veien KT, et al. Optical coherence tomography guided percutaneous coronary intervention with Nobori stent implantation in patients with non-ST-segment-elevation myocardial infarction (OCTACS) trial. Circ Cardiovasc Interv 2015;8:e002446.
[75] Imola F, Occhipinti M, Biondi-Zoccai G, Di Vito L, Ramazzotti V, Manzoli A, et al. Association between proximal stent edge positioning on atherosclerotic plaques containing lipid pools and postprocedural myocardial infarction (from the CLI-POOL study). Am J Cardiol 2013;111: 526–31.
[76] Imanaka T, Hao H, Fujii K, Shibuya M, Fukunaga M, Miki K, et al. Analysis of atherosclerosis plaques by measuring attenuation coefficients in optical coherence tomography: thin-cap fibroatheroma or foam cells accumulation without necrotic core? Eur Heart J 2013;34:P5482.
[77] Adriaenssens T, Ughi GJ, Dubois C, De Cock D, Onsea K, Bennett J, et al. STACCATO (Assessment of Stent sTrut Apposition and Coverage in Coronary ArTeries with Optical coherence tomography in patients with STEMI, NSTEMI and stable/unstable angina undergoing everolimus vs. biolimus A9-eluting stent implantation): a randomised co. EuroIntervention 2016;11: e1619–26.
[78] Prati F, Romagnoli E, Burzotta F, Limbruno U, Gatto L, La Manna A, et al. Clinical impact of OCT findings during PCI: the CLI-OPCI II study. JACC Cardiovasc Imag 2015;8:1297–305.
[79] Kubo T, Shinke T, Okamura T, Hibi K, Nakazawa G, Morino Y, et al. Optical frequency domain imaging vs. intravascular ultrasound in percutaneous coronary intervention (OPINION trial): study protocol for a randomized controlled trial. J Cardiol 2016. http://dx.doi.org/10.1016/j.jjcc.2015.11.007.
[80] Maehara A, Ben-Yehuda O, Ali Z, Wijns W, Bezerra HG, Shite J, et al. Comparison of stent expansion guided by optical coherence tomography versus intravascular ultrasound. JACC Cardiovasc Interv 2015;8:1704–14.
[81] Ughi GJ, Adriaenssens T, Onsea K, Kayaert P, Dubois C, Sinnaeve P, et al. Automatic segmentation of in-vivo intra-coronary optical coherence tomography images to assess stent strut apposition and coverage. Int J Cardiovasc Imag 2012;28:229–41.
[82] Nam HS, Kim CS, Lee JJ, Song JW, Kim JW, Yoo H. Automated detection of vessel lumen and stent struts in intravascular optical coherence tomography to evaluate stent apposition and neointimal coverage. Med Phys 2016;43:1662–75.

[83] Ughi GJ, Wang H, Gerbaud E, Gardecki JA, Fard AM, Hamidi E, et al. Clinical characterization of coronary atherosclerosis with dual-modality OCT and near-infrared autofluorescence imaging. JACC Cardiovasc Imag 2016. http://dx.doi.org/10.1016/j.jcmg.2015.11.020.
[84] Tsantis S, Kagadis GC, Katsanos K, Karnabatidis D, Bourantas G, Nikiforidis GC. Automatic vessel lumen segmentation and stent strut detection in intravascular optical coherence tomography. Med Phys 2012;39:503.
[85] Stone PH, Saito S, Takahashi S, Makita Y, Nakamura S, Kawasaki T, et al. Prediction of progression of coronary artery disease and clinical outcomes using vascular profiling of endothelial shear stress and arterial plaque characteristics: the PREDICTION study. Circulation 2012;126: 172–81.
[86] Vergallo R, Papafaklis MI, Yonetsu T, Bourantas CV, Andreou I, Wang Z, et al. Endothelial shear stress and coronary plaque characteristics in humans combined frequency-domain optical coherence tomography and computational fluid dynamics study. Circ Cardiovasc Imag 2014;7: 905–11.
[87] Ughi GJ, Van Dyck CJ, Adriaenssens T, Hoymans VY, Sinnaeve P, Timmermans JP, et al. Automatic assessment of stent neointimal coverage by intravascular optical coherence tomography. Eur Heart J Cardiovasc Imag 2014;15:195–200.
[88] Lu H, Gargesha M, Wang Z, Chamie D, Attizzani GF, Kanaya T, et al. Automatic stent detection in intravascular OCT images using bagged decision trees. Biomed Opt Express 2012;3:2809–24.
[89] Bonnema GT, Cardinal KO, Williams SK, Barton JK, Bonnema GT, Cardinal KO, et al. An automatic algorithm for detecting stent endothelialization from volumetric optical coherence tomography datasets. Phys Med Biol 2008;53:3083–98.
[90] Ughi GJ, Adriaenssens T, Desmet W, D'hooge J. Fully automatic three-dimensional visualization of intravascular optical coherence tomography images: methods and feasibility in vivo. Biomed Opt Express 2012;3:3291–303.
[91] Ughi GJ, Adriaenssens T, Larsson M, Dubois C, Sinnaeve PR, Coosemans M, et al. Automatic three-dimensional registration of intravascular optical coherence tomography images. J Biomed Opt 2012;17:026005.
[92] De Cock D, Bennett J, Ughi GJ, Dubois C, Sinnaeve P, Dhooge J, et al. Healing course of acute vessel wall injury after drug-eluting stent implantation assessed by optical coherence tomography. Eur Heart J Cardiovasc Imag 2014;15:800–9.
[93] Ughi GJ, Dubois C, Desmet W, D'hooge J, Adriaenssens T. Provisional side branch stenting: presentation of an automated method allowing online 3D OCT guidance. Eur Heart J Cardiovasc Imag 2013;14:715.
[94] Farooq V, Gogas BD, Okamura T, Heo JH, Magro M, Gomez-Lara J, et al. Three-dimensional optical frequency domain imaging in conventional percutaneous coronary intervention: the potential for clinical application. Eur Heart J 2013;34:875–85.
[95] Wang A, Nakatani S, Eggermont J, Onuma Y, Garcia-Garcia HM, Serruys PW, et al. Automatic detection of bioresorbable vascular scaffold struts in intravascular optical coherence tomography pullback runs. Biomed Opt Express 2014;5:3589–602.
[96] Gnanadesigan M, van Soest G, White S, Scoltock S, Ughi GJ, Baumbach A, et al. Effect of temperature and fixation on the optical properties of atherosclerotic tissue: a validation study of an ex-vivo whole heart cadaveric model. Biomed Opt Express 2014;5:1038.
[97] Wang Z, Kyono H, Bezerra HG, Wang H, Gargesha M, Alraies C, et al. Semiautomatic segmentation and quantification of calcified plaques in intracoronary optical coherence tomography images. J Biomed Opt 2010;15:061711.
[98] Wang Z, Chamie D, Bezerra HG, Yamamoto H, Kanovsky J, Wilson DL, et al. Volumetric quantification of fibrous caps using intravascular optical coherence tomography. Biomed Opt Express 2012;3:1413–26.
[99] Zahnd G, Karanasos A, van Soest G, Regar E, Niessen W, Gijsen F, et al. Quantification of fibrous cap thickness in intracoronary optical coherence tomography with a contour segmentation method based on dynamic programming. Int J Comput Assist Radiol Surg 2015;10:1383–94.
[100] Jang SJ, Park HS, Song JW, Kim TS, Cho HS, Kim S, et al. ECG-triggered, single cardiac cycle, high-speed, 3D, intracoronary OCT. JACC Cardiovasc Imag 2016;9:623–5.

[101] Wang T, Pfeiffer T, Regar E, Wieser W, van Beusekom H, Lancee CT, et al. Heartbeat OCT: in vivo intravascular megahertz-optical coherence tomography. Biomed Opt Express 2015;6: 5021–32.

[102] Wang T, Pfeiffer T, Regar E, Wieser W, van Beusekom H, Lancee CT, et al. Heartbeat OCT and motion-free 3D in vivo coronary artery microscopy. JACC Cardiovasc Imag 2016;9:622–3.

[103] de Boer JF, Milner TE, van Gemert MJC, Nelson JS. Two-dimensional birefringence imaging in biological tissue by polarization-sensitive optical coherence tomography. Opt Lett 1997; 22:934.

[104] Saxer CE, de Boer JF, Park BH, Zhao Y, Chen Z, Nelson JS. High-speed fiber based polarization-sensitive optical coherence tomography of in vivo human skin. Opt Lett 2000;25:1355–7.

[105] Oh WY, Yun SH, Vakoc BJ, Shishkov M, Desjardins AE, Park BH, et al. High-speed polarization sensitive optical frequency domain imaging with frequency multiplexing. Opt Express 2008;16:1096–103.

[106] Villiger M, Zhang EZ, Nadkarni S, Oh WY, Bouma BE, Vakoc BJ. Artifacts in polarization-sensitive optical coherence tomography caused by polarization mode dispersion. Opt Lett 2013; 38:923.

[107] Villiger M, Zhang EZ, Nadkarni SK, Oh WY, Vakoc BJ, Bouma BE. Spectral binning for mitigation of polarization mode dispersion artifacts in catheter-based optical frequency domain imaging. Opt Express 2013;21:16353–69.

[108] Nadkarni SK, Pierce MC, Park BH, de Boer JF, Whittaker P, Bouma BE, et al. Measurement of collagen and smooth muscle cell content in atherosclerotic plaques using polarization-sensitive optical coherence tomography. J Am Coll Cardiol 2007;49:1474–81.

[109] Doradla P, Villiger M, Tshikudi DM, Bouma BE, Nadkarni SK. Assessment of atherosclerotic plaque collagen content and architecture using polarization-sensitive optical coherence tomography. In: Choi B, Kollias N, Zeng H, Kang HW, Wong BJF, Ilgner JF, et al. editors. International Society for Optics and Photonics (Conference Presentation, 96893L), 2016. http://dx.doi.org/10.1117/12.2218191.

[110] van der Sijde JN, Karanasos A, Villiger M, Bouma BE, Regar E. First-in-man assessment of plaque rupture by polarization-sensitive optical frequency domain imaging in vivo. Eur Heart J 2016;ehw179. http://dx.doi.org/10.1093/eurheartj/ehw179.

[111] Villiger M, Karanasos A, Ren J, Lippok N, Shishkov M, Daemen J, et al. First clinical pilot study with intravascular polarization sensitive optical coherence tomography. In: Choi B, Kollias N, Zeng H, Kang HW, Wong BJF, Ilgner JF, et al. editors. International Society for Optics and Photonics (Conference Presentation, 96892O), 2016. http://dx.doi.org/10.1117/12.2211463.

[112] Virmani R, Burke AP, Willerson JT, Farb A, Narula J, Kolodgie FD. The vulnerable atherosclerotic plaque: strategies for diagnosis and management. Oxford, UK: Blackwell Publishing; 2007. p. 19–36. http://dx.doi.org/10.1002/9780470987575.ch2.

[113] Li BH, Leung ASO, Soong A, Munding CE, Lee H, Thind AS, et al. Hybrid intravascular ultrasound and optical coherence tomography catheter for imaging of coronary atherosclerosis. Catheter Cardiovasc Interv 2013;81:494–507.

[114] Li X, Li J, Jing J, Ma T, Liang S, Zhang J, et al. Integrated IVUS-OCT imaging for atherosclerotic plaque characterization. IEEE J Sel Top Quantum Electron 2014;20:7100108.

[115] Li J, Li X, Mohar D, Raney A, Jing J, Zhang J, et al. Integrated IVUS-OCT for real-time imaging of coronary atherosclerosis. JACC Cardiovasc Imag 2014;7:101–3.

[116] Granada JF, Wallace-Bradley D, Win HK, Alviar CL, Builes A, Lev EI, et al. In vivo plaque characterization using intravascular ultrasound-virtual histology in a porcine model of complex coronary lesions. Arterioscler Thromb Vasc Biol 2007;27:387–93.

[117] Thim T, Hagensen MK, Wallace-Bradley D, Granada JF, Kaluza GL, Drouet L, et al. Unreliable assessment of necrotic core by virtual histology intravascular ultrasound in porcine coronary artery disease. Circ Cardiovasc Imag 2010;3:384–91.

[118] Kang SJ, Mintz GS, Pu J, Sum ST, Madden SP, Burke AP, et al. Combined IVUS and NIRS detection of fibroatheromas: histopathological validation in human coronary arteries. JACC Cardiovasc Imag 2015;8:184–94.

[119] Fard AM, Vacas-Jacques P, Hamidi E, Wang H, Carruth RW, Gardecki JA, et al. Optical coherence tomography-near infrared spectroscopy system and catheter for intravascular imaging. Opt Express 2013;21:30849–58.
[120] Yoo H, Kim JW, Shishkov M, Namati E, Morse T, Shubochkin R, et al. Intra-arterial catheter for simultaneous microstructural and molecular imaging in vivo. Nat Med 2011;17:1680–4.
[121] Vinegoni C, Botnaru I, Aikawa E, Calfon MA, Iwamoto Y, Folco EJ, et al. Indocyanine green enables near-infrared fluorescence imaging of lipid-rich, inflamed atherosclerotic plaques. Sci Transl Med 2011;3:84ra45.
[122] Jaffer FA, Calfon MA, Rosenthal A, Mallas G, Razansky RN, Mauskapf A, et al. Two-dimensional intravascular near-infrared fluorescence molecular imaging of inflammation in atherosclerosis and stent-induced vascular injury. J Am Coll Cardiol 2011;57:2516–26.
[123] Verjans J, Osborn E, Ughi GJ, Calfon M, Hamidi E, Anotniadis A, et al. Targeted near-infrared fluorescence imaging of atherosclerosis: clinical and intracoronary evaluation of indocyanine green. JACC Cardiovasc Imag 2016;9(9):1087–95.
[124] Hara T, Ughi GJ, McCarthy JRJ, Erdem SS, Mauskapf A, Lyon SC, et al. Intravascular fibrin molecular imaging improves the detection of unhealed stents assessed by optical coherence tomography in vivo. Eur Heart J 2015;7:1081–92.
[125] Wang H, Gardecki JA, Ughi GJ, Jacques PV, Hamidi E, Tearney GJ. Ex vivo catheter-based imaging of coronary atherosclerosis using multimodality OCT and NIRAF excited at 633 nm. Biomed Opt Express 2015;6:1363.
[126] Psaltis PJ, Nicholls SJ. Imaging: focusing light on the vulnerable plaque. Nat Rev Cardiol 2016;13:253–5.
[127] Wang B, Su JL, Amirian J, Litovsky SH, Smalling R, Emelianov S. Detection of lipid in atherosclerotic vessels using ultrasound-guided spectroscopic intravascular photoacoustic imaging. Opt Express 2010;18:4889–97.
[128] Jansen K, van Soest G, van der Steen AFW. Intravascular photoacoustic imaging: a new tool for vulnerable plaque identification. Ultrasound Med Biol 2014;40:1037–1048.
[129] Zhang EZ, Povazay B, Laufer J, Alex A, Hofer B, Pedley B, et al. Multimodal photoacoustic and optical coherence tomography scanner using an all optical detection scheme for 3D morphological skin imaging. Biomed Opt Express 2011;2:2202–15.
[130] Kataoka Y, Puri R, Hammadah M, Duggal B, Uno K, Kapadia SR, et al. Cholesterol crystals associate with coronary plaque vulnerability in vivo. J Am Coll Cardiol 2015;65:630–2.
[131] Wang PP, Ma T, Slipchenko MN, Liang S, Hui J, Shung KK, et al. High-speed intravascular photoacoustic imaging of lipid-laden atherosclerotic plaque enabled by a 2-kHz barium nitrite Raman laser. Sci Rep 2014;4:6889.
[132] Yeager D, Karpiouk A, Wang B, Amirian J, Sokolov K, Smalling R, et al. Intravascular photoacoustic imaging of exogenously labeled atherosclerotic plaque through luminal blood. J Biomed Opt 2012;17:106016.
[133] Liu L, Gardecki JA, Nadkarni SK, Toussaint JD, Yagi Y, Bouma BE, et al. Imaging the subcellular structure of human coronary atherosclerosis using micro-optical coherence tomography. Nat Med 2011;17:1010–4.
[134] Chu KK, Ughi GJ, Liu L, Tearney GJ. Toward clinical μOCT—a review of resolution-enhancing technical advances. Curr Cardiovasc Imag Rep 2014;7:1–8.
[135] Yin B, Chu KK, Liang CP, Singh K, Reddy R, Tearney GJ. μOCT imaging using depth of focus extension by self-imaging wavefront division in a common-path fiber optic probe. Opt Express 2016;24:5555.

SECTION III

Vascular Biomechanics and Modeling

CHAPTER 9

Vascular Hemodynamics with Computational Modeling and Experimental Studies

S. Beier*, J. Ormiston†, M. Webster‡, J. Cater*, S. Norris*, P. Medrano-Gracia*, A. Young*, B. Cowan*

*The University of Auckland, Auckland, New Zealand
†Auckland Heart Group, Auckland, New Zealand
‡Auckland City Hospital, Auckland, New Zealand

Chapter Outline

Computing and Visualization for Intravascular Imaging and Computer-Assisted Stenting
http://dx.doi.org/10.1016/B978-0-12-811018-8.00009-6

1. VASCULAR HEMODYNAMICS AND ATHEROSCLEROSIS

Atherosclerotic cardiovascular disease causes 31% of global fatalities annually [1], and this is expected to increase due to the aging of the world's population and rising obesity rates. Atherosclerosis is referred to as the "21st century disease" [2], and there is a pressing need for a better understanding of its causes and improved treatments. The most common type of cardiovascular disease is coronary artery disease (CAD) (43%), which alone causes an estimated 7.4 million deaths worldwide each year [1].

1.1 A Brief Description of Coronary Artery Disease

CAD is characterized by plaque build-up in the subendothelium (the innermost layer) of the arterial wall, leading to narrowing, reduced blood supply, and a reduction in oxygen and nutrient delivery to the beating heart muscle. The clinical consequences of this include cardiac ischemia manifested as chest pain or angina, myocardial infarction (commonly known as a "heart attack"), and sudden death [1].

Blood flow in arteries induces shear stresses on the vessel wall. In some locations, these stresses can influence or even injure the endothelial cells lining the wall [3]. Vessel branching or bifurcations have complex shapes which create complex flow regions with a higher likelihood of adverse stresses, making them more susceptible to atherosclerotic disease [4, 5].

The severity of arterial narrowing is determined by measuring the pressure before and after the narrowed lesion, and calculating the resulting pressure gradient. This technique is referred to as fractional flow reserve (FFR). If the pressure gradient is above a given threshold, then the lesion is commonly treated with the implant of stents (wire mesh tube which scaffolds the vessel open) [6]. However, some stent patients experience complications, and flow alternations induced by the presence of the stent are known to be a major cause [7]. Understanding blood flow can therefore aid in the understanding of disease and treatment complications, and potentially improve stent design and deployment techniques, particularly for bifurcations.

1.2 Study of Coronary Artery Flow

One cornerstone in coronary blood flow research is the investigation of the hemodynamics, such as wall shear stress, wall shear stress gradient, and oscillatory shear stress at the arterial wall. A coronary hemodynamic assessment can yield important information on the localization and progression of the arterial disease [3]. Coronary flow assessment with hemodynamic quantification can be achieved by using either computational modeling or experimental studies, or a combination of the two.

Computational modeling with computational fluid dynamics (CFD) is a well-established technique which provides numerical prediction of velocity, pressure, and stress in the fluid domain of interest. With an accurate description of the boundary conditions, it provides insights into the effects of stent design [8] and deployment [9], and vessel shape effects on blood flow [5]. Vessel regions exposed to adverse stress can be identified, which enables predictions of sites for plaque development and evolution [10]. In order to make these computational efforts feasible, simplifications and assumptions need to be made and CFD simulations are only as good as these underlying assumptions. These limitations should always be considered during application [11].

Artery flow can also be investigated in vivo, with noninvasive phase contrast magnetic resonance imaging (PC-MRI) currently being the gold standard technology [12]. However, PC-MRI application in vivo is limited to larger arteries such as the aorta [13], since it has a spatial resolution (approximately $1 \times 1 \times 1$ mm) which may only visualize a coronary artery with 2–3 voxels. Relatively long acquisition times are also common and respiratory and cardiac motion can be difficult to manage [13, 14]. These limitations have led to the use of in vitro studies, an established concept to overcome temporal measurement challenges [15].

The applicability of in vitro PC-MRI flow quantification of larger vessels has been demonstrated for medium-size vascular stents [16, 17]. Time-resolved PC-MRI flow measurements of normal and moderately stenosed carotid bifurcation models were described by Marshall et al. [18], with measured in- and outflows being used to prescribe the boundary conditions for comparison with CFD simulations. There was a good agreement of velocities between PC-MRI and CFD, but the geometries were generally oversimplified which potentially resulted in less complex flow zones and likely improved correspondence. The differences between PC-MRI and CFD occurred particularly in regions of low velocity. Overall, the value of combining CFD and PC-MRI was demonstrated for future studies.

The potential for PC-MRI phantom studies has been further demonstrated with detailed measurements of the hemodynamics of the aortic arch [19, 20] and intracranial aneurysms [21]. The PC-MRI images agreed well with 3D printed phantoms ($r^2 = .99$, $p < .001$) [21], including coherent flow patterns [19], and a close correlation ($r^2 > .9$) with in vivo mean velocities [20].

While in vitro experiments can overcome temporal limitations in larger vessels, spatial restrictions remain for the small coronary arteries. To overcome this, the scaling of vascular replicas based on the principle of dynamic similarity was first suggested by Friedman et al. [22]. The Law of Dynamic Similarity ensures the development of identical flow patterns in scaled-up and true-scale cases by preserving the relationship between geometric scales and fluid forces with the constant ratios of forces expressed as nondimensional numbers such as the Reynolds number [23]. This is a common engineering procedure [24], and was recently applied to the problem of coronary artery flow, where a close correlation for true-scale computational fluid dynamics (CFD) and scaled-up PC-MRI flow was demonstrated for patient-specific arteries [5].

This leads to the suggestion that CFD simulations (and in particular their boundary conditions) could be defined based on in vitro PC-MRI data [19, 20] to improve the accuracy of the computed results [11]. This may enable a more detailed flow analysis than currently available, particularly useful for the assessment of stenting techniques [25] and patient-specific profiling [11].

2. VESSEL GEOMETRY

In order to computationally model and experimentally investigate coronary flow, the coronary artery geometry, including any deployed stents, must be specified. Anatomical accuracy is highly desired as it influences local blood flow and thus hemodynamics significantly [26].

2.1 Coronary Artery Geometry

The definition of the anatomy or geometry of the coronary arteries may be made from images acquired by a number of technologies (Table 9.1).

A combination of technologies, such as OCT and CT, or IVUS and biplanar XR, would also be possible and largely depends on availability. In 2002, Antiga et al. [27] were the first to reconstruct arterial bifurcation geometries from computed tomography angiography (CTA), and to demonstrate the speed, reliability, and reproducibility of this technique. Many studies have substantially adopted this methodology to generate 3D models of the coronary arteries [28–30], and even a comprehensive atlas [31].

Table 9.1 Advantages and disadvantages of modalities for geometrical coronary artery assessment

	CCA	CTA	MRI	IVUS	OCT
Spatial resolution (μm)	100–200	500–600	1000	80–120	10–20
Type of radiation	X-rays	X-rays	Radio waves	Ultrasound	Near-IR light
Radiation exposure	++	+++	–	–	–
Invasiveness	+++	+	+	+++	+++
Coronary assessment	+++	++	+	+++	+++
Stenosis detection	✓	✓	✓	✓	✓
Plaque composition detection	++	–	+	✓	✓ [a]
Applicability (ease of use)	+++	+++	++	+	+
Cost (patient, hospital, state)	++	+	+	++	++

[a] Limited capability due to the restricted depth of penetration of OCT enabling only endothelial and superficial plaques to be imaged with estimation of intima-media thickness growth in the early stages of atherosclerosis, but not discrimination between the four classical plaque types.

CCA, conventional coronary angiogram; *CTA*, computed tomography angiogram; *MRI*, magnetic resonance imaging; *IVUS*, intravascular ultrasound; *OCT*, optical coherence tomography; grading from + (least) to +++ (most); ✓ indicates ability, – indicates lack of capability.

Modified from Morlacchi S, Colleoni SG, Cardenes R, Chiastra C, Diez JL, Larrabide I, et al. Patient-specific simulations of stenting procedures in coronary bifurcations: two clinical cases. Med Eng Phys 2013.

This image-based vessel reconstruction technique requires the definition of the inner vessel surface (or lumen) by means of manual or automatic image segmentation. A combination of all of the registered and segmented 2D images of the vessels cross section then creates a 3D representation of the vessel's luminal surface [27]. For straight (or gently curving) vessels, this can be achieved by longitudinally connecting many cross sectional splines defining the vessel circumferences at each location. For bifurcations, the merging of the branches is accomplished using computer-aided drawing software and boundary patch tools [11, 32, 33].

In Beier et al. [5], a cohort of more than 100 CTA nondiseased patients was studied. The imaging was performed using a multidetector CT scanner using retrospective ECG gating following administration of a contrast medium. Segmentation was performed by an experienced analyst using the OsiriX CMIV CTA plug-in [34] to generate vessel centerlines. The vessel's luminal meshes were subsequently created using the software MIALite [35], which had previously been validated using the Rotterdam Coronary Artery Algorithm Evaluation Framework [36, 37]. An atlas of the left main coronary bifurcation was then created, with a principal component analysis revealing that the first mode of shape variation was the bifurcation angle between the left anterior descending (LAD) and the circumflex (LCX) arteries [38]. Three specific patient cases were studied, with bifurcation angles representative of the $\pm$1SD of the population spread.

2.2 Stented Vessels

Stented vessels can be reconstructed in computer-aided drawing software directly [11, 39], using virtual, idealized insertion techniques [32], or geometry information can be based on micro-CT images of experimentally deployed stents in phantoms [8]. It is possible to perform a structural analysis to generate a virtually deformed stented vessel shape which accounts for the stent-induced vessel deformation. These methods typically use a mechanical model to create the starting geometry for subsequent CFD analysis [40]. This approach was used for idealized straight vessels [41, 42] and generic bifurcations [43], but has also been applied to image-based nonbifurcating vessels [32].

Three main limitations remain for these studies: (1) the inability to account for the angioplasty balloon, which has been demonstrated to produce different results in stent deployment [9, 44], (2) the use of load-controlled boundary conditions, which prevents direct comparison to inflation pressure in clinical settings, and (3) the assumption of wall homogeneity due to a lack of knowledge of the wall composition [11]. Further research is needed to address these assumptions in order to provide more realistic models of the stent deployment process.

Using an idealized representation has the advantage of eliminating lesion specific characteristics (stent overlap, local deformation depending on tissue characteristics, etc.) and may be preferable outside the scope of patient-specific profiling [5, 8].

3. COMPUTATIONAL (CFD) MODELING

CFD modeling of cardiovascular flow has developed into a large research field due to the cost, limitations, and often invasive nature of in vivo flow measurement studies [45]. The inability of experimental studies to measure fluid motion at high resolution in time and space makes CFD a suitable complementary approach [19]. CFD simulations have a predictive value, and parameters can easily be changed to investigate different scenarios quickly and conveniently once the basic simulation model has been developed. In this context, CFD allows analysis of blood flow at many times in the cardiac cycle, coronary flow under varying physiological conditions, and estimation of induced vessel wall stresses.

3.1 Governing Equations and Modeling Assumptions

Given a geometry and definition of the boundary conditions, CFD is able to compute a solution for the flow domain by solving a set of partial differential equations representing the laws of physics, to determine velocity, pressure, stress, and other parameters. The equations of motion, also known as the Navier-Stokes equations, represent a force balance between pressure, acceleration, and viscous and gravitational forces

$$\rho\frac{\partial \mathbf{u}}{\partial t} + \rho(\mathbf{u} \bullet \nabla\bullet)\mathbf{u} = \mu\nabla^2\mathbf{u} + f - \nabla p \tag{9.1}$$

where $\mathbf{u}$ is the 3D velocity field, f is a body force (e.g., gravity), p is the scalar pressure field, ρ is the fluid density, and μ is the dynamic viscosity. For incompressible fluids such as blood, these equations can be closed with the conservation of mass, which can be expressed as [46]

$$\nabla \bullet \mathbf{u} = 0 \tag{9.2}$$

considered a constraint on the set of admissible velocities.

3.1.1 Flow Characterization

The two most important dimensionless numbers for blood flow are the Reynolds and Womersley numbers.

The Reynolds number *Re* is the ratio between the inertial and viscous forces, and is defined as

$$Re = \frac{\rho U D_h}{\mu} \tag{9.3}$$

where D_h is the hydraulic diameter of the vessel (for noncircular shapes this is $D_h = 4\frac{A}{P}$, where A is the cross-sectional area and P is the perimeter). The value of the Reynolds number can vary from a few thousand in the aorta to a few hundred in smaller arteries, with higher values (>2000) indicative of turbulent flow, and lower values of laminar flow.

The Womersley number characterizes the relative unsteadiness of an oscillating flow and is defined as

$$\alpha = \frac{D_h}{2}\sqrt{\frac{\omega\rho}{\mu}} \tag{9.4}$$

where ω is the oscillation frequency corresponding to the heartbeat. For the aorta, the Womersley number is approximately 15, and for the left main coronary artery it is approximately 3 [47]. Flow pulsatility and inertia are as important as the viscous forces because all of the respective terms in the governing equations are of a similar order of magnitude and therefore cannot be neglected [47]. The flow regime is laminar in most arteries, with exceptions in the larger vessels such as the aorta. The flow in severely narrowed vessels may cause accelerating jets, which create instability and can initiate transitional or turbulent flow. Real flows are also highly three-dimensional, and strong secondary flows are common in diseased blood vessels.

3.1.2 Simplifying Assumptions

While the study of wall pulsation is critical for plaque evolution or rupture, for coronary flow assessment, pulsation appears not to be significant. Cardiac motion has been found to have a relatively small hemodynamic effect in the coronary arteries [48]. Similarly, compliance of the coronary arteries has a minimal effect on flow [49, 50], with computational studies that compare rigid and compliant models finding only minor differences in hemodynamics [40]. Therefore, for simplicity, coronary arterial walls are often assumed to be rigid and stationary, particularly when modeling diseased or stented cases.

It is possible to perform a fluid-structure-interaction (FSI) analysis to account for compliance and motion. Ongoing limitations with this approach include the lack of heart motion including translational stretching, bending, and twisting, which are still a subject of study [40].

3.2 Mesh Discretization

After creating a digital geometry of the region of interest, discretization into finite volumes must occur for numerical analysis. The mesh type is selected depending on the topology of the vessel. Complex 3D volumes, such as stented arteries, are commonly discretized using unstructured tetrahedral meshes as they are easy to generate [33, 40, 51]. Good-quality meshes can generally be produced [52], but mesh convergence should always be validated to ensure the accuracy of the solution [53].

Several commercial and freely available meshing software packages exist, such as ANSYS Mesher or ICEM CFD (ANSYS Inc., Canonsburg, USA), or Hypermesh (Altair, Troy, US). Verification can be performed using a sensitivity analysis where the parameter(s) of interest are generated for different mesh densities for spatial convergence, and different time step sizes for temporal convergence testing. Widely accepted and recommended testing procedures for estimating the discretization error are Richardson extrapolation, and the "GCI method" (Grid Convergence Index) [54].

3.3 Boundary and Initial Conditions

After creation of the fluid domain, the definition of appropriate boundary conditions is an important modeling choice in CFD. These commonly comprise the inlet velocity flow profile, the pulsatile waveform describing the cardiac cycle, the definition of the outlet conditions, and the boundary conditions at the lumen walls.

3.3.1 Inlet Boundary Conditions

In vivo measurements from either a series of Doppler wire (a system allowing instantaneous peak velocity measurements) [55], or ultrasound measurements [13], have previously been used to prescribe the coronary inlet flow. The inherent limitation of these methods is that they influence the flow field they measure. As a consequence, the left main inlet flow has often been defined as a standard parabolic inlet profile [5, 8, 29, 41, 56–58].

3.3.2 Outlet Boundary Conditions

With the inlet flow defined, it is common practice to define the outlet boundary either using a constant pressure condition [59], or specifying the flow split ratio between the two daughter branches at a bifurcation [10, 60]. The latter approach is usually based on Murray's study in 1926, who used Poiseuille's law to determine the vessel radius minimizing energy expenditure [61]. This led to the following formula

$$\frac{q_{D2}}{q_{D1}} = \left(\frac{d_{D2}}{d_{D1}}\right)^3 \tag{9.5}$$

where the volume flow rates in each daughter branch q_{D1} and q_{D2} are related to their respective diameters d_{D1}, d_{D2}. For the coronaries, it was later demonstrated that flow ratios are better described using exponent values of less than three [62]. The presence of obstructive disease or stents can influence the flow further. For this reason and also stability, many CFD studies define a constant outlet pressure instead [41, 56, 59, 62, 63].

3.3.3 Shear-Thinning Behavior

It is well known that the assumption of Newtonian behavior does not apply to low shear rates flows such as in the coronaries [64, 65], and it has been demonstrated that wall shear stresses may be underestimated by up to 10% if this non-Newtonian behavior is ignored [66–68]. This difference may be even greater in regions of complex flow such as at bifurcations [69, 70], and for unsteady flow [71]. Although the significance of non-Newtonian behavior for coronary hemodynamics is well established, many investigators have neglected it [45]. A number of different mathematical viscosity models have been proposed to describe the non-Newtonian behavior of blood, whereas the "Carreau-Yasuda" model was found most accurate [72].

4. EXPERIMENTAL STUDIES

While CFD currently provides the highest spatial and temporal resolution for blood flow quantification, a complementary experimental approach can help to overcome some of its inherent limitations.

PC-MRI is a technology capable of providing velocity data in a noninvasive manner, and has previously been utilized in vivo in larger vessels such as the pulmonary arteries [73], and the thoracic aorta [19, 20]. In vitro flow acquisition was performed in 3D vascular replicas of the carotid arteries [74], the thoracic aorta [20], and for the coronary arteries using dynamic scaling [5].

4.1 Dynamic Scaling

Dynamic similarity is a well-established concept in fluid mechanics. It allows the faithful reproduction of the same flow pattern in a phantom (which may be scaled-up or down) as real-scale by matching the relevant dimensionless numbers. In the case of the small coronaries, an enlarged phantom is desirable due to the limitations of spatial resolution with PC-MRI.

In incompressible steady flow without surface tension, the Reynolds number is the only dimensionless number of interest (Eq. 9.3). By scaling the phantom diameter, the dynamic viscosity, velocity, or density can then be changed to obtain the full-scale Reynolds number in the phantom. This reproduces a flow field in the scaled-up phantom which is identical to the true-scale coronary artery.

This concept was first applied and validated for steady flow in three patient-specific left main coronary bifurcations using PC-MRI (Fig. 9.1) [5].

4.1.1 Steady State

In Beier et al. [5], matching of the Reynolds number was achieved by dynamically scaling the fluid viscosity, density, and velocity to compensate for the larger phantom diameter. The desired viscosity was estimated at the inlet. The target flow rate was taken from in vivo flow rates from the literature [75] (Table 9.2). The scaling varied slightly depending on the true-scale vessel diameter of the patient with all phantoms having a 25 mm inlet diameter for connection to the in vitro MRI setup (Fig. 9.2).

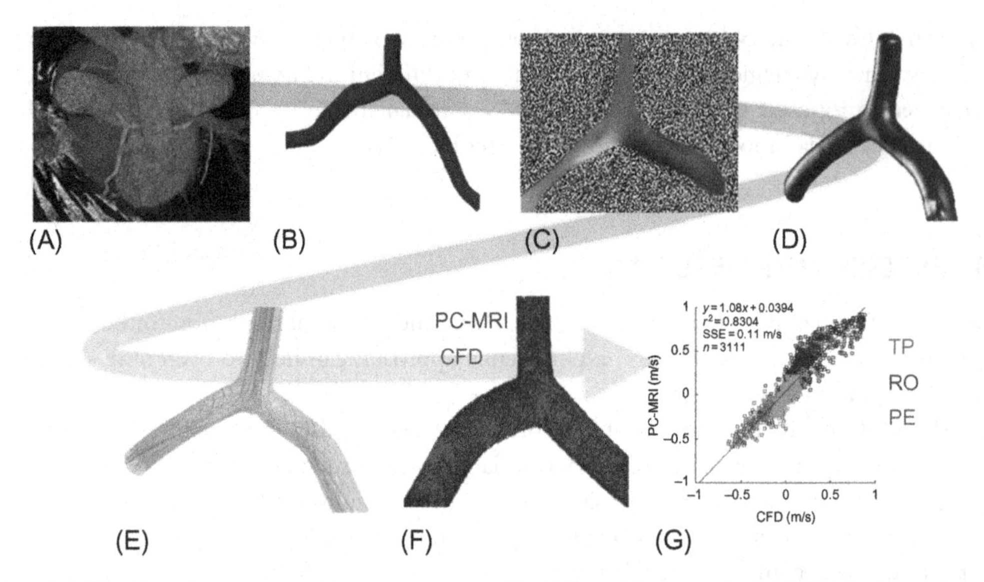

Fig. 9.1 Workflow for coronary flow assessment with CFD and large-scale PC-MRI. Workflow: from (A) CTA scans, (B) computer aided drawings (CAD) of vessels (digital geometry), (C) large-scale PC-MRI, (D) PC-MRI image segmentation, (E) true-scale computational fluid dynamics (CFD), (F) co-registration of PC-MRI and CFD for (G) analysis for the through-plane (TP), phase-encoded (PE), and read-out (RO) directions. *(From Beier S, Ormiston JA, Webster MW, Cater JE, Norris SE, Medrano-Gracia P, et al. Dynamically scaled phantom phase contrast MRI compared to true-scale computational modeling of coronary artery flow. J Magn Reson Imaging 2016;44(4):983–92.)*

Table 9.2 Scaling parameters to achieve a matching Reynolds number through dynamic similarity

		True-scale CFD	Scaled-up PC-MRI
Mean velocity (m/s)	v_{mean}	0.40	0.77
Inlet diameter (mm)	d	4.7	25
Density (kg/m^3)	ρ	1060	989
Viscosity (kg/ms)	μ_{mean}[a]	0.0035	0.0350
Resulting Reynolds number (–)	Re	540	540

[a] μ_{mean} is the mean viscosity estimated at the inlet.
Data from Beier S, Ormiston J, Webster M, Cater J, Norris S, Medrano-Gracia P, et al. Hemodynamics in idealized stented coronary arteries: important stent design considerations. Ann Biomed Eng 2015;1–15.

The local velocity varied by vessel shape with high velocity regions proximally, and low velocity regions in the distal daughter branches. The local velocity influences the MRI signal strength, which in turn affects the signal-to-noise ratio (SNR_v)

$$SNR_v = \left(\frac{\pi}{\sqrt{2}}\right)\left(\frac{v}{VENC}\right)SNR_{Img}. \tag{9.6}$$

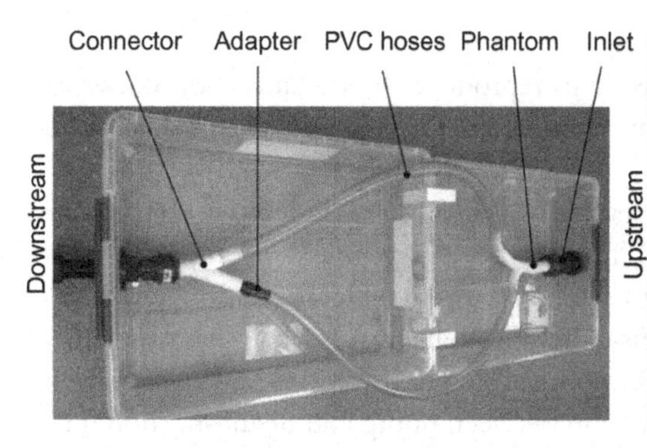

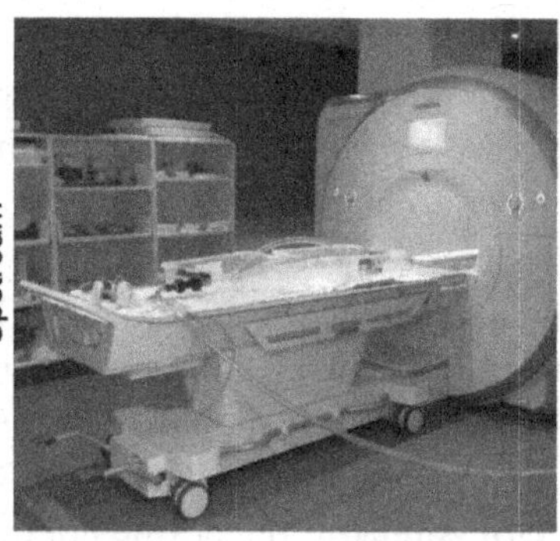

Fig. 9.2 Flow circuit for in vitro PC-MRI. In vitro PC-MRI with a dynamically scaled anatomically accurate coronary phantom incorporated into a flow circuit. *(From Beier S, Ormiston JA, Webster MW, Cater JE, Norris SE, Medrano-Gracia P, et al. Dynamically scaled phantom phase contrast MRI compared to true-scale computational modeling of coronary artery flow. J Magn Reson Imaging 2016;44(4):983–92.)*

where SNR_{Img} is the image SNR, $VENC$ is the selected Velocity ENCoding parameter determining the dynamic range of the grey scale intensity of the acquired PC-MRI images, and v is the local velocity [76].

The VENC should be selected so that the maximum velocity does not cause signal aliasing, and if this is high due to a wide range of velocities, then the local velocity SNR_v can be low in regions of slow flow [76]. For this reason, it is important to select the VENC just high enough to capture the highest peak velocities, but not higher to avoid regions of low SNR [5]. In case the region of interest has a high dynamic range of velocities, then the imaging subregions with different VENCs may be necessary to achieve a good SNR for the whole area.

Beier et al. demonstrated that more complex geometries (e.g., more tortuous vessels) may have larger regions of low velocities (and therefore lower SNR), which introduces greater errors in the PC-MRI measurements [5].

Achieving good velocity measurements is a trade-off between lowering the circuit flow rate and increasing the phantom size (to assure dynamic similarity with matching Reynolds numbers), because a larger phantom improves the effective spatial resolution but requires a slower flow rate, which reduces the SNR. Therefore, compared to Ref. [5] it is possible to increase the velocities to improve the image SNR by using lower phantom scaling factors, but this then causes a lower effective spatial resolution. Current 4D MRI can achieve a practical spatial resolution of $1 \times 1 \times 1$ mm, with 2D techniques achieving a greater in-plane resolution of 0.6×0.6 mm [13]. This may be useful when analyzing specific complex regions only, or in conjunction with a 3D whole volume acquisition. While better spatial resolution can be achieved, it comes at the expense of lower SNR or longer scan-times.

4.1.2 Transient Considerations

For transient experiments, it is necessary to reproduce the pulsatile blood flow through the cardiac cycle. Commercial pump systems are available to produce in vitro flow waveforms to match in vivo data.

Pressure loss associated with the length of the tubing connecting the pump to the phantom needs to be considered. This may be compensated by integrating a valve downstream of the phantom in order to adjust the resistance of the flow circuit.

Waveform degradation may be caused by wave reflections, which has previously been addressed in part by installing an MRI compatible pump on the MRI scanner's patient table [20]. Similarly, the hosing connection between pump and phantom should be rigid and straight for transient studies.

If dynamic scaling is performed for unsteady flow, then the Womersley number must be considered. This means that the pulsatile frequency needs to be adjusted according to the scaling factor chosen (Eq. 9.4, α). To the author's knowledge, this has not been reported for coronary artery flow investigations.

4.2 Phantoms

Phantoms have previously been produced using rubber casts from postmortem arteries [22], but the development of rapid prototyping, also called 3D printing, has more recently enabled the fast, accurate, and effective production of rigid phantoms [19, 20]. These can be anthropomorphically realistic when manufactured from 3D reconstructions of medical imaging data from patients. A resolution as fine as a 16 μm slice with 25–50 μm in-plane resolution is feasible using a modern 3D printing system [5]. These can even capture the stent wires within the vessel [77]. The material is usually a rigid plastic.

To achieve compliant phantoms, either commercially available rubber-like material can be used for PolyJet rapid prototyping [78], or rigid negative molds can be 3D printed and then a compliant material cast into them [79].

One research group has pursued efforts to generate a compliant phantom resembling the complex composition of the three individual arterial vessel wall layers [80]. The phantom was molded using an injection-like casting sequence with a three-layer construction to reflect more accurately the atherosclerotic vessel. The material used was silicone, an agar-based compound with water, glycerol, and cellulose particles, and the composition varied for each of the three layers to adjust their compliance. The phantom was then embedded in an air-chamber allowing outside pressure to be adjusted to study vessel compliance. As the geometry was only "inspired" by IVUS images [80], it is questionable how much this complex three-layer silicone construction actually represented arterial wall properties, which are very complex (anisotropic and nonhomogeneous), and still poorly quantified [50].

4.3 Compliance and Cardiac Motion

4.3.1 Coronary Compliance

The maximum pressure and velocity peak reduces with increasing distance from the heart resulting in a reduced pulsatile change in coronary diameter more distally in the coronary arteries [81]. Clinical observations show that diameter change also reduces with advancing disease as the vessel becomes stiffer, and even completely subsides in the presence of calcification [50]. The most recent computational fluid-structure-interaction comparisons show similar trends, concluding that the rigid-wall assumption for CFD appears reasonable [40].

One study at variance with these findings [82] reported flow measurements in elastic straight tubes with reduced differences in instantaneous and time-averaged shear and wall shear stresses of up to 30%. How well an elastic tube represents the coronary artery wall remains unknown. On the balance of evidence, and considering the complexities associated with modeling walls that move in response to pressure, a rigid vessel assumption is usually made when computationally modeling coronary blood flow.

4.3.2 Coronary Motion

To date, only one study has investigated the impact of coronary vessel motion resulting from cardiac motion on flow, demonstrating that the (axial) pulsatile flow dominates flow structure, and outweighs any minor bulk motion effects [48]. Based on these observations, a simplified model is usually used, where the coronaries are assumed to be motionless [11, 83].

4.4 Newtonian Versus Non-Newtonian Fluids

In larger vessels, non-Newtonian fluid properties do not need to be considered. However, for smaller vessels, such as the coronary arteries, modeling the blood as non-Newtonian is essential to accurately predict hemodynamics (Section 3.3.3) [84]. Wall shear stress measures, which are used to predict disease location or for stent design considerations, vary significantly between non-Newtonian and Newtonian fluids [66].

Dynamic scaling of non-Newtonian flow has previously been demonstrated [85] and applied to large-scale patient PC-MRI acquisition [5]. Previous blood-mimicking fluids include a mixture of 60% distilled water and 40% glycerol [20], various aqueous solutions of glycerol and xanthan gum [84], and aqueous solutions of sodium chloride and xanthan gum [5].

4.4.1 Xanthan Gum as Blood-Mimicking Fluid

A commonly used blood-mimicking solution with similar shear thinning properties is a xanthan gum-water solution. Xanthan gum is a natural, high-molecular weight polysaccharide produced by the microorganism *Xanthomonas campestris* by microbial fermentation. A solution of 0.05 w/w% exhibits shear thinning behavior similar to blood. The advantages of xanthan gum mixtures are that they are easy to acquire, clean, and safe to use. The gum's disadvantage is that it has a strong tendency to agglomerate into lumps when added to water, which can adversely affect the generation of the desired shear properties. A high shear mixer can be used to ensure a homogeneous mixture [5].

4.4.2 Viscosity Properties

Viscosity measurements of the final mixtures are recommended to confirm that the desired target viscosity is achieved [77]. For xanthan gum, increases in temperature and addition of sodium chloride (to prevent mold) did not have any effect on the measured viscosity [77].

4.5 Experimental Design with PC-MRI

4.5.1 Circuit Design

It is recommended to have a recirculating flow circuit to avoid flow stagnation, which can either be completely MR compatible (containing no magnetic material) and located within the scan room, or include non-MR compatible components which may be placed outside by passing pipe connections through the scan room wave guides.

The phantom should be positioned at the isocenter of the MRI scanner where the magnetic field homogeneity is greatest to minimize image distortion. A pump is used to circulate the fluid, with a reservoir to collect fluid, a flow meter to monitor flow rate, and a control valve to regulate the flow.

It should be noted that most pipes and connections come in standard sizes. To fit the phantom to commercially available fittings, a thread can be 3D printed, or a flexible PVC pipe may be fitted around the outside of the phantom and secured with a plastic hose clip (Fig. 9.2).

Rigid pipes are recommended for transient studies. The pump should be placed close to the region of interest without bends between the pump and phantom to minimize wave reflections. For steady-state studies, pipes may be soft and a rigid entrance length is only required if a fully developed parabolic profile is desired (Fig. 9.3).

4.5.2 MRI Scan Protocol

The use of contrast agent such as Gd-BOPTA (Multihance, Bracco) may be considered to increase signal strength and therefore improve signal-to-noise ratio. However, this may be costly as a sufficient quantity for a large 100L flow circuit would be required. Experiments without contrast have previously been found sufficient [5].

Importance of non-Newtonian considerations have often been addressed however, and recent experiments successfully used a xanthan gum - water mixture as blood

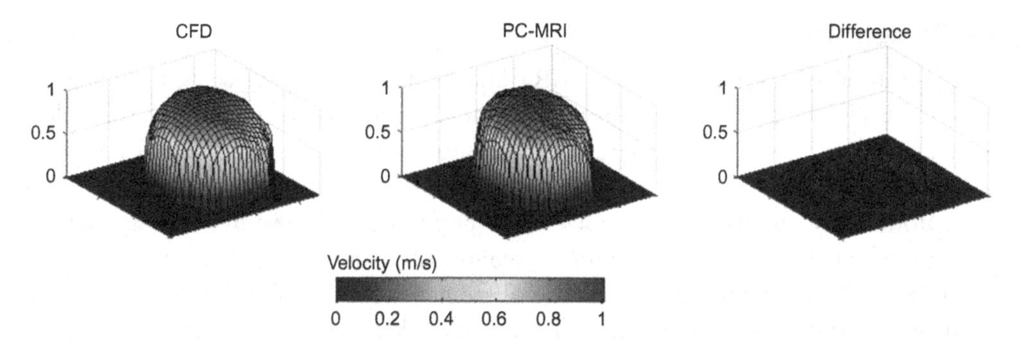

Fig. 9.3 Scaled-up PC-MRI and true-scale CFD velocity profiles. Inlet velocity profile for true-scale CFD, large-scale experimental PC-MRI and their difference (*left-to-right*) for a patient-specific bifurcation where PC-MRI's measured profile was applied to the CFD solution. *(From Beier S, Ormiston JA, Webster MW, Cater JE, Norris SE, Medrano-Gracia P, et al. Dynamically scaled phantom phase contrast MRI compared to true-scale computational modeling of coronary artery flow. J Magn Reson Imaging 2016.)*

mimicking fluid [5]. In case non-Newtonian properties are ignored for simplification, then water can be used for the experiments [20].

Three-directional velocity data can be acquired where the intensity of each pixel is decoded using an appropriate VENC for each of the through-plane (TP), phase-encoding (PE), and read-out (RO) directions. The selected VENCs may vary from phantom to phantom in order to optimize the dynamic range and achieve optimal SNR (Eq. 9.6). This can be achieved by choosing a VENC close to the expected local maximum velocity for each encoding direction as discussed earlier.

5. DATA POSTPROCESSING, CO-REGISTRATION, AND COMPARISON

Experimental flow acquisition with PC-MRI and haemodynamic CFD simulations combined can overcome each techniques shortcomings and improve insights. To achieve this, the acquired data need to be segmented, co-registered, and statistically analyzed.

5.1 Segmentation of the Imaging Data

Medical imaging data, in vitro or in vivo needs image segmentation. Several different techniques exist, such as thresholding, histogram-based methods, edge detection, region-growth methods, and many more [86]. The technique should be selected according to the images acquired. When segmenting an in vitro phantom MRI, segmentation is easier than in vivo due to the more clearly defined interfaces, and more advanced methods may not be necessary. Using a histogram-based approach was found to be effective, to identify the intensity cluster location and therefore enabled an easy selection of an appropriate threshold [5].

5.2 Co-Registration

Co-registration of the CFD and PC-MRI images may be performed using intensity-based methods or feature-based approaches. The latter allows the user to spatially select corresponding points in both coordinate systems for an automated alignment calculation (a single value decomposition is used to determine a transformation matrix). Lines or contours may also be used to establish a point-by-point feature correspondence. Intensity-based techniques use an entire image or subimages for registration.

Other forms of registration include coherent point drift algorithms [87], which optimize the distance between the surface point cloud data sets after segmentation [5]. These are based on Gaussian mixture models and are suitable for different sizes as required for dynamic scaling [88]. These lead to a transformation matrix, which provides an affine or nonrigid registration between both sets of topological images, with scale, rotation, translation for the former, and an additional shear component for the latter. For dynamic scaling, a simple affine transformation is usually adequate [5].

5.3 Statistical Analysis

After co-registration, the results of both CFD and PC-MRI technique can be directly compared, even when PC-MRI was scaled in vitro. Previous efforts used interpolation methods to perform a point-by-point velocity comparison [5, 19]. Linear interpolation is not recommended for velocity data close to the vessel walls where there is a strong nonlinear spatial gradient (boundary layer), as this would introduce errors. More sophisticated methods using Gaussian processes are a possibility, but a simpler natural-neighbor approach has previously been found sufficient [5].

To understand the correlation between the data sets, a quality of fit should be calculated. General correlation may not always be suitable, as a higher functional relationship may return a zero correlation (meaning not correlated) even when the shapes are clearly correlated. Isotropic covariance σ^2 (mm^2) has been used in the past and can be calculated as

$$\sigma^2 = \frac{1}{DMN} \sum_{n=1}^{N} \sum_{m=1}^{M} ||x_n - y_m||^2 \tag{9.7}$$

where x and y are the point sets of MRI and CFD respectively, and M, N are the number of dimensions of the points D [5, 88].

In order to evaluate the differences between the CFD and MRI flow field, the median and maximum velocity norm differences can be calculated. PC-MRI flow quantification errors are typically on the order of 5–10% [89], and thus an error <15% may be deemed acceptable [5].

The Spearman rank coefficient is a nonparametric measure of statistical dependence between two variables and may be used to further quantify the velocity norm differences. It assesses how well the relationship between two variables can be described using a monotonic function. If there are no repeated data values, a perfect Spearman correlation of +1 or −1 occurs when each of the variables is a perfect monotonic function of the other [77].

The linear, parametric Pearson correlation can be applied to assess correlations between individual vector components. Other possible assessment tools include Bland-Altman plots to determine the mean variation according to the velocity norm range [77].

6. ACCURACY AND RELIABILITY

6.1 Validation with Experimental Data

The experimental modeling of transient flow is challenging. In Ref. [90], good correlation was found close to the inlet between the velocities of true-scale in vitro MRI to CFD of an identical thoracic aorta. However, discrepancies rapidly increased downstream through the phantom. It was found that the experimental setup introduced a pulsatile waveform reflection which was not modeled by the CFD, causing secondary forward flow. The correlation between the two flows reduced from an r^2 of .88 in the first cross-sectional slice (by the inlet boundary) to only .47 in the last slice close to the exit boundary, with virtually no similarity between the velocity profiles.

In other work, the importance of having an appropriate CFD inlet boundary condition (e.g., an experimentally determined velocity profile) has been demonstrated, in order to gain a good agreement with the experiment [8, 20].

A more recent in vitro PC-MRI flow acquisition was performed using scaled-up coronary phantoms of patient-specific coronary geometries. While the study only examined steady flow, correlation to identical but true-scale CFD (driven with PC-MRI measured inlet velocity profiles) showed good agreement throughout the phantom (2–8% difference and $\rho > .72$ in velocity norms with $r^2 > .82$ in the velocity vector direction, Table 9.3) [5].

Table 9.3 Comparison between dynamically scaled PC-MRI and true scale CFD 3D flow fields for a patient-specific coronary bifurcation

		Patient-specific
Surface correlation (mm^2)	σ^2	3.5×10^{-5}
Pearson correlation(–)	r^2	0.87
Intercept (m/s)	–	0.946
Slope (–)	–	0.025
Spearmen rank coefficient (–)	ρ	0.79
Median difference velocity magnitudes (%)	–	2

From Beier S, Ormiston JA, Webster MW, Cater JE, Norris SE, Medrano-Gracia P, et al. Dynamically scaled phantom phase contrast MRI compared to true-scale computational modeling of coronary artery flow. J Magn Reson Imaging 2016.

Overall flow features were captured by both methods even in complex flow regions (Fig. 9.4), but small discrepancies were found in slow or oscillating flow zones. This can be explained by the lower SNR in slow flow regions, which increases the PC-MRI's acquisition error [5, 89]. These zones were bifurcation shape dependent, and more complex bifurcation shapes, such as those with high tortuosity, introduced larger discrepancies. The dynamic scaling reduced the PC-MRI's inherent spatio-temporal limitation by providing an effectively higher spatial resolution, which likely contributed to the better agreement.

The same method was tested for stented phantoms, revealing that a dynamically scaled in vitro PC-MRI was able to detect small flow-alterations induced by the presence of the stent [91], which has not been previously achieved (Fig. 9.5) [77].

6.2 Validation with In Vivo Data

Previous true-scale 3D MR flow measurements have been compared to CFD for the thoracic aorta, and while the flow patterns where somewhat similar, the velocity correlation was poor and did not reflect in vivo conditions [20]. Spatio-temporal resolution and intrinsic MRI measurement errors were identified as possible causes.

In vivo 3D MRI velocity measurements in three intracranial aneurysms were more promising when compared to patient-specific CFD models [90]. The flow patterns correlated somewhat with the CFD, but velocity norms and shear stresses had poor agreement (Spearman's coefficient ρ of .56 and .48–.59, respectively). Again, the MRI's

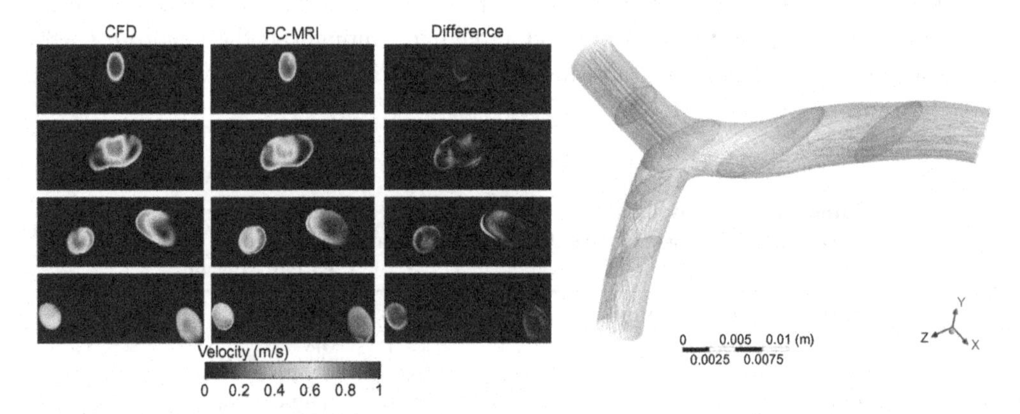

Fig. 9.4 Scaled-up PC-MRI and true-scale CFD comparison. Cross-sectional velocity contour comparison between true-scale CFD and large-scale PC-MRI and their difference (*left*) for the patient-specific bifurcation shown as 3D flow field volume with streamlines (*right*). *(From Beier S, Ormiston JA, Webster MW, Cater JE, Norris SE, Medrano-Gracia P, et al. Dynamically scaled phantom phase contrast MRI compared to true-scale computational modeling of coronary artery flow. J Magn Reson Imaging 2016.)*

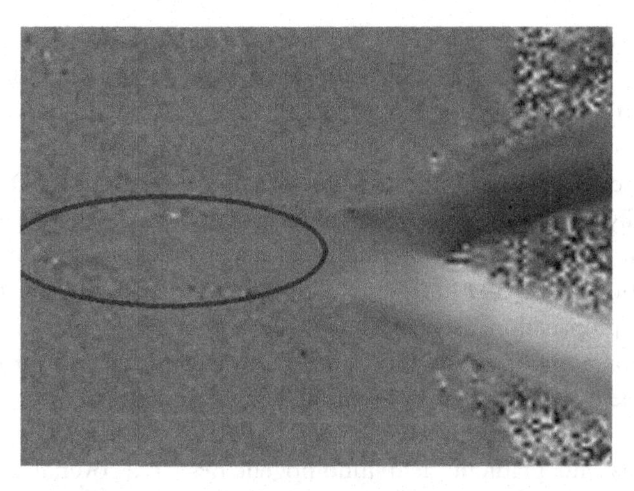

Fig. 9.5 Scaled-up PC-MRI image of a stented phantom. Left-to-right read out direction for a stented phantom PC-MRI acquisition image where the lattice of stent induced flow alternations can be seen (circled). *(Image courtesy of Susann Beier.)*

spatial resolution was identified as the most important source of discrepancies, followed by the temporal resolution.

Recently, in vivo efforts to generate patient-specific boundary conditions have been pursued for medium-size vessels such as the pulmonary arteries [73]. These provided boundary data for CFD and were compared to the standard of care technique (QFlow, Medis Inc., Leiden, the Netherlands). Although larger differences (22.6%) were found in regions with lower flow rates due to excessive flow regurgitation, the potential of a combined MRI-to-CFD again approach was demonstrated.

6.3 Ongoing Limitations

The definition of the accurate CFD boundary conditions is crucial for meaningful simulation results [20]. Previous research derived CFD boundary conditions from in vitro phantom acquisitions [5, 19, 20], or from accessible in vivo vessels [73, 90, 92]. For the coronary arteries, detailed boundary conditions remained unavailable to date however, with invasive technologies changing the measured flow profile (IVUS), and noninvasive PC-MRI providing maximum velocity when in vivo only, rather than a detailed profile [13].

While steady-state efforts are promising [5], and even enable stent-induced flow detection when dynamically scaled [93], transient tests remain unsuccessful [20].

MRI's limited spatio-temporal resolution inhibits detailed noninvasive flow acquisition. The use of in vitro steady-state studies can eliminate temporal challenges [19], and dynamically scaling such experiments can provide higher effective spatial resolution [5], making even small flow patterns accessible, such as stent-induced changes [93].

In vitro transient models have been unsatisfactory in the past [20], and using dynamic scaling for transient models may be an option but has not been explored to date. It is possible however, that the challenges of in vitro reproduction of the transient pulse wave, namely wave reflection within the experimental flow circuit, outweigh the discrepancies introduced when initiating a transient CFD simulation with an experimentally validated steady-state solution only [5, 20]. Thus, an initial steady-state description of the boundary conditions may deliver an acceptable subsequent transient prediction with CFD [94], due to confidence in the mathematical description of pulsatile wave propagation. Besides there is a lack of knowledge of in vivo parameters, such as local lesion composition, which affect the wave propagation through the arterial tree and make them therefore hard to reproduce experimentally.

Other sources of discrepancies may be MRI's limited measurement accuracy, caused by eddy currents and gradient field inhomogeneities. The overall accuracy of PC-MRI flow measurements is expected to be approximately 5–10% [89]. PC-MRI also systematically overestimates near-wall-velocities, which can be attributed to partial voluming effects (errors due to variation in velocity within one voxel), and are most significant near vessel/phantom walls [89].

Most importantly, the link between in vitro studies and in vivo observation is still weak, such that it remains uncertain to what extent the experimentally observed flow patterns reflect the true vascular hemodynamics. The potential for in vitro efforts has, however, been demonstrated in the past, and the continuation of this pursuit is warranted.

7. CURRENT DEVELOPMENTS

Although in vivo studies are desirable, they do not offer the possibility of predicting hemodynamic effects of vascular alterations related to surgical interventions to date.

The use of 5:1 scaled-up phantoms allows PC-MRI acquisition with approximately five times better spatial resolution, which is otherwise not accessible with in vivo or true-scale in vitro studies. Dynamically scaled in vitro PC-MRI flow experiments provide accurate flow measurements and enable measurement of stent-induced flow patterns that were previously unresolvable.

It is believed that in vitro studies will ultimately be useful for surgical planning to test interventional strategies on a patient-specific basis (patient profiling) or to increase the understanding of hemodynamic effects of different intervention strategies for patient cohorts.

In the future, the study of pathological patient vessels would benefit from a dynamically scaled in vitro profiling approach as the narrowed vessel lumen can be effectively enlarged, enabling detailed flow analysis in the lesion region. Advanced disease affects the local blood flow dramatically such that the flow profile can change from naturally laminar zones to turbulent, jet-like flow structures associated with adverse pathological effects (endothelial cell injury, etc.).

Modern medical research increasingly considers patient-specific treatment. For coronary artery disease, FFR remains the current standard technique in coronary catheterization to determine if a patient will undergo surgical intervention or medical treatment. This technique is associated with risks for the patients, and research efforts today pursue noninvasive alternatives using either CFD [95, 96] or 4D PC-MRI flow [97]. While these approaches seem feasible, further validation is needed to prove in vivo correlation of the underlying assumptions for the former, and the limited spatio-temporal resolution of the latter. Using dynamically scaled in vitro PC-MRI can overcome spatio-temporal limitations, deliver reliable pressure measurements [97], and improve the fidelity of CFD models through the provision of accurate boundary condition data [5]. Therefore, dynamic scaling of in vitro PC-MRI studies may enable patient-specific profiling for FFR, wall shear stress, and flow pattern assessment, and provide a validation platform for CFD studies.

This can also be relevant for hemodynamic stent studies such as testing of bifurcation stenting techniques, stent design, and variability between surgeons for interventional guidance and assessment. Further studies are warranted, however, as the applicability for transient considerations have not been demonstrated, replication of in vivo boundary conditions needs to be achieved, and the method is so far not applicable for routine clinical practice. Comparison to in vivo measurements is desirable.

ACKNOWLEDGMENTS

SB was funded by the Auckland Heart Group Charitable Trust. PMG was supported by the Green Lane Research and Educational Fund. The authors wish to acknowledge the Centre for eResearch (http://www.eresearch.auckland.ac.nz) and NeSI high-performance computing facilities (https://www.nesi.org.nz) for their support of this research. New Zealand's national facilities are provided by the New Zealand eScience Infrastructure and funded jointly by NeSI's collaborator institutions through the Ministry of Business, Innovation & Employment Research Infrastructure.

REFERENCES

[1] Alwan A. Global status report on noncommunicable diseases 2010. Geneva: World Health Organization; 2011.

[2] George SJ, Johnson J. Atherosclerosis: molecular and cellular mechanisms. New York: John Wiley & Sons; 2010.

[3] Ku DN. Blood flow in arteries. Annu Rev Fluid Mech 1997;29(1):399–434.

[4] Cecchi E, Giglioli C, Valente S, Lazzeri C, Gensini GF, Abbate R, et al. Role of hemodynamic shear stress in cardiovascular disease. Atherosclerosis 2011;214(2):249–56.

[5] Beier S, Ormiston JA, Webster MW, Cater JE, Norris SE, Medrano-Gracia P, et al. Dynamically scaled phantom phase contrast MRI compared to true-scale computational modeling of coronary artery flow. J Magn Reson Imaging 2016;44(4):983–92.

[6] Sels JEM, Tonino WAL, Pijls NHJ. Fractional flow reserve. In: Catheter-based cardiovascular interventions. Berlin: Springer; 2013. p. 349–61.

[7] Baber U, Kini AS, Sharma SK. Stenting of complex lesions: an overview. Nat Rev Cardiol 2010;7(9):485–96.
[8] Beier S, Ormiston J, Webster M, Cater J, Norris S, Medrano-Gracia P, et al. Hemodynamics in idealized stented coronary arteries: important stent design considerations. Ann Biomed Eng 2015;1–15.
[9] De Beule M, Mortier P, Carlier SG, Verhegghe B, Van Impe R, Verdonck P. Realistic finite element-based stent design: the impact of balloon folding. J Biomech 2008;41(2):383–9.
[10] Malvé M, Gharib AM, Yazdani SK, Finet G, Martínez MA, Pettigrew R, et al. Tortuosity of coronary bifurcation as a potential local risk factor for atherosclerosis: CFD steady state study based on in vivo dynamic CT measurements. Ann Biomed Eng 2015;43(1):82–93.
[11] Morlacchi S, Colleoni SG, Cardenes R, Chiastra C, Diez JL, Larrabide I, et al. Patient-specific simulations of stenting procedures in coronary bifurcations: two clinical cases. Med Eng Phys 201335(9):1272–81.
[12] Markl M, Frydrychowicz A, Kozerke S, Hope M, Wieben O. 4D flow MRI. J Magn Reson Imaging 2012;36(5):1015–36.
[13] Torii R, Keegan J, Wood NB, Dowsey AW, Hughes AD, Yang G-Z, et al. MR image-based geometric and hemodynamic investigation of the right coronary artery with dynamic vessel motion. Ann Biomed Eng 2010;38(8):2606–20.
[14] Johnson K, Sharma P, Oshinski J. Coronary artery flow measurement using navigator echo gated phase contrast magnetic resonance velocity mapping at 3.0 T. J Biomech 2008;41(3):595–602.
[15] Perktold K, Hofer M, Rappitsch G, Loew M, Kuban BD, Friedman MH. Validated computation of physiologic flow in a realistic coronary artery branch. J Biomech 1997;31(3):217–28.
[16] Holton A, Walsh E, Anayiotos A, Pohost G, Venugopalan R. Comparative MRI compatibility of 316 L stainless steel alloy and nickel-titanium alloy stents: original article technical. J Cardiovasc Magn Reson 2002;4(4):423–30.
[17] Walsh EG, Holton AD, Brott BC, Venugopalan R, Anayiotos AS. Magnetic resonance phase velocity mapping through NiTi stents in a flow phantom model. J Mag Reson Imaging 2005;21(1):59–65.
[18] Marshall I. Targeted particle tracking in computational models of human carotid bifurcations. J Biomech Eng 2011;133(12):124501.
[19] Stalder AF, Frydrychowicz A, Russe MF, Korvink JG, Hennig J, Li K, et al. Assessment of flow instabilities in the healthy aorta using flow sensitive MRI. J Magn Reson Imaging 2011;33(4):839–46.
[20] Canstein C, Cachot P, Faust A, Stalder AF, Bock J, Frydrychowicz A, et al. 3D MR flow analysis in realistic rapid-prototyping model systems of the thoracic aorta: comparison with in vivo data and computational fluid dynamics in identical vessel geometries. Magn Reson Med 2008;59(3):535–46.
[21] Anderson JR, Thompson WL, Alkattan AK, Diaz O, Klucznik R, Zhang YJ, et al. Three-dimensional printing of anatomically accurate, patient specific intracranial aneurysm models. J Neurointerven Surg 2016;8:517–20.
[22] Friedman MH, Kuban BD, Schmalbrock P, Smith K, Altan T. Fabrication of vascular replicas from magnetic resonance images. J Biomech Eng 1995;117(3):364–66.
[23] Langhaar HL. Dimensional analysis and theory of models, vol. 2. New York: Wiley; 1951.
[24] Sedov LI. Similarity and dimensional methods in mechanics. Boca Raton: CRC Press; 1993.
[25] Katritsis DG, Theodorakakos A, Pantos I, Gavaises M, Karcanias N, Efstathopoulos EP. Flow patterns at stented coronary bifurcations: computational fluid dynamics analysis. Circ Cardiovasc Interv 2012;5:530–9.
[26] Beier S, Ormiston J, Webster M, Cater J, Norris S, Medrano-Gracia P, et al. Impact of bifurcation angle and other anatomical characteristics on blood flow–A computational study of non-stented and stented coronary arteries. J. Biomech. 2016;49(9):1570–82.
[27] Antiga L, Ene-Iordache B, Caverni L, Cornalba GP, Remuzzi A. Geometric reconstruction for computational mesh generation of arterial bifurcations from CT angiography. Comput Med Imaging Graph 2002;26(4):227–35.
[28] Dvir D, Marom H, Assali A, Kornowski R. Bifurcation lesions in the coronary arteries: early experience with a novel 3-dimensional imaging and quantitative analysis before and after stenting. EuroIntervention 2007;3(1):95–9.

[29] Frauenfelder T, Boutsianis E, Schertler T, Husmann L, Leschka S, Poulikakos D, et al. In-vivo flow simulation in coronary arteries based on computed tomography datasets: feasibility and initial results. Eur Radiol 2007;17(5):1291–300.

[30] Galassi AR, Tomasello SD, Capodanno D, Seminara D, Canonico L, Occhipinti M, et al. A novel 3-D reconstruction system for the assessment of bifurcation lesions treated by the mini-crush technique. J Interven Cardiol 2010;23(1):46–53.

[31] Medrano-Gracia P, Ormiston J, Webster M, Beier S, Young A, Ellis, C., et al., A computational atlas of normal coronary artery anatomy. EuroIntervention 2016;12(7):845–54.

[32] Gijsen FJH, Migliavacca F, Schievano S, Socci L, Petrini L, Thury A, et al. Simulation of stent deployment in a realistic human coronary artery. Biomed Eng Online 2008;7(1):23.

[33] van der Giessen AG, Schaap M, Gijsen FJH, Groen HC, van Walsum T, Mollet NR, et al. 3D fusion of intravascular ultrasound and coronary computed tomography for in-vivo wall shear stress analysis: a feasibility study. Int J Cardiovasc Imaging 2010;26(7):781–96.

[34] Wang C, Frimmel H, Persson A, Smedby Ö. An interactive software module for visualizing coronary arteries in CT angiography. Int J Comput Assist Radiol Surg 2008;3(1–2):11–18.

[35] Wang C, Frimmel H, Smedby Ö. Fast level-set based image segmentation using coherent propagation. Med phys 2014;41(7):073501.

[36] Schaap M, Metz CT, van Walsum T, van der Giessen AG, Weustink AC, Mollet NR, et al. Standardized evaluation methodology and reference database for evaluating coronary artery centerline extraction algorithms. Med Image Anal 2009;13(5):701–14.

[37] Kirişli HA, Schaap M, Metz CT, Dharampal AS, Meijboom WB, Papadopoulou SL, et al. Standardized evaluation framework for evaluating coronary artery stenosis detection, stenosis quantification and lumen segmentation algorithms in computed tomography angiography. Med Image Anal 2013;17(8):859–76.

[38] Medrano-Gracia P, Ormiston J, Webster M, Beier S, Ellis C, Wang C, et al. Construction of a coronary artery atlas from CT angiography. Med Image Comput Comput Assist Interv 2014;17:513–20.

[39] Morlacchi S, Keller B, Arcangeli P, Balzan M, Migliavacca F, Dubini G, et al. Hemodynamics and in-stent restenosis: micro-CT images, histology, and computer simulations. Ann Biomed Eng 2011;39(10):2615–26.

[40] Chiastra C, Migliavacca F, Martinez MA, Malve M. On the necessity of modelling fluid-structure interaction for stented coronary arteries. J Mech Behav Biomed Mater 2014;34:217–30.

[41] Balossino R, Gervaso F, Migliavacca F, Dubini G. Effects of different stent designs on local hemodynamics in stented arteries. J Biomech 2008;41(5):1053–61.

[42] Martin DM, Murphy EA, Boyle FJ. Computational fluid dynamics analysis of balloon-expandable coronary stents: influence of stent and vessel deformation. Med Eng Phys 2014;36(8):1047–56.

[43] Morlacchi S, Chiastra C, Cutri E, Zunino P, Burzotta F, Formaggia L, et al. Stent deformation, physical stress, and drug elution obtained with provisional stenting, conventional culotte and Tryton-based culotte to treat bifurcations: a virtual simulation study. EuroIntervention 2014;9(12):1441–53.

[44] De Beule M, Van Impe R, Verhegghe B, Segers P, Verdonck P. Finite element analysis and stent design: reduction of dogboning. Technol Health Care 2005;14(4–5):233–41.

[45] Lewis G. Materials, fluid dynamics, and solid mechanics aspects of coronary artery stents: a state-of-the-art review. J Biomed Mater Res B Appl Biomater 2008;86(2):569–90.

[46] Ferziger JH, Peric M. Computational methods for fluid dynamics. Berlin: Springer Science & Business Media; 2012.

[47] Rayz VL, Berger SA. Computational modeling of vascular hemodynamics. In: Computational modeling in biomechanics. New York: Springer; 2010. [chapter 5].

[48] Zeng D, Ding Z, Friedman MH, Ethier CR. Effects of cardiac motion on right coronary artery hemodynamics. Ann Biomed Eng 2003;31(4):420–29.

[49] Zeng D, Boutsianis E, Ammann M, Boomsma K, Wildermuth S, Poulikakos D. A study on the compliance of a right coronary artery and its impact on wall shear stress. J Biomech Eng 2008;130(4):041014.

[50] Mattace-Raso F, Van Der Cammen T, Hofman A, Van Popele N, Bos ML, Schalekamp M, et al. Arterial stiffness and risk of coronary heart disease and stroke the Rotterdam study. Circulation 2006;113(5):657–63.

[51] Chiastra C, Morlacchi S, Gallo D, Morbiducci U, Cardenes R, Larrabide I, et al. Computational fluid dynamic simulations of image-based stented coronary bifurcation models. J R Soc Interface 2013;10(84):20130193.
[52] Wang E, Nelson T, Rauch R. Back to elements-tetrahedra vs. hexahedra. In: Proceedings of the 2004 international ANSYS conference; 2004.
[53] De Santis G, Mortier P, De Beule M, Segers P, Verdonck P, Verhegghe B. Patient-specific computational fluid dynamics: structured mesh generation from coronary angiography. Med Biol Eng Comput 2010;48(4):371–80.
[54] Celik IB, Ghia U, Roache PJ. Procedure for estimation and reporting of uncertainty due to discretization in CFD applications. J Fluids Eng Trans ASME 2008;130(7):078001.
[55] Samady H, Eshtehardi P, McDaniel MC, Suo J, Dhawan SS, Maynard C, et al. Coronary artery wall shear stress is associated with progression and transformation of atherosclerotic plaque and arterial remodeling in patients with coronary artery disease. Circulation 2011;124(7):779–88.
[56] Jiménez JM, Davies PF. Hemodynamically driven stent strut design. Ann Biomed Eng 2009;37(8):1483–4.
[57] He Y, Duraiswamy N, Frank AO, Moore Jr JE. Blood flow in stented arteries: a parametric comparison of strut design patterns in three dimensions. J Biomech Eng 2005;127(4):637–47.
[58] Hsiao H, Chiu Y, Lee K, Lin C. Computational modeling of effects of intravascular stent design on key mechanical and hemodynamic behavior. Comput Aided Des 2012;44(8):757–65.
[59] Chaichana T, Sun Z, Jewkes J. Haemodynamic analysis of the effect of different types of plaques in the left coronary artery. Comput Med Imaging Graph 2013;37(3):197–206.
[60] Chaichana T, Sun Z, Jewkes J. Computation of hemodynamics in the left coronary artery with variable angulations. J Biomech 2011;44(10):1869–78.
[61] Murray CD. The physiological principle of minimum work: I. The vascular system and the cost of blood volume. Proc Natl Acad Sci USA 1926;12(3):207.
[62] Huo Y, Finet G, Lefevre T, Louvard Y, Moussa I, Kassab GS. Which diameter and angle rule provides optimal flow patterns in a coronary bifurcation? J Biomech 2012;45(7):1273–79.
[63] Pant S, Bressloff NW, Forrester AIJ, Curzen N. The influence of strut-connectors in stented vessels: a comparison of pulsatile flow through five coronary stents. Ann Biomed Eng 2010;38(5):1893–907.
[64] Johnston BM, Johnston PR, Corney S, Kilpatrick D. Non-Newtonian blood flow in human right coronary arteries: steady state simulations. J Biomech 2004;37(5):709–20.
[65] Johnston BM, Johnston PR, Corney S, Kilpatrick D. Non-Newtonian blood flow in human right coronary arteries: transient simulations. J Biomech 2006;39(6):1116–28.
[66] Mejia J, Mongrain R, Bertrand OF. Accurate prediction of wall shear stress in a stented artery: Newtonian versus non-Newtonian models. J Biomech Eng 2011;133(7):074501.
[67] Seo T, Schachter LG, Barakat AI. Computational study of fluid mechanical disturbance induced by endovascular stents. Ann Biomed Eng 2005;33(4):444–56.
[68] Benard N, Perrault R, Coisne D. Computational approach to estimating the effects of blood properties on changes in intra-stent flow. Ann Biomed Eng 2006;34(8):1259–71.
[69] Liepsch D, Moravec S. Pulsatile flow of non-Newtonian fluid in distensible models of human arteries. Biorheology 1984;21(4):571–86.
[70] Pohl M, Wendt MO, Werner S, Koch B, Lerche D. In vitro testing of artificial heart valves: comparison between Newtonian and non-Newtonian fluids. Artif Organs 1996;20(1):37–46.
[71] Chen J, Lu X. Numerical investigation of the non-Newtonian pulsatile blood flow in a bifurcation model with a non-planar branch. J Biomech 2006;39(5):818–32.
[72] Razavi A, Shirani E, Sadeghi MR. Numerical simulation of blood pulsatile flow in a stenosed carotid artery using different rheological models. J Biomech 2011;44(11):2021–30.
[73] Das A, Wansapura JP, Gottliebson WM, Banerjee RK. Methodology for implementing patient-specific spatial boundary condition during a cardiac cycle from phase-contrast MRI for hemodynamic assessment. Med Image Anal 2015;19(1):121–36.
[74] Marshall I, Zhao S, Papathanasopoulou P, Hoskins P, Xu XY. MRI and CFD studies of pulsatile flow in healthy and stenosed carotid bifurcation models. J Biomech 2004;37(5):679–87.
[75] Anderson HV, Stokes MJ, Leon M, Abu-Halawa SA, Stuart Y, Kirkeeide RL. Coronary artery flow velocity is related to lumen area and regional left ventricular mass. Circulation 2000;102(1):48–54.

[76] Taylor CA, Draney MT. Experimental and computational methods in cardiovascular fluid mechanics. Ann Rev Fluid Mech 2004;36:197–231.
[77] Beier S. Haemodynamic assessment of coronary flow with CFD and phase-contrast MRI [Biomedical Sciences thesis]. The University of Auckland; 2015.
[78] Biglino G, Verschueren P, Zegels R, Taylor AM, Schievano S. Rapid prototyping compliant arterial phantoms for in-vitro studies and device testing. J Cardiovasc Magn Reson 2013;15(1):1.
[79] Kim GB, Lee S, Kim H, Yang DH, Kim Y-H, Kyung YS, et al. Three-dimensional printing: basic principles and applications in medicine and radiology. Korean J Radiol 2016;17(2):182–97.
[80] Brunette J, Mongrain R, Laurier J, Galaz R, Tardif JC. 3D flow study in a mildly stenotic coronary artery phantom using a whole volume PIV method. Med Eng Phys 2008;30(9):1193–200.
[81] Truskey GA, Yuan F, Katz DF. Transport phenomena in biological systems. Upper Saddle River, NJ: Pearson-Prentice Hall; 2004. p. 6–12.
[82] Eguchi T, Watanabe S, Takahara H, Furukawa A. Development of pulsatile flow experiment system and PIV measurement in an elastic tube. Mem Fac Eng Kyushu Univ 2003;63(3):161–72.
[83] Murphy J, Boyle F. Predicting neointimal hyperplasia in stented arteries using time-dependent computational fluid dynamics: a review. Comput Biol Med 2010;40(4):408–18.
[84] Anastasiou AD, Spyrogianni AS, Koskinas KC, Giannoglou GD, Paras SV. Experimental investigation of the flow of a blood analogue fluid in a replica of a bifurcated small artery. Med Eng Phys 2012;34(2):211–18.
[85] Gray JD, Owen I, Escudier MP. Dynamic scaling of unsteady shear-thinning non-Newtonian fluid flows in a large-scale model of a distal anastomosis. Exp Fluids 2007;43(4):535–46.
[86] Narkhede HP. Review of image segmentation techniques. Int J Sci Mod Eng 2013;1(8):54–61.
[87] Medrano-Gracia P, Ormiston J, Webster M, Beier S, Ellis C, Wang C, et al. A statistical model of the main bifurcation of the left coronary artery using coherent point drift. In: Joint MICCAI workshops on computing and visualisation for intravascular imaging and computer-assisted stenting; 2015.
[88] Myronenko A, Song X. Point set registration: coherent point drift. IEEE Trans Pattern Anal Mach Intell 2010;32(12):2262–75.
[89] Jiang J, Kokeny P, Ying W, Magnano C, Zivadinov R, Mark Haacke E. Quantifying errors in flow measurement using phase contrast magnetic resonance imaging: comparison of several boundary detection methods. Magn Reson Imaging 2015;33(2):185–93.
[90] Boussel L, Rayz V, Martin A, Acevedo-Bolton G, Lawton MT, Higashida R, et al. Phase-contrast magnetic resonance imaging measurements in intracranial aneurysms in vivo of flow patterns, velocity fields, and wall shear stress: comparison with computational fluid dynamics. Magn Reson Med 2009;61(2):409–17.
[91] Ormiston JA, Beier S, Cowan B, Webster MWI, Why similar stent designs cause new clinical issues reply. JACC Cardiovasc. Interv. 2012;5(3):362–3.
[92] Isoda H, Ohkura Y, Kosugi T, Hirano M, Alley MT, Bammer R, et al. Comparison of hemodynamics of intracranial aneurysms between MR fluid dynamics using 3D cine phase-contrast MRI and MR-based computational fluid dynamics. Neuroradiology 2010;52(10):913–20.
[93] Beier S, Ormiston J, Webster M, Cater J, Norris S, Medrano-Gracia P, et al. Overcoming spatio-temporal limitations using dynamically scaled in vitro PC-MRI-A comparison to real scale CFD of idealised, stented and patient left main geometries. In: Joint MICCAI workshops on computing and visualisation for intravascular imaging and computer-assisted stenting; 2015.
[94] Ramaswamy SD, Vigmostad SC, Wahle A, Lai Y-G, Olszewski ME, Braddy KC, et al. Fluid dynamic analysis in a human left anterior descending coronary artery with arterial motion. Ann Biomed Eng 2004;32(12):1628–41.
[95] Zarins CK, Taylor CA, Min JK. Computed fractional flow reserve (FFTCT) derived from coronary CT angiography. J Cardiovasc Transl Res 2013;6(5):708–14.
[96] Taylor CA, Fonte TA, Min JK. Computational fluid dynamics applied to cardiac computed tomography for noninvasive quantification of fractional flow reserve: scientific basis. J Am Coll Cardiol 2013;61(22):2233–41.
[97] Deng Z, Fan Z, Xie G, He Y, Natsuaki Y, Jin N, et al. Pressure gradient measurement in the coronary artery using 4D PC-MRI: towards noninvasive quantification of fractional flow reserve. J Cardiovasc Magn Reson 2014;16(1):1.

CHAPTER 10

Arterial Flow Impact on Aneurysmal Hemodynamics

H.G. Morales, O. Bonnefous
Medisys-Philips Research, Paris, France

Chapter Outline

Computing and Visualization for Intravascular Imaging and Computer-Assisted Stenting
http://dx.doi.org/10.1016/B978-0-12-811018-8.00010-2

1. INTRODUCTION

1.1 Cerebral Aneurysms: The Pathology

A cerebral aneurysm is a cardiovascular disease consisting of a balloon-like dilatation of arteries as shown in Fig. 10.1. Aneurysms are mainly located near or at arterial bifurcations of those vessels forming the Circle of Willis, which is located at the brain base [1]. Around 2% of the general population harbors cerebral aneurysms [2], many of those remaining asymptomatic and harmless (between 50% and 80%) [1]. However, a spontaneous rupture of an aneurysm is the main cause (around 85%) of the nontraumatic subarachnoid hemorrhage (SAH), which is lethal between 25% and 50% of the time [3].

Aneurysm rupture produces an increase in the intracranial pressure due to blood leakage in the subarachnoid space. This leakage may injure the brain parenchyma [4] and can obstruct and stagnate the cerebrospinal fluid circulation, producing hydrocephalus [3]. Moreover, SAH can also induce vasospasm on the affected arteries, triggering critical ischemia in the supplied brain regions after one or two weeks of the bleeding [3].

1.2 Cerebral Aneurysms: The Clinical Problem

There are two ways to diagnose cerebral aneurysms and their rupture: by symptoms or by imaging.

The most common symptom of aneurysm rupture and subsequent SAH is a strong and sudden headache developed in seconds, described as the worst in one's lifetime by those affected. Usually, this headache is followed by a decrease in consciousness and by other neurological symptoms, such as paresis, vomiting, or neck stiffness [3, 5]. Some patients have either no symptoms or a minor headache, and can be misdiagnosed. Nonetheless, a sudden death could occur for the most unfortunate cases.

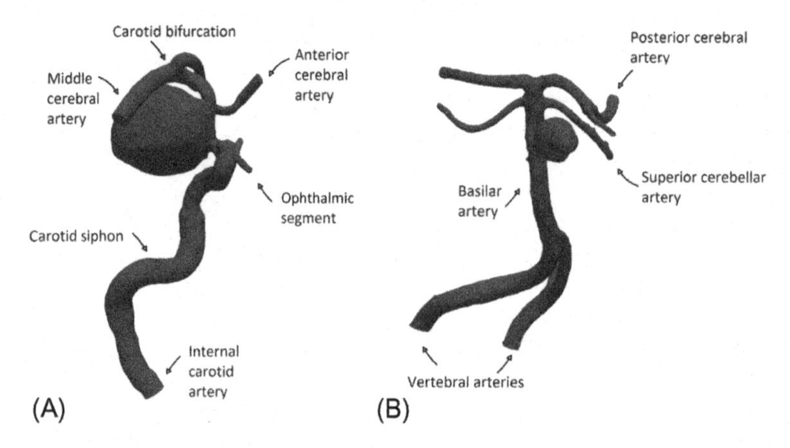

Fig. 10.1 Saccular aneurysms: (A) lateral aneurysm in the internal carotid artery (anterior circulation); (B) lateral trunk aneurysm in the Basilar artery (posterior circulation).

To diagnose cerebral aneurysms, there are three types of imaging modalities: magnetic resonance angiography, computed tomography angiography, and digital subtraction angiography (DSA). The use of these modalities depends on several factors, including patient medical record, age, aneurysm stage, spatial resolution of the modality, and costs.

Once a cerebral aneurysm is diagnosed by any of these means (symptoms or imaging), the clinical question is whether this patient should be monitored or treated. If monitored, there is the risk of later aneurysm rupture and potential patient death. If the aneurysm is treated, then the physicians need to decide which therapy should be used, considering the full patient condition, long-term success of the treatment, as well as the risk of the intervention [6]. Therefore, the physicians need to balance the risk of monitoring versus the risk of the best treatment option, for the specific patient at a given time. This decision is not easy to make since neither aneurysm rupture causes nor treatment outcomes are completely understood.

Several factors including genetics, gender, age, lifestyle habits (such as smoking, drug and alcohol consumptions among others), chemical and biomechanical factors have been considered and correlated with the stages of aneurysm (initiation, development, and potential rupture). Among biomechanical factors, blood flow has an enormous importance on the arterial wall behavior [7]. Although one of the fundamental roles of blood flow is to provide nutrients to the organs and to transport their wastes away, it also affects vascular remodeling [7] by supporting healing processes and by participating in degenerative progressions of the arterial wall [7–10].

In the case of treatment outcome, hemodynamics play a key role after the placement of endovascular devices such as coils or flow diverters [11, 12]. Endovascular therapies seek to modify locally the intra-aneurysmal hemodynamics and to trigger the coagulation cascade inside the aneurysm. Hopefully, the posttreatment hemodynamics will produce a stable thrombus inside the aneurysm sac, excluding the arterial malformation from blood circulation. This intra-saccular aneurysm thrombus will therefore reduce the risk of aneurysm rupture.

1.3 Cerebral Aneurysms: Hemodynamics

The need to understand the role of cardiovascular hemodynamics is essential to assess the causes of aneurysm initiation, development, rupture, and potential treatment outcome. Several review articles have been published in recent years in order to expose, discuss, and consolidate the main findings around hemodynamics at different aneurysm stages [13, 14]. One of the most commonly measured hemodynamic quantities is wall shear stress (WSS). WSS is the frictional force per surface unit that blood flow exerts on the arterial wall. The importance of WSS relies on its association with the stimuli and response of the innermost arterial layer in contact with blood, which is formed by endothelial cells [7, 9]. Both high and low WSS values have been associated with different pathological mechanisms that lead to aneurysm rupture [15–20]. Elevated WSS has been linked to aneurysm rupture through a pure acute mechanical failure of the aneurysm wall, while, low WSS and slow blood flow recirculation has been associated with a degenerative process of the arterial wall and blood coagulation.

From an engineering perspective, the study of hemodynamics can be analyzed using three complementary approaches: (1) by pure theoretical descriptions of the physical problem, where an analytic study of the governing flow equations are evaluated under certain hypotheses; (2) by experimental setups using in vitro, ex vivo, or in vivo data, where the real world is exposed, observed and measured; and (3) by numerical modeling where full control, reproducibility, cheap experimentations, and the potential of testing unfeasible scenarios, such as aneurysm rupture or several therapeutic options, are available.

This chapter exemplifies how a numerical model can be used to understand cardiovascular hemodynamics using image-based vascular models. In particular, it is focused on the impact of arterial flow conditions on intra-aneurysmal hemodynamics. For that, computational fluid dynamics (CFD) simulations were performed. The numerical tools and methodologies presented here have been extensively used in this field to investigate hemodynamics in cerebral aneurysms [21, 22].

2. MODELING ANEURYSM HEMODYNAMICS

Due to the scale of the studied system (arterial caliber >1 mm), blood flow can be modeled as a continuous matter. This assumption leads to the formulation of the continuity and Navier-Stokes equations as the physical foundations of our problem. These equations are complemented with additional assumptions and formulations such as non-Newtonian viscous models, incompressibility of the fluid, and wall rigidity among others. The set of equations that finally describes the studied system should be solved by numerical schemes (such as finite-volumes, finite-elements, or finite differences, for instance) since the morphology of the system is rather complex. Finally, to adapt these equations and numerical solvers to model patient-specific hemodynamics, the studied system needs to be defined in terms of morphology and boundary conditions.

In terms of morphology, the advances in image acquisition, quality, and postprocessing tools have enormously increased the availability of 3D surface models of cerebral arteries. The accessibility of these models has opened the gate to patient-specific analysis using computational tools. For this purpose, a clear workflow from medical data to 3D vascular flow model has been defined and used [21, 22]. Basically, this workflow starts with the medical acquisition using the above-mentioned imaging modalities. Afterwards,

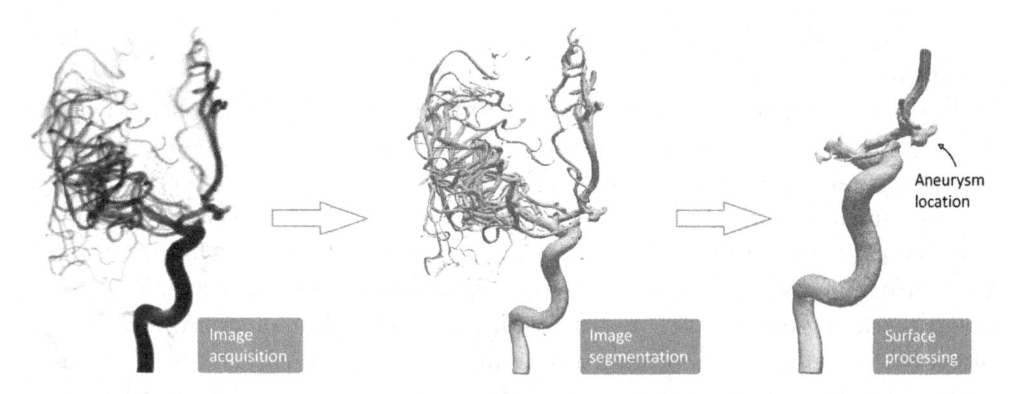

Fig. 10.2 Workflow scheme to extract a 3D vascular model from medical images.

the images are processed and a surface model is extracted using segmentation techniques. Finally, manual corrections are required in order to reduce the region of interest, remove undesirable artifacts, or fill surface holes that were produced during the segmentation process. An example of this processing pipeline is presented in Fig. 10.2.

In terms of boundary conditions, the availability of functional information is harder to obtain compared to morphological data. The main reason is that current medical examinations aimed at anatomical features of the diagnosed aneurysm, since those are well defined in clinical guidelines for prognosis, therapeutic planning, and treatments of cerebral aneurysms [6, 23, 24]. Additionally, image modalities that could extract flow information in cerebral arteries are either technically limited or excluded from the required clinical protocols. For example, the Doppler ultrasound (US) is relatively cheap and has been extensively used to measure flow in carotid arteries and heart chambers. Nevertheless, the presence of the skull and the depth of the organs make US unfeasible for flow measurements in cerebral arteries. An image modality that overcomes these limitation is phase-contrast magnetic resonance (pc-MRI). However, it is not usually available in hospitals (more expensive than US) and pc-MRI has limited spatial resolution [25]. Only particular studies at the academic level use pc-MRI to extract flow information [26]. Finally, it has been recently shown that the use of modulated contrast agent from DSA can be used to extract functional information of blood flow in cerebral arteries [27, 28]. These recent developments could certainly provide the missing part to set properly patient-specific numerical simulations.

3. CONTRIBUTIONS OF THIS CHAPTER

3.1 General Aspects

This chapter is divided into two parts, which are based on two main publications: [29, 30]. Both parts are meant to serve as examples of how CFD simulation can be used to understand intra-aneurysmal hemodynamics better using image-based vascular models.

In the first part of this chapter, temporal variations of aneurysmal hemodynamics in both, parent artery and aneurysms, were studied. In particular, a thorough analysis of peak-systolic hemodynamics is covered since peak systole is generally used as the time instance during the cardiac cycle where high WSS occurred [16, 31, 32]. Peak-systolic hemodynamics was also compared against the time instance where the actual maximum WSS occurred. A proper quantification of the maximum WSS is clinically relevant due to its correlation with the likelihood of aneurysm rupture [15]. Moreover, the implications of normalizing the intra-aneurysmal hemodynamics using arterial segments are also analyzed and discussed. Hemodynamic normalization is done to compensate for different arterial flow conditions, when a case comparison is performed.

The clinical importance to quantify properly the maximum aneurysm stresses and their time instance during the cardiac cycle is due to their correlations with the potential rupture of the aneurysm wall.

Since the analysis intra-aneurysmal is complex due to both morphology and flow conditions, the second part of this chapter aims to unravel the relationship between the arterial flow and aneurysmal hemodynamics. Quantitative means to compare several arterial flow conditions in a simple manner are proposed. For this part, an analysis of the approximation of steady-state solutions of spatiotemporal average results was conducted. Afterwards, simple characterizations of aneurysmal hemodynamic variables as functions of the arterial flow rate were proposed.

The clinical importance of this analysis is to provide a simple way for quantitative characterization of aneurysm hemodynamics under any arterial flow condition for patient stratification and comparison.

In both parts, the results were based on pulsatile flow CFD simulations. Therefore, a brief explanation of the general methodology is first introduced here. Afterwards, the specific data for each part, such as the number of patients, aneurysms, and their respective analyses are covered.

3.2 Image Acquisition

Medical images were acquired with a 3D rotational angiographic X-ray system (Allura Xper FD20 system of Philips Healthcare) in patients with cerebral aneurysms (see Fig. 10.2). Only aneurysms located in the internal carotid artery (ICA), between the carotid siphon and downstream bifurcation (included), were considered (see Fig. 10.1). Although only a portion of cerebral aneurysm incidence is located in this arterial segment (25% for posterior communicating artery and 7.5% for ICA bifurcation), the reason to discard other locations, such as the middle cerebral artery (20% of aneurysm incidence) and the anterior communication artery (30% of aneurysm incidence), is actually to control arterial flow rate range for all simulations, which might not be the case if several arteries are taken into account. Details of aneurysm occurrence can be found in Ref. [1].

3.3 Surface and Volumetric Mesh Generations

After image acquisition and case selection based on their location, 3D surface models of the arteries and aneurysms were extracted using manual thresholding (see Fig. 10.2). This segmentation is not optimal for CFD simulation since holes, kissing vessels, and incomplete arteries of small branches can be obtained in the resulting triangular mesh. To correct these surface imperfections, the software Remesh was used [33]. Afterwards, to eliminate surface wrinkles, the Taubin smoothing algorithm of MeshLab was applied [34] (see final image in Fig. 10.2). To reduce the uncertainties of parent artery truncation, the largest arterial segment upstream the aneurysm available from the images was preserved [35, 36]. Finally, the inlet of the model was extruded and morphed to a circular cross section using the Vascular Modeling Toolkit (VMTK) [37] to facilitate the setup of the parabolic velocity profile (Poiseuille's law) as an inlet boundary condition.

The SnappyHexMesh utility of OpenFoam [38] was used to generate structured hexahedral meshes inside the 3D surface models after closing each inlet and outlet of all cases. This meshing step is required since the governing equations were numerically solved with a finite-volume numerical scheme. Cell size at the arterial wall was approximately 0.2 mm (size level 2) and 0.8 mm elsewhere (size level 0), with four elements between cell size level. Afterwards, three prism layers were added near the wall to improve the accuracy of velocity gradients in the wall normal direction. The

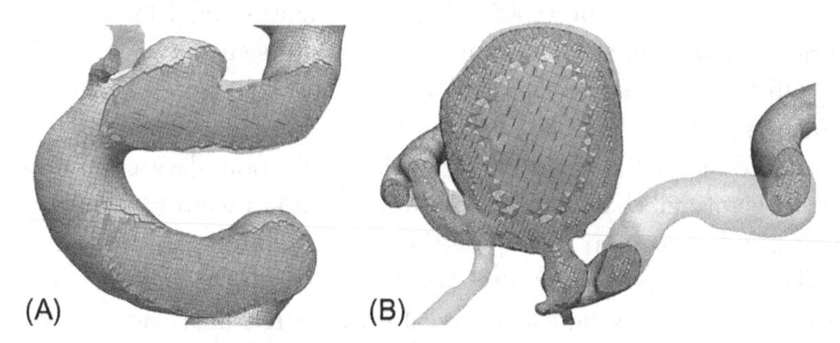

Fig. 10.3 Volumetric meshes for (A) case 4 and (B) case 5. Meshes were clipped with a plane to visualize the internal cells.

volumetric meshes for two cases are illustrated in Fig. 10.3. This meshing configuration was defined after obtaining grid-independent CFD simulations (differences in velocity, WSS, and pressure below 3% between consecutive mesh refinements). The need for this mesh test is to ensure that the results are not dependent on the cell size and their quality.

3.4 Morphological Descriptor

To support the postsimulation data analysis, the aneurysm size was taken as morphological descriptor of each aneurysm. The aneurysm size was manually measured as the largest distance inside the aneurysm cavity, parallel to the parent artery longitudinal orientation. Aneurysms were classified depending of their sizes as either small, medium, or large (<3 mm; between 3 and 5 mm; and >5 mm, respectively).

3.5 CFD Modeling

Several parameters can be modified in a numerical simulation to study relevant problems. For example, the impact of vascular morphology (arterial diameter, vessel tortuosity, narrowing) on aneurysmal hemodynamics can be analyzed by modeling arteries with different morphological characteristics.

Other variations include different viscous models (Carreau, Casson, Herschel-Bulkley, etc.), segmentation strategies, image modality or the arterial flow rate waveform, among others. This work analyzed the impact of the mean arterial flow rate on aneurysm hemodynamics, and therefore all other parameters were unaltered. If the reader is interested on the impact of changing these parameters on aneurysm hemodynamics, the following references are recommended [21, 39–43].

3.5.1 Common CFD Configuration

CFD simulations were performed using OpenFoam v2.2.1 [38]. Since the fluid being simulated mimics human blood, it was therefore considered as an incompressible Newtonian fluid (viscosity of 0.0035 Pa s) with density of 1060 kg/m^3. Although blood is a non-Newtonian shear-thinning fluid, it has been shown through numerical experiments that the increase in blood viscosity as shear strain rates decrease is negligible inside aneurysms [42].

Walls were considered as rigid bodies with no slip boundary condition. This was assumed since the overall characteristics of the WSS do not seem to change considerably when moving walls are used [44]. For the outlet(s), zero-pressure conditions were imposed. The impact of different resistances downstream of the aneurysms was not covered here, which is expected to be small only on terminal aneurysms where flow split occurs and null on lateral aneurysms [41].

A parabolic profile was imposed at the inlet of each model as a boundary condition. This is an acceptable simplification of the fully developed velocity profile in straight tubes for pulsatile flow, called the Womersley profile. This was done, following the recommendation of previous studies where the Womersley profile will be an unrequired complication of the simulation setup if image-based models are used with sufficient entrance length [45, 46]. This parabolic profile was changed over time to obtain pulsatile flow simulations, following a flow rate curve (FRC) that was derived from optical flow measurements performed on patient data [27, 47]. To standardize this waveform for the studied population, a scaling factor was applied to obtain a mean flow rate of $\overline{Q} = 4.08\,mL/s$. This $\overline{Q}$ corresponds to the physiological mean arterial flow rate at the ICA, measured in general population [48]. The scaled waveform was denoted as FRC_8 and was used in all the cases. The index 8 is clarified in the next section.

As initial conditions, velocity vector and pressure fields were set with zero values. Time stepping of $4.16 \times 10^{-4}\,s$ was used for all simulations. Two cardiac cycles were computed in all cases, with the exception of case 5 for FRC_2 ($\overline{Q} = 2\,mL/s$), where three cycles were required.

The need to compute several cycles is due to the zero velocity and pressure values that were used to initialize each simulation. Indeed, the CFD solver propagates the values from the boundaries to each cell inside through time. It was found that the initial part of the first computed cardiac cycle is affected by the initialization. Nevertheless to be sure that the results are not affected by the initialization, they were recovered from the second cycle. In the particular situation of case 5, an extra cardiac cycle was required since the aneurysm was too big, and to propagate the boundary conditions everywhere inside the aneurysm cavity, the additional cycle was required.

3.5.2 Arterial Flow Rate Curve Generation

To analyze the effects of different arterial flows on intra-aneurysmal and arterial hemodynamics in the studied cases, ten waveforms were generated from the previous one (FRC_8) at different $\overline{Q}$ (see Table 10.1 and Fig. 10.4). These additional waveforms were produced using Eq. (10.1), which relates the new FRC_i as a function of its mean flow rate $\overline{Q}_i$, the known FRC_8 and $\overline{Q}_8$ ($= 4.08\,mL/s$). This is a simplified equation of the transformation that Geers et al. implemented [49], since neither pulsatility nor heart rate were changed in this study [49].

$$FRC_i = FRC_8 \cdot \frac{\overline{Q}_i}{\overline{Q}_8} \tag{10.1}$$

The index i is intended to order the FRCs according to their $\overline{Q}$. This means that when comparing two FRCs, the one with the lower index has the lower $\overline{Q}$. Four sequential steps were performed to determine the new waveforms, which are mentioned in order of execution. The implementation of these particular steps to generate the waveforms is mainly because these approaches are commonly used in the literature.

Table 10.1 Inlet model radius and mean flow rates ($\overline{Q}$) for each case

Case	Radius (mm)	Mean flow rates ($\overline{Q}$) (mL/s) for:										
		FRC_1	FRC_2	FRC_3	FRC_4	FRC_5	FRC_6	FRC_7	FRC_8	FRC_9	FRC_{10}	FRC_{11}
1	1.64	0.52[a]	1.00	1.52[b]	2.00	2.50	3.08	3.58	4.08	4.58	5.08	6.30
2	2.65	0.50	1.00	1.50	2.00	2.50	3.08	3.58	4.08	4.58	5.08	6.30[b]
3	2.19	0.50	1.00	1.47[a]	2.00	2.50	3.08	3.58	4.08	4.58	5.08	6.30
4	2.52	0.50	1.00	1.50	2.00	2.47[a]	3.08	3.58	4.08	4.58	5.08	5.36[b]
5	2.19	0.50	1.00	1.49[a]	2.00	2.50	3.08	3.58	4.08	4.58	5.08	6.30
6	2.05	0.50	1.17[a]	1.50	2.00	2.50	3.08	3.58	4.08	4.58	5.08	6.30
7	1.80	0.50	0.72[a]	1.50	1.95[b]	2.50	3.08	3.58	4.08	4.58	5.08	6.30

[a] These values are derived from Eq. (10.2) (flow-area relationship).
[b] These values are derived from Eq. (10.3) (1.5 Pa of WSS at inlet of the models.)
$\overline{Q}_6$ to $\overline{Q}_{10}$ were extracted from the mean ± standard deviation reported by Hoi et al. [48].
From Morales HG, Bonnefous O. Peak systolic or maximum intra-aneurysmal hemodynamic condition? Implications on normalized flow variables. J Biomech 2014;47(10):2362–70.

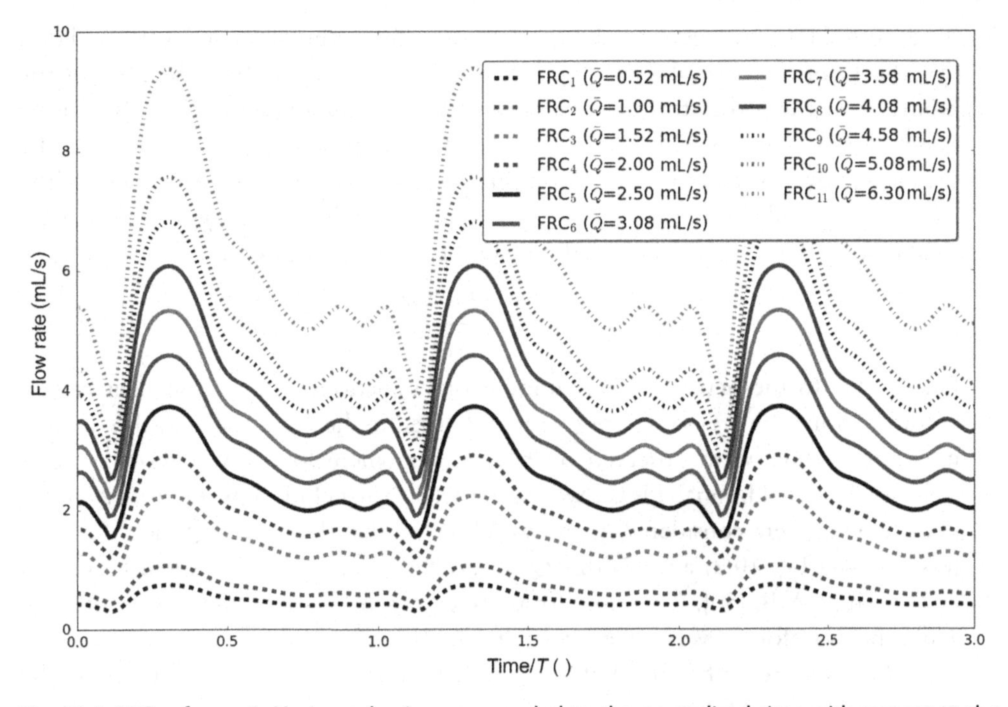

Fig. 10.4 FRCs of case 1. Horizontal axis corresponded to the normalized time with respect to the cardiac period ($T = 1.1$ s). *(From Morales HG, Bonnefous O. Peak systolic or maximum intra-aneurysmal hemodynamic condition? Implications on normalized flow variables. J Biomech 2014;47(10):2362–70.) For interpretation of the color in this figure, please see the online version.*

1. In Hoi et al. [48], a mean arterial flow rate at the ICA of 4.08 ±1 mL/s is reported. Using this information, two waveforms above this mean value, FRC_9 and FRC_{10} with respective $\overline{Q}_9 = 4.58$ mL/s and $\overline{Q}_{10} = 5.08$ mL/s were produced. In the same way, two waveforms below this mean value, FRC_6 with a $\overline{Q}_6 = 3.08$ mL/s and FRC_7 with a $\overline{Q}_7 = 3.58$ mL/s, were generated. These five FRCs were imposed in all cases.
2. Cebral et al. [50] proposed a relationship between the inlet area of the model (in centimeters) and the mean arterial flow. This relationship is presented in Eq. (10.2). Using this equation, a sixth waveform was generated. In this equation, the coefficients $k = 48.21$ and $n = 1.84$ were experimentally found. Since this approach is dependent on the inlet of each case, it is not possible to set a unique index for the arterial flow in all cases since the resulting $\overline{Q}$ can vary. Moreover, if the resulting $\overline{Q}$ from Eq. (10.2) is within the range $\overline{Q}_6$ and $\overline{Q}_{10}$ (generated from the previous step), then the FRC that can be generated by Eq. (10.2) was not included.

$$\overline{Q} = k \cdot \text{Area}^n \tag{10.2}$$

3. A seventh FRC was generated for each case, from the physiological condition that the $\overline{Q}$ of the arterial flow should produce a WSS = 1.5 Pa at the inlet of the model. A mathematical formulation of this relation is presented in Eq. (10.3) using Poiseuille's law. A WSS = 1.5 Pa is considered as the physiological stress exerted by the blood flow on the arterial wall [8]. Since this condition is also dependent of the inlet of each model, it is not possible to use a unique index for all cases and the FRC produced by this step was discarded if it was between $\overline{Q}_6$ and $\overline{Q}_{10}$.

$$\overline{Q} = \frac{\pi \cdot r^3 \cdot \mathrm{WSS}}{4 \cdot \mu} \tag{10.3}$$

4. Finally, due to the heterogeneous amount of simulations among cases that can be produced using the previous steps (commonly used in the literature) and also because interesting results were observed on those cases where low flow rates (0.5 mL/s) were available, additional FRCs were produced. To include low flow waveforms, two elements were considered: (1) It was observed that the lowest $\overline{Q}$ was 0.52 mL/s (case 1 using Eq. 10.2) and (2) this $\overline{Q}$ is near the lowest value measured by Cebral et al. with pc-MR [50]. Therefore for those cases where lower FRCs were missing, additional waveforms were added starting from 0.5 mL/s and every 0.5 mL/s. These extra FRCs were only added if they were not too close (±0.25 mL/s) to the previously derived FRCs.

   In the same fashion, an upper FRC was established knowing that (1) the highest $\overline{Q}$ using the previous steps was 6.30 mL/s (case 2 with Eq. 10.3) and (2) this $\overline{Q}$ is a high flow rate compared to the measured data of by Cebral et al. [50]. With this information, only one extra simulation at 6.3 mL/s was included if there was not a FRC already above 5.08 mL/s in each of the cases.

These four steps provided 11 $\overline{Q}$s for each case that applying Eq. (10.1), generated 11 pulsatile waveforms per case in total. Table 10.1 presents the imposed $\overline{Q}$ for seven of the studied cases and Fig. 10.4, the FRCs for case 1 as example. This setup was used in both parts of this chapter.

4. PART 1: PEAK-SYSTOLIC AND MAXIMUM HEMODYNAMIC CONDITION

The purpose of this section is to understand if peak systole actually produces the maximum hemodynamic condition inside aneurysms. For that, the impact on changing the arterial flow rate is taken into account as well as the implication of aneurysm hemodynamic normalization. The clinical importance of a proper quantification of the maximum aneurysm hemodynamics relies on its correlation with the potential rupture of the aneurysm.

Normalization of aneurysm hemodynamics is meant to standardize the variables inside the aneurysm such as WSS when different arterial flow conditions are used. In the case of WSS, this is done by taking an arterial segment upstream the aneurysm and calculating its spatial averaged WSS. Then, the aneurysmal WSS is divided by the arterial WSS. This normalization relies on the assumption that the impact of the arterial flow can be canceled if the aneurysmal hemodynamic quantity is divided by its equivalent quantity in a given arterial segment. Moreover, since the aneurysm could have abnormal WSS quantities, the normalized WSS should be able to indicate if it is below or above arterial WSS values for the given flow condition. Several studies have implemented this strategy, allowing a comparison among cases [16, 19, 51, 52].

In this part of the chapter, the spatial-averaged of hemodynamic variables were analyzed as essential quantity in 11 aneurysm models. Other studies provide further and deeper analysis of aneurysmal hemodynamics by considering local values in space, such as maximum or minimum WSS [20, 53]. This means that for each case, every location in the aneurysm sac is inspected in time. This type of analysis is not possible in this study due to the large amount of simulated data that is generated by changing the FRCs.

4.1 Data Analysis

For each case, at least three regions were defined: three arterial segments upstream from the aneurysm and the aneurysm(s) sac(s). The arterial segments were denoted as ICA segments 1, 2, and 3. The first two have a length of one arterial diameter approximately, while segment 3 is a larger region that may include the previous small segments. An example of the arterial segments and aneurysm region is presented in Fig. 10.5. In each of these regions, the spatial-averaged velocity magnitude and WSS were calculated during the latest computed cardiac cycle.

4.2 Arterial Hemodynamics

First, the velocity and WSS in the arterial segments were analyzed. To provide a visual representation of the results, Fig. 10.6 shows these variables during a cardiac cycle for three cases (2, 5, and 7) at different arterial flow rates. In each plot, five curves are illustrated: the inlet (given boundary condition), the ICA segment 3 (larger arterial region), the mean $\pm$SD of the three arterial segments (during the second cardiac cycle), and two curves for the aneurysm region (first and second cardiac cycles), which will be discussed in the following section. The aneurysm curves of Fig. 10.6 show the importance of computing an additional cardiac cycle. The initial condition (zero velocity) demonstrates that first-cycle aneurysm velocity curve rises from zero and then follows the second-cycle aneurysm velocity curve. The stabilization time depends on each case (aneurysm size) and arterial flow rate.

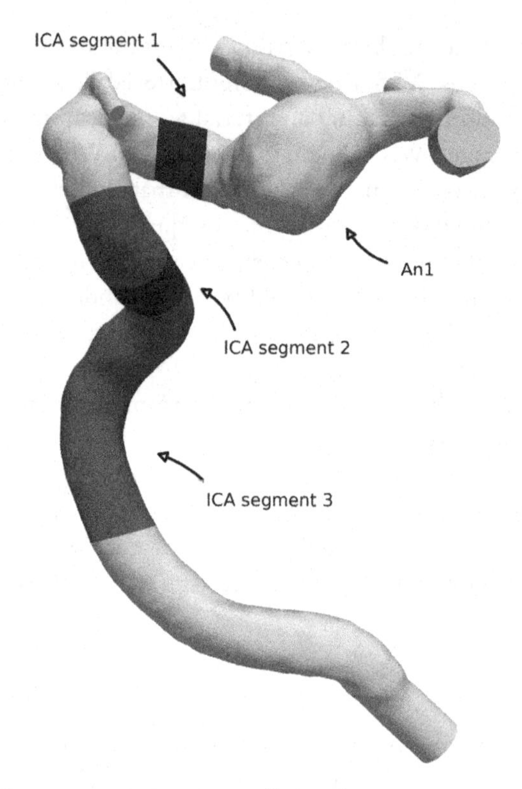

Fig. 10.5 Example of ICA segments and aneurysm (An) cavity.

In Fig. 10.6, it can be observed that during the cardiac cycle, both the arterial velocity and WSS changed over time due to the flow pulsatility, with maximum values at peak systole ($t_{\text{systole}} = 0.818$ in a normalized cardiac cycle t/T, with $T = 1.1$ s). Moreover, both velocity and WSS are dependent on the arterial segment as can be appreciated in the mean ±SD deviation curves. These curves also indicates that the larger variabilities of these quantities among the arterial segments occurred near peak systole. To quantify the variability at peak systole, Table 10.2 presents WSS values for each arterial segment for two arterial flow rates. There, it is observed that the differences in WSS are also dependent on the arterial flow rate.

4.3 Aneurysm Hemodynamics

During the cardiac cycle, it was observed that peak systole is the time instance where neither the maximum velocity nor maximum WSS occurred inside the aneurysm cavity. As examples, maximum values of these variables are indicated with the arrows in plots of Fig. 10.6. This maximum can occur before or after peak systole, depending on the case

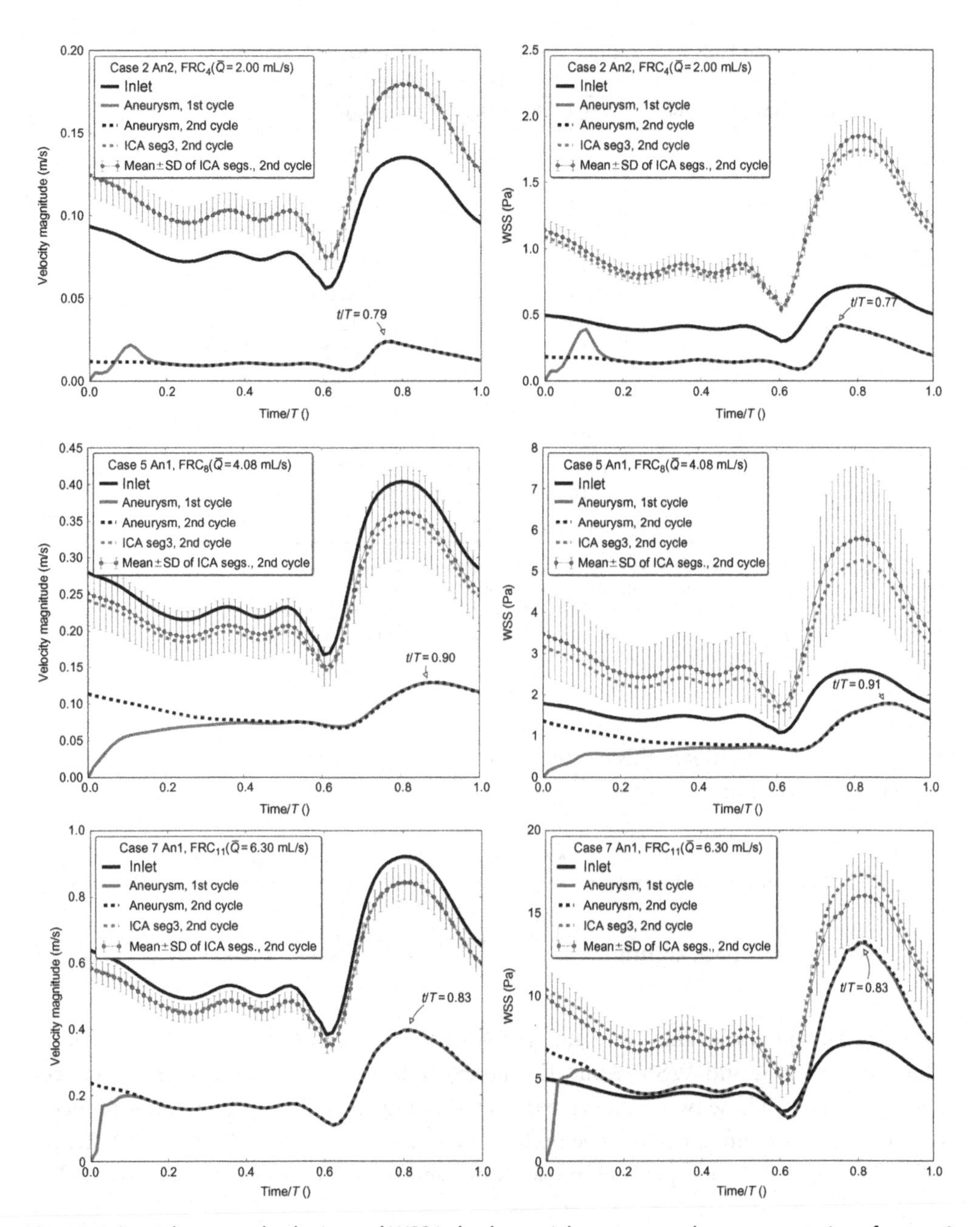

Fig. 10.6 Spatial averaged velocity and WSS in both arterial segment and aneurysm regions for case 2 (An2), case 5 (An1), and case 7 (An1) for $\overline{Q}_4 = 2.0$ mL/s, $\overline{Q}_8 = 4.08$ mL/s, and $\overline{Q}_{11} = 6.30$ mL/s, respectively. T is the cardiac period equal to 1.1 s. Arrows indicate the normalized time instance (t/T) where the maximum velocity and WSS occurred inside the aneurysm. For the aneurysms, two consecutive cardiac cycles are plotted to show the cyclic numerical solution. To illustrate the results for ICA segments (seg.), segment 3 (the largest one) and the mean ± SD using all three segments are shown. (*Modified from Morales HG, Bonnefous O. Peak systolic or maximum intra-aneurysmal hemodynamic condition? Implications on normalized flow variables. J Biomech 2014;47(10):2362–70.*)

Table 10.2 Peak systolic WSS at ICA segments for FRC_1 and FRC_8

Case	FRC	WSS at ICA segments			
		seg1 (Pa)	seg2 (Pa)	seg3 (Pa)	% Diff.
1	1	0.363	0.437	0.428	16.80
	8	4.965	8.211	7.842	39.52
2	1	0.349	0.248	0.286	28.89
	8	5.727	4.864	4.877	15.07
3	1	0.493	0.464	0.523	7.79
	8	8.582	8.755	8.870	1.97
4	1	0.235	0.107	0.129	54.54
	8	3.757	2.641	2.682	29.69
5	1	0.419	0.213	0.280	49.05
	8	8.168	3.928	5.228	51.91
6	1	0.816	0.353	0.598	56.67
	8	13.07	7.935	11.94	39.30
7	1	0.458	0.575	0.553	20.43
	8	6.659	10.01	9.274	33.49

Percentage differences (% Diff.) were calculated as | seg1 − seg2 | / Max(seg1, seg2) × 100. *Max*, maximum; *Min*, minimum; *seg*, segment.
From Morales HG, Bonnefous O. Peak systolic or maximum intra-aneurysmal hemodynamic condition? Implications on normalized flow variables. J Biomech 2014;47(10):2362–70.

and flow rate. To investigate this matter, two additional indices were calculated using the following formulations:

$$\text{Percentage difference} = \frac{\text{Maximum value} - \text{peak value}}{\text{peak value}} \times 100 \tag{10.4}$$

$$\text{Normalized peak shifting} = \frac{\text{Time}_{\text{Maximum}} - \text{Time}_{\text{peak}}}{T} \times 100 \tag{10.5}$$

Eq. (10.4) quantifies the difference between the maximum value and peak systolic values of velocities and WSS, while Eq. (10.5) indicates the time difference between these two events. These two indices are depicted in Fig. 10.7. In this figure, it is observed that the maximum differences can be up to ~35% for velocity and ~60% of WSS. Larger differences occurred in larger aneurysms compared to small aneurysms and lower flow rates.

The second row of Fig. 10.7 depicts the normalized peak shifting, where negative and positive times are presented. In most of the cases, positive-shifted times occurred for all flow rates, meaning that maximum velocity and WSS occurred after peak systole, with a recurrence for positive values of 84.3% for velocity and 89.3% for WSS (vs. 15.7% for velocity and 10.7% for WSS for negative times). A negative time means that

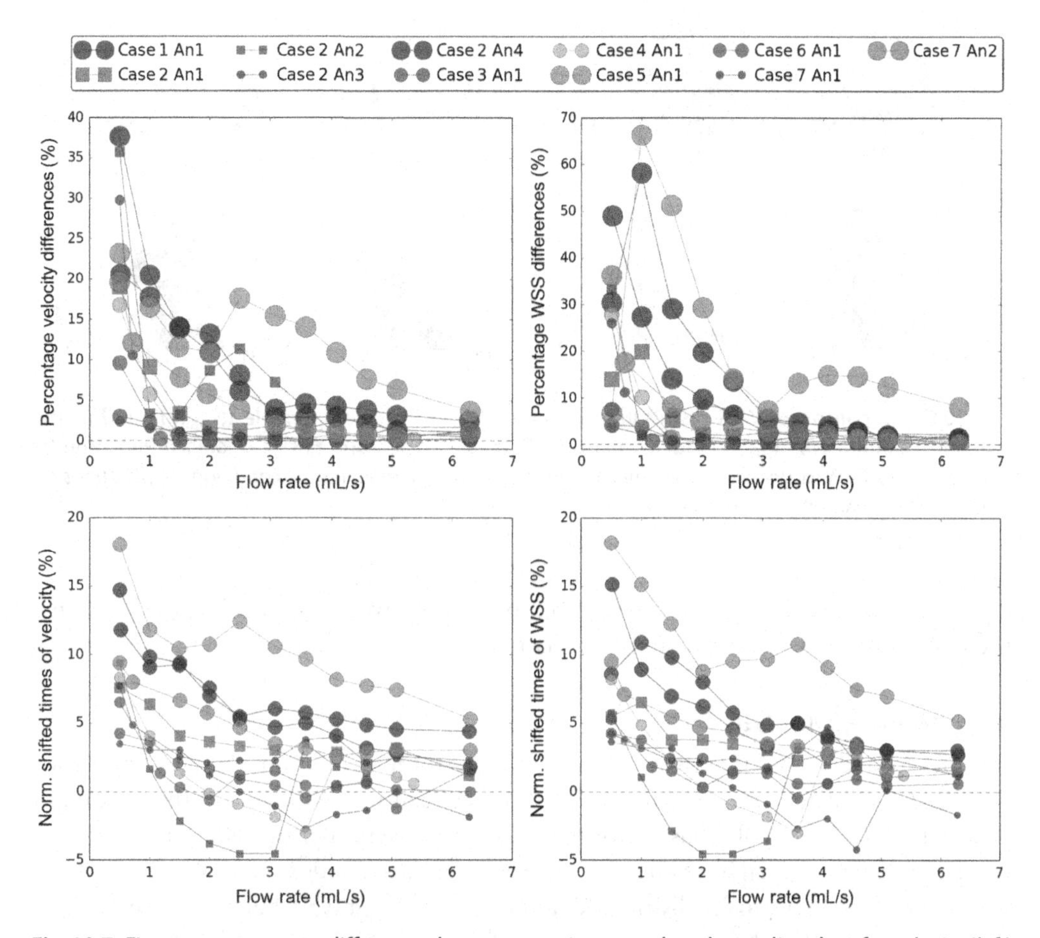

Fig. 10.7 *First row*: percentage differences between maximum and peak systolic values for velocity (*left*) and WSS (*right*). *Second row*: normalized peak shifted times for velocity and WSS. Marker size represents the aneurysm size (small, medium, and large), while *circles* and *squares* signify lateral and terminal aneurysms, respectively. (*From Morales HG, Bonnefous O. Peak systolic or maximum intra-aneurysmal hemodynamic condition? Implications on normalized flow variables. J Biomech 2014;47(10):2362–70.*) *For interpretation of the color in this figure, please see the online version.*

the maximum value occurred earlier. In magnitude, positive shifted times were larger (18.02% for velocity and WSS) than negative times (4.55% for velocity and WSS).

To complement these results, WSS distributions are shown in Fig. 10.8 for these two time instances (t_{peak}/T and t_{max}/T). For case 2 An2, t_{max}/T is lower than t_{peak}/T (= 0.818) for FRQ_4, because the impingement flow jet moved away from the aneurysm neck during flow acceleration. Finally, Eq. (10.5) was modified by replacing the time$_{maximum}$ for time$_{minimum}$ and time$_{systole}$ for time$_{diastole}$ to evaluate the relation between the minimum value of these variables and end-diastole. With this modified

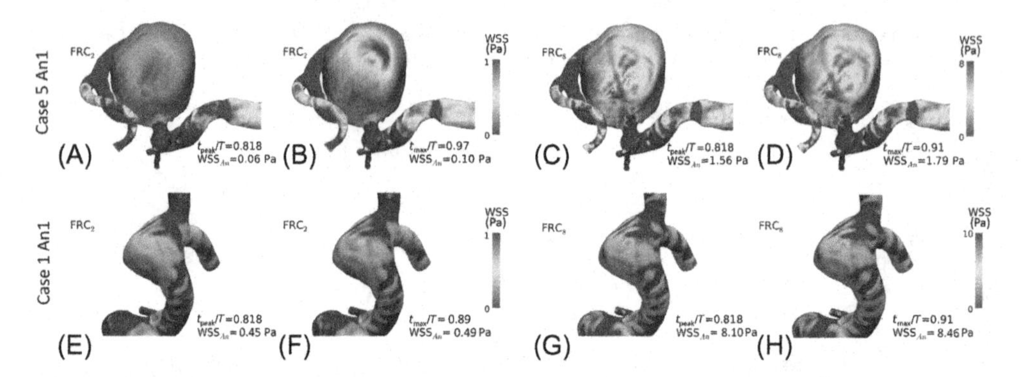

Fig. 10.8 Comparison of WSS distribution for cases 5 and 1 at peak systole (t_{peak}/T) and at the time of maximum WSS (t_{max}/T) for two arterial flow rates (FRC_2 and FRC_8). (*Modified from Morales HG, Bonnefous O. Peak systolic or maximum intra-aneurysmal hemodynamic condition? Implications on normalized flow variables. J Biomech 2014;47(10):2362–70.*)

equation, only positive shifted times were observed between 0% and 7%, meaning that the minimum value occurred after end-diastole.

4.3.1 Statistical Analysis

In order to understand if there is a relation between the maximum and peak difference (Eq. 10.4) population and the time difference between peak systole and the maximum value (Eq. 10.5) populations, two statistical tests were performed. The first test corresponded to a Shapiro-Wilk test on each separated index (and variable) to see whether they were normally distributed. p-Values of this test were below .005, indicating that it is unlikely that these populations came from normal distributions.

Having these results for index/variable, a Spearman rank-order test was performed as a second statistical analysis to evaluate if the differences correlated or not with the shifted times. As a result, this test revealed statistical significant correlations (p-value $<.001$) between variable difference and times for velocity and WSS ($\rho_s = .84$ for velocity and $\rho_s = .82$ for WSS). This result means, e.g., that if larger differences between peak systole and maximum velocity are found, it is most likely that the time delay between these two events is also relatively large; in the same way, smaller differences correlated with shorted shifted time.

Finally, two additional statistical tests were performed to compare the shifted time of velocity and WSS (second row of Fig. 10.7). First, a Spearman statistical test showed that there is a significant correlation between these two populations ($\rho_s = .97$, p-value $<.001$). Afterwards, a Wilcoxon signed-rank test for matched paired data indicated that both populations of shifted times were not significantly different ($p = .65$). This last result means that there is no evidence that the medians of these two populations differ.

4.4 Discussion of Part 1

As mentioned, this part of the chapter was meant to investigate the temporal variations of intra-aneurysmal hemodynamics in both parent artery and aneurysms. Particular attention was paid to the peak-systolic velocity and WSS, which were compared with the maximum value of these variables during the cardiac cycle. Two main results can be discussed: maximum versus peak-systolic value and the normalization strategy using an arterial segment.

4.4.1 Maximum Versus Peak-Systolic Value

As Fig. 10.6 shows, the maximum value of both velocity and WSS does not necessarily occur at peak systole inside the aneurysm cavity. In the studied population, the differences between peak systole and maximum values could be up to ~35% for velocity and ~65% for WSS. These differences are due to both morphological and hemodynamic factors. In general, small aneurysms have smaller differences that larger aneurysms. Lower arterial flow rate produces larger differences. Aneurysm size and arterial flow rate introduce positive shifted time as a delayed response of the peak systolic arterial flow. Indeed, when peak systole arrives, the arterial hemodynamics (velocity and WSS) is at its maximum. However, to accelerate the blood inside the aneurysm cavity, the peak systolic fluid needs time to go from the aneurysm neck and to move all around the aneurysm cavity. If the aneurysm is small, then the distance that peak systolic fluid needs to cover is small compared to a large aneurysm. Additionally, if the arterial flow rate is faster, then the same distance will be covered in a shorter amount of time.

Negative times were observed in Fig. 10.7, which are not intuitive as positive shifted times, since they mean that the maximum value occurred before peak systole. Some explanations were found to understand these negative times in the studied cases. For example, during flow acceleration, the impingement jet is shifted away of the aneurysm neck, and therefore the velocity and WSS inside the aneurysm are lower at peak systole compared to the earlier instance. These shifted times can be a quantitative way to evaluate flow stabilities in cerebral aneurysms instead of using qualitative assessments. Moreover, they have a direct and indirect implication in the normalization strategy that it is commonly used.

4.5 Direct Implication

Peak systolic WSS has been associated with aneurysm rupture risk in several studies with large populations [17, 20, 32, 54]. This is generally done as a mean to obtain the maximum stress at which the aneurysm wall would be exposed. However, our results indicate that hemodynamic variables at peak systole are not necessarily in their maximum inside the aneurysm, although maximum values at the artery occurred at this time instance. The shifting in time depends on several morpho-hemodynamic factors such as aneurysm size, flow rate, surrounding vasculature, and the stabilities of the flow patterns. The combination of all these factors makes it difficult to extrapolate from peak-systolic values to the maximum variable. Therefore, when peak systole value is used to compare cases, some of them will not be properly evaluated (especially large aneurysms). This finding has great importance since it has been shown that large aneurysms tend to have lower peak systolic WSS and the latest has been associated with aneurysm rupture risk [32]. Therefore, if the maximum WSS needs to be studied, e.g., it is important to calculate the actual time instance when it happens instead of the peak systolic value.

In the same way, if the lowest value is required, then end-diastole should not be used and the minimum value of the variable should be sought, although lower differences between lower and end diastole were calculated in our population (the larger differences was only 7%). As a final comment of this finding, it is important to highlight the evaluation of treatment performance as coils or flow diverter stents using computational tools [11, 55]. These devices change both the morphology of the aneurysm and local hemodynamics, which potentially can modify the time instance where the maximum value occurred in the cardiac cycle.

4.6 Indirect Implication

This finding is also related to the recurrent strategy of normalizing the intra-aneurysmal variables, like peak WSS using an arterial segment, which is presented in Eq. (10.6). Here, the problem is related to the fact that peak systolic values in the aneurysm are used, but those values can be far from the maximum value at which the aneurysm is actually exposed. In a sense this can be seen as an underestimation calculation of the numerator in Eq. (10.6).

$$\text{Normalized variable} = \frac{\text{Variable}_{\text{aneurysm at systole}}}{\text{Variable}_{\text{arterial segment at systole}}} \times 100 \tag{10.6}$$

As previously indicated, this part of the chapter also investigated the use of hemodynamic variables at several arterial segments, which has implications in the denominator of Eq. (10.6).

4.7 Normalization Strategy Using an Arterial Segment

The normalization of hemodynamics variables is meant to standardize them when different boundary conditions are used. To provide a representative quantity of the artery, as presented in Eq. (10.6), a segment upstream from the aneurysm is used. The selection of the arterial segment can be manually or automatically done, but in both cases, it is arbitrary chosen without considering if the segment is representative of arterial hemodynamics.

According to the findings presented here, both velocity and WSS vary among the arterial segments that were defined. According to Table 10.2, differences in WSS can be higher than 50% among segments. These differences are particularly large near peak systole, where larger variations were observed compared to the rest of the cardiac cycle (see Fig. 10.6).

5. PART 2: CHARACTERISTIC CURVES OF INTRA-ANEURYSMAL HEMODYNAMICS

In the previous section, one of the generic messages that can be extracted is that temporal variations produce an enormous variability of hemodynamic quantities such as WSS, velocity, or pressure. This hinders the comparison among aneurysms using hemodynamic descriptors, since many elements (morphological and functional) play a role in time-dependent hemodynamic characterizations. Having this in mind, the purpose of this section is to see whether generic expressions of aneurysmal hemodynamics can be obtained, those independent on both time variations and normalization strategies using an arterial segment. To achieve this goal, the spatial averaged quantities that were computed in the previous section were now averaged over time (spatiotemporal average). These quantities were used in this section as the essential quantity for analysis. In addition, the number of studied aneurysm models increased from 11 to 15.

5.1 Data Analysis

To evaluate the arterial hemodynamics, a user-independent approach was followed this time using some of the available tools of VMTK [37]. These tools were the centerline extraction and arterial cross-section generation. Cases 4 and 6 were considered for this analysis, and their selection was only based on their differences in arterial caliber. Only the arterial centerline from the inlet to the one of the middle cerebral arteries was computed. This centerline was then discretized every 0.1 mm and the arterial cross-section on each point was computed. Afterwards in each cross-section, its area, the spatiotemporal-averaged velocity magnitude ($\overline{\mathrm{vel}}_{\mathrm{sa}}$), and the maximal radius of the inscribed sphere (denoted here as radius r) were computed. Having these variables for each cross-section and both blood density ρ and viscosity μ, the computation of the Reynolds number (Re) was calculated as follows:

$$Re = \frac{\rho \cdot 2r \cdot \overline{\mathrm{vel}}_{\mathrm{sa}}}{\mu} \tag{10.7}$$

In the case of the aneurysm ($\overline{\mathrm{vel}}_{\mathrm{sa}}$) the spatiotemporal-averaged WSS ($\overline{\mathrm{WSS}}_{\mathrm{sa}}$) and spatiotemporal-averaged pressure ($\overline{\mathrm{pre}}_{\mathrm{sa}}$) were calculated. Additionally, the oscillatory

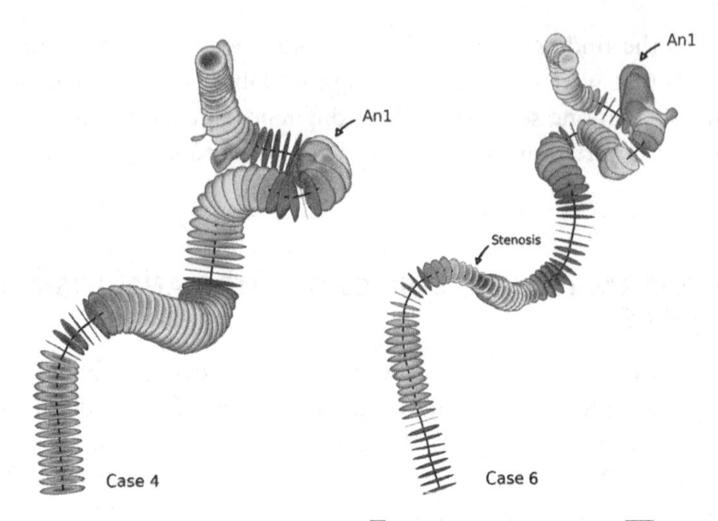

Fig. 10.9 Arterial cross-sections for cases 4 and 6 at $\overline{Q} = 3.08$ mL/s colored by $\overline{\text{vel}}_{\text{sa}}$.

shear Index (*OSI*), the relative residence time (*RRT*) and maximum spatial-averaged WSS max($\overline{\text{vel}}_{\text{sa}}$) were also computed. The interest for *OSI* and *RRT* lies in their relationship with the initiation and promotion of atherosclerosis [56, 57].

5.2 Spatiotemporal Arterial Hemodynamics

The calculated cross-section along the arteries for cases 4 and 6 are presented in Fig. 10.9 and measured variables in each section along the discretized centerline are shown in Fig. 10.10. It is interesting to note in this figure how *Re* varies along the artery and how much it depends on the arterial flow rate. Additionally, the stenosis in case 6 produced an increase in $\overline{\text{vel}}_{\text{sa}}$.

To compare $\overline{\text{vel}}_{\text{sa}}$ with the calculated arterial radius, this variable was divided by the mean arterial flow rate, $\overline{Q}$, to obtain an inverse of the cross-section area (units of m^{-2}). As can be observed on the second row of Fig. 10.10 (for both cases), all $\overline{\text{vel}}_{\text{sa}}$ curves become a unique representative line when divided by $\overline{Q}$. The percentage differences between $\overline{\text{vel}}_{\text{sa}} \backslash \overline{Q}$ and the cross-section area are shown in Fig. 10.10 along the centerline. Large percentage differences are due to the aneurysm (cross-section enlargement) and downstream bifurcation where inappropriate cuts occurred. Keep in mind that the used tool to generate the cross sections was not designed to identify bifurcations and to cut one of the branches without touching the other one. The small differences along the artery (where the cross-sections were well defined) were mainly due to the calculation of $\overline{\text{vel}}_{\text{sa}}$, which considers all the vector components of the velocity field instead of only the normal component.

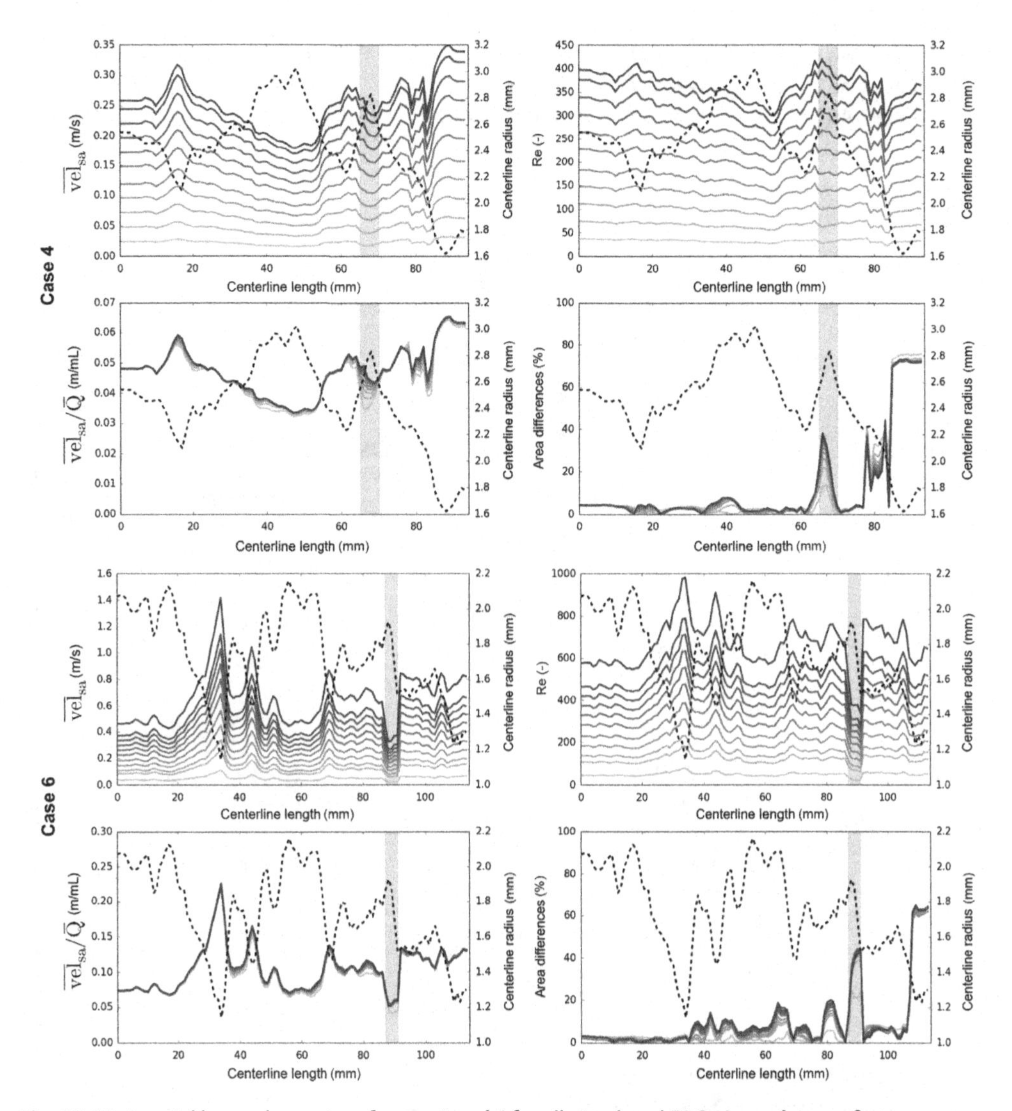

Fig. 10.10 Arterial hemodynamics of cases 4 and 6 for all simulated FRCs. In each case, *first row* present $\overline{vel}_{sa}$ (*left*) and *Re* (*right*) along the discretized arterial centerline. *Second row*: norm $\overline{vel}_{sa}$ (*left*) and area percentage difference. Arterial radii are represented by the *dashed line* in each plot. The *vertical grey rectangle* indicates the aneurysm location along the centerline.

5.3 Spatiotemporal Aneurysmal Hemodynamics

Fig. 10.11 presents the dependency of spatiotemporal-averaged variables ($\overline{vel}_{sa}$, $\overline{WSS}_{sa}$ and $\overline{pre}_{sa}$), max$\overline{WSS}_{sa}$, *OSI*, and *RRT* to $\overline{Q}$. Higher flow rates increased the spatiotemporal-averaged variables and max$\overline{WSS}_{sa}$. *OSI* was almost constant in all cases,

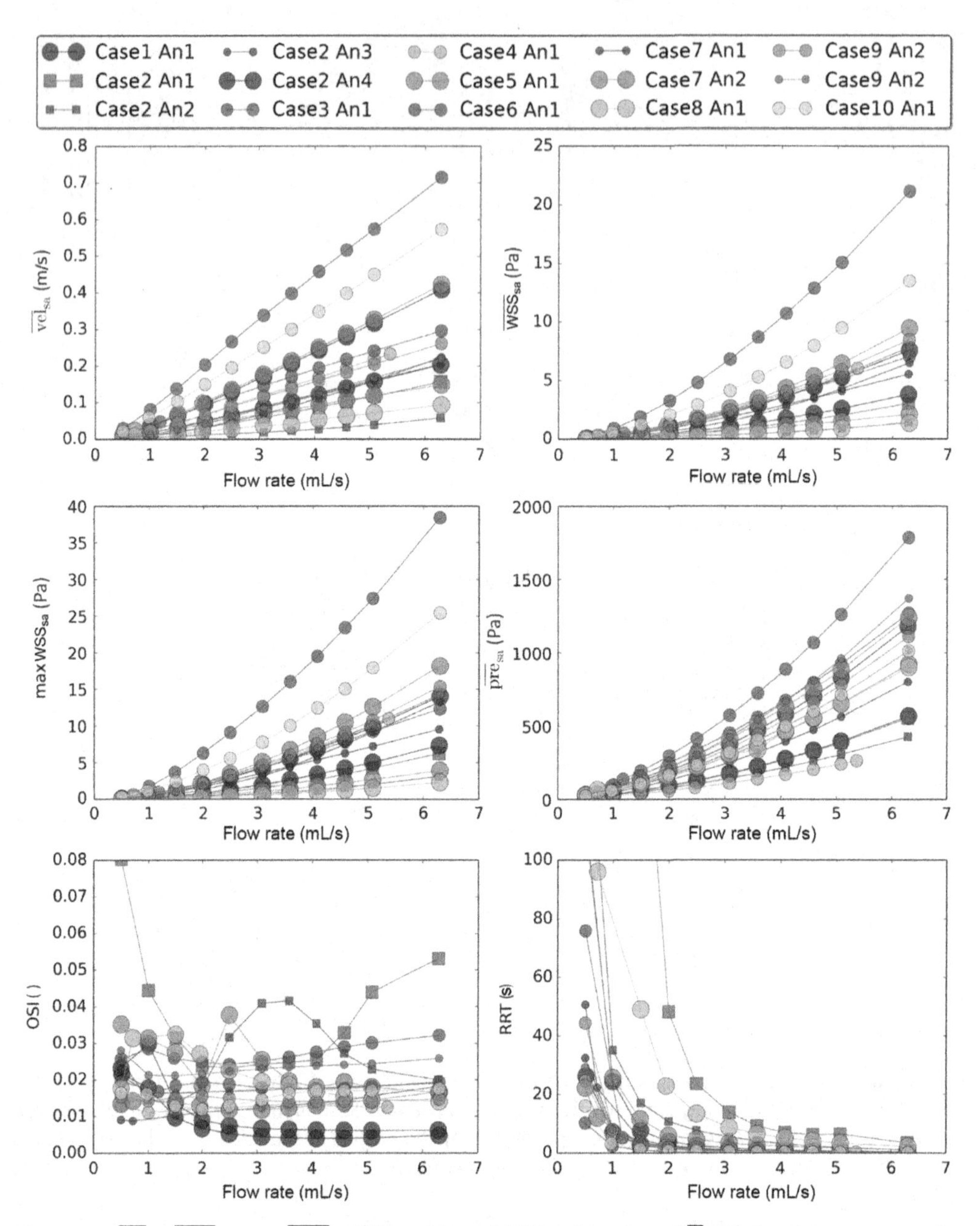

Fig. 10.11 $\overline{\text{vel}}_{\text{sa}}$, $\overline{\text{WSS}}_{\text{sa}}$, max $\overline{\text{WSS}}_{\text{sa}}$, $\overline{\text{pre}}_{\text{sa}}$, *OSI*, and *RRT* as function of $\overline{Q}$. *Marker size* represents the aneurysm size label (small, medium, or large). *Squares* correspond to terminal aneurysms and circles to lateral ones. (*From Morales HG, Bonnefous O. Peak systolic or maximum intra-aneurysmal hemodynamic condition? Implications on normalized flow variables. J Biomech 2014;47(10):2362–70.*) *For interpretation of the color in this figure, please see the online version.*

ranging from 0.004 to 0.08. Lateral aneurysms exhibited less variations in *OSI* than terminal aneurysms (see aneurysms of case 2, for instance). *RRT* rapidly decreased with increasing $\overline{Q}$.

5.4 Characteristic Curves

Using these spatiotemporal descriptions of velocity, WSS, and pressure ($\overline{vel}_{sa}$, $\overline{WSS}_{sa}$, and $\overline{pre}_{sa}$), linear and quadratic regressions were applied to quantify their relationship with the mean arterial flow rate, $\overline{Q}$, in each case. For intra-aneurysmal flow velocities, linear regression models were imposed in the form of:

$$\overline{vel}_{sa} = m \cdot (\overline{Q} - \overline{Q}_{min}) \tag{10.8}$$

where m is the slope of the linear model and $\overline{Q}_{min}$ is the defined as the intersection of the regression models with the x-axis (i.e., flow rates).

Quadratic regressions were applied to WSS and pressure in the following form:

$$\overline{WSS}_{sa} = A \cdot (\overline{Q} - \overline{Q}_{min}) \cdot (\overline{Q} - B) \tag{10.9}$$

$$\overline{pre}_{sa} = A \cdot \overline{Q} \cdot (\overline{Q} - B) \tag{10.10}$$

In these formulations, m, A, $\overline{Q}_{min}$, and B are constants. Note that the regression models were independently applied on each set of data (variable-flow rate) and therefore, these coefficients define a particular equation for each aneurysm. Concerning the pressure variable, the quadratic regressions were set to intercept the origin, i.e., $\overline{Q}_{min} = 0$. This was done since there is always a nonzero pressure within the aneurysm sac and it should be larger than the lowest pressure imposed at the outlets, which was zero. These regression models constitute the characteristic curves of $\overline{vel}_{sa}$, $\overline{WSS}_{sa}$, and $\overline{pre}_{sa}$ as functions of the mean flow rate $\overline{Q}$ for each aneurysm.

The linear and quadratic coefficients of the applied regression models are displayed in Table 10.3. In the quadratic regressions of WSS, $\overline{Q}_{min}$ is the positive root of the polynomial and B is the negative one. Nevertheless, the best quadratic regression of $\overline{WSS}_{sa}$ for case 2 An2 does not intersect the x-axis. For this case only, a zero value for $\overline{Q}_{min}$ was set, which is the lower possible flow rate at which WSS should be zero. All regressions exhibited an excellent fitting, where the minimum $R^2 > .963$ for velocity, $R^2 > .996$ for WSS and $R^2 > .997$ for pressure, considering all cases.

5.5 Generalization

An important aspect of the analysis of these variables and the proposed characteristic curves is to know their dependency on the chosen waveform shape (frequency, pulsatility, mean flow rate). If these curves are independent of the waveform, a very strong generalization will be possible and meaningful. Inspired by Geers et al. [49], who

Table 10.3 Linear and quadratic regressions for $\overline{vel}_{sa}$, $\overline{WSS}_{sa}$, and $\overline{pre}_{sa}$

Model		$\overline{vel}_{sa} = m \cdot (\overline{Q} - \overline{Q}_{min})$			$\overline{WSS}_{sa} = A \cdot (\overline{Q} - \overline{Q}_{min}) \cdot (\overline{Q} - B)$				$\overline{pre}_{sa} = A \cdot \overline{Q} \cdot (\overline{Q} - B)$		
Case	**An**	m	$\overline{Q}_{min}$	R^2	A	B	$\overline{Q}_{min}$	R^2	A	B	R^2
1	1	0.069	0.529		0.134	-3.644	0.561		10.273	−0.407	
2	1	0.026	0.712		0.071	−1.25	0.573		9.9882	−0.394	
	2	0.009	0.684		0.032	−0.460	0.0		6.8255	−0.272	
	3	0.034	0.493		0.071	−7.633	0.589		14.553	−0.401	
	4	0.034	0.520		0.08	−2.238	0.659		21.335	−0.394	
3	1	0.119	0.282		0.316	−5.335	0.522		17.904	−0.199	
4	1	0.047	0.502		0.141	−3.562	0.531		5.8083	−0.327	
5	1	0.024	0.214	>0.963	0.03	−6.169	0.649	>0.996	20.564	−0.299	>0.997
6	1	0.049	0.151		0.092	−6.976	0.470		31.363	−0.357	
7	1	0.037	0.917		0.164	−0.628	0.541		18.550	−0.412	
	2	0.070	0.482		0.192	−2.242	0.508		16.037	−0.347	
8	1	0.015	0.487		0.033	−0.9	0.515		14.200	−0.262	
9	1	0.044	0.446		0.155	−3.357	0.583		20.349	−0.412	
	2	0.038	0.738		0.156	−2.693	0.642		25.456	−0.435	
10	1	0.095	0.385		0.231	−3.704	0.431		18.141	−0.385	
Mean ± SD		0.048 ±0.028	0.50 ±0.196	–	0.127 ±0.078	−3.386 ±2.201	0.518 ±0.152	–	16.756 ±6.644	−0.354 ±0.066	–
[Min, Max]		[0.015, 0.119]	[0.151, 0.917]	–	[0.03 0.316]	[−7.633 −0.462]	[0.0, 0.649]	–	[5.808 31.363]	[−0.435, −0.199]	–

Q_{min} for pressure equation was zero.
From Morales HG, Bonnefous O. Peak systolic or maximum intra-aneurysmal hemodynamic condition? Implications on normalized flow variables. J Biomech 2014;47(10):2362–70.

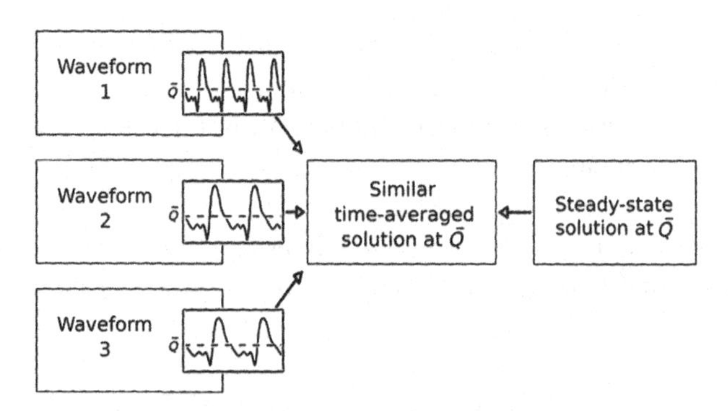

Fig. 10.12 Different waveforms with the same mean flow rate (blocks on the left) have a similar spatiotemporal-averaged solution. This solution can be obtained through a steady-state simulation at $\overline{Q}$ [49]. *(From Morales HG, Bonnefous O. Peak systolic or maximum intra-aneurysmal hemodynamic condition? Implications on normalized flow variables. J Biomech 2014;47(10):2362–70.)*

reported that $\overline{WSS_{sa}}$ does not depend on the shape of the pulsatile waveform of the input arterial flow curve, but only on $\overline{Q}$, a comparison of the spatiotemporal-averaged variables ($\overline{vel_{sa}}$, $\overline{WSS_{sa}}$, and $\overline{pre_{sa}}$) with steady-state simulations was performed. To do so, three steady-state simulations were performed in six aneurysm models. The chosen arterial flow rates were 3.0, 4.0, and 5.0 mL/s, which corresponded to physiological measurements at the ICA [48, 58]. The characteristic curves previously computed on those aneurysm models were then evaluated on these specific $\overline{Q}$.

As a result of these additional simulations, the percentage differences between the steady-state and the characteristic curves evaluated on these $\overline{Q}$ were computed. The mean ±SD of these differences were 1.16±1.28% for $\overline{vel_{sa}}$, 2.67±1.90% for $\overline{WSS_{sa}}$, and 2.89±0.07% for $\overline{pre_{sa}}$. This first mean that steady state simulation does approximate the spatiotemporal averaged simulation as reported by Geers et al. for $\overline{WSS_{sa}}$ [49], but it also means that these averaged values are unique and independent of the temporal variations. In other words, during the pulsatile flow simulation, values above $\overline{Q}$ are canceled out by the flow conditions below $\overline{Q}$ and only time-averaged quantities are preserved. This means that if another pulsatile curve is used, the same situation will occur and this unique value of time-averaged characterization will appear. A generalization of these results is described in the scheme of Fig. 10.12.

5.6 Discussion of Part 2

In this part, the effect of the arterial flow rates on both arterial and intra-aneurysmal hemodynamics was studied by looking at spatiotemporal averaged variables. An increase of the arterial flow rate produces an increase of $\overline{vel_{sa}}$, $\overline{WSS_{sa}}$, and $\overline{pre_{sa}}$. Moreover, these hemodynamic quantities are related to $\overline{Q}$ through simple polynomial functions, which do not depend on the pulsatile shape of the input arterial waveform (frequency, pulsatility, or mean arterial value). The polynomial coefficients of these curves are specific of each morphology (artery and aneurysm).

$\overline{vel}_{sa}$ varies linearly with respect to $\overline{Q}$, whose relationship can be characterized by two coefficients: the slope m and a minimum arterial flow $\overline{Q}_{min}$. $\overline{WSS}_{sa}$ and $\overline{pre}_{sa}$ exhibit a quadratic behavior with respect to $\overline{Q}$, which is determined only by three or two parameters, respectively. These characteristic curves contain a condensed representation of the intra-aneurysmal hemodynamics where the full range of flow rates is naturally considered.

5.7 Spatiotemporal Averaged Variables and Mean Arterial Flow Rate

The noncrossing zero linear relationship between $\overline{Q}$ and $\overline{vel}_{sa}$ indicates that there is a minimum flow rate to induce a detectable motion within the aneurysm. This is not plausible in reality, since no matter the arterial flow rate, viscous forces will create aneurysm flow motion. These viscous forces will be relatively more important at low *Re*. There is effectively motion within the aneurysm, due to the moving arterial fluid layer closest to the aneurysm ostium, even for very slow arterial flow. $\overline{Q}_{min}$ was derived from the regression models applied on our set of experiments, and not from the CFD simulations using very low flow rate or from fluid dynamics. The need to perform simulations at lower flow rates to investigate deeper into this matter is arguable since lower flow rates are outside the range of physiological flows. Moreover, other physiological issues may arise at such low flow rates as platelets activation and blood coagulation.

The values of $\overline{Q}_{min}$, considering all cases, was 0.5 ± 0.19 mL/s. Although coherent among cases, these values remain to be studied in the future, considering the influence of the aneurysm shapes and arterial curvatures, for instance. The slope m from Eq. (10.8) would benefit from the same analysis. A study including the morphology of the vascular models could lead to even more generic characteristic curves, relating shape and hemodynamic values.

Concerning $\overline{WSS}_{sa}$, a quadratic behavior was found with respect to $\overline{Q}$. The minimum arterial flow rate necessary to induce an observable WSS was higher on $\overline{WSS}_{sa}$ curves (0.56 ± 0.06 mL/s) than on $\overline{vel}_{sa}$ curves, and occurred in more cases (73.3%). This is consistent with the fact that the WSS must be zero if the aneurysm velocity is zero. For the remaining cases (26.6%), the discrepancies with this hypothesis could be due to how regressions were calculated and their accuracy at low flow rates.

5.8 Waveform-Dependent Variables

He and Ku [56] and Himburg et al. [57] proposed *OSI* and *RRT* as hemodynamic indexes to discriminate the likelihood of aneurysm rupture due to their link with atherosclerosis initiation and promotion. Max$\overline{WSS}_{sa}$ was also studied here as a way of evaluating the maximum possible shear stress at which the aneurysm wall would be affected. The latest is a better measurement compared to peak-systolic $\overline{\text{WSS}}_{\text{sa}}$, since they are different quantities, as demonstrated in the previous part of this chapter. For these three variables, polynomial regression models were not applied since they are waveform-dependent. Indeed, max$\overline{\text{WSS}}_{\text{sa}}$, *OSI*, and *RRT* are highly dependent of the waveform pulsatility and frequency.

In previous studies, a single-flow condition is used to obtain these three variables, and therefore, the impact of changing either the arterial flow rates or the shape of the waveform itself was not considered so far. Fig. 10.11 showed that the different mean flow rates modify the magnitude of these waveform-dependent variables. Moreover, it is still required to see how these variables can be affected by changes in frequency and pulsatility of the waveform. This may be the reason why statistically significant correlations between OSI and aneurysm rupture have been either found [56] or not [51]. However, further investigation is required to clarify these discrepancies among studies.

5.9 Clinical Applications of the Curves

5.9.1 Hemodynamic Comparison at Any Flow Condition

The relationship between arterial flow-aneurysm hemodynamics is still not well understood. The literature describes a wide range of possible flow conditions (patient-specific, generic, area-dependent, physiological as discussed previously). This huge variety in the used flow conditions hinders the comparison among studies, both experimental and clinical. For example, in the field of functional imaging, medical images are compared without taking into account the possible different hemodynamic conditions related to the specific acquisition time and circumstances.

Pre- and posttreatment hemodynamics can be compared to quantify the performance of a deployed endovascular device, like flow diverter stents or coils. If the pretreatment characteristic curves are available, any posttreatment measurement performed in a single arterial flow rate can be compared and can be related to its respective untreated flow rate condition. If this posttreatment measurement is done from medical images, using optical flow techniques, e.g., [27, 28, 47], an actual assessment of the device performance will be done. If CFD is used instead to extract those posttreatment measurements, then a predictive assessment will be conducted (see Fig. 10.13A). In this case, if explicit representation of the virtual devices is used [11, 12] which is computationally expensive, then the need for a single simulation will be sufficient (and not several to cover the full range or arterial flow rates).

5.9.2 Physiological Value or Range

As it is known, WSS is a key factor in the physiological response of cerebral arteries [8, 59] and associated with aneurysm rupture likelihood [15, 60]. In numerical

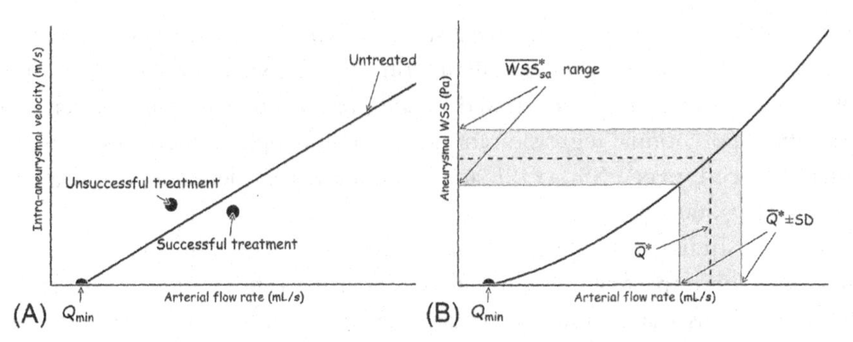

Fig. 10.13 Clinical use of the proposed characteristic curves. (A) Velocity and (B) WSS curve as function of arterial flow rate. In (A), successful (unsuccessful) treatment can be identified when an aneurysm measurement of velocity lies above or under the untreated velocity curve. Here, the treatment success refers to an actual reduction on intra-aneurysmal velocity. (B) $\overline{Q}^*$ represents the most probable flow rate of a patient, knowing his/her clinical history, age, gender, etc. The range $\overline{Q}^* \pm SD$ corresponds to the most probable range of arterial flow rate, providing a $\overline{WSS}^*_{sa}$ range. (*From Morales HG, Bonnefous O. Peak systolic or maximum intra-aneurysmal hemodynamic condition? Implications on normalized flow variables. J Biomech 2014;47(10):2362–70.*)

modeling, a common practice to estimate WSS is through a single-flow condition. Nevertheless, arterial flow rate changes in the short and long terms due to patients' activities, thus affecting aneurysmal quantities [26, 61, 62]. Therefore when estimated aneurysm WSS, the full physiological range of arterial flow rates of the patient should be taken into account and not only a single-flow condition (see Fig. 10.13B). The study of a range instead of a single value has to be done, even if the functional information comes from medical examinations, since those acquisitions are also dependent on the time instance when they were acquired.

Having access to the characteristic curve for WSS, the full range of stresses can be available for all possible arterial flow rates. The range can be personalized and narrowed if patient data are used. In any case, the potential use of the curves can be seen as follows: (1) the analysis of the coefficients describing its quadratic shape and its association with aneurysm rupture likelihood and (2) the identification of a harmful WSS range that can be generated by the physiologic arterial flow rate of the patient.

5.9.3 Normalization

A normalization of aneurysmal variables is usually performed to deal with the uncertainties in the arterial flow as previously discussed to compare among cases. For this reason, when investigating hemodynamic stimuli in aneurysms, reference arterial values are defined, like the WSS at the inlet or at any arterial segment [16]. Nevertheless, these hemodynamic variables depend on the selected arterial segment, but they also vary both with the flow rate and within the cardiac cycle. Therefore, this usual normalization approach has to be reconsidered. A way to allow the comparison among cases without the need of normalization is to use the characteristic curves, since they provide the time-averaged result for any physiological flow rate condition that is needed for comparison.

5.9.4 Patient-Specific Flow Condition

Geers et al. recently analyzed the effect of varying flow rate, heart rate, and pulsatility [49]. They found that the heart rate and the pulsatility index do not affect $\overline{WSS_{sa}}$ for a given $\overline{Q}$. These results mean that the shape of the FRC does not matter concerning $\overline{WSS_{sa}}$, but the mean arterial flow $\overline{Q}$ does. Using steady-state simulations, they found that $\overline{WSS_{sa}}$ can be approximated within an error of 4.3%. This error is relatively small compared to other potential sources, like the image modality (44.2% reported by Geers et al. [40]) or the level of surface smoothing (18.0% reported by Gambaruto et al. [39]).

These results are confirmed and extended in this study to the $\overline{vel_{sa}}$ and $\overline{pre_{sa}}$, using steady state simulations. Here, characteristic curves obtained with pulsatile conditions were confronted with steady state simulations using same $\overline{Q}$s. The difference was lower than 3% in average. These results prove that the proposed characteristic curves can be considered as generic and independent from any patient condition that affects the flow waveform. These curves are then more meaningful than any flow measurement performed in the patient themselves, since those values are not unique and vary during time.

5.10 Limitations and Further Comments

This numerical study is not exempt of limitations. First, the mechanical properties of the wall were not taken into account and rigid walls were forced to be imposed, even though arteries vary their diameter due to arterial pressure chances. However, according to Dempere-Marco et al. [44], the hemodynamic characteristics might not be greatly affected when using these constraints. Secondly, a parabolic profile was imposed at the inlet, although a better condition will be to impose a Womersley-based velocity profile due to the pulsatile nature of the flow. To reduce potential errors, the vascular models were not truncated and the largest available arterial segments upstream of the aneurysms were kept from the medical images [35]. Additionally, since image-based vascular models are used in this study, the need for a Womersley profile might be an unrequired complication of the simulation setup [45, 46].

At the outlets, a zero pressure condition was imposed in the CFD simulations to build the characteristic curves. As reported by Evju et al. [41], aneurysms located further downstream a bifurcation will be greatly affected by changes in arterial flow rates due to different pressure conditions (pulsatile or constant pressure value) that alter flow split. However, the studied cases were mostly lateral aneurysms located upstream from the main ICA bifurcation and only two terminals were investigated. In those two aneurysms, the changes in pressure condition will not affect WSS [41].

We proposed characteristic curves describing key hemodynamic variables as functions of the arterial flow rate. To complement this study, morphology descriptors (such as aneurysm size, aspect ratio, and the presence of blebs) and specific flow patterns (impingement zone, low WSS regions, etc.) should be added to this analysis [18, 24, 61]. A combination of all these elements should be considered in the future to build a unique and more comprehensive picture of the aneurysm hemodynamic status when varying the arterial flow rate.

6. CONCLUSIONS

This chapter shows how numerical modeling can be used to understand vascular hemodynamics. Particularly, it is meant to investigate how aneurysmal hemodynamics is affected by the arterial flow rate and its possible physiological variations. The main contributions of this study are as follows:

- The maximum hemodynamic condition inside the aneurysm does not necessarily occur at peak systole, where the arterial hemodynamics are at its highest. Several factors influence the shifted time between peak systole and the maximum condition, including aneurysm size, arterial flow rate, and arterial morphology, among others.
- The common strategy of aneurysm variable normalization using arterial information should be revised. WSS and velocity inside the artery are strongly dependent on the chosen arterial segment and their variation is larger near peak systole.
- When spatiotemporal variables are considered to characterize aneurysmal hemodynamics, such as $\overline{vel}_{sa}$, $\overline{WSS}_{sa}$, and $\overline{pre}_{sa}$, it is possible to obtain simple relationships with respect to the mean arterial flow rate. It was found that $\overline{vel}_{sa}$ linearly depends on $\overline{Q}$. $\overline{WSS}_{sa}$ and $\overline{pre}_{sa}$ can be represented by quadratic functions of $\overline{Q}$.
- The proposed aneurysm characteristic curves are independent of the waveform shape, since they can also be obtained from steady-state simulations.
- These curves go beyond any patient-specific temporal-dependent flow conditions. They provide a complete view for comparison of both experimental and clinical studies under any physiological flow condition since the full flow rate range is included.

ACKNOWLEDGMENTS

The authors would like to thank the team of the Department of Medical Imaging and Information Sciences, Interventional Neuroradiology Unit, University Hospitals of Geneva, Switzerland, for providing the medical images from which vascular models were extracted.

REFERENCES

[1] Brisman JL, Song JK, Newell DW. Cerebral aneurysms. N Engl J Med 2006;355:928–39.
[2] Rinkel GJE, Djibuti M, Algra A, van Gijn J. Prevalence and risk of rupture of intracranial aneurysms: a systematic review. Stroke 1998;29(1):251–6.
[3] van Gijn J, Rinkel GJE. Subarachnoid haemorrhage: diagnosis, causes and management. Brain 2001;124(Pt 2):249–78.
[4] Grote E, Hassler W. The critical first minutes after subarachnoid hemorrhage. Neurosurgery 1988;22(4):654–61.
[5] Menghini VV, Brown RD Jr, Sicks JD, O'Fallon WM, Wiebers DO. Clinical manifestations and survival rates among patients with saccular intracranial aneurysms: population-based study in Olmsted County, Minnesota, 1965 to 1995. Neurosurgery 2001;49(2):251–8.
[6] Connolly ES, Rabinstein AA, Carhuapoma JR, Derdeyn CP, Dion J, Higashida RT, et al. Guidelines for the management of aneurysmal subarachnoid hemorrhage: a guideline for healthcare professionals from the American Heart Association/American Stroke Association. Stroke 2012;43(6):1711–37.
[7] Wootton DM, Ku DN. Fluid mechanics of vascular systems, diseases, and thrombosis. Ann Rev Biomed Eng 1999;1:299–329.
[8] Reneman RS, Arts T, Hoeks APG. Wall shear stress—an important determinant of endothelial cell function and structure—in the arterial system in vivo. Discrepancies with theory. J Vasc Res 2006;43(3):251–69.
[9] Lasheras JC. The biomechanics of arterial aneurysms. Annu Rev Fluid Mech 2007;39(1):293–319.
[10] Sadasivan C, Fiorella DJ, Woo HH, Lieber BB. Physical factors effecting cerebral aneurysm pathophysiology. Ann Biomed Eng 2013;41(7):1347–65.
[11] Morales HG, Kim M, Vivas EE, Villa-Uriol MC, Larrabide I, Sola T, et al. How do coil configuration and packing density influence intra-aneurysmal hemodynamics? Am J Neuroradiol 2011;32(10): 1935–41.
[12] Larrabide I, Aguilar ML, Morales HG, Geers AJ, Kulcsár Z, Rüfenacht DA, et al. Intra-aneurysmal pressure and flow changes induced by flow diverters: relation to aneurysm size and shape. Am J Neuroradiol 2013;34(4):816–22.
[13] Sforza DM, Putman CM, Cebral JR. Hemodynamics of cerebral aneurysms. Annu Rev Fluid Mech 2009;41(2008):91–107.
[14] Castro MA. Understanding the role of hemodynamics in the initiation, progression, rupture, and treatment outcome of cerebral aneurysm from medical image-based computational studies. ISRN Radiol 2013;2013:1–17.
[15] Meng H, Tutino VM, Xiang J, Siddiqui AH. High WSS or low WSS? Complex interactions of hemodynamics with intracranial aneurysm initiation, growth, and rupture: toward a unifying hypothesis. Am J Neuroradiol 2014;35(7):1254–62.
[16] Jou LD, Lee DH, Morsi H, Mawad ME. Wall shear stress on ruptured and unruptured intracranial aneurysms at the internal carotid artery. Am J Neuroradiol 2008;29(9):1761–7.
[17] Cebral JR, Mut F, Weir J, Putman CM. Association of hemodynamic characteristics and cerebral aneurysm rupture. Am J Neuroradiol 2011;32(2):264–70.
[18] Cebral JR, Mut F, Weir J, Putman C. Quantitative characterization of the hemodynamic environment in ruptured and unruptured brain aneurysms. Am J Neuroradiol 2011;32(1):145–51.
[19] Miura Y, Ishida F, Umeda Y, Tanemura H, Suzuki H, Matsushima S, et al. Low wall shear stress is independently associated with the rupture status of middle cerebral artery aneurysms. Stroke 2013;44(2):519–21.
[20] Shojima M, Oshima M, Takagi K, Torii R, Hayakawa M, Katada K, et al. Magnitude and role of wall shear stress on cerebral aneurysm: computational fluid dynamic study of 20 middle cerebral artery aneurysms. Stroke 2004;35(11):2500–5.
[21] Cebral JR, Castro MA, Appanaboyina S, Putman CM, Millan D, Frangi AF. Efficient pipeline for image-based patient-specific analysis of cerebral aneurysm hemodynamics: technique and sensitivity. IEEE Trans Med Imaging 2005;24(4):457–67.
[22] Villa-Uriol MC, Larrabide I, Pozo JM, Kim M, Camara O, De Craene M, et al. Toward integrated management of cerebral aneurysms. Philos Trans A Math Phys Eng Sci 2010;368(1921):2961–82.

[23] Ujiie H, Tachibana H, Hiramatsu O, Hazel AL, Matsumoto T, Ogasawara Y, et al. Effects of size and shape (aspect ratio) on the hemodynamics of saccular aneurysms: a possible index for surgical treatment of intracranial aneurysms. Neurosurgery 1999;45(1):119–29; Discussion 129–30.
[24] Ujiie H, Tamano Y, Sasaki K, Hori T. Is the aspect ratio a reliable index for predicting the rupture of a saccular aneurysm? Neurosurgery 2001;48(3):495–503.
[25] Markl M, Kilner PJ, Ebbers T. Comprehensive 4D velocity mapping of the heart and great vessels by cardiovascular magnetic resonance. J Cardiovasc Magn Reson 2011;13:7.
[26] Marzo A, Singh PK, Larrabide I, Radaelli AG, Coley S, Gwilliam M, et al. Computational hemodynamics in cerebral aneurysms: the effects of modeled versus measured boundary conditions. Ann Biomed Eng 2011;39(2):884–96.
[27] Bonnefous O, Pereira VM, Ouared R, Brina O, Aerts H, Hermans R, et al. Quantification of arterial flow using digital subtraction angiography. Med Phys 2012;39(10):6264–75.
[28] Pereira VM, Ouared R, Brina O, Bonnefous O, Satwiaski J, Aerts H, et al. Quantification of internal carotid artery flow with digital subtraction angiography: validation of an optical flow approach with Doppler ultrasound. Am J Neuroradiol 2014;35(1):156–63.
[29] Morales HG, Bonnefous O. Peak systolic or maximum intra-aneurysmal hemodynamic condition? Implications on normalized flow variables. J Biomech 2014;47(10):2362–70.
[30] Morales HG, Bonnefous O. Unraveling the relationship between arterial flow and intra-aneurysmal hemodynamics. J Biomech 2015;48(4):585–91.
[31] Hassan T, Timofeev EV, Saito T, Shimizu H, Ezura M, Matsumoto Y, et al. A proposed parent vessel geometry-based categorization of saccular intracranial aneurysms: computational flow dynamics analysis of the risk factors for lesion rupture. J Neurosurg 2005;103(4):662–80.
[32] Valencia AA, Morales H, Rivera R, Bravo E, Galvez M. Blood flow dynamics in patient-specific cerebral aneurysm models: the relationship between wall shear stress and aneurysm area index. Med Eng Phys 2008;30(3):329–40.
[33] Remesh; 2011. http://remesh.sourceforge.net.
[34] MeshLab; 2012. http://meshlab.sourceforge.net.
[35] Pereira VM, Brina O, Marcos Gonzales A, Narata AP, Bijlenga P, Schaller K, et al. Evaluation of the influence of inlet boundary conditions on computational fluid dynamics for intracranial aneurysms: a virtual experiment. J Biomech 2013;46(9):1531–39.
[36] Valen-Sendstad K, Piccinelli M, KrishnankuttyRema R, Steinman DA. Estimation of inlet flow rates for image-based aneurysm CFD models: where and how to begin? Ann Biomed Eng 2015;43(6): 1422–31.
[37] Antiga L, Steinman DA. Vascular modeling toolkit; 2009. http://www.vmtk.org/.
[38] OpenFOAM; 2013. http://www.openfoam.org.
[39] Gambaruto AM, Janela Ja, Moura A, Sequeira A. Sensitivity of hemodynamics in a patient specific cerebral aneurysm to vascular geometry and blood rheology. Math Biosci Eng 2011;8: 409–23.
[40] Geers AJ, Larrabide I, Radaelli AG, Bogunović H, Kim M, Gratama van Andel HAF, et al. Patient-specific computational hemodynamics of intracranial aneurysms from 3D rotational angiography and CT angiography: an in vivo reproducibility study. Am J Neuroradiol 2011;32(3): 581–6.
[41] Evju Ø, Valen-Sendstad K, Mardal KA. A study of wall shear stress in 12 aneurysms with respect to different viscosity models and flow conditions. J Biomech 2013;46(16):2802–8.
[42] Morales HG, Larrabide I, Geers AJ, Aguilar ML, Frangi AF. Newtonian and non-Newtonian blood flow in coiled cerebral aneurysms. J Biomech 2013;46(13):2158–64.
[43] Fisher C, Rossmann JS. Effect of non-Newtonian behavior on hemodynamics of cerebral aneurysms. J Biomech Eng 2009;131(9):091004.
[44] Dempere-Marco L, Oubel E, Castro M, Putman C, Frangi A, Cebral J. CFD analysis incorporating the influence of wall motion: application to intracranial aneurysms. Med Image Comput Comput Assist Interv 2006;9(Pt 2):438–45.
[45] Marzo A, Singh PK, Reymond P, Stergiopulos N, Patel UJ, Hose DR. Influence of inlet boundary conditions on the local haemodynamics of intracranial aneurysms. Comput Meth Biomech Biomed Eng 2009;12(4):431–44.

[46] Moyle KR, Antiga L, Steinman DA. Inlet conditions for image-based CFD models of the carotid bifurcation: is it reasonable to assume fully developed flow? J Biomech Eng 2006;128(3):371–9.
[47] Pereira VM, Bonnefous O, Ouared R, Brina O, Stawiaski J, Aerts H, et al. A DSA-based method using contrast-motion estimation for the assessment of the intra-aneurysmal flow changes induced by flow-diverter stents. Am J Neuroradiol 2013;34(4):808–15.
[48] Hoi Y, Wasserman BA, Xie YJ, Najjar SS, Ferruci L, Lakatta EG, et al. Characterization of volumetric flow rate waveforms at the carotid bifurcations of older adults. Physiol Meas 2010;31(3):291–302.
[49] Geers AJ, Larrabide I, Morales HG, Frangi AF. Approximating hemodynamics of cerebral aneurysms with steady flow simulations. J Biomech 2014;47(1):178–85.
[50] Cebral JR, Castro MA, Putman CM, Alperin N. Flow-area relationship in internal carotid and vertebral arteries. Physiol Meas 2008;29(5):585–94.
[51] Xu J, Yu Y, Wu X, Wu Y, Jiang C, Wang S, et al. Morphological and hemodynamic analysis of mirror posterior communicating artery aneurysms. PLoS One 2013;8(1):e55413.
[52] Huang Q, Xu J, Cheng J, Wang S, Wang K, Liu JM. Hemodynamic changes by flow diverters in rabbit aneurysm models: a computational fluid dynamic study based on micro-computed tomography reconstruction. Stroke 2013;44:1936–41.
[53] Sforza DM, Putman CM, Cebral JR. Computational fluid dynamics in brain aneurysms. Int J Numer Methods Biomed Eng 2012;28(6/7):801–8.
[54] Li C, Wang S, Chen J, Yu H, Zhang Y, Jiang F, et al. Influence of hemodynamics on recanalization of totally occluded intracranial aneurysms: a patient-specific computational fluid dynamic simulation study. J Neurosurg 2012;117(2):276–83.
[55] Cavazzuti M, Atherton M, Collins MW, Barozzi G. Non-Newtonian and flow pulsatility effects in simulation models of a stented intracranial aneurysm. Proc Inst Mech Eng H 2011;225(6):597–609.
[56] He X, Ku DN. Pulsatile flow in the human left coronary artery bifurcation: average conditions. J Biomech Eng 1996;118(1):74–82.
[57] Himburg HA, Grzybowski DM, Hazel AL, LaMack JA, Li XM, Friedman MH. Spatial comparison between wall shear stress measures and porcine arterial endothelial permeability. Am J Physiol Heart Circ Physiol 2004;286(5):H1916–22.
[58] Ford MD, Alperin N, Lee SH, Holdsworth DW, Steinman DA. Characterization of volumetric flow rate waveforms in the normal internal carotid and vertebral arteries. Physiol Meas 2005;26(4): 477–88.
[59] Gao L, Hoi Y, Swartz DD, Kolega J, Siddiqui AH, Meng H. Nascent aneurysm formation at the basilar terminus induced by hemodynamics. Stroke 2008;39(7):2085–90.
[60] Xiang J, Natarajan SK, Tremmel M, Ma D, Mocco JD, Hopkins LN, et al. Hemodynamic-morphologic discriminants for intracranial aneurysm rupture. Stroke 2011;42(1): 144–52.
[61] Jiang J, Strother CM. Computational fluid dynamics simulations of intracranial aneurysms at varying heart rates: a "patient-specific" study. J Biomech Eng 2009;131(9):091001.
[62] McGah PM, Levitt MR, Barbour MC, Morton RP, Nerva JD, Mourad PD, et al. Accuracy of computational cerebral aneurysm hemodynamics using patient-specific endovascular measurements. Ann Biomed Eng 2014;42(3):503–14.

CHAPTER 11

Toward a Mechanical Mapping of the Arterial Tree: Challenges and Potential Solutions

R.L. Maurice*, K.Y.H. Chen†,‡, D. Burgner†,‡,§, L.B. Daniels¶, L. Vaujois*, N. Idris‖, J.-L. Bigras*, N. Dahdah*

*University of Montreal, Montreal, QC, Canada
†Murdoch Childrens Research Institute, Parkville, VIC, Australia
‡University of Melbourne, Parkville, VIC, Australia
§Monash University, Clayton, VIC, Australia
¶University of California San Diego, La Jolla, CA, United States
‖University of Indonesia/Cipto Mangunkusumo General Hospital, Jakarta, Indonesia

Chapter Outline

1. OVERVIEW AND OBJECTIVES

Cardiovascular disease (CVD) is the leading cause of mortality worldwide, despite major advances in healthcare and therapeutic measures [1]. Much effort is invested in preventive care in order to improve health outcomes further. In this respect, vascular physiology modulation is a key element for early diagnosis and risk stratification. Our long-term objective is to facilitate identification in early life of individuals at increasing risk of CVD. This would allow primary prevention, prior to the onset of clinical disease.

Computing and Visualization for Intravascular Imaging and Computer-Assisted Stenting
http://dx.doi.org/10.1016/B978-0-12-811018-8.00011-4

CVD arises from cumulative inflammatory arterial damage. Arterial stiffness is an independent predictor of CVD. Our group has developed an ultrasound-based elastography method that is able to detect vascular functional changes (arterial wall elasticity/stiffness) prior to the onset of anatomical modification (such as the intima-media thickening). The aim of this work is to establish a *longitudinal* mechanical mapping of the arterial tree in healthy individuals. Using our proprietary noninvasive imaging-based biomarker (ImBioMark) algorithm, we address the challenges and potential solutions related to quantifying the complex mechanics of the cardiovascular system.

ImBioMark is an ultrasound-based elastography method. It stems from the premises that the speckle kinematics observed from a time-sequence of echographic images reproduces the underlying tissue kinematics and that tissue motion can be inferred from speckle tracking. This implicitly assumes that speckle pattern is a material property that can be tracked with respect to time and space. ImBioMark then expresses the motion of such a material property in terms of total derivative, also known as optical flow (OF) equations. The method was implemented through solving a nonlinear minimization problem, allowing the assessment of the complete 2D-strain tensor.

The theoretical framework is here supported and illustrated with clinical data reported on the common carotid artery (CCA) in Section 4.1, abdominal aorta artery (AAA) in Section 4.2, and brachial artery (BA) in Section 4.3. We will also address the qualitative, quantitative, and physiological relevance of ImBioMark. These preliminary data provide a foundation for CVD research in children and adults.

2. ARTERIAL PATHOPHYSIOLOGY, MECHANICS, AND STIFFNESS ASSESSMENT

2.1 Arterial Pathophysiology

The arterial wall is made up of the intima, media, and adventitia layers. The media, mostly consisting of lamellae of elastic material, layers of vascular smooth muscle cells, and collagen fibers, is largely responsible for arterial compliance [2]. Collagen, which is at least two orders of magnitude stiffer than elastin and smooth muscle cells, predominates in peripheral arteries, leading to increasing vascular stiffness from the proximal to the distal arterial compartments [3].

Arterial stiffness increases with age [4] and with cardiovascular risk factors [5]. As tissue ages, elastin is either abnormally remodeled or simply not replaced. To compensate for elastin loss, collagen is produced under the influence of angiotensin as suggested in in vitro models [6]. An unbalanced collagen-to-elastin ratio results in an extracellular matrix that distends and recoils less under a given load, collagen being much stiffer than elastin [3, 6]. As the arterial elastic properties regress, their absorbing capacity diminishes, leading to an increasing pulse pressure (PP), aortic impedance, and left ventricular wall tension, all of which increase the workload of the left ventricle, thereby increasing the risk for CVD [7]. Arterial stiffness is then considered to be an independent predictor of cardiovascular mortality [8].

2.1.1 "Occult" Arteriopathies Undermining Health Status

Elevated blood pressure and hypertension are well-established CVD risk factors causing early mortality, increasing overall morbidity, and exacerbating comorbidities such as kidney disease and neurovascular events [4]. For many years now, the focus of prevention worldwide has been on the "traditional" cardiovascular risk factors, namely sedentary lifestyle, obesity, arterial hypertension, diabetes, hypercholesterolemia, and smoking.

However, there are poorly recognized situations that undermine the health status of many individuals very early in life. Kawasaki disease (KD), for instance, is an acute vasculitis typically associated with subclinical myocarditis occurring during the preschool years [9]. Whereas the long-term consequences of KD are typically associated with the extent and severity of the related coronary artery complications (aneurysms), little is known about the long-term consequences on the vascular dynamics status [10].

Another potential cardiovascular risk factor is intrauterine growth restriction (IUGR). The effects of IUGR and premature birth on vascular health status were initially put forth based on epidemiological studies from England, commonly known nowadays as the Barker theory [11]. This investigation created a trend in epidemiological, clinical, and basic research.

Marfan syndrome [12, 13], Williams Buren syndrome [14, 15], and Horton syndrome [16, 17] are additional less-known pathologies that are believed to be potential precursors of CVDs. It is worth emphasizing the fact that, similar to hypertension, those clinical disease models (KD, IUGR) and *occult* syndromes (Marfan, Williams Buren, and Horton) are phenotyped by the alteration of the arterial wall mechanics.

2.1.2 The Importance of Early Detection of CVD Risk

The importance of early detection of vascular dysfunction in various disease models has multiple implications. One essential key role is risk stratification and prognostic assessments. Identification of the classical cardiovascular risk is a great example in modern medicine, since lifestyle modification and early medical intervention have dramatically improved longevity and reduced related morbidity on both individual and population levels. Identifying new risk factors with the possibility of detecting them very early in life, such as we propose to do, would further enhance our knowledge and our potential to protect better against preventable diseases.

2.2 Arterial Mechanics and Stiffness Assessment

The role of large conduit arteries is to minimize pulsatility caused by the flow pattern of left ventricular stroke volume upon ejection. The determinants of such pulsatility are the cushioning capacity of arteries and the timing and intensity of wave reflections. This cushioning capacity, which is usually expressed in terms of compliance and distensibility, is influenced by the arterial stiffness.

2.2.1 Indirect Measurement of Stiffness

As surrogates to stiffness, the ultrasound-based assessment of the arterial biophysical profile (ABP) has been proposed. The ABP can be seen as a set of biometric measurements representative of vascular structure dynamic changes to cardiac work. The ABP parameters, namely the arterial diameter modulation (ADM) with cardiac cycle, the Peterson's elastic modulus (E_P), the β-stiffness index (β_i), and the pulse wave velocity (PWV), are believed to be predictors of CVD [18]. Other indirect measurement parameters that are used for the diagnosis/prognosis of CVD are distension, characteristic impedance (Z_C), and input impedance (Z_I). However, these later parameters all are, to some extent, derived from ADM, E_P, and PWV. In addition, the intima-media thickness (IMT) measurement is a widely used biomarker of subclinical atherosclerosis in clinical practice and research [19, 20]. Table 11.1 lists parameters and formulas conventionally used for the indirect measurement of stiffness.

2.2.2 Ultrasound Elastography for Direct Measurement of Stiffness

Ophir et al. introduced elastography, which is defined as biological tissue elasticity (stiffness) imaging [21]. Primary objectives of elastography were to complement B-mode ultrasound imaging as a screening method to detect hardened areas in the breast [22]. Basically, the tissue under inspection is externally compressed and the displacement between pairs of pre- and postcompression radio frequency (RF) lines (also called "raw" data; that is data before processing such as beam forming, log amplification, and compression) is estimated using cross-correlation analysis. The strain profile in the tissue

Table 11.1 List of parameters/formulas conventionally used for the indirect measurement of stiffness

Pulse pressure (PP) (mmHg) = $P_{SYST} - P_{DIAST}$
Artery diameter modulation (ADM) (%) = $100^*(D_S - D_D)/D_D$
Peterson's elastic modulus (EP) (mmHg) = PP/(ADM/100)
Distension (per mmHg) = 1/EP
β-stiffness index = $\ln(P_{SYST}/P_{DIAST})/(\text{ADM}/100)$
Pulse wave velocity (PWV) (cm/s) = L_W/TT
Artery peak flow (P_F (cm^3/s)) = Peak velocity × CSA
Characteristic impedance (Z_C (dyne × s/cm^5)) = PP/P_F
Input impedance (Z_I (dyne × s/cm^5)) = PWV × ρ/CSA

Notes: ρ, blood density; CSA, cross-section area; L_W, distance covered by the blood flow; TT, transit time calculated with Doppler mode.

is determined from the gradient of the displacement field. It is worth recalling that the display of the axial (equivalent lateral) displacement distribution is known as an axial (lateral) displacement elastogram. Similarly, the display of the axial (equivalent lateral) strain distribution is known as an axial (lateral) strain elastogram.

2.2.3 Noninvasive Vascular Elastography

Vascular ultrasound elastography has primarily been developed as a catheter-based modality to evaluate coronary arteries [23–26] and as a noninvasive way (NIVE) of studying superficial arteries in humans [27–32]. Time sequences of RF data are transcutaneously recorded to assess the vascular tissue motion induced by the blood flow pulsation. As illustrated in Fig. 11.1 for the carotid, in NIVE, the ultrasound probe is placed on the region of interest and dynamic sequences of RF images are recorded.

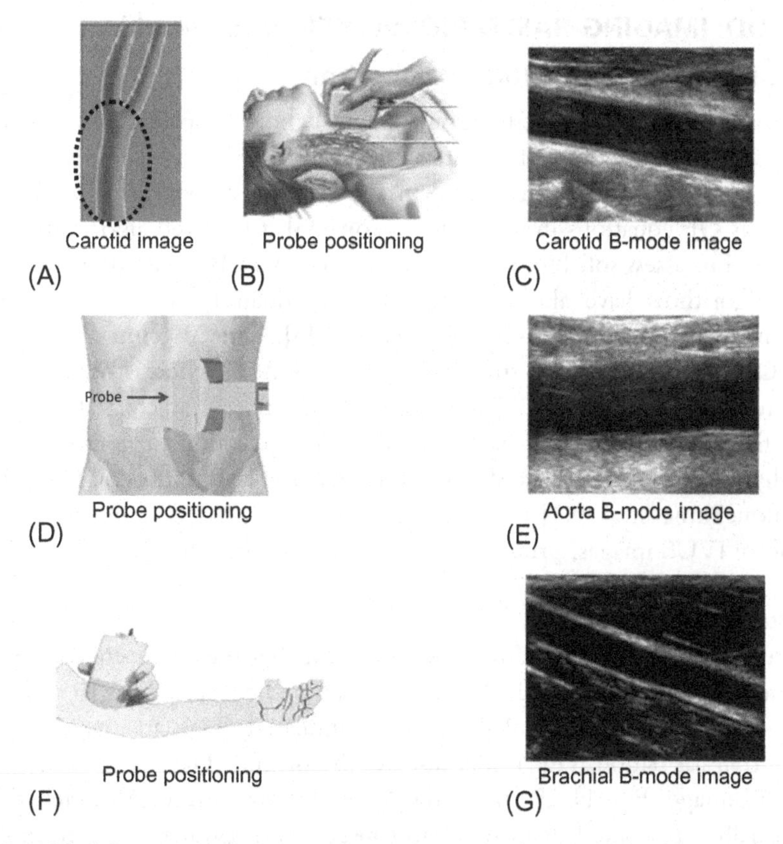

Fig. 11.1 Simplified illustration of NIVE acquisition data procedure: (A–C) for the carotid artery; (D, E) for the abdominal aorta; and (F, G) for the brachial artery. In summary, the ultrasound probe is placed transcutaneously on the region of interest and dynamic sequences of RF (equivalently B-mode) images are recorded.

2.2.4 B-Mode Imaging-Based NIVE

Most current elastographic methods employ RF data (also known as raw data, i.e., prior to any processing) to compute motion estimates, mainly for the purpose of improving elastogram resolution. In B-mode images (i.e., envelop-detected and logarithmic-compressed data, prior to displaying and archiving), for example, signal demodulation may cause loss of motion information. However, such high accuracy in the motion estimation is not required to investigate the arterial wall remodeling in plaque-free conditions. In such a context, it would be convenient to use B-mode data for at least two reasons: (1) they have the advantage of being less bulky, and thus allow significant improvement of the computation time and (2) instead of RF data, they are readily available under the DICOM format available on any clinical ultrasound machine, allowing an application to be easily implementable.

3. METHOD: IMAGING-BASED BIOMARKER (ImBioMark)

3.1 Optical Flow-Based B-Mode Elastography

B-mode images are made up of textures, that is, echographic textures, which can be associated with the definition of a material property [33]. It is convenient to express the motion of a material property in terms of material derivative or total derivative, also known as OF equations in computer vision [33]. OF-based algorithms have then been proposed to assess soft biological tissue motion with B-mode data. By extension, OF-based algorithms have also been advanced to quantify tissue motion from RF data. Meunier and Bertrand [34], Behar et al. [35], and Angelini and Gerard [36] proposed OF-based methods to study myocardial motion, whereas other groups used OF-based techniques for breast [37] and prostate [38] cancer characterizations. Our team very recently introduced an OF-based method for application in noninvasive vascular elastography (NIVE) [39, 40] and for intravascular ultrasound (IVUS) elastography [26]. Danilouchkine et al. [41] recently proposed an OF-based approach for the rigid registration of IVUS images, prior to motion estimation in palpography.

3.1.1 Tissue-Motion Model

Within small measurement windows, tissue motion between two consecutive images recorded at a short-time interval "δt" can be assumed to be affine. To introduce the "tissue-motion" model, Fig. 11.2 illustrates a simulated Gaussian image subjected to such affine transformation. Fig. 11.2A displays the original Gaussian shape, that is, the "premotion" image. Fig. 11.2B shows the "postmotion" image. Although the whole image is usually represented in an Eulerian (observer's) coordinate system (x, y), a sub-ROI (namely, the Gaussian in this example) can have its own coordinates, known as Lagrangian coordinates (x', y'). Fig. 11.2C presents lateral translation of the Gaussian, namely T_1, whereas axial translation is given as T_2 in Fig. 11.2D. Fig. 11.2C and D

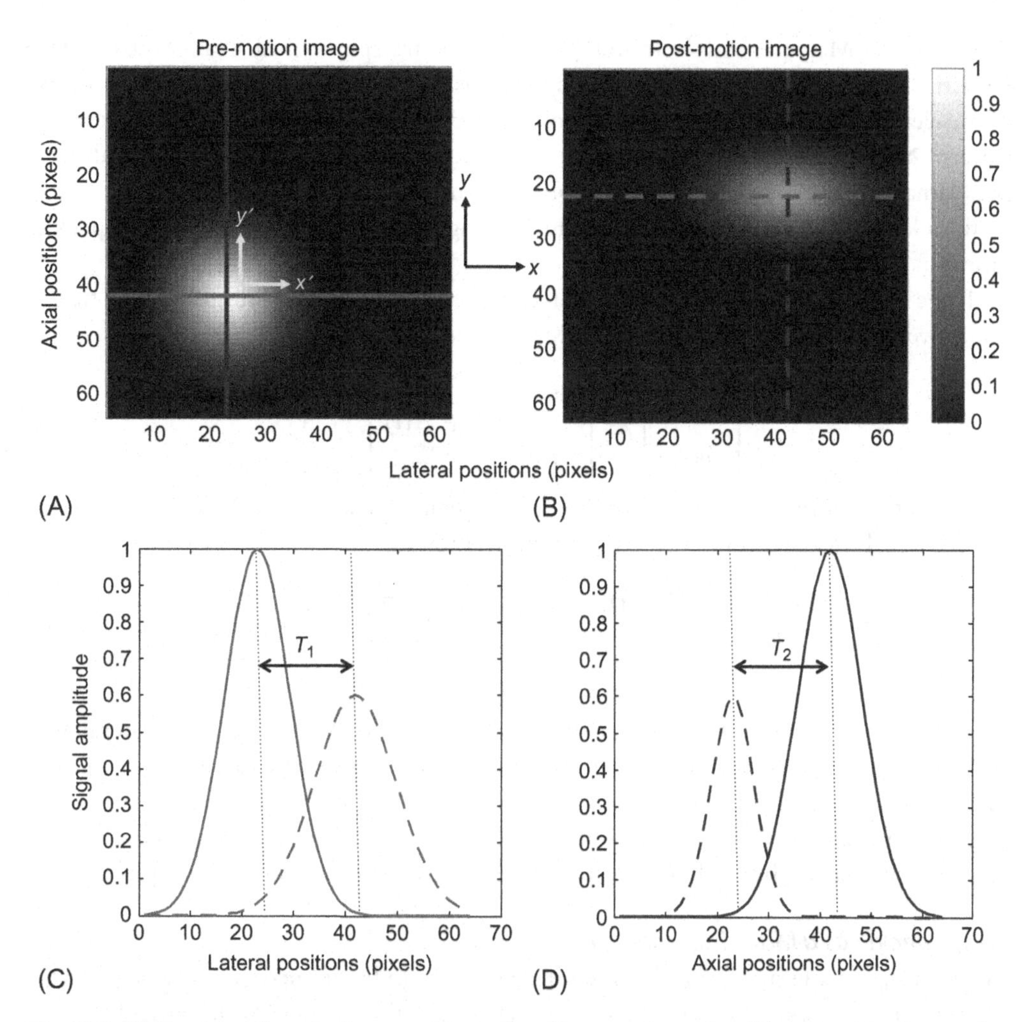

Fig. 11.2 (A) Simulated Gaussian image subjected to affine transformation; (B) the resulting image after application of affine transformation; (C–D) lateral and axial plots as given by ROIs delineated in (A) and (B), respectively. The parameters T_1 and T_2 give lateral and axial translations, respectively. Lateral expansion and axial compression of Gaussian shape, along with a change in image intensity (signal amplitude), can be observed in C and D.

also exhibit lateral expansion and axial compression of the Gaussian shape, respectively. In addition, a change in image intensity (signal amplitude) can be observed during motion.

When a tissue is subjected to complex movements, such as rotation and nonrigid motion (shear, compression/expansion, etc.), changes occurring in speckle pattern may result from the motion of speckles and also from alterations of their morphology.

Referring to Maurice and Bertrand [42], speckle morphological changes associated with tissue motion originate from modifications in the relative amplitude and phase of wavelets backscattered by individual points, leading to modulation of the interference pattern. Namely, rotation and nonrigid motion would produce such effects. In addition, in 2D imaging, out-of-plane tissue motion alters the recorded signal pattern. The change in the Gaussian shape intensity, illustrated in Fig. 11.2, aims to model such speckle morphological alterations.

If we now assume that $u(t)$ and $v(t)$ express lateral and axial displacement fields, respectively, affine tissue motion can be formulated as follows:

$$\begin{bmatrix} u(t) \\ v(t) \end{bmatrix} = \begin{bmatrix} T_1(t) \\ T_2(t) \end{bmatrix} + \Delta(t) \begin{bmatrix} x \\ y \end{bmatrix} \tag{11.1}$$

As presented in Fig. 11.2, T_1 and T_2 give lateral and axial translation parameters, respectively. From this equation, $\Delta(t)$ can be written as:

$$\Delta(t) = \begin{bmatrix} \Delta_{xx}(t) & \Delta_{xy}(t) \\ \Delta_{yx}(\mathrm{t}) & \Delta_{yy}(t) \end{bmatrix} = \begin{bmatrix} \frac{\partial u(t)}{\partial x} & \frac{\partial u(t)}{\partial y} \\ \frac{\partial v(t)}{\partial x} & \frac{\partial v(t)}{\partial y} \end{bmatrix} \tag{11.2}$$

where Δ_{xx} is the lateral strain, Δ_{yy} is the axial strain, and Δ_{xy} and Δ_{yx} are lateral and axial shear parameters, respectively. It is important to emphasize that the optical flow-based B-mode elastography (OFBE) implementation reported below does not explicitly compute the derivative of the displacement fields, as given in Eq. (11.2), to assess the 2D deformation matrix, but rather consists of solving a nonlinear minimization problem.

3.1.2 Speckle as a Material Property

The Gaussian-made sub-ROI presented in Fig. 11.2 can be defined as a *material property*. By extension, speckle pattern can be seen as a continuum of material property. It is convenient to express the motion of a material property in terms of material derivative or total derivative, also known as OF equations in computer vision [33]:

$$\frac{dI\left(x(t), y(t)\right)}{dt} = \frac{\partial I}{\partial x}\frac{dx}{dt} + \frac{\partial I}{\partial y}\frac{dy}{dt} + \frac{\partial I}{\partial t} = I_x u + I_y v + I_t \tag{11.3}$$

In this equation, $I(x(t), y(t))$ is the material property, and dI/dt is the total derivative that expresses the rate of change for $I(x(t), y(t))$ of a "material point" (x, y) as it moves to $(x+\delta x, y+\delta y)$ in $[t, t+\delta t]$ time interval. On the other hand, the partial derivative I_t gives the rate of change for $I(x(t), y(t))$ at a fixed observation point (x, y). Substituting Eq. (11.1) in Eq. (11.3) yields:

$$\frac{dI\left(x(t), y(t)\right)}{dt} = I_x\left(\Delta_{xx}x + \Delta_{xy}y + T_1\right) + I_y\left(\Delta_{yx}x + \Delta_{yy}y + T_2\right) + I_t \tag{11.4}$$

More explicitly, let us consider the material property $(I(x(t), y(t)))$ after movement, that is, $I(x(t+\delta t), y(t+\delta t))$. Under the assumption that speckles reproduce underlying tissue motion, OFBE consists of computing the affine transformation that allows the best match between pre- and postmotion material properties, that is:

$$\underset{\text{AT}}{MIN} \quad \left\| I\left(x(t), y(t)\right) - I\left(x(t+\delta t), y(t+\delta t)\right) \right\|^2 \tag{11.5}$$

In Eq. (11.5), AT defines affine transformation, that is, lateral and axial translations (T_1, T_2), and the 2D deformation matrix (Δ).

3.1.3 Imaging-Based BioMarker (ImBioMark)

We define (ImBioMark) as a stepwise implementation of Eq. (11.5). As a first iteration, lateral and axial translations were assessed as:

$$\underset{\vec{T}r}{MIN} \left\| I\left(x(t), y(t)\right) - I\left(x(t+\delta t), y(t+\delta t)\right) \right\|^2 \tag{11.6}$$

In this equation, $\vec{T}r$, which is the translation vector (T_1, T_2), allows rigid registration. Such a step is particularly important when investigating the aorta artery that is subjected to very substantial motion artifacts, due to its proximity to the heart. It is important to emphasize that, unlike other groups who proposed to compensate for nonrigid movements, namely scaling, by stretching postmotion signals [43] and the 2D companding method [44, 45], here we suggest rigid registration (i.e., compensation for translation) to improve pre- and postmotion signal coherence.

In Eqs. (11.3)–(11.6), $I(x(t), y(t))$ can be seen as being equivalent to a measurement window. As reported by Mercure et al. [40], assuming $dI(x(t), y(t)) = 0$, Eq. (11.6) can be written in a discrete form, according to which the translation vector, $\vec{T}r = (T_1, T_2)$, can be computed as the least squares solution of the following equation:

$$\begin{aligned}
&I_{x_n} T_1 + I_{y_n} T_2 = -I_{t_n}, \quad n = 1, \ldots, p \times q, \quad \text{with:} \\
&I_x = \frac{1}{2}\left[\frac{\partial}{\partial x} I\left(x(t), y(t)\right) + \frac{\partial}{\partial x} I\left(x(t+\delta t), y(t+\delta t)\right)\right] \\
&I_y = \frac{1}{2}\left[\frac{\partial}{\partial y} I\left(x(t), y(t)\right) + \frac{\partial}{\partial y} I\left(x(t+\delta t), y(t+\delta t)\right)\right] \\
&I_t = I\left(x(t+\delta t), y(t+\delta t)\right) - I\left(x(t), y(t)\right)
\end{aligned} \tag{11.7}$$

where "$p \times q$" is the number of pixels within the measurement window, which is sized p-lines (axial samples) $\times$ q-columns (B-mode lines). The spatial derivatives were calculated using finite differences.

In the second iteration, tissue deformation (Δ) is assessed as:

$$\underset{\Delta}{MIN} \left\| I\left(x(t), y(t)\right) - I\left(x(t+\delta t) - [T_1], y(t+\delta t) - [T_2]\right) \right\|^2 \tag{11.8}$$

where $[T_i]$ rounds T_i to the nearest integer, $i = 1, 2$. In other words, the premotion ROI is compared to the postmotion ROI compensated for translations (T_1, T_2). Similarly to Eq. (11.7), the deformation matrix (Δ) can be found by solving this equation:

$$I_{x_n} x_n \Delta_{xx} + I_{x_n} y_n \Delta_{xy} + I_{y_n} x_n \Delta_{yx} + I_{y_n} y_n \Delta_{yy} = -I_{t_n}, \quad n = 1, \ldots, p \times q \tag{11.9}$$

In ImBioMark applications in the carotid, brachial, and aorta arteries, the measurement window is typically set to 40×20 pixels (i.e., "$p = 20$" axial samples $\times$ "$q = 40$" B-mode lines), with 90% axial and 90% lateral overlaps, respectively. As an example, that is roughly close to 750 μm × 1500 μm, for the carotid. These OF-based implementations of displacement (Eq. 11.7) and deformation (Eq. 11.9) assessments do

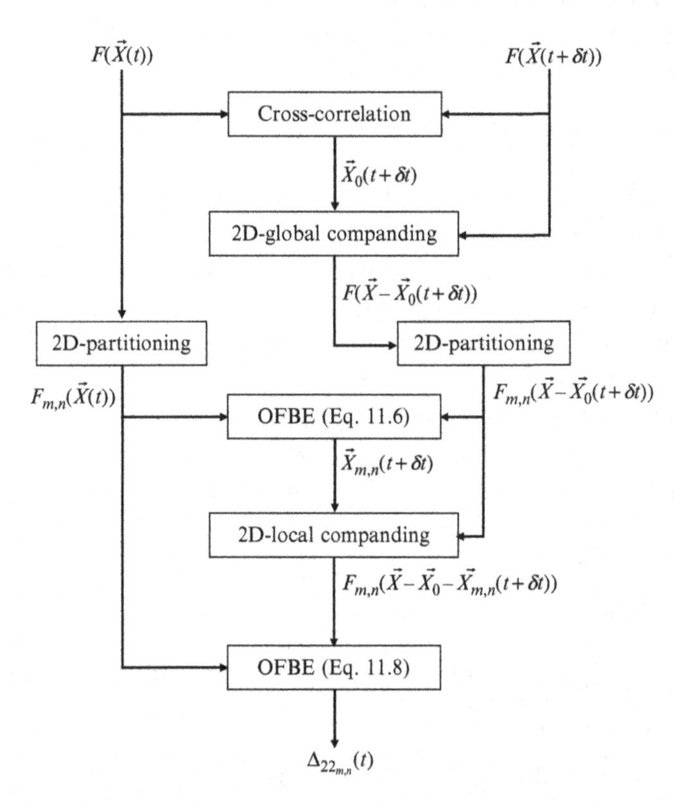

Fig. 11.3 Simplified block-diagram illustrating the new implementation of OFBE as it was adapted to study the abdominal aorta artery. Without lost of generality, the notation in this illustration slightly differs from Eqs. (11.6), (11.8). $F(\vec{X}(t))$ and $F(\vec{X}(t+\delta t))$ here define B-mode images of a cine-loop, while $F_{m,n}(\vec{X}(t)) = I(\vec{X}(t))$ and $F_{m,n}(\vec{X}(t+\delta t)) = I(\vec{X}(t+\delta t))$ represent pre- and postmotion measurement windows and $\vec{X}_{m,n}(\vec{X}(t+\delta t)) = \vec{T}_r$, respectively.

not require any iterative computation, and the data are not interpolated. To summarize, Fig. 11.3 is a flow chart of this OFBE implementation.

3.1.4 Qualitative and Quantitative Illustrations of ImBioMark

For qualitative illustration purposes, we present typical B-mode images (Fig. 11.4A–C), diastolic axial displacement elastograms (Fig. 11.4D–F), and systolic axial displacement elastograms (Fig. 11.4G–I) for the CCA, AAA, and BA, respectively. In this configuration, positive displacement values indicate downward wall motion and inversely for negative displacement values. Strain was assessed, at the bottom wall, in longitudinal segments typically >1 cm length. In this context, Δ_{yy} was averaged over 3-pixel wall thickness starting from the blood/intima interface through the adventitia as a rough approximation of the IMT.

For quantitative illustration purposes, Fig. 11.4J–L exhibits instantaneous strain curves computed over three consecutive cardiac cycles for the CCA, AAA, and BA, respectively. The systolic phase of the cardiac cycle is associated with positive strain values and inversely for diastole. In addition, Fig. 11.4M–O exhibits cumulative strain curves for the CCA, AAA, and BA, respectively.

In the context of this study, we averaged cumulated systolic ($\bar{\Delta}_{yy}^{syst}$) and cumulated diastolic ($\bar{\Delta}_{yy}^{diast}$) strains, respectively, over at least three cardiac cycles. $\bar{\Delta}_{yy}$ was calculated as the average in absolute values of $\bar{\Delta}_{yy}^{syst}$ and $\bar{\Delta}_{yy}^{diast}$. ImBioMark elastic moduli (E_{IBM}), for a given subject's CCA, AAA, and BA, were calculated as the ratio between the PP ($\Delta P \equiv$ peak-systole blood pressure − nadir-diastole blood pressure) and $\bar{\Delta}_{yy}$, such as

$$E_{IBM} = \frac{\Delta P}{\bar{\Delta}_{yy}} \tag{11.10}$$

4. ImBioMark: APPLICATIONS ON CAROTID, BRACHIAL, AND AORTA ARTERIES

4.1 Common Carotid Artery Study

4.1.1 Study Population

Subjects examined in this study were recruited as controls for other investigations conducted in four institutions worldwide, namely the Murdoch Childrens Research Institute Melbourne, the Universitas Indonesia/Cipto Mangunkusumo General Hospital, the University of California at San Diego, and the CHU Sainte-Justine at Montreal. Exclusion criteria were diabetes, known atherosclerotic CVDs, treatment for hypertension and/or hyperlipidaemia, or chronic inflammatory conditions requiring previous or ongoing treatment. The studies were approved by respective ethics committees, and written consent was obtained from the parents of the children involved in this investigation.

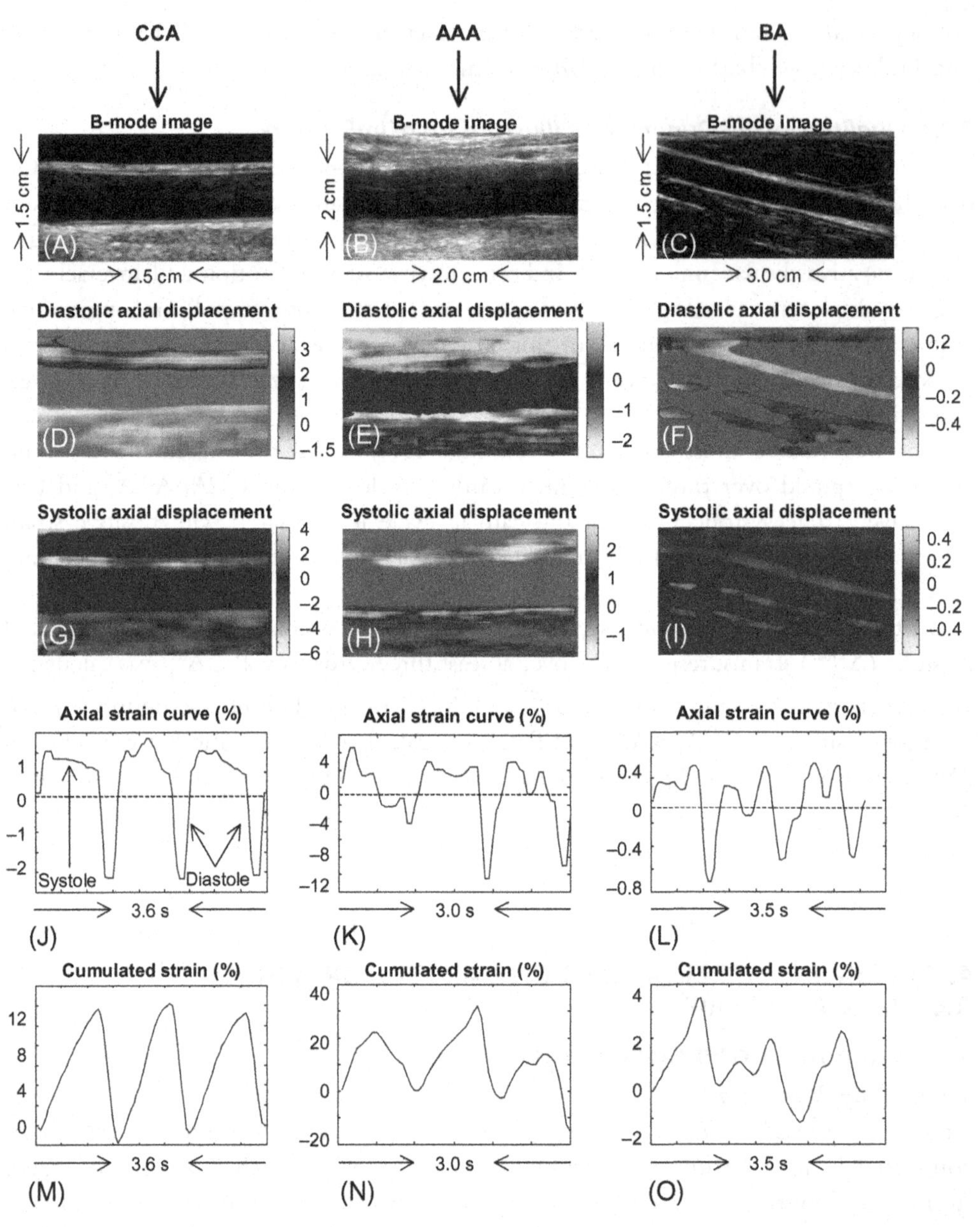

Fig. 11.4 (A–C) B-mode images of the CCA, AAA, and BA, respectively; (D–F) diastolic axial displacement elastograms computed with ImBioMark for the CCA, AAA, and BA, respectively; (G–I) systolic axial displacement elastograms computed with ImBioMark for the CCA, AAA, and BA, respectively; (J–L) instantaneous axial strain curves (for three consecutive cardiac cycles) computed with ImBioMark for the CCA, AAA, and BA, respectively; (M–O) cumulative strain curves computed with ImBioMark for the CCA, AAA, and BA, respectively. *For interpretation of the color in this figure, please see the online version.*

4.1.2 Materials

Depending on the institution, B-mode data of longitudinal segments of the right CCA (Fig. 11.4A) were recorded with either an iE33 Philips (Philips, Andover, MA) echography machine, a Vivid-i ultrasound machine (General Electronics Medical Systems, Tirat Harcarmel, Israel), or a MyLab One (Esaote), at 1 cm proximal to the carotid bulb. Typically, probes in a range of 8–11 MHz were used. The frame rate was generally close to 40 Hz depending on the depth, which was typically 4 cm with the focus positioning in the middle of the CCA. Loops of seven to eight beats were recorded serially. We then analyzed, offline with ImBioMark, four to five complete consecutive cardiac cycles.

For each subject, blood pressure was measured with automated sphygmomanometers at the beginning of imaging recording. ECG signals were simultaneously recorded for appropriate cardiac cycle determination as well as proper identification of systole and diastole. In addition to cardiovascular ultrasound and physiological parameters, somatic data such as weight and height were also obtained.

4.1.3 CCA Stiffness as a Function of Aging

The right CCA of 64 male subjects, in a range of 3–43 years old, was examined. As illustrated in the histogram in Fig. 11.5, there were 22 children aged 3–9 years, 25 adolescents aged 10–17 years, and 17 subjects aged 18–43 years. Fig. 11.6 shows

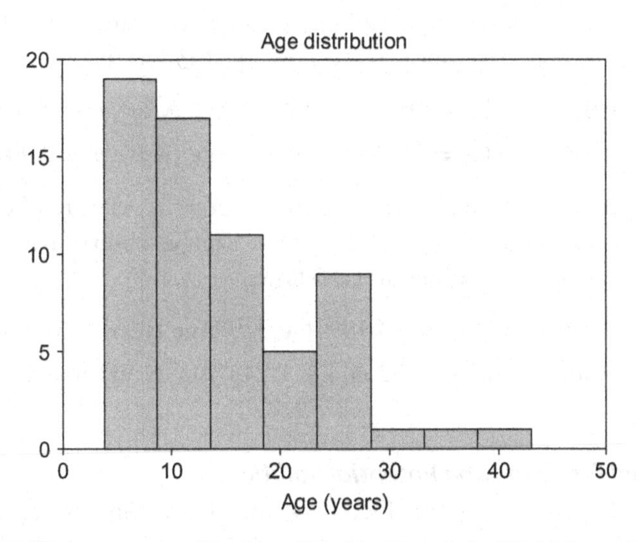

Fig. 11.5 The right CCA was examined for 64 male subjects; namely: 32 children ∈ [3, 12[years old, 15 adolescents ∈ [13, 18[years old, and 17 young adults and adults ∈ [18, 43[years old.

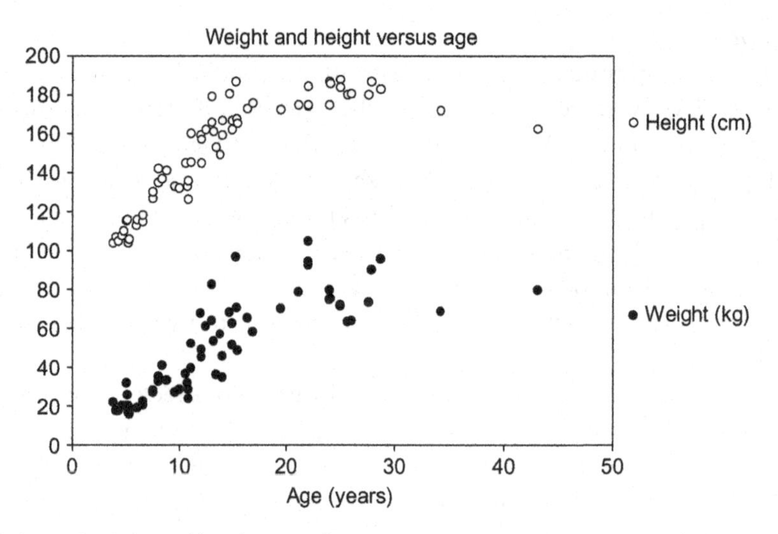

Fig. 11.6 Subjects' weight and height according to age.

an increase in somatic parameters such as weight and height up until the age of 22 approximately; with respective linear regression equations as follows:

$$\text{Weight (kg)} = 15.777 + 2.426^{*}\text{Age (year)}; p < 0.001$$
$$\text{Height (cm)} = 112.760 + 2.660^{*}\text{Age (year)}; p < 0.001.$$

Fig. 11.7 shows an increase in physiologic parameters, namely systolic and diastolic blood pressures, with respective linear regression equations as follows:

$$\text{Syst BP (mmHg)} = 98.597 + 0.927^{*}\text{Age (year)}; p < 0.001$$
$$\text{Diast BP (mmHg)} = 55.399 + 0.598^{*}\text{Age (year)}; p < 0.001.$$

Fig. 11.8 shows a moderate increase in PP, whereas a very significant increase in stiffness (elastic moduli in a range of 22–90 kPa) can be observed in Fig. 11.9. The respective linear regression equations are as follows:

$$\text{PP (mmHg)} = 43.199 + 0.329^{*}\text{Age (year)}; p = 0.022$$
$$\text{Elastic modulus (kPa)} = 32.464 + 1.248^{*}\text{Age (year)}; p < 0.001.$$

4.1.4 Anticipated Problems and Potential Solutions

As reported above and illustrated in Fig. 11.4, the elastograms display maps of strain, shear, and displacement distributions, including near and far vessel walls, lumen, and surrounding tissues. Based on phenomenological knowledge, the far wall is more appropriate to investigate the carotid artery and BA. In contrast, the near wall is more

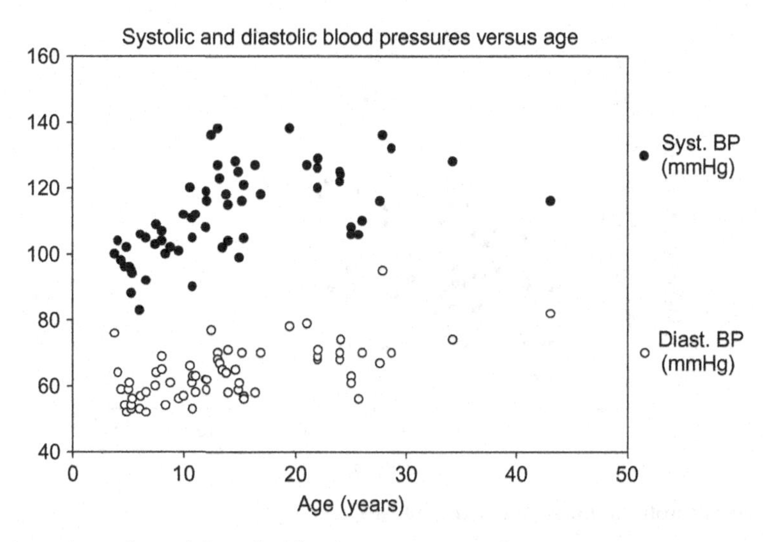

Fig. 11.7 Subjects' systolic and diastolic blood pressures according to age.

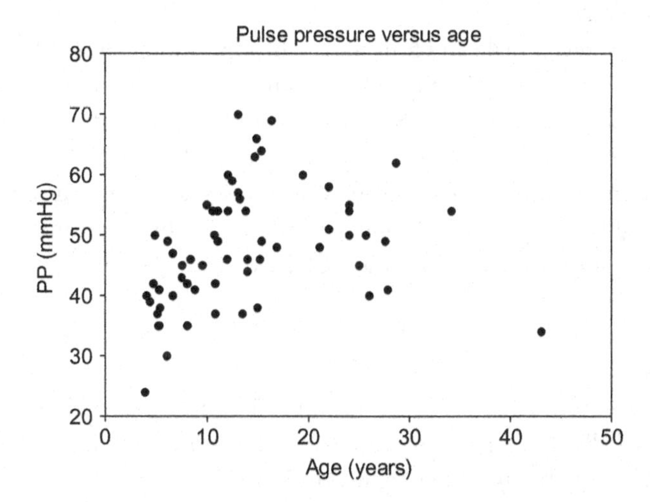

Fig. 11.8 Subjects' pulse pressure according to age.

adequate for the aorta. We take advantage of this assumption and assess the carotid stiffness for a region of interest selected at the bottom wall.

The CCA wall (including intima, media, and adventitia) is typically as large as 1.2 mm [46, 47]. In addition, the axial image resolution is around 150 μm, depending on the image acquisition set-up. In this regard, a "750 μm × 1500 μm" measurement window is quite appropriate to provide reliable motion estimates.

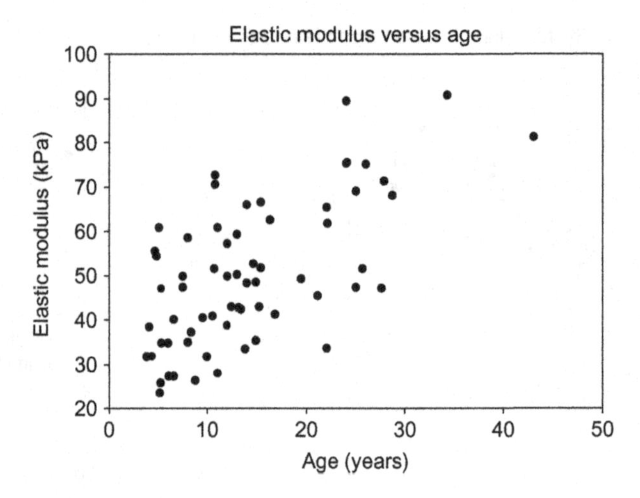

Fig. 11.9 Subjects' carotid stiffness (elastic modulus) according to age.

4.2 Brachial Artery Study

4.2.1 Study Population

We examined BA in seven male and four female healthy adolescents, 14.4±1.2 years old, who were recruited as control subjects for a clinical research study involving children with IUGR, conducted at the CHU Ste-Justine. Study subjects were assessed at the ages of 13–15 years old according to a standardized protocol; they were all free of known cardiovascular comorbidities or risk factors. Cardiovascular ultrasounds, blood pressures, weights, and heights were obtained. The institutional ethics committee approved the study and written informed consent was obtained from the subjects' parents for this investigation.

4.2.2 Materials

Blood pressure was measured with an automated sphygmomanometer (Welch Allyn Inc., Skaneateles Falls, NY) prior to vascular ultrasound imaging recording. B-mode data of longitudinal segments of the right-arm BA (Fig. 11.4C) were recorded with an iE33 Philips (Philips) echography machine, using an 11 MHz probe. The frame rate was generally close to 40 Hz depending on the depth, which was typically 4 cm with the focus positioning in the middle of the artery. Three loops of seven to eight beats were recorded serially. ECG signals were simultaneously recorded for appropriate cardiac cycle determination as well as proper identification of systole and diastole.

4.2.3 BA Stiffness Assessment

BA exhibited higher E_{IMB} compared to CCA for these adolescents (14.4 ± 1.2 years old), 129.73 ± 25.67 kPa (range 78–168) versus 49.55 ± 14.75 kPa (range 31–71),

$p < 0.001$ (paired T-test) [48]. As illustrated in Fig. 11.10, it was observed that there was an apparent increasing linear relationship between the two arteries with a slope of +0.36 and an intercept of 111.62 (R^2 0.045).

4.2.4 Anticipated Problems and Potential Solutions

The BA is deeply located and runs parallel to the humerus. As observed in Fig. 11.4C, it can be difficult to image horizontal longitudinal wall segments with neat delineation between the intima-media layer and the lumen. First, the deviation of the arterial wall from the horizontal plane may induce a bias in motion estimates requiring compensation as reported by Mercure et al. [49]. Second, *speckle* patterns of BA B-mode images result from multiple interferences due to the proximity of this artery to various surrounding tissues (muscles, nerves, bone, etc.), which can generate motion artifacts [50]. The latter phenomenon causes *speckle* decoherence, which impacts on the delineation between the intima-media layer and the lumen. Although the potential drawbacks reported here for BA were not a major concern in the current investigation, it could be convenient to implement a segmentation algorithm to delimit the lumen and intima interface prior to ImBioMark assessment for broader applications.

4.3 Abdominal Aorta Artery Study

4.3.1 Study Population

We examined AAA in six male and seven female healthy adolescents, 12.9 ± 2.5 years old, who were recruited as control subjects for a research study evaluating vascular

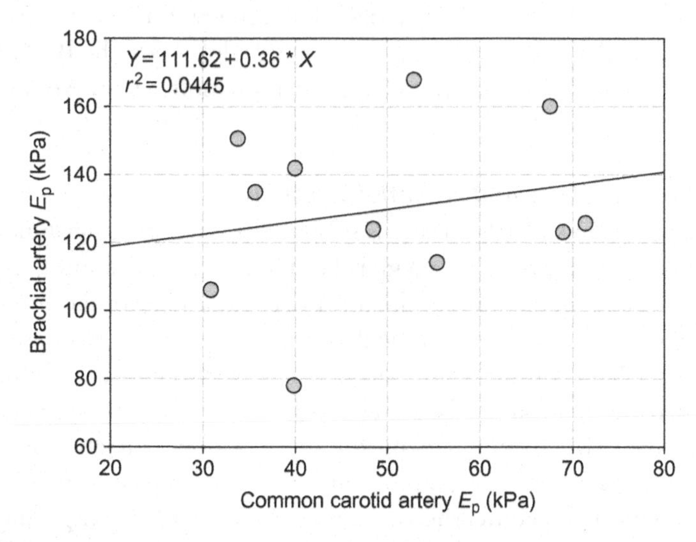

Fig. 11.10 Brachial artery elastic modulus (E_{IMB}) correlated positively with that of the common carotid artery ($Y = 111.62 + 0.36 * X$).

health following KD. Control subjects were recruited from the outpatient clinics of the Royal Children's Hospital Melbourne, and from children of staff members and friends of KD patients. Exclusion criteria were pregnancy, diabetes, known atherosclerotic CVDs, treatment for hypertension and/or hyperlipidaemia, and chronic inflammatory conditions requiring previous or ongoing treatment. The study was approved by the Royal Children's Hospital Human Research Ethics Committee and written consent was obtained from the parents.

4.3.2 Materials

After a minimum 6-h fast, participants attended Murdoch Childrens Research Institute for a one-off appointment at which anthropometric and adiposity measurements (BC 418, Tanita, Tokyo, Japan), automated blood pressure (SphygmoCor XCEL, AtCor Medical Pty Ltd, NSW, Australia), fasting lipid profile and blood glucose (Vitros 5600, Ortho-Clinical Diagnostics, Rochester, NY), and high sensitivity C-reactive protein (CRP Vario, Abbott Laboratories Inc., Abbott Park, IL) were obtained.

B-mode ultrasound images of the AAA (Fig. 11.4B) were obtained using a Vivid-i ultrasound machine (General Electronics Medical Systems) with an 8 MHz linear probe and continuous electrocardiography grating. Five to fifteen cardiac cycles consisting of longitudinal images of the AAA just proximal to the femoral bifurcation were recorded for offline analysis. Because of body habitus, the depth of the AAA varied between 4 and 8 cm and the frame rates ranged from 14 to 27 frames/s.

4.3.3 AAA Stiffness Assessment

AAA stiffness for those 12.9 $\pm$ 2.5-year-old adolescents was assessed at 31 $\pm$ 6 kPa. In summary, as illustrated in Fig. 11.11 for a 14-year-old adolescent, the BA (130 $\pm$ 26 kPa) was stiffer than CCA (49 $\pm$ 16 kPa), which in turn was stiffer than AAA (31 $\pm$ 6 kPa); $p < 0.001$.

4.3.4 Anticipated Problems and Potential Solutions

The aorta, being proximal to the heart, is subjected to complex dynamics. While the cyclic movement of the heart induces a global displacement of the aorta, the blood flow pulsation induces cyclic translations of the top and bottom walls of the aorta (i.e., lumen dilation/compression). This makes it difficult to record high signal-to-noise ratio images, that is with neat delineation between the intima-media layer and the lumen. We have recently modeled the motion artifacts that can be induced by the complex dynamics of the aorta artery [51]. To compensate for such motion artifacts, we implemented an iterative stepwise rigid registration procedure. The first step consists of a correlation-based rigid registration procedure; this is equivalent to the global companding procedure introduced by Chaturvedi et al. [44]. The second step consists of an OF-based rigid registration with the purpose of compensating for local translations of the arterial walls;

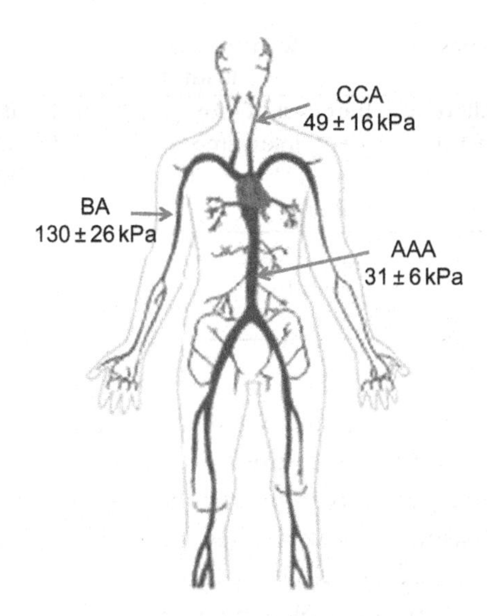

Fig. 11.11 Preliminary mechanical mapping of the main upper body arterial branches for a 14-year-old adolescent: AAA (31 ± 6 kPa), CCA (49 ± 16 kPa), and BA (130 ± 26 kPa).

this is equivalent to the local companding procedure introduced by Chaturvedi et al. [45]. An additional alternative could be the implementation of a segmentation algorithm to improve the delineation between the intima-media layer and the lumen.

5. DISCUSSION

An accurate assessment of arterial stiffness may help early detection of vascular diseases as well as stratification of the severity of the disease affecting the vessels. Whereas IMT was developed to measure anatomical changes to the carotid arteries, ImBioMark measures dynamic responses of the arterial vascular wall to distension and recoil forces as they are generated throughout systole and diastole. This assessment is unique and gives a new functional perspective of the arterial wall structure. It is worth noting that ImBioMark was designed to assess mechanical properties of large- and medium-sized vessels that are transcutaneously accessible, namely the carotid artery, AAA, and BA in this study. The algorithm is applicable to clinically available ultrasound imaging modalities, namely DICOM format data recorded with GE Vivid-i, Philips iE33, and Esaote MyLab One ultrasound machines in the current studies. Table 11.2 summarizes the different echographs that were used with respect to the institution and the arteries investigated.

Table 11.2 Ultrasound machines with respect to institutions and arteries

	Institutions			
	Murdoch Childrens Research Institute (Melbourne)	**Indonesia/Cipto Mangunkusumo General Hospital (Jakarta)**	**University of California (San Diego)**	**CHU Sainte-Justine (Montreal)**
Echographs	GE Vivid-i	Esaote MyLab One	Philips iE33	Philips iE33
Arteries	Carotid Abdominal aorta	Carotid	Carotid	Carotid Brachial

We reported data showing a neat and significant increase in the CCA stiffness with aging ($p < 0.001$). Elastic moduli were measured in a range of 22–90 kPa for healthy subjects aged 3–43 years old. We also compared the arterial stiffness of the AAA, CCA, and BA in healthy adolescents. Distal arteries were stiffer than proximal arteries, in keeping with the predominance of collagen over elastin in peripheral arteries [3]. Here we show for the first time, in an adolescent population, the progressive increase of stiffness from AAA to CCA and BA. This work is an additional step toward extensive vascular mapping of arterial wall elastic properties. Although we now assume BA stiffness of 130 ± 26 kPa, CCA stiffness of 49 ± 16 kPa, and AAA stiffness of 31 ± 6 kPa for a 14-year-old adolescent, we are investigating a larger cohort of healthy subjects in order to allow for age-stratification and to include additional arteries such as the pulmonary, the renal, and the femoral.

6. CONCLUSION

One essential key role of early detection of vascular dysfunction is risk stratification and prognostic assessments of CVDs. ImBioMark enables the possibility of detecting such a dysfunction early in life; this would further enhance our knowledge and our potential to protect better against CVDs.

The in vivo data presented here compare, for the first time, stiffness between the abdominal aorta, common carotid, and brachial arteries in the same adolescent subjects. As expected, due to the predominance of collagen in peripheral arteries, arterial stiffness was observed to increase from the proximal to the distal compartments.

We also corroborated the arterial stiffness increase with aging in the CCA for a population ranging from young children to adults. In summary, these results suggest that *longitudinal* mechanical mapping of the arterial tree in healthy individuals is feasible, thanks to ImBioMark. In the longer term, our research may facilitate identification of individuals at increased risk of CVD earlier in the life course, thereby allowing primary and primordial prevention, prior to the onset of clinical diseases.

ACKNOWLEDGMENTS

Special thanks go to Ramona for her kind support and encouragement to pursue this research.

REFERENCES

[1] http://www.who.int/mediacentre/factsheets/fs317/en/.
[2] Benetos A, Bouaziz H, Albaladejo P, Guez D, Safar ME. Carotid artery mechanical properties of Dahl salt-sensitive rats. Hypertension 1995;25:272–7.
[3] Li JKJ. The arterial circulation: physical principles and clinical applications. Totowa: Human Press; 2000. p. 271.
[4] Benetos A, Waeber B, Izzo J, Mitchell G, Resnick L, Asmar R, et al. Influence of age, risk factors, and cardiovascular and renal disease on arterial stiffness: clinical applications. Am J Hypertens 2002;15(12):1101–8.
[5] Liao J, Farmer J. Arterial stiffness as a risk factor for coronary artery disease. Curr Atheroscler Rep 2014;16(2):387.
[6] Wagenseil JE, Mecham RP. Elastin in large artery stiffness and hypertension. J Cardiovasc Transl Res 2012;5:264–73.
[7] Patrianakos AP, Parthenakis FI, Karakitsos D, Nyktari E, Vardas PE. Proximal aortic stiffness is related to left ventricular function and exercise capacity in patients with dilated cardiomyopathy. Eur J Echocardiogr 2009;10:425–32.
[8] Laurent S, Boutouyrie P, Asmar R, Gautier I, Laloux B, Guize L, et al. Aortic stiffness is an independent predictor of all-cause and cardiovascular mortality in hypertensive patients. Hypertension 2001;37:1236–41.
[9] Dahdah N, Fournier A. Natriuretic peptides in Kawasaki disease: the myocardial perspective. Diagnostics 2013;3:1–12.
[10] McCrindle BW, McIntyre S, Kim C, Lin T, Adeli K. Are patients after Kawasaki disease at increased risk for accelerated atherosclerosis? J Pediatr 2007;151(3):244–8.
[11] Barker DJ, Osmond C, Simmonds SJ, Wield GA. The relation of small head circumference and thinness at birth to death from cardiovascular disease in adult life. BMJ 1993;306(6875):422–6.
[12] Adams JN, Brooks M, Redpath TW, Smith FW, Dean J, Gray J, et al. Aortic distensibility and stiffness index measured by magnetic resonance imaging in patients with Marfan's syndrome. Br Heart J 1995;73(3):265–9.
[13] Abraham PA, Perejda AJ, Carnes WH, Uitto J. Marfan syndrome: demonstration of abnormal elastin in aorta. J Clin Investig 1982;70(6):1245–52.
[14] Aggoun Y, Sidi D, Levy B, Lyonnet S, Kachaner J, Bonnet D. Mechanical properties of the common carotid artery in Williams syndrome. Heart 2000;84(3):290–3.
[15] Kozel BA, Danback J, Waxler J, Knutsen RH, Fuentes L, Reusz GS, et al. Williams syndrome predisposes to vascular stiffness modified by anti-hypertensive use and copy number changes in NCF1. Hypertension 2013;63(1):74–9.
[16] Lurati A, Bertani L, Re KA, Marrazza M, Bompane D, Scarpellini M. Successful treatment of a patient with giant cell vasculitis (Horton arteritis) with tocilizumab a humanized anti-interleukin-6 receptor antibody. Case Rep Rheumatol 2012;2012:639612.
[17] Chebbi W, Jerbi S. Aortite thoraco-abdominale révélant une maladie de horton. Pan Afr Med J 2014;18:139.
[18] Vaujois L, Dallaire F, Maurice RL, Fournier A, Houde C, Therrien J, et al. The biophysical properties of the aorta are altered following Kawasaki disease. JASE 2013;26(12):1388–96.
[19] Lee SJ, Ahn HM, You JH, Hong YM. Carotid intima-media thickness and pulse wave velocity after recovery from Kawasaki disease. Korean Circ J 2009;39:264–9.
[20] Kadono T, Sugiyama H, Hoshiai M, Osada M, Tan T, Naitoh A, et al. Endothelial function evaluated by flow-mediated dilatation in pediatric vascular disease. Pediatr Cardiol 2005;26:385–90.
[21] Ophir J, Cespedes EI, Ponnekanti H, Yazdi Y, Li X. Elastography—a quantitative method for imaging the elasticity of biological tissues. Ultrason Imaging 1991;13(2):111–34.

[22] Garra BS, Céspedes EI, Ophir J. Elastography of breast lesions: initial clinical results. Radiology 1997;202:79–86.
[23] de Korte CL, van der Steen AFW, Céspedes EI, Pasterkamp G, Carlier SG, Mastik F, et al. Characterization of plaque components and vulnerability with intravascular ultrasound elastography. Phys Med Biol 2000;45(6):1465–75.
[24] de Korte CL, Sierevogel MJ, Mastik F, Strijder C, Schaar JA, Velema E, et al. Identification of atherosclerotic plaque components with intravascular ultrasound elastography in vivo: a Yucatan pig study. Circulation 2002;105:1627–30.
[25] Maurice RL, Fromageau J, Brusseau E, Finet G, Rioufol G, Cloutier G. On the potential of the Lagrangian speckle model estimator for endovascular ultrasound elastography: in vivo human coronary study. Ultrasound Med Biol 2007;33(8):1199–2005.
[26] Maurice RL, Fromageau J, Doyley M, Demuinck E, Robb J, Cloutier G. Characterization of atherosclerotic plaques and mural thrombi with intravascular ultrasound elastography: a potential method evaluated in an aortic rabbit model and a human coronary artery. IEEE Trans Inf Technol Biomed 2008;12(3):290–8.
[27] Mai JJ, Insana MF. Strain imaging of internal deformation. Ultrasound Med Biol 2002;28(11–12):1475–84.
[28] Bang J, Dahl T, Bruinsma A, Kaspersen JH, Hernes TA, Myhre HO. A new method for analysis of motion of carotid plaques from RF ultrasound images. Ultrasound Med Biol 2003;29(7):967–76.
[29] Kanai H, Hasegawa H, Ichiki M, Tezuka F, Koiwa Y. Elasticity imaging of atheroma with transcutaneous ultrasound: preliminary study. Circulation 2003;107(24):3018–21.
[30] Kim K, Weitzel WF, Rubin JM, Xie H, Chen X, O'Donnell M. Vascular intramural strain imaging using arterial pressure equalization. Ultrasound Med Biol 2004;30(6):761–71.
[31] Trahey GE, Palmeri ML, Bentley RC, Nightingale KR. Acoustic radiation force impulse imaging of the mechanical properties of arteries: in vivo and ex vivo results. Ultrasound Med Biol 2004;30(9):1163–71.
[32] Dumont D, Behler RH, Nichols TC, Merricks EP, Gallippi CM. ARFI imaging for noninvasive material characterization of atherosclerosis. Ultrasound Med Biol 2006;32(11):1703–11.
[33] Horn BKP. Robot vision. New York: McGraw-Hill; 1986. p. 278–98.
[34] Meunier J, Bertrand M. Echographic image mean gray level change with tissue dynamics: a system-based model study. IEEE Trans Biomed Eng 1995;42(4):403–10.
[35] Behar V, Adam D, Lysyansky P, Friedman Z. Improving motion estimation by accounting for local image distortion. Ultrasonics 2004;43(1):57–65.
[36] Angelini ED, Gerard O. Review of myocardial motion estimation methods from optical flow tracking on ultrasound data. Conf Proc IEEE Eng Med Biol Soc 2006;1:1537–40.
[37] Pellot-Barakat C, Frouin F, Insana MF, Herment A. Ultrasound elastography based on multiscale estimations of regularized displacement fields. IEEE Trans Med Imag 2004;23(2):153–63.
[38] Lorenz A, Schäpers G, Sommerfeld HJ, Garcia-Schürmann M, Philippou S, Senge T, et al. On the use of a modified optical flow algorithm for the correction of axial strain estimates in ultrasonic elastography for medical diagnosis. In: Proceedings of Forum Acusticum. Berlin, Germany; 1999. p. 4aBB1–4.
[39] Maurice RL, Soulez G, Giroux MF, Cloutier G. Non-invasive vascular elastography for carotid artery characterization on subjects without previous history of atherosclerosis. Med Phys 2008;35(8): 3436–43.
[40] Mercure E, Cloutier G, Schmitt C, Maurice RL. Performance evaluation of different implementations of the Lagrangian speckle model estimator for non-invasive vascular ultrasound elastography. Med Phys 2008;5(37):3116–26.
[41] Danilouchkine MG, Mastik F, van der Steen AFW. Improving IVUS palpography by incorporation of motion compensation based on block matching and optical flow. IEEE Trans Ultrason Ferroelectr Freq Control 2008;55(11):2392–404.
[42] Maurice RL, Bertrand M. Lagrangian speckle model and tissue motion estimation—theory. IEEE Trans Med Imaging 1999;18(7):593–603.
[43] Ophir J, Alam SK, Garra B, Kallel F, Konofagou E, Krouskop T, et al. Elastography: ultrasonic estimation and imaging of the elastic properties of tissues. Proc Inst Mech Eng H 1999;213(3):203–33.

[44] Chaturvedi P, Insana MF, Hall TJ. 2-D companding for noise reduction in strain imaging. IEEE Trans Ultrason Ferroelectr Freq Control 1998;45(1):179–91.
[45] Chaturvedi P, Insana MF, Hall TJ. Testing the limitations of 2-D companding for strain imaging using phantoms. IEEE Trans Ultrason Ferroelectr Freq Control 1998;45(4):1022–31.
[46] Kazmierski R, Watala C, Lukasik M, Kozubski W. Common carotid artery remodeling studied by sonomorphological criteria. J Neuroimaging 2004;14:258–64.
[47] Krejza J, Arkuszewski M, Kasner SE, Weigele J, Ustymowicz A, Hurst RW, et al. Carotid artery diameter in men and women and the relation to body and neck size. Stroke 2006;37:1103–5.
[48] Maurice RL, Vaujois L, Dahdah N, Nuyt A-M, Bigras J-L. Comparing carotid and brachial artery stiffness: a first step toward mechanical mapping of the arterial tree. Ultrasound Med Biol 2015;41(7):1808–13.
[49] Mercure E, Deprez JF, Fromageau J, Basset O, Soulez G, Cloutier G, et al. A compensative model for angle-dependence of motion estimates in non-invasive vascular elastography. Med Phys 2011;38(2):727–35.
[50] Maurice RL, Bertrand M. Speckle motion artifact under tissue shearing. IEEE Ultrason Ferroelectr Freq Control 1999;46(3):584–94.
[51] Maurice RL, Dahdah N. Characterization of aortic remodeling following Kawasaki disease: toward a fully-developed automatic bi-parametric model. Med Phys 2012;30(10):6104–10.

SECTION IV

Computer-Assisted Stenting

CHAPTER 12

Computerized Navigation Support for Endovascular Procedures

P. Fallavollita*, S. Demirci†
*University of Ottawa, Ottawa, ON, Canada
†Technical University of Munich, Munich, Germany

Chapter Outline

Chapter points

- This chapter gives an insight into computer-aided endovascular procedures.
- It provides an historical overview of simulators for procedural training.
- The chapter concludes by investigating existing medical imaging techniques for navigation support during endovascular procedures.

1. INTRODUCTION

The field of cardiovascular and vascular surgery has seen a tremendous evolution in the past decades. The care of patients with related diseases such as cardiac arrhythmias or the direct repair of lesions of the vascular tree has been governed by the excellence of the surgical team. The first use of medical images (X-ray) as an adjunct to surgery, reported only a few months after the discovery of X-rays in 1895 [1], marked the dawn of the *image guided surgery* (IGS) era. The subsequent rapid progress of IGS was predominantly made possible by the development of various medical imaging

Computing and Visualization for Intravascular Imaging and Computer-Assisted Stenting
http://dx.doi.org/10.1016/B978-0-12-811018-8.00012-6

technologies including endoscopic and laparoscopic imaging, ultrasound, and molecular imaging. For a complete review on the history of IGS, the reader is referred to Vaezy and Zderic [2].

In the late 1960s, Charles Dotter, a radiologist, pioneered the concept of *endovascular intervention* by being able to dilate stenotic atherosclerotic lesions in the iliac arteries using a series of catheter of increasing diameter. Very soon, *catheter-based therapy* became a popular alternative for treating certain vessel lesions. The most recent development has been a hybrid of a limited surgical exposure and the catheter-based introduction of a stent graft to treat aneurysmal disease. Taking their name from the English dentist Charles Stent who developed a thermoplastic material for taking impressions of toothless mouths in 1856 [3], stent grafts became the major impetus for endovascular procedures excluding an aneurysm, closing an arteriovenous fistula, and reconstructing the central lumen of a dissected vessel.

In contrast to conventional open surgery, *endovascular interventions* are performed minimally invasive, that is, only a small incision is needed to insert instruments such as catheters and guide wires. Location and navigation of these instruments is then performed under fluoroscopic and angiographic image guidance.

As *endovascular intervention* became competitive with and (in many cases) more desirable than direct vascular surgery, the traditional role of surgeons was challenged. To emphasize the fact that intraarterial intervention was another form of surgery and, hence, should be included in the repertoire of a vascular surgeon, the term "endovascular surgery" was invented as an alternative to "endovascular intervention."

> *There must be a perfect coordination between the eyes on the monitor and the foot on the fluoro pedal, interconnectedby a brain that thinks interventionally!*

This quotation, found in a textbook for medical students [4], points out very nicely the challenges that a vascular surgeon has to face nowadays. Besides deep knowledge about the history of vascular diseases and extensive training in conventional treatment methods, they have to acquire certain endo-skills and habits including remote catheter-mediated actions, indirect visualization, interventional mindset, and catheter and imaging skills. This chapter investigates the role and current techniques of computer-assisted navigation during minimally invasive procedures.

We begin by providing historical insights and reasons for the development and use of simulators for procedural training. The literature demonstrates that today's training and education processes for resident surgeons are positively affected by modern simulations, although there are still problems like the associated cost of simulators. Another point to look into is how the skill level of students or trainees is evaluated when practicing surgical tasks. There are different metrics, such as procedure time, and different processes, such as the Simulator for Testing and Rating Endovascular Skills for measuring their performance. These measurements show that skill levels are clearly distinguishable.

We conclude by investigating existing medical imaging techniques for navigation support during endovascular procedures. Besides the need for invasive imaging, the state-of-the-art comprises high-level medical image processing methods such as image registration and instrument detection and tracking. First, in cardiovascular interventions such as arrhythmia ablations, detecting and tracking the position of catheters relative to the patient anatomy is vital for several reasons. They include (i) heart motion compensation, (ii) facilitating positioning, (iii) planning the ablation procedure, and (iv) facilitating navigation by registration of intraoperative catheter positions to preoperative data. Similarly, in endovascular stenting procedures, stent detection and tracking is important since subsequent image-based detection of the stent graft in the interventional image and its simultaneous recovery in 3D makes it possible to visualize the device further in its current position within the preoperative volume. This fusion and appropriate visualization helps improve the difficult situation for the interventionalist as well as increase the chance of a cure for the patient.

2. SIMULATION FOR TRAINING

Over the last couple of decades, the use of simulators has greatly increased due to new technological possibilities as well as the need to optimize the training and practicing process. The need for it is due to recent increases in working hours restricting the availability of expert surgeons for training. Globally, simulations take place in a lot of different professions where precise handling is needed and physical tasks in high-risk environments are performed [5]. These simulations are a standard procedure for teaching in some occupations, for example, in the aviation industry, where commercial airline pilots are trained for 5000 h and astronauts for 12,000 h, which of course includes practice in real environments [6]. Although studies in nonmedical fields have shown that high fidelity simulations are effective teaching tools and enable repetitive practice of tasks in a range of conditions, and can improve learner success and satisfaction along with task safety, simulations have not yet been consistently established in medical education and residency training programs [5]. This is primarily a result of a simulator's high cost and missing organization and personnel, for example, training centers. What follows is a look into the range of simulators for (cardio)vascular surgery as well as education and skill assessment processes.

2.1 Overview of Simulators

Medical simulation means "a person, device, or set of conditions which attempts to present [education and] evaluation problems authentically" [7]. That means that a student or trainee has to perform a simple task, like making a knot, or complete a more difficult procedure like a surgery in a staged environment. In this way, possible mistakes do not risk the life of a real patient, but instead act as education. Another advantage of simulators is that the trainee can get feedback in real-time and a detailed evaluation at the end of the task (depending on the simulator's capabilities). Although the first simulators in the flight and military field became popular in the 1930s with the Link Trainer, other areas, such as in medicine were not quite ready because of technological barriers and a lack of knowledge about the human body. The first medical simulators were Resusci Anne, a manikin to perform CPR, or Harvey, "a part task trainer for heart diseases that used heart sounds, pulse waveforms, precordial impulses, peripheral pulses, and blood pressure of many cardiac conditions" [8]. They were initially used in the late 1960s.

Today, there are a lot of types of simulators being used (Fig. 12.1). They range from pure computer programs, to part task trainers, to full body manikins with complex anatomy. Further, there are different input devices and tools such as real instruments that are used in a real operating room, simplified instruments that are designed for a special simulator, or simply a mouse and keyboard. All of these options are combinable and can form very complex and expensive models like the VIST-Lab from Mentice, which uses a body-sized manikin, real instruments with a haptic system, and computer programs to simulate X-ray images that are displayed on several screens. Conversely, there are extremely simple simulators, for example, "using nothing more than copper tubing, plastic bird guards, a drain and some miscellaneous items such as wood screws and two-by-fours" [9]. They are really inexpensive to build (about $5) but still capable of teaching basic skills.

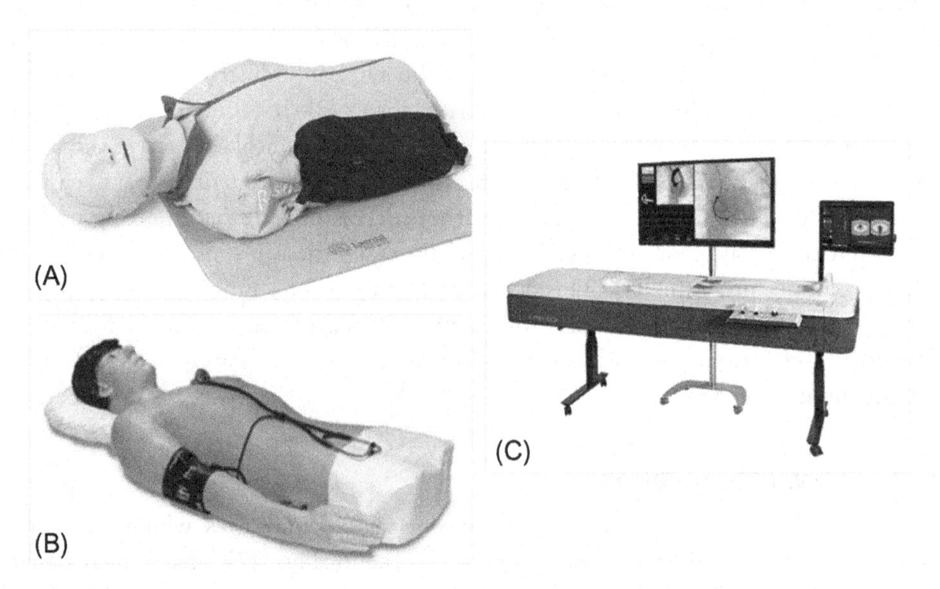

Fig. 12.1 (A and B) Example manikin simulators such as the Resusci Anne and Harvey. (C) The VIST-lab vascular simulator.

Table 12.1 Overview of sample commercial simulators and their descriptions

Simulator	Manufacturer	Simulated procedure	Type	Haptic feedback	Price
The replicator	Vascular simulations LLC	Different endovascular procedures	Hardware	Yes	On request
Different devices	Limbs and things	Arterial procedures; abdominal aortic aneurysm repair; carotid endarterectomy; femoro-peroneal anastomosis; femoral popliteal bypass	Hardware	Yes	$500–$3000
SimVascular	Funded by NSF SSI	Blood flow	Software	No	Open source
Pontresina	Vascular International	Open aortic/carotid surgery; open peripheral vascular surgery; shunt-surgery	Hardware	Yes	$2900–$6700
Angio Mentor	Simbionix	More than 19 different endovascular procedures and over 130 patient scenarios	Hardware/software	Yes	On request
VIST-Lab (stationary), VIST-C (portable)	Mentice	Variety of training modules for endovascular surgery	Hardware/software	Yes	On request
Coro3D	CATHI GmbH	Interactive 3D learning program	Software	No	Basic version 80 *e* first add-on 155 *e*
CATHIS H CATHIS M	CATHI GmbH	Different endovascular procedures	Hardware/software	Yes	On request
3mensio CT	Esaote	Mitral valve	Software	No	On request
CathLabVR	CAE Healthcare	Percutaneous coronary intervention; percutaneous peripheral interventions; cardiac surgery	Hardware/software	Yes	$90,000–$250,000
Simantha	Medical Simulation Corporation	Acute myocardial infarction; basic endovascular skills; cardiac cath lab orientation; carotid course	Hardware/software	Yes	$90,000–$250,000

Several simulators for (cardio)vascular interventions have been designed and fabricated. Table 12.1 provides a basic overview of what exists commercially as well as a detailed description about them.

2.2 Training and Education

The training through simulators is facing a number of challenges. The acceptance of virtual reality training has only slowly been progressing due to an initial lack of evidence validating the training and a lack of data showing benefits to patients. This view has been changed by a number of studies validating the use of virtual reality simulators in surgical training in cardiovascular and vascular interventions. In Dawson et al. [10], residents from vascular surgery programs were invited to a series of 2-day training programs using the simulator SimSuite, which provides a high degree of realism. Each participant completed an average of 9.5 simulated cases. Their procedural skills improved significantly after finishing the training program. The total procedure time decreased by 54%, volume of contrast decreased 44%, and fluoroscopy time decreased by 48%. The study concludes by affirming the improvement in key performance metrics but leaves open whether simulation training affects following clinical performance. A nonnegligible part of the training with vascular simulators is realized through workshop-based training [11]. This can be handled in many cases by local training, as a great number of institutions providing postgraduate surgical training have a skills laboratory. To offer basic vascular surgical training, "benchtop jigs, surgical instruments, sutures and grafts" are required [11]. Such laboratories can be used for hands-on teaching or self-directed practice to enhance clinical practice.

With the use of better and more specific simulators, the cost for laboratories increases dramatically. Therefore training needs to be centralized in order to remain efficient. For example, the London Postgraduate School of Surgery offers a fully funded monthly half-day of teaching to all general and vascular surgical trainees across London with a generic program in the first year and a specialty program in subsequent years [11]. By having this training centralized at a professional institution, all trainees have access to more sophisticated simulators then they would have in their respective local institutions. Procedures not commonly performed can often be carried out during training in nationally based workshops, especially if the expertise is mainly at a single institution. In the case of vascular trainees, a prime example is Spain, where all vascular trainees are invited by the Simuvasc group to take part in their training course, which is funded by the industry [11].

2.3 Skills Assessment

There is a wide variety of studies assessing the skills of trainees before, during, and after performing surgical tasks with different simulators. There are different approaches in quantifying the quality of a surgery performed with a simulator. A number of studies [12, 13] rely on the opinion of two or more independent experts for evaluating the technical quality of the surgery. Other measurements that are often used include the safe handling of surgical tools and the time needed to complete the task. In a study by the

European Vascular Workshop, the effectiveness of workshop training on simulators was assessed [12]. For the study, 15 participants of the workshop in 2003 and 2004 were given the task to perform a proximal anastomosis on a commercial abdominal aortic aneurysm simulator. After a 3-day training on a sophisticated model, the task was reassessed. The technical skill of the task was measured in four ways: the generic skill, procedural skill, a five-point technical rating of the anastomosis, and the procedure time. Analysis of the results showed a significant increase in technical performance and the procedure time, when comparing the results from before and after the 3-day course.

The paper by Willems et al. [13] demonstrates how to distinguish between novice, intermediate, and expert trainees. Eighteen candidates from all three groups (eight novices, five intermediate, five experts) participated in the study. Participants in the novice group were recruited from surgical residents of the hospital. They had no specialized cardiovascular or vascular skills. The intermediate and expert participants were specialized surgeons grouped by their skill level. Two tasks were given to assess the skills of the candidates. The first assignment was to access the left renal artery from the right groin, while the second assignment was to access the right renal artery from the left groin. The second assignment is the main and more difficult task used for the actual assessment, while the first assignment is mainly for candidates to get used to the simulator. All candidates were given a total score between 0 and 100, based on a combination of operation specific and global assessments similar to the checklist derived from the Imperial College Evaluation of Procedure Specific Skill (ICEPS) scoring system for the task. Additionally the time needed for each assignment was recorded.

Table 12.2 clearly shows a sharp distinction between the mean total scores and the mean procedural time of the different groups of candidates. Therefore the simulator is suitable for rating resident trainee and expert skills.

2.4 Outlook and Conclusion

In recent years, the quality of simulators has greatly improved. Nevertheless further research will be needed to increase the closeness to interventional reality of the

Table 12.2 Results of the study by Willems et al. [13]

	Novice (n = 8)	Intermediate (n = 5)	Expert (n = 5)
Mean total score (0–100)	42.75	65.6	82.8
Mean procedure time (min)			
First assignment	7.0	3.1	2.0
Second assignment	8.3	6.2	5.5
Sum of procedure time for both assignments	15.3	9.2	7.5

Notes: The mean total score and the procedural time used per group per assignment are displayed.

simulators. With future progress in building simulators, it is hoped that the price will decrease and therefore the availability for students and resident trainees will increase. Also, new and better methods to evaluate and assess the skill level constantly have to be developed. Another interesting part of research would be to investigate at which level of already obtained skills the simulator training has the most impact on further learning. This would be particularly interesting for cardiovascular and vascular surgeons with a very low or very high skill level.

Although the use of simulators started slowly, important progress has been made in the recent years. Recent studies have shown the usefulness and efficiency of both cardiovascular and vascular simulators. A vast majority of the participants in studies achieved significantly better results after simulator training than before. All simulators provide a safe way of learning new surgical procedures and training individuals without endangering patients' health and lives. In contrast to other educational methods, feedback can be given in real-time. This trend of using simulators in education is likely to continue as even more sophisticated simulators are being developed. There will also probably be an increase in workshops as more and better simulators become available.

3. INTERVENTIONAL NAVIGATION SUPPORT

Endovascular IGS predominantly depends on accurate and robust navigation of cardiovascular instruments. The field of computer-aided surgery and intervention has proposed a variety of solutions to support a surgeon's navigation further by information mapping. Here, conventional 2D image information is enhanced with 3D models of the patient and navigation of instruments is essentially performed in a 3D virtual space. In order to yield a correct mapping of the 3D scenery and the action therein, besides optimal image alignment methods (complete reviews on respective methods are provided by Markelj et al. [14] and Matl et al. [15]), accurate and robust detection and tracking of interventional instruments, such as guide wires, catheters, and stents, have to be ensured. Whereas this section focuses on curvilinear structures (catheters and guide wires), detection and tracking of specific shapes such as stents will be described subsequently.

3.1 Tracking and Detection in Electrophysiology

Sudden cardiac death (SCD) is among the leading causes of death in Western countries, with a worldwide incidence of 4–5 million cases per year [16]. Advanced technology called radio frequency (RF) catheter ablation can help treat related disorders such as atrial or ventricular arrhythmia that may eventually lead to SCD. This procedure involves inserting a catheter inside the heart and delivering RF currents through the catheter tip in order to ablate the arrhythmogenic site. The aim is to achieve scarring or destroying heart tissue that triggers the abnormal heart rhythm and hence preventing abnormal electrical signals from traveling through the heart, eventually causing the arrhythmia to disappear.

Electroanatomic mapping (EAM) that creates a 3D chamber geometry and displays the ablation catheter position is often helpful during the entire procedure. The technology refers to point-by-point contact mapping combined with the ability to display the location of each mapping point in 3D. In the 1990s, EAM systems were translated into clinical practice for the treatment of several cardiac arrhythmias. Early technologies such as the LocaLisa system had the capacity of tracking intracardiac electrodes and tagging points in a 3D electrical field [17]. LocaLisa reduced X-ray image acquisitions for all kinds of ablations [18]. Nevertheless, the navigation system has been cast aside due to its inability to build surface geometries and to create electrophysiological maps. This was followed by the Ensite Array, which used a mesh of 64 coated wires spanned over a balloon forming >3000 unipolar electrodes. Later generations of the LocaLisa system has been integrated so that array and ablation catheter localization is achieved via impedance-based tracking. Thus, more than 3000 noncontact far-field signals are simultaneously superimposed on a 3D cardiac surface reconstruction, allowing for high-density electrophysiological maps from single heartbeats. Yet the system did not prevail in clinical routine because of its complexity, costs, and other limitations [19]. Twenty years later, two EAM systems have become the standard in the treatment of cardiac arrhythmias: the EnsiteNavX system and the Carto system. Since their translation into clinical practice, these two EAM systems have seen continuous technical development in various aspects, such as (multiple) catheter visualization, creation of chamber models, image integration, and mapping modalities [20]. What follows is a synopsis of the state-of-the-art features and derived strategies of existing commercial EAM systems in the context of cardiac ablation interventions.

The foundation of established EAM systems consists of a nonfluoroscopic visualization of mapping catheters and a 3D reconstruction of the anatomy of interest (e.g., the heart atria and ventricles) that are created by the manipulation of the mapping catheter by the electrophysiologists during cardiac ablation procedures. The resulting 3D reconstruction is displayed as a shell (i.e., convex hull) representing the cardiac structure of interest and can become more precise in its representation by increasing the number of sampled points. The latest EAM systems come with tools that enable fast and automated multipoint model creation during the mapping catheter movement and positioning. Special multipolar catheters support rapid chamber 3D reconstruction and mapping of individual anatomical locations such as the valvular annuli or His bundle. Another aspect of current EAM systems is the simultaneous display of various diagnostic catheters and devices in real-time in relation to the 3D geometry [20]. The most popular mapping techniques include color-coded displays of the referenced electrical activation sequence known as an activation map, and mapping of unipolar/bipolar electrograms as part of fractionation mapping and voltage mapping on the model surface [21]. Optional features allow electrophysiologists to measure and display electrical contact/contact force at the catheter electrode heart tissue interface. Today, the two most widely applied contemporary contact EAM systems for fibrillation ablation are Carto3 and EnSiteNavX.

The original Carto system is principally based on three active and weak magnetic fields generated by a three-coil location pad placed underneath the patient during ablation procedures. The magnetic field strength is measured with small sensors embedded in the catheter tip on a continuous basis providing information about the exact position and orientation of the sensor in real-time [22]. However, during an intervention, patient movement or dislocation of the location pad may lead to uncorrectable map shifts. Modern versions such as the Carto3 integrate a hybrid of magnetic and current-based catheter localization technology that enables visualization of multiple catheters simultaneously without the need for X-ray radiation via C-arm fluoroscopy. To achieve this, a total of six electrode patches are affixed on the patient's back and chest; these continuously screen the current emitted at a unique frequency from different catheter electrodes. Precise localization of the nonmagnetic electrodes can be calibrated by the detection of the magnetic sensor within the coordinate system in order to overcome distortions from nonuniform intrathoracic resistances. New versions of the Carto3 system allow for fast anatomical mapping, which offers detailed anatomical shells/convex hulls with all electrodes of the mapping catheter by moving it around in the heart chamber of interest. These representations of end-diastolic surfaces provide a much better reconstruction than the traditional point-by-point maps in the first Carto versions [20]. Finally, a new way of improving the accuracy of the 3D surface reconstructions is accomplished by a unique type of respiratory gating in which varying thoracic impedances are measured throughout the respiratory cycle. Such developments in the current Carto3 system have already been shown to be beneficial in terms of reducing C-arm fluoroscopy requirements [23].

The other EAM system commonly used today is the EnsiteNavX. This system is based on the LocaLisa technology and uses six skin electrode patches to create high-frequency electric fields in three orthogonal planes. The 3D localization of traditional electrophysiology catheters is calculated based on an impedance gradient in relation to a reference electrode [17]. Calculations can be complicated by the body's nonlinear impedance, which is correctable by a process called field scaling [24]. To account for cardiac and respiratory motion artifacts, intracardiac reference catheters such as catheters placed at the relatively stable proximal coronary sinus or septal wall are usually preferred over an extra-cardiac reference electrode. It is an advantage of the EnSiteNavX system over the Carto system that it is architecturally open and allows for visualization of multiple catheters from different manufacturers [25]. Further, all electrodes of all catheters can be used simultaneously for a relatively quick reconstruction

Table 12.3 Sample of commercially available electroanatomic mapping system

Electroanatomic mapping system	Manufacturer	Example visualization
EnSite NavX	St. Jude Medical	Image by Estner et al. [26]
Carto 3	Biosense Webster	Image by Dickfeld et al. [27]
CartoSound	Biosense Webster	Image by Dickfeld et al. [27]
Sensei X Robotic System	Hansen Medical	Image by Beasley [28]
Rhythmia	Boston Scientific	Image by Boston Scientific [29]

of the cardiac chamber of interest, providing not only 3D anatomical information but also electrophysiological mapping data in color-code.

Table 12.3 provides a basic overview of existing commercially available EAM system as well as a depiction of the visualization output capabilities.

While the EAM systems provide a wealth of data and reduce C-arm fluoroscopy times and radiation exposure, they cannot replace careful interpretation of X-ray data and strict adherence to electrophysiological principles. Although these benefits are achieved at a greater cost, there may be long-term benefits to the community and catheter laboratory staff. There are several reasons why detecting and tracking the position of catheters in X-ray images relative to the patient anatomy is important. They include (i) heart motion compensation, (ii) facilitating positioning, (iii) planning the ablation procedure, and (iv) facilitating navigation by registration of intraoperative catheter positions to preoperative data. To achieve detection and tracking, there are two primary options available: (a) hardware solutions that include mechanical and optical systems or the EAM systems and (b) software and image-based solutions that make use of intraoperative images such as C-arm fluoroscopy.

We observe notable differences when comparing mapping systems with image-based solutions. First, although mapping solutions are expensive, they provide high accuracy and robustness, and their clinical usage is getting popular. However, there is a notable learning curve for clinicians to get used to this technology. As for image-guided solutions, these require dealing with high data variation and involve the acquisition of a large amount of X-ray images. Their usage is still very limited since there is a high performance requirement for a practical clinical solution. Whether using EAM systems or conventional RF ablation techniques, clinicians still rely on X-ray images to position and guide catheters to complete the ablation procedures. Thus, exploiting X-ray image information is crucial for providing additional information to clinicians during cardiac ablation procedures. There has been a trend in the past several years where researchers began investigating image-based solutions, since these would also provide inexpensive and simple assistance to clinicians and could alleviate some of the burdens involved with the commercially available expensive technologies [30]. What follows is a synopsis of the current literature related to the detection, tracking, and subsequent reconstruction of catheters in X-ray images.

A first approach for the detection of catheters was proposed by Franken et al. [31]. The motivation of the approach was to detect automatically all catheter tips in X-ray images. The results lacked robustness and when trying to extend the very same algorithm to detect the entire catheter sheath, the success rate dropped significantly. A second approach for catheter detection was proposed by Yatziv et al. [32]. A drawback of their approach is that the algorithm cannot be applied to single X-ray images and requires the analysis of temporal image sequences to compute a mean image representing the background. A random image from the temporal sequence is then selected and the background image is subtracted to obtain moving structures such as catheters including their sheaths. Beginning from a point marked by the user, all potential catheter sheaths are traced. Nevertheless, this information is not processed further to detect the full catheter sheath, but rather used to reduce the search space when detecting all catheter tip electrodes. A third method for catheter detection is based on traditional blob detectors and was proposed by Ma et al. [33]. A unique detection method relies on the fusion of hypotheses generated by learning-based detectors in a Bayesian framework as presented in Wu et al. [34]. The premise of the algorithm focuses on catheter electrodes only, limiting a subsequent full catheter sheath reconstruction to those catheter segments that carry electrodes. Consequently, the detection of catheters not carrying electrodes (e.g., cryoballoon catheters), is not possible. A fifth detection approach was described by Cazalas et al. [35], whereas a technique focusing on detecting and separating overlapping catheters was presented by Milletari et al. [36]. All these approaches focus solely on the detection of catheter tip electrodes and not the entire catheter sheath. Without detection and 3D reconstruction of the full catheter sheath, little information about the overall catheter shape can be provided to the electrophysiologists during cardiac ablation procedures.

Aside from catheters as instruments used during intervention, guide wires are commonly used by the interventionist to facilitate the deployment of catheters. In X-ray images, guide wires are imaged as thin, homogeneously dense objects with little variance in appearance. A method was proposed to detect curvilinear structures in X-ray images by Wang et al. [37]. Here, a graph-based solution was developed requiring the user to mark the start and endpoints of the curvilinear object. To increase robustness, additional points can be manually added. Although the method can be adapted to the detection of ablation catheters, it has only been evaluated for guide wires. Several interesting works on guide wire detection have been proposed [38–41]. Yet again, these methodologies cannot be transferred for the detection of ablation catheters directly. This is because catheters have a thicker sheath compared to their thinner counterparts. They also differ in contrast, especially in the regions close to the electrodes.

As mentioned briefly previously, catheter detection can be accomplished temporally across C-arm fluoroscopy sequences. The ambition for this is to achieve catheter tracking. To arrive at a robust solution, catheter tracking requires prior information, such as an initialization at the beginning or the result of a previous frame. Extending this notion, if a 3D structure is tracked using images from different C-arm fluoroscope viewpoints, it is also related to 3D reconstruction. Tracking of ablation catheters in 3D can be performed by altering the 3D structure until its projections fit the 2D images [42, 43], or by 2D tracking in both X-ray images followed by a calculation of the 3D position based on the 2D objects. This approach was first proposed by Baert et al. [44] for tracking of guide wires with low curvature. Brost et al. [45] focused on curved objects but their approach is limited to the elliptical part of the catheter. Papalazarou et al. [46] proposed a similar approach handling only rigid objects. Biplane C-arm fluoroscopy systems are capable of scanning the patient from two views simultaneously, but they are costly and not very common in clinical practice. They have been used by Bender et al. [47] for catheter reconstruction. However, their technique lacked automation and real-time performance. Deformable reconstruction using single-plane C-arm fluoroscopy imaging is described in Refs. [48, 49]; both methods were applied to guide wires inside catheter sheaths. All the mentioned methods rely on preoperative 3D volumes to constrain the solution space. To alleviate some of these limitations, Baur et al. [50] proposed a novel and robust method for automatic reconstruction of 3D catheters in nonsynchronized multiple X-ray images. The method employs a probabilistic graphical model that optimizes all catheter electrode correspondences and their 3D reconstructions. Alternatively, Hoffmann et al. [51] detect the catheter of interest in multiple X-ray views after manual initialization using a graph-search method. The detection results are then used to reconstruct a full 3D model of the catheter sheath based on automatically determined point pairs for triangulation.

In conclusion, future work should rely on the development of real-time detection, tracking, and reconstruction of mapping catheters visible in X-ray images during ablation procedures. The result of this can be easily fused with the EAM systems visualization outputs to strengthen further the precision of the catheter deployment and subsequent and successful ablation of the arrhythmia.

3.2 Computer-Assisted Stenting

In the last decade, endovascular stenting procedures have gained acceptance as an alternative to open surgical repair, with reduced risks. These procedures provide substantial clinical benefits for the patient [52], such as decreased use of the intensive care unit, diminished length of hospital stay, and early return to normal activities [53]. However, not all patients are suitable for these type of procedure.

Although many people associate the word "stent" with a prosthetic device used in human medicine, it goes back to the English dentist Charles Stent who invented a material to form dental impression compounds in 1856. The plastic surgeon Jan Esser used the exact same material to craft forms for facial reconstruction and named it "Stent's material" after its inventor [54]. It is difficult to follow an exact line through history, but the expression "stent" ended up being a synonym for a special prosthesis replacing or supporting parts of the body structure [55].

Medical stents are implanted into various anatomical structures and organs. Their design changes with the requirements a supporting prosthesis must fulfill for certain anatomical locations. The most widely known is the vascular stent placed inside a blood vessel to prevent, or counteract, a disease-induced, localized flow constriction.

Vascular stents and *stent grafts* exist in various shapes and configurations, depending on the type of blood vessel and kind of disease. Whereas coronary and carotid stents are solely made from metal webs or meshes, (endovascular) stent grafts are composed of a synthetic fabric tube (graft) supported by a rigid structure (stent).

Depending on the anatomical location and severity of the lesion, it is common to deploy combinations of different stents or stent grafts plugged into each other and thereby forming various shapes. This is particularly the case for so-called bifurcation stents or stent grafts as used for endovascular repair of the abdominal aorta. Deployed stent graft combinations usually consist of at least two parts: a trunk component (covering the abdominal aorta and one iliac artery branch) and a contra-lateral leg component (covering the other iliac artery branch). If necessary, additional tubular stent grafts can be attached to the endpoints of this stent graft configuration. The components are folded up in order to fit into delivery catheters, and are capable of expanding to the preestablished diameter when placed and released in the artery.

The insertion of a prosthesis inside the aneurysmatic aorta requires accurate treatment planning. Physicians need to choose carefully the appropriate stent graft from a variety of different models and producers, each of them being unique in material and shape. Depending on the anatomical location and type of lesion, important measurements include diameter, length, and angulation of the proximal and distal landing zones, various distance measurements in between branching arteries and/or bifurcations, diameter measurements at different locations inside the vessel, presence of and location of thrombus material, quality of landing zones and potential access routes in terms of calcification or atheroma, and presence of vascular anomalies (multiple renal arteries, early bifurcations, venous anomalies) [56].

Due to the unbeaten advantages of a CTA scan in terms of resolution, most clinicians do the measurements in this scan of the patient. Nowadays, angiography suites are equipped with modified stationary C-arms that are able to acquire CT-like slice images following the same main technical principles as CT. Although this new technology aims at improving the interventional situation by allowing 3D reconstructions during the intervention, recent studies [57, 58] have also examined its feasibility for preoperative scans. Although resulting images have lower contrast resolution than conventional CTA scans and artifacts are more likely to occur, it has been shown that radiologists are able to do accurate measurements. This could be timesaving in particular for acute cases, avoiding patient transfer to a CT laboratory.

3.2.1 Intraoperative Challenges

The implantation of a stent (graft) inside the human vasculature is a minimally invasive procedure for the treatment of aortic aneurysms and aortic dissections. After the insertion of a pigtail catheter and guide wires, a shaft catheter including a folded stent or stent graft is placed to cover the lesion. Before unfolding the stent (graft), the physician must ensure that it is positioned correctly inside the vessel tree. In the case of covered stent grafts, misplacements can lead to partial or total cut-offs of blood supply of vitally important organs. Another critical complication is presented by an endoleak where blood still enters in between the stent graft and the aneurysm sac [47].

In the current clinical workflow, the intraoperative deployment of the stent graft into the aorta is performed under continuous 2D imaging. Using either general or regional anesthesia, one or both femoral arteries are exposed depending on the type of stent (graft) that is required. A needle followed by a guide wire is then placed in the femoral artery, and the guide wire is extended up the lesion under continuous fluoroscopy imaging. An angiography or DSA sequence is then acquired in order to provide a roadmap for placing the device. The stent (graft) catheter delivery system is passed up, over the guide wire, and positioned across the aneurysm. Acquiring several angiography sequences from different view angles, the stent (graft) is deployed immediately below the renal arteries. A balloon within the catheter delivery system is then positioned across the attachment site and expanded in order to seat the hooks into the wall of the aorta. A completion angiography or DSA sequence is then obtained to make certain that the stent (graft) is properly seated and there is no evidence of flow between the graft and the aneurysm.

The entire interventional catheter navigation is done under 2D angiography imaging where the physician is missing the important 3D information. As the catheter and stent position are only visualized in 2D, more image acquisitions are needed during the fine positioning of the stent graft before unfolding. This means an increase in radiation dose and used contrast agent at the same time as branching vessels need to be made visible in the images. As conventional planar angiography is not able to detect all stent (graft)-related anomalies, it has been suggested to use CBCT instead to ensure clinical success [57, 60, 61]. Despite improved 3D depth visualizations of the anatomy, recent studies have shown a significant increase in radiation exposure to the patient as well as the surgical team when employing interventional CBCT in comparison to 2D angiography [62]. In addition, it has been rated as too time-consuming for interventional usage [63].

Over the last decades, endovascular interventions have seen a rapid increase, replacing more and more conventional open surgery techniques and allowing an ever-larger number of patients to benefit from minimally invasive procedures. The increasing complexity of stent (graft) configurations used, however, goes hand in hand with a higher risk of stent graft-related complications such as occlusion of visceral arteries due to stent (graft) displacement and endoleaks. Disadvantages related to conventional intraoperative imaging modalities, such as fluoroscopy and angiography, include prolonged procedure duration, poor visualization in some regions of the aorta, and navigational limitations.

3.2.2 Detection and Tracking of Stents and Stent Grafts

Despite the vast variety of approaches detecting and tracking catheter and guide wire tips for various applications and purposes as presented in Section 3, these techniques cannot be applied for stents (grafts) or their delivery device due to their inhomogeneous appearance and rather large size. Moreover, tracking only the tip of this device, as done for EP applications, has turned out to be less useful as the tip may not always be visible during the procedure. In particular, for complex navigation tasks with different combinations of stents (grafts), it is far more important for the physician to focus on the plug locations connecting different stents (grafts).

Given the variety of stents and stent grafts available, different algorithms have been proposed for the tracking and detection of such structures. In addition to conventional approaches attaching electromagnetic sensors to the stent delivery devices and thereby tracking the insertion of the device via an external tracking system [64, 65], solely image-based solutions have been presented. Most commonly, such methods take advantage from the fact that stents (grafts) as well as related instruments (such as delivery devices) show tiny little markers that are being used by interventionalists to correctly align the structure with respect to the used imaging system. Stent (graft) detection is then being performed by employing blob detection filters in combination with local maxima selection [66]. More advanced methods have employed marginal space learning in combination with the Viterbi algorithm to detect the markers [67].

In this section, we present a novel algorithm to match a 3D model of the stent graft to an intraoperative 2D image showing the device [68, 69]. The method is fully automatic and does therefore not interrupt the medical workflow. By choosing a global-to-local approach, we are able to abandon any user interaction and still meet the required robustness. The complexity of our registration scheme is further reduced by a semisimultaneous optimization scheme and by including constraints that correspond to the geometry of the stent graft.

We define the stent segment model to be the curve,

$$\mathcal{M}_i(\mathbf{x}) = (a_i(\mathbf{x}), b_i(\mathbf{x}), c_i(\mathbf{x})) \tag{12.1}$$

consisting of the set of parametric equations

$$a_i(\mathbf{x}) = r_{\mathcal{M},i}(\mathbf{x}) \cos(\mathbf{x}) \tag{12.2}$$

$$b_i(\mathbf{x}) = r_{\mathcal{M},i}(\mathbf{x}) \sin(\mathbf{x}) \tag{12.3}$$

$$c_i(\mathbf{x}) = A_i \sin(p_i\mathbf{x} + s_i) \tag{12.4}$$

with amplitude A_i specifying the height of the segment, period p_i equal to the number of peaks, and phase shift s_i merely shifting the starting point. The radius of the stent segment model is calculated by

$$r_{\mathcal{M},i}(\mathbf{x}) = \left(1 - \frac{\sin(p_i\mathbf{x} + 1)}{2}\right) r_i^t + \frac{\sin(p_i\mathbf{x}) + 1}{2} r_i^b \tag{12.5}$$

with r_i^t and r_i^b representing the upper and lower radius as visualized in Fig. 12.2.

In our notation, an entire stent graft $\mathcal{M} = \{\mathcal{M}_1, \ldots, \mathcal{M}_l\}$ is defined as the set of l stent segments.

Automatic Candidate Region and Feature Extraction

The method is summarized in Fig. 12.3. At first, d frames of an EVAR operation are acquired in a real-time fashion. Each frame is viewed as a matrix of m rows and n

columns. This first block of frames is filtered using the *Frangi* filter [70] and reshaped into a matrix D of dimensions $m \cdot n \times d$.

Employing *Robust Principal Component Analysis* and its more efficient formulation *Principal Component Pursuit*, matrix D can be decomposed into a sum of two matrices: a low rank part A representing the background part of the X-ray images, and a sparse part E corresponding to its foreground, that is, the stent graft:

$$\begin{aligned} \underset{A,E}{\text{minimize}} \quad & \|A\|_\star + \lambda\|E\|_1 \\ \text{subject to} \quad & A + E = D \end{aligned} \tag{12.6}$$

Parameter λ establishes a trade-off between rank and sparsity. A detailed overview of how this parameter behaves for different values is presented in Volpi et al. [71]. This work also gives details on how to solve Eq. (12.6) numerically, employing the *augmented Lagrange multiplier* algorithm [72].

Finally, the rows of matrix A are first averaged yielding to a single vector of length $m \cdot n$. This vector is reshaped in a single image with the original dimensions $m \times n$ that represents a filtered background and stored in memory as matrix B. This mask B is then subtracted from the filtered versions of the original image frames to yield candidate pixel locations for the stent (graft) and its delivery device (cf. Fig. 12.3).

Conventionally, *robust principal component analysis* (Eq. 12.6) performs retroactive data separation making an clinical application for real-time tracking and detection impossible. To solve this issue, Volpi et al. [71] have introduced *online robust principal component analysis* that essentially performs RPCA in a block-wise manner. While the first d frames are being decomposed, subsequent d frames are captured and filtered. The system is calibrated in such a way that the computation takes slightly less time than the time required for the acquisition of d new frames. Hence, the new block will be later used to compute another parallel execution of RPCA.

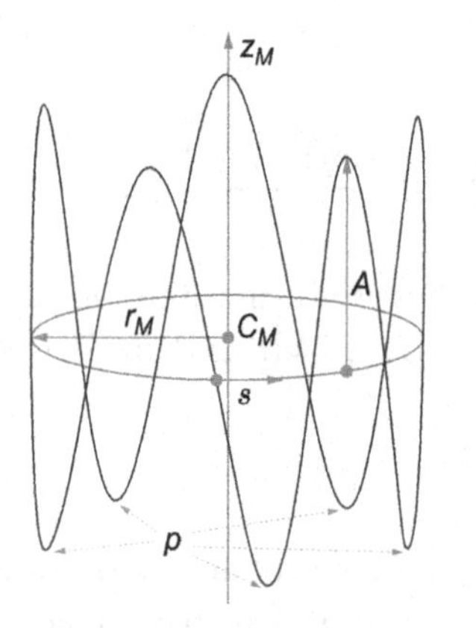

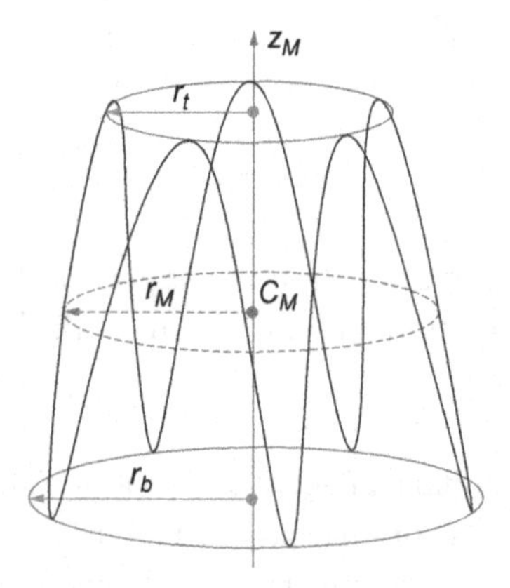

Fig. 12.2 Stent segment model with $r^t = r^b$ (*left*) and $r^t < r^b$ (*right*).

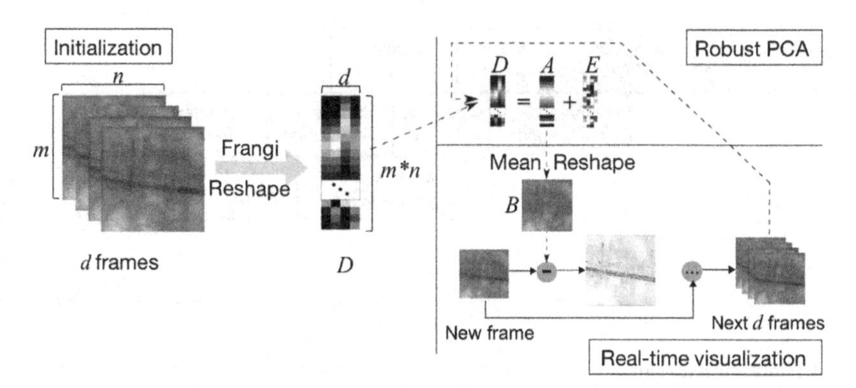

Fig. 12.3 Overview of the system.

Registration Algorithm

In order to place the model in the 2D image space of the interventional image I, a projection is necessary to map the stent model $\mathcal{M}$ to the image coordinate system. Similar to 2D-3D image registration, the projection transformation $\mathbf{P} = K[R|t]$ consists of the 6-DOF extrinsic parameters $[R|t]$ for rotation and translation of the 3D volume and the 4-DOF intrinsic camera parameter K of the *pinhole camera model* [73]. For the following considerations, we assume the camera matrix K to be given by the interventional angio system.

Using all transformation and model parameters together as

$$\mathbf{p} = \{\mathbf{p}_1, \ldots, \mathbf{p}_l\}, \quad \mathbf{p}_i = \{R_i, t_i, r_i^t, r_i^b, A_i\} \tag{12.7}$$

the registration problem for the entire stent model can be formalized as

$$\hat{\mathbf{p}} = \arg\min_{\mathbf{p}} \sum_{i=1}^{l} \sum_{x \in \mathcal{M}_i} D_{I_f}(T_{\mathbf{p}_i}(x)) \tag{12.8}$$

where $T_{\mathbf{p}_i}(x)$ is a projection of point x of the 3D segment model $\mathcal{M}_i$ using parameters $\mathbf{p}_i$ into 2D image space. This formalization equals a simultaneous registration of all stent segment models introducing a parameter space of dimension $l \times 9$. Considering that conventional abdominal aortic stent grafts consist of more than 10 segments, the cost for optimization increases rapidly. Another drawback is introduced by many local minima in the cost function plot, each of them belonging to one stent segment that is displayed in the image. The cost function for the registration of the whole stent model consisting of l stent segments is

$$E(\mathbf{p}) = \sum_{i=1}^{l} \sum_{\mathcal{M}_i} I(T_{\mathbf{p}_i}(\mathcal{M}_i)) \tag{12.9}$$

where $\mathbf{p} = \{\mathbf{p}_1, \ldots, \mathbf{p}_l\}$.

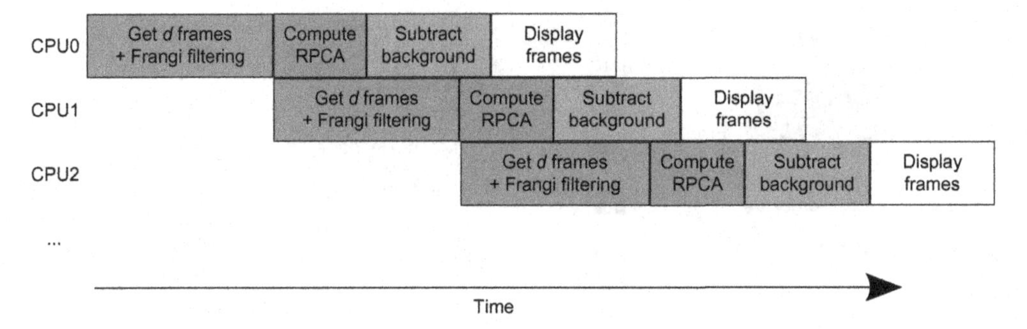

Fig. 12.4 Parallelization for *online robust principal component analysis.*

A sequential fitting of each segment of a conventional aortic stent graft leads to parameter space of dimension multiple of nine. In order to reduce the complexity in our registration procedure, we use a semisimultaneous optimization framework introduced by Sidorov et al. [74]. Instead of optimizing all parameters for all segments at once, we optimize the parameters of one segment for a certain number of iterations and then move to the next randomly chosen segment. By applying this strategy, we implicitly make use of the tubular appearance of the stent graft and constrain the pose change in between neighboring segments.

We can smooth the cost function plot by the inclusion of prior knowledge about the stent graft to be implanted: mean diameter, mean amplitude of the stent graft, and distance d_i between consecutive segments (see Fig. 12.4). This information is default for all samples of a certain model by a certain producer and can be delivered by the vendor (i.e., in an xml file). Having this information available and setting the remaining parameters of each segment to initial values, we can model the implanted stent graft approximately and divide our registration problem in two steps.

Global Registration Here, we solve for the overall orientation of all segments in order to be very close for the local calculations. The global pose of the entire stent graft model is defined by the global parameters $K, R_{\text{global}}, t_{\text{global}}$. The angle γ_{global} for rotation around the camera's z-axis as well as the translation $t^x_{\text{global}}, t^y_{\text{global}}$ can be estimated from the stent region $\mathcal{S}$ via *principal component analysis* and center of mass detection. Therefore, only the rotation around the camera's x- and y-axis and translation in along z-axis need to be optimized, and hence we define $\mathbf{p}_{\text{global}} = \{\alpha_{\text{global}}, \beta_{\text{global}}, t^z_{\text{global}}\}$.

Accordingly, let now $\mathbf{p}_i = \{\mathbf{v}_i, \mathbf{t}_i, r^t_i, r^b_i, A_i\}$ define the set of remaining parameters for each single segment i, where $\mathbf{v}_i = [\alpha_i, \beta_i, \gamma_i]^T$ represents the vector containing the three rotation angles that form rotation matrix $\mathbf{R}_i$. Setting parameter vectors $\mathbf{p}_i$ to initial values corresponding to the definitions from the xml file starting with no rotation and translation, the registration problem can be formalized as

$$\hat{\mathbf{p}}_{\text{global}} = \arg\min_{\mathbf{p}_{\text{global}}} \sum_{i=1}^{l} \sum_{x \in \mathcal{M}_i} D_{I_f}(T_{\{\gamma_{\text{global}}, t^x_{\text{global}}, t^y_{\text{global}}\} \circ \mathbf{p}_{\text{global}} \circ \mathbf{p}_i}(x)) \tag{12.10}$$

The optimal $\hat{\mathbf{p}}_{\text{global}}$ leads us to the approximated position of the stent graft in the interventional image. In the next step, we will refine the shape and position of each single segment of the stent graft.

Local Registration We aim at finding the correct values for each $\mathbf{p}_i$. Similar to Eq. (12.8) for the simultaneous registration, we define our costfunction for each segment i $(i = 1, \ldots, l)$ as

$$E(\mathbf{p}_i) = \sum_{x \in \mathcal{M}_i} D_{I_f}\left(T_{\{\gamma_{\text{global}}, t^x_{\text{global}}, t^y_{\text{global}}\} \circ \hat{\mathbf{p}}_{\text{global}} \circ \mathbf{p}_i}(x)\right) \cdot \phi(\mathbf{p}_i) \cdot \phi(\mathbf{p}_{i+1}) \tag{12.11}$$

with

$$\phi(\mathbf{p}_i) = \left\| \begin{matrix} |\mathbf{t}_i - \mathbf{t}_{i-1}| - \mathbf{t}_i^{\Delta} \\ |\mathbf{v}_i - \mathbf{v}_{i-1}| - \mathbf{v}_i^{\Delta} \\ |r_i^t - r_{i-1}^b| - r_i^{\Delta} \end{matrix} \right\| + \lambda \tag{12.12}$$

penalizing the change of translation, rotation, and radius in between neighboring segments. This change is constrained by the graft material fixing the stent wires (Fig. 12.4), which is not stretchable. Therefore translation, rotation radii, and the radii of the curves depend on the predefined distance d_i between the segments:

$$\mathbf{t}_i^{\Delta} = \begin{bmatrix} \frac{d_i}{2} \\ \frac{d_i}{2} \\ \frac{A_i}{2} + d_i \end{bmatrix} \quad \mathbf{v}_i^{\Delta} = \begin{bmatrix} \frac{d_i}{\sqrt{1+d_i}} \\ \frac{d_i}{\sqrt{1+d_i}} \\ \frac{s_i}{2} \end{bmatrix} \quad r_i^{\Delta} = |r_i^t - r_i^b|$$

In order to account for small measurement errors, an additional parameter λ was added to the penalization equation (12.12).

3.3 Outlook and Conclusion

The purpose of interventional navigation support for endovascular procedures is mainly the correct location and identification of instruments within the patient's anatomy. Here, physicians can greatly benefit from 3D visualizations showing the patient's anatomy together with the interventional instrument. This aspect is depicted in Fig. 12.5 where the recovered shape of a partially unfolded stent graft is shown within a 3D patient's anatomy model extracted from the respective preoperative CTA scan. On the downside, this visualization does not account for any vascular deformations that happen during the procedure due to the insertion of medical instruments. Some initial results in this direction have been proposed by Toth et al. [75] and need to be advanced further for practical application.

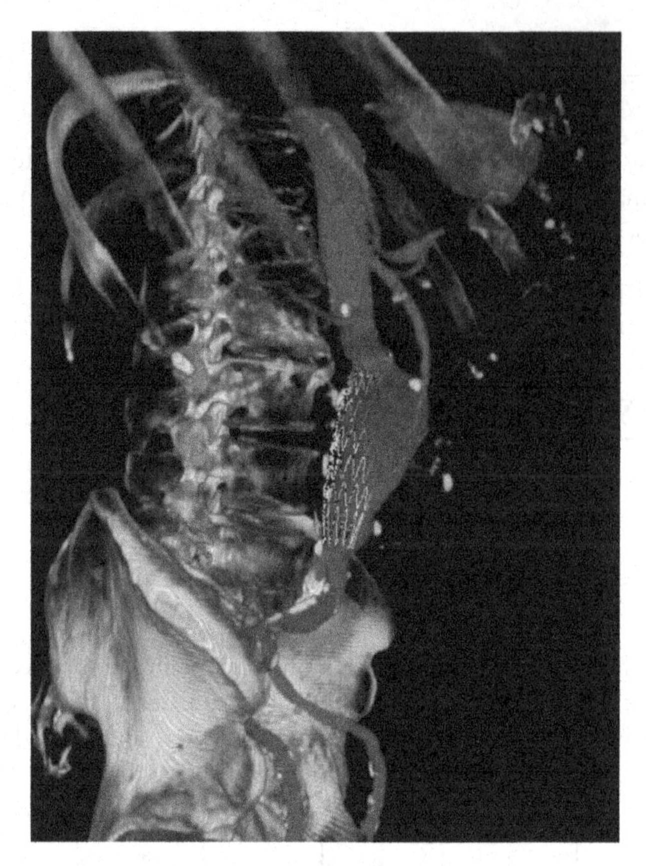

Fig. 12.5 Visualization of the recovered shape of a partially unfolded stent graft within a 3D patient's anatomy model extracted from the respective preoperative CTA scan.

The field of endovascular interventions is rapidly changing over time. The continuous introduction of more and more optimized and customized stent grafts allow for endovascular treatment of even more complex diseases and degeneration of the vessel wall. Computer assistance systems have to follow this evolution closely in order to provide intelligent and optimized navigation support for highly complex stent graft configurations.

Although customized stent grafts that are being fabricated for each patient's anatomy specifically bear superior outcome results and long-term effects, it seems that major stent graft producers are more and more moving toward off-the-shelf stent grafts that can be easily plugged together for realizing almost all possible anatomical configurations. This, however, requires a much more precise and reliable procedural planning in order to define stent graft sizes accurately. In particular, for cases with complex anatomical degenerations such as kinking of the aorta and combination of aneurysm and dissection, an optimal planning of stent graft configuration and sizes may require interventional image data in addition to the regular preoperative CTA scan in order to simulate the interaction accurately between stent grafts and vessel wall. The research field of virtual stenting has presented some initial solutions for simulating the effect of a stent graft being inserted into a patient's vasculature relying solely on image information extracted from respective preoperative CTA scans [76, 77]. In the future, this may be accompanied with interventional image data revealing an insight into how likely the vasculature is to be deformed by inserted instruments, as already initially shown by Toth et al. [75], and integrating this information into the simulation process.

REFERENCES

[1] Peters TM. Image-guided surgery. In: Kim Y, Horii SC, editors. Display and PACS, Handbook of medical imaging, vol. 3. SPIE; 2000. p. 103–53.
[2] Vaezy S, Zderic V. Image-guided therapy systems (Engineering in medicine & biology), Artech House; 2009.
[3] Venbrux AC, Chen HK, Tren GD, Gagarin DA. A brief history of image-guided therapy: endovascular milestones and nonvascular interventions. In: Mauro MA, Murphy K, Thomson K, Zollikofer CL, editors. Image-guided interventions, vol. 1. Saunders; 2008. p. 5–16.
[4] Criado FJ. Fundamental skills and "endo-habits". In: Endovascular intervention: basic concepts & techniques. Futura Publishing; 1999.
[5] Green SM, Klein AJ, Pancholy S, Rao SV, Steinberg D, Lipner R, et al. The current state of medical simulation in interventional cardiology: a clinical document from the Society for Cardiovascular Angiography and Intervention's (SCAI) Simulation Committee. Catheter Cardiovasc Interv 2014;83(1):37–46.
[6] Klass D, Tam MD, Cockburn J, Williams S, Toms AP. Training on a vascular interventional simulator: an observational study. Eur Radiol 2008;18(12):2874–8.
[7] Bismuth J, Donovan MA, O'Malley MK, El Sayed HF, Naoum JJ, Peden EK, et al. Incorporating simulation in vascular surgery education. J Vasc Surg 2010;52(4):1072–80.
[8] Lake CL. Simulation in cardiothoracic and vascular anesthesia education: tool or toy? Semin Cardiothorac Vasc Anesth 2005;9(4):265–73.
[9] Eckhard M. Surgical resident designs low-cost vascular simulator. Vanderbilt University Medical Center's Weekly Newspaper; 2012.
[10] Dawson DL, Meyer J, Lee ES, Pevec WC. Training with simulation improves residents' endovascular procedure skills. J Vasc Surg 2007;45(1):149–54.
[11] Pandey VA, Wolfe JHN. Expanding the use of simulation in open vascular surgical training. J Vasc Surg 2012;56(3):847–52.
[12] Pandey VA, Black SA, Lazaris AM, Allenberg JR, Eckstein HH, Hagmuller GW, et al. Do workshops improve the technical skill of vascular surgical trainees? Eur J Vasc Endovasc Surg 2005;30(4):441–7.
[13] Willems MCM, van der Vliet JA, Williams V, Kool LJS, Bergqvist D, Blankensteijn JD. Assessing endovascular skills using the simulator for testing and rating endovascular skills (STRESS) machine. Eur J Vasc Endovasc Surg 2009;37(4):431–6.
[14] Markelj P, Tomaževič D, Likar B, Pernuš F. A review of 3D/2D registration methods for image-guided interventions. Med Image Anal 2012;16(3):642–61.
[15] Matl S, Brosig R, Baust M, Navab N, Demirci S. Vascular image registration techniques: a living review. Med Image Anal 2016;35:1–17.
[16] Tomaselli GF. Introduction to a compendium on sudden cardiac death: epidemiology, mechanisms, and management. Circ Res 2015;116(12):1883–6.

[17] Wittkampf FHM, Wever EFD, Derksen R, Wilde AAM, Ramanna H, Hauer RNW, et al. LocaLisa new technique for real-time 3-dimensional localization of regular intracardiac electrodes. Circulation 1999;99(10):1312–7.
[18] Kirchhof P, Loh P, Eckardt L, Ribbing M, Rolf S, Eick O, et al. A novel nonfluoroscopic catheter visualization system (LocaLisa) to reduce radiation exposure during catheter ablation of supraventricular tachycardias. Am J Cardiol 2002;90(3):340–3.
[19] Earley MJ, Abrams DJR, Sporton SC, Schilling RJ. Validation of the noncontact mapping system in the left atrium during permanent atrial fibrillation and sinus rhythm. J Am Coll Cardiol 2006;48(3): 485–91.
[20] Rolf S, Hindricks G, Sommer P, Richter S, Arya A, Bollmann A, et al. Electroanatomical mapping of atrial fibrillation: review of the current techniques and advances. J Atrial Fibrillation 2015;7:57–68.
[21] Knackstedt C, Schauerte P, Kirchhof P. Electro-anatomic mapping systems in arrhythmias. Europace 2008;10(Suppl. 3):iii28–34.
[22] Gepstein L, Hayam G, Ben-Haim SA. A novel method for nonfluoroscopic catheter-based electroanatomical mapping of the heart in vitro and in vivo accuracy results. Circulation 1997;95(6): 1611–22.
[23] Scaglione M, Biasco L, Caponi D, Anselmino M, Negro A, Donna PD, et al. Visualization of multiple catheters with electroanatomical mapping reduces X-ray exposure during atrial fibrillation ablation. Europace 2011;13(7):955–62.
[24] Govil A, Calkins H, Spragg DD. Fusion of imaging technologies: how, when, and for whom? J Interv Card Electrophysiol 2011;32(3):195–203.
[25] Eitel C, Hindricks G, Dagres N, Sommer P, Piorkowski C. EnSite Velocity™ cardiac mapping system: a new platform for 3D mapping of cardiac arrhythmias. 2014;7(2):185–92.
[26] Estner HL, Hessling G, Ndrepepa G, Wu J, Reents T, Fichtner S, et al. Electrogram-guided substrate ablation with or without pulmonary vein isolation in patients with persistent atrial fibrillation. Europace 2008;10(11):1281–7.
[27] Dickfeld T, Tian J, Ahmad G, Jimenez A, Turgeman A, Kuk R. MRI-guided ventricular tachycardia ablation integration of late gadolinium-enhanced 3D scar in patients with implantable cardioverter-defibrillators. Circ Arrhythm Electrophysiol 2011;4(2):172–84.
[28] Beasley RA. Medical robots: current systems and research directions. J Robot 2012;2012:1–14. Article ID 401613.
[29] Boston Scientific. Rhythmia™ mapping system—mapping, redefined; 2016. April, http://www.bostonscientific.com/.
[30] Fallavollita P. Cardiac arrhythmias—mechanisms, pathophysiology, and treatment, Detection, tracking and related costs of ablation catheters in the treatment of cardiac arrhythmias. Intech Open Science; 2014. doi:10.5772/57424.
[31] Franken E, Rongen P, van Almsick M, ter Haar Romeny B. Detection of electrophysiology catheters in noisy fluoroscopy images. In: Medical image computing and computer-assisted intervention (MICCAI 2006). Springer; 2006. p. 25–32.
[32] Yatziv L, Chartouni M, Datta S, Sapiro G. Toward multiple catheters detection in fluoroscopic image guided interventions. IEEE Trans Inf Technol Biomed 2012;16(4):770–81.
[33] Ma Y, Gogin N, Cathier P, Housden RJ, Gijsbers G, Cooklin M, et al. Real-time X-ray fluoroscopy-based catheter detection and tracking for cardiac electrophysiology interventions. Med Phys 2013;40(7):071902.
[34] Wu W, Chen T, Barbu A, Wang P, Strobel N, Zhou SK, et al. Learning-based hypothesis fusion for robust catheter tracking in 2D X-ray fluoroscopy. In: Proceedings of IEEE conference on computer vision and pattern recognition. IEEE; 2011. p. 1097–104.
[35] Cazalas M, Bismuth V, Vaillant R. An image-based catheter segmentation algorithm for optimized electrophysiology procedure workflow. In: Functional imaging and modeling of the heart. Berlin: Springer; 2013. p. 182–90.
[36] Milletari F, Navab N, Fallavollita P. Automatic detection of multiple and overlapping EP catheters in fluoroscopic sequences. In: Medical image computing and computer-assisted intervention (MICCAI 2013). Springer; 2013. p. 371–9.

[37] Wang P, Liao WS, Chen T, Zhou SK, Comaniciu D. Graph based interactive detection of curve structures in 2D fluoroscopy. In: Medical image computing and computer-assisted intervention (MICCAI 2010). Springer; 2010. p. 269–77.
[38] Barbu A, Athitsos V, Georgescu B, Boehm S, Durlak P, Comaniciu D. Hierarchical learning of curves application to guidewire localization in fluoroscopy. In: IEEE conference on computer vision and pattern recognition. 2007. p. 1–8.
[39] Spiegel M, Pfister M, Hahn D, Daum V, Hornegger J, Struffert T, et al. Towards real-time guidewire detection and tracking in the field of neuroradiology. In: Medical imaging 2009: visualization, image-guided procedures, and modeling. SPIE, vol. 726105; 2009.
[40] Wang P, Chen T, Zhu Y, Zhang W, Zhou SK, Comaniciu D. Robust guidewire tracking in fluoroscopy. In: IEEE conference on computer vision and pattern recognition. 2009. p. 691–8.
[41] Bismuth V, Vaillant R, Talbot H, Najman L. Curvilinear structure enhancement with the polygonal path image-application to guide-wire segmentation in X-ray fluoroscopy. Med Image Comput Comput Assist Interv 2012;15(Pt 2):9–16.
[42] Cañero C, Vilariño F, Mauri J, Radeva P. Predictive (un)distortion model and 3-D reconstruction by biplane snakes. IEEE Trans Med Imaging 2002;21(9):1188–201.
[43] Schenderlein M, Stierlin S, Manzke R, Rasche V, Dietmayer K. Catheter tracking in asynchronous biplane fluoroscopy images by 3D B-snakes. In: Medical imaging 2010: visualization, image-guided procedures, and modeling. SPIE; 2010. p. 76251U.
[44] Baert SAM, van de Kraats EB, van Walsum T, Viergever MA, Niessen WJ. Three-dimensional guide-wire reconstruction from biplane image sequences for integrated display in 3-D vasculature. IEEE Trans Med Imaging 2003;22(10):1252–8.
[45] Brost A, Liao R, Strobel N, Hornegger J. Respiratory motion compensation by model-based catheter tracking during EP procedures. Med Image Anal 2010;14(5):695–706.
[46] Papalazarou C, Rongen PMJ, de With PHN. Surgical needle reconstruction using small-angle multi-view X-ray. In: IEEE international conference on image processing (ICIP). IEEE; 2010. p. 4193–6.
[47] Bender HJ, Männer R, Poliwoda C, Roth S, Walz M. Reconstruction of 3D catheter paths from 2D X-ray projections. In: Medical image computing and computer-assisted intervention (MICCAI'99). Springer; 1999. p. 981–9.
[48] van Walsum T, Baert SAM, Niessen WJ. Guide wire reconstruction and visualization in 3DRA using monoplane fluoroscopic imaging. IEEE Trans Med Imaging 2005;24(5):612–23.
[49] Brückner M, Deinzer F, Denzler J. Temporal estimation of the 3D guide-wire position using 2D X-ray images. In: Medical image computing and computer-assisted intervention (MICCAI 2009). Springer; 2009. p. 386–93.
[50] Baur C, Milletari F, Belagiannis V, Navab N, Fallavollita P. Automatic 3D reconstruction of electrophysiology catheters from two-view monoplane C-arm image sequences. Int J Comput Assist Radiol Surg 2016;11(7):1319–28.
[51] Hoffmann M, Brost A, Koch M, Bourier F, Maier A, Kurzidim K, et al. Electrophysiology catheter detection and reconstruction from two views in fluoroscopic images. IEEE Trans Med Imaging 2016;35(2):567–79.
[52] Hill BH, Wolf YG, Lee WA, Arko FR, Olcott IVC, Schubart PJ, et al. Open versus endovascular AAA repair in patients who are morphological candidates for endovascular treatment. J Endovasc Ther 2002;9:255–61.
[53] Zarins CK, White RA, Diethrich EB, Hodgson KJ, Fogarty TJ. AneuRx stent graft versus open surgical repair of abdominal aortic aneurysms: multicenter prospective clinical trial. J Vasc Surg 1999;29(2):292–305.
[54] Ring ME. How a dentist's name became a synonym for a life-saving device: the story of Dr. Charles Stent. J Hist Dent 2001;49(2):77–80.
[55] Hedin M. The origin of the word stent. DynaCT during EVAR—a comparison with multidetector CT. Acta Radiol 1997;38:937–9.
[56] Geller SC, The Members of the Society of Interventional Radiology Device Forum. Imaging guidelines for abdominal aortic aneurysm repair with endovascular stent grafts. J Vasc Interv Radiol 2003;14(9 Pt 2):S263–4.

[57] Eide K, Ødegård A, Myhre HO, Lydersend S, Hatlinghus S, Haraldseth O. DynaCT during EVAR—a comparison with multidetector CT. Eur J Vasc Endovasc Surg 2009;37(1):23–30.
[58] Nordon IM, Hinchliffe RJ, Malkawi AH, Taylor J, Holt PJ, Morgan R, et al. Validation of DynaCT in the morphological assessment of abdominal aortic aneurysm for endovascular repair. J Endovasc Ther 2010;17(2):183–9.
[59] Chaikof EL, Blankensteijn JD, Harris PL, White GH, Zarins CK, Bernhard VM, et al. Reporting standards for endovascular aortic aneurysm repair. J Vasc Surg 2002;35(5):1048–60.
[60] Binkert CA, Alencar H, Singh J, Baum RA. Translumbar type II endoleak repair using angiographic CT. J Vasc Interv Radiol 2006;17(8):1349–53.
[61] Biasi L, Ali T, Hinchliffe R, Morgan R, Loftus I, Thompson M. Intraoperative DynaCT detection and immediate correction of a type 1a endoleak following endovascular repair of abdominal aortic aneurysm. Cardiovasc Intervent Radiol 2009;32(3):535–8.
[62] Schulz B, Heidenreich R, Heidenreich M, Eichler K, Thalhammer A, Naeeme NNN, et al. Radiation exposure to operating staff during rotational flat-panel angiography and C-arm cone beam computed tomography (CT) applications. Eur J Radiol 2012;81(12):4138–42.
[63] Eide KR, Ødegård A, Myhre HO, Haraldseth O. Initial observations of endovascular aneurysm repair using Dyna-CT. J Endovasc Ther 2007;14:65–8.
[64] Pujol S, Cinquin P, Pecher M, Bricault I, Viorin D. Minimally invasive navigation for the endovascular treatment of abdominal aortic aneurysm: preclinical validation of the endovax system. In: Medical image computing and computer-assisted intervention (MICCAI 2003), Springer; 2003. p. 231–8.
[65] Manstad-Hulaas F, Tangen GA, Gruionu LG, Aadahl P, Hernes TAN. Three-dimensional endovascular navigation with electromagnetic tracking: ex vivo and in vivo accuracy. J Endovasc Ther 2011;18(2):230–40.
[66] Bismuth V, Vaillant R, Funck F, Guillard N, Najman L. A comprehensive study of stent visualization enhancement in X-ray images by image processing means. Med Image Anal 2011;15(4):565–76.
[67] Zheng Y, Comaniciu D. Marginal space learning for medical image analysis. New York: Springer; 2014.
[68] Demirci S, Bigdelou A, Wang L, Wachinger C, Baust M, Tibrewal R, et al. 3D stent recovery from one X-ray projection. In: Medical image computing and computer-assisted intervention (MICCAI 2011), Springer; 2011. p. 178–85.
[69] Demirci S, Manstad-Hulaas F, Navab N. Interventional 2D-3D registration in the presence of occlusion. In: XIII Mediterranean conference on medical and biological engineering and computing 2013 (MEDICON 2013), September 25–28, 2013, Seville, Spain. Cham: Springer International Publishing; 2014. p. 277–80.
[70] Frangi AF, Niessen WJ, Vincken KL, Viergever MA. Multiscale vessel enhancement filtering. In: Wells WM, Colchester A, Delp S, editors. Proceedings of first international conference on medical image computing and computer-assisted interventation (MICCAI'98), October 11–13, 1998, Cambridge, MA, USA, Berlin, Heidelberg: Springer Berlin Heidelberg; 1998. p. 130–7.
[71] Volpi D, Sarhan MH, Ghotbi R, Navab N, Mateus D, Demirci S. Online tracking of interventional devices for endovascular aortic repair. Int J Comp Assist Radiol Surg 2015;10(6):773–81.
[72] Lin Z, Chen M, Ma Y. The augmented Lagrange multiplier method for exact recovery of corrupted low-rank matrices. UIUC technical report UILU-ENG-09-2215, 2009.
[73] Hartley R, Zisserman A. Multiple view geometry in computer vision. 2nd ed. New York, NY, USA: Cambridge University Press; 2003.
[74] Sidorov KA, Richmond S, Marshall D. An efficient stochastic approach to groupwise non-rigid image registration. In: IEEE conference on computer vision and pattern recognition (CVPR 2009), June, 2009. p. 2208–13.
[75] Toth D, Pfister M, Maier A, Kowarschik M, Hornegger J. Adaption of 3D models to 2D X-ray images during endovascular abdominal aneurysm repair. In: Medical image computing and computer-assisted intervention (MICCAI 2015), Springer International Publishing; 2015. p. 339–46.
[76] Egger J, Grosskopf S, Nimsky C, Kapur T, Freisleben B. Modeling and visualization techniques for virtual stenting of aneurysms and stenoses. Comput Med Imaging Graph 2012;36(3):183–203.
[77] Larrabide I, Kim M, Augsburger L, Villa-Uriol MC, Rüfenacht D, Frangi AF. Fast virtual deployment of self-expandable stents: method and in vitro evaluation for intracranial aneurysmal stenting. Med Image Anal 2012;16(3):721–30.

CHAPTER 13

Interventional Quantification of Cerebral Blood Flow

S. Demirci[*], M. Kowarschik[†]
[*] Technical University of Munich, Munich, Germany
[†] Siemens Healthineers, Forchheim, Germany

Chapter Outline

Chapter points

- This chapter gives an insight into the medical relevance of blood flow quantification.
- It provides an overview of the state-of-the-art in cerebral blood flow assessment and introduces several novel concepts of flow quantification using angiographic imaging in 2D and 3D.

1. INTRODUCTION TO THE CLINICAL VALUE OF BLOOD FLOW QUANTIFICATION

1.1 Blood Flow and Perfusion

A major purpose of blood flow is to transport both oxygen and nutrients to tissue of various types (e.g., organ tissue, muscle tissue). A variety of diseases are thus related to abnormal flow of blood in vessels or abnormal perfusion of capillary tissue. Potentially, such diseases can be related to all regions of the human body.

For example, the blood flow in coronary arteries and the corresponding myocardial perfusion play an important role in the assessment of a patient's cardiac condition. As another example, a malignant tumor in a patient's liver is typically characterized by enhanced perfusion due to hypervascularization. Finally, elderly patients with a relevant

Computing and Visualization for Intravascular Imaging and Computer-Assisted Stenting
http://dx.doi.org/10.1016/B978-0-12-811018-8.00013-8

cardiovascular risk profile often have blood flow and perfusion deficits in their periphery, i.e., their legs and feet. Generally speaking, the treatment of all of these pathologies can benefit from accurate measurements of blood flow and tissue perfusion.

Minimally invasive procedures are gaining importance due to shorter patient recovery times and lower complication rates, enhanced clinical workflows, and cost reductions. Such procedures cover endovascular treatments, where interventional devices such as guidewires and catheters are advanced through the patient's vasculature to the lesion to be treated, e.g., via the transfemoral route. We refer to Refs. [1, 2] for detailed overviews of the growing field of image-guided interventions and minimally invasive therapies.

The clinical focus of this chapter is on brain imaging and thus on applications in interventional neuroradiology. Therefore, our primary motivation for the assessment of blood flow and perfusion is based on patients suffering from cerebral vascular disorders. In particular, these pathologies cover ischemic and hemorrhagic strokes, arteriovenous malformations (AVMs), dural arteriovenous fistulas (DAVFs), and aneurysms, which will be described briefly in the following section.

1.2 Selected Diseases Related to Abnormal Flow and Perfusion Patterns in the Brain

Stroke. According to the World Health Organization (WHO), stroke is the second leading cause of death, behind ischemic heart disease. In 2012, 11.9% of all deaths worldwide were caused by stroke.[1]

It is worth noting that there are two types of stroke: ischemic and hemorrhagic [3]. An ischemic stroke is caused by an occlusion of a cerebral artery by an embolus, whereas a hemorrhagic stroke is related to the spontaneous rupture of an aneurysm or an AVM, for instance. Ischemic strokes account for approximately 85% of all strokes. An overview of today's imaging strategies for assessing stroke patients is given in Ref. [4]. Generally speaking, the high morbidity of stroke results from the interplay between neurological impairment, the emotional and also the social consequences of that neurological impairment, and the additional high recurrence risk of stroke [5].

> *In patients experiencing a typical large vessel acute ischemic stroke, 120 million neurons, 830 billion synapses, and 714 km (447 miles) of myelinated fibers are lost each hour. In each minute, 1.9 million neurons, 14 billion synapses, and 12 km (7.5 miles) of myelinated fibers are destroyed.*

This quotation by Saver [6] highlights several impressive quantities regarding the impact of ischemic stroke events. It thus motivates the need for the fastest, most appropriate treatment. In patients suffering from acute ischemic stroke, the assessment of blood flow and brain tissue perfusion can help to identify both the infarct core and the so-called penumbra region (i.e., the tissue at risk) that may benefit from revascularization.

[1]See http://www.who.int.

Since ischemic stroke is caused by vessel occlusion, its therapy is based on systemic intravenous thrombolysis and—for improved revascularization rates—on catheter-based procedures comprising both intra-arterial thrombolysis and mechanical recanalization of the occluded vessel, e.g., using clot-retrieving devices. Mechanical recanalization of a blocked vessel is also referred to as thrombectomy. We refer to Refs. [3, 4, 7–9] and the references to additional clinical publications provided therein for more detailed information.

Therefore, the assessment of blood flow and cerebral perfusion can support risk stratification and help the interventionalist to decide upon the appropriate treatment strategy, which ranges from leaving the patient untreated to drug-based or mechanical recanalization of the blocked artery, also beyond the established time window. For a comprehensive definition and classification of ischemic stroke events and for a list of respective treatment recommendations, we refer to the latest release of the stroke prevention and treatment guidelines by the American Heart Association/American Stroke Association [5].

Arteriovenous Malformation (AVM). A cerebral AVM is characterized by an abnormal network of blood vessels that commonly do not supply any capillary bed. This network is typically referred to as the nidus of the AVM. The nidus is supplied by a number of arteries (feeders) and drained by a number of veins. AVMs exhibit a high morphologic variety. So far, the causes of such malformations are not well understood. They are assumed to be based on head injuries or genetic disorders. The point prevalence of AVMs in adults is reported to be about 18 in 100,000.

A comprehensive discussion of brain AVMs including a review of their frequency and their prognosis can be found in Ref. [10]; a clinically widespread classification of AVMs was proposed by Spetzler and Martin [11].

Cerebral AVMs may cause various symptoms and issues, ranging from neurological disorders to bleedings, i.e., hemorrhagic strokes. For the case of an unruptured AVM, it is essential to assess its risk of rupture by analyzing the number of feeding arteries and the blood flow therein, the angioarchitecture of its nidus, as well as the number and the structure of its draining veins and the blood flow therein.

A common treatment approach of AVMs covers embolization using endovascular techniques and their successive resection, based on either open surgery or radiosurgery [12, 13].

Dural Arteriovenous Fistula (DAVF). The general term "fistula" refers to an abnormal connection, e.g., between organs or vessels. DAVFs are fistulas connecting the branches of dural arteries to dural veins or a venous sinus. In particular, AVMs may contain DAVFs. According to Ref. [14], the incidence of DAVFs is unknown. While many DAVFs remain clinically silent, some of them may cause symptoms similar to AVMs, including neurological disorder or hemorrhage, among others. Similar to AVMs, the treatment of DAVFs is based on endovascular embolization or surgical clipping.

We refer to Ref. [15] for a widely established classification of DAVFs and again to Ref. [14] for details on DAVFs and their medical management.

Aneurysm. Cerebral aneurysms are balloon-like dilatations of arterial vessel walls and may occur at a variety of different locations in the brain. Their reported incidence rates in the adult population range from 1% to 5% [16]. According to configuration, size, and location, aneurysms may cause various clinical symptoms. Aneurysmal rupture represents the most severe event and causes subarachnoid hemorrhage, which corresponds to a hemorrhagic stroke and represents a major cause of morbidity and mortality throughout the world [17]. However, most brain aneurysms will remain asymptomatic.

Today's options of treating both ruptured and unruptured aneurysms cover surgical as well as endovascular therapies [18–20]. In either case, it is important to assess cerebral flow before, during, and after the treatment in order to protect the parent artery and to react to periprocedural complications.

2. BLOOD FLOW ASSESSMENT USING ANGIOGRAPHIC X-RAY IMAGING

2.1 Overview of Angiographic X-Ray Imaging

Due to their flexibility, angiographic C-arm devices represent today's most commonly used systems in interventional imaging. Their name is derived from their architecture, which is characterized by a C-shaped arm that has an X-ray source mounted on one end and a flat-panel detector attached to the other. Meanwhile, C-arm systems based on image intensifier technology have been widely replaced by devices using flat-panel detectors, primarily due to the resulting improvements in image quality and the less bulky design of the latter.

C-arm systems primarily perform a wide spectrum of endovascular diagnostic and therapeutic procedures. Today's scanners can be used to perform both 2D and 3D imaging. When referring to 3D imaging using C-arm systems, the term "cone-beam CT (CBCT) imaging" is often used in order to emphasize the cone-shaped geometry of the X-ray beam that is emitted by the X-ray source. We refer to Ref. [21] for an introduction to angiographic C-arm systems.

Compared to a mono-plane system consisting of just one single C-arm, a biplane system comprises of two independent C-arms such that two views of the patient's anatomy and the interventional procedure can be acquired simultaneously. Such biplane systems are used in interventional neuroradiology, for instance, due to the geometric complexity of the cerebral vasculature.

Radio-opaque X-ray contrast agents (also referred to as dye) are commonly based on iodine and need to be used since the X-ray attenuation of blood is too low to enable the identification of blood vessels in native X-ray images. We refer to Ref. [21] for discussions of contrast agents for X-ray, MR, and ultrasound imaging. Due to the nephrotoxicity of iodine- and gadolinium-based contrast media, a general objective is to keep the contrast agent load as low as reasonable, particularly in patients with renal insufficiency.[2]

The assessment of blood flow in arterial and venous vessels plays an important role in the diagnosis and treatment of vascular disorders. The term "blood flow assessment" is rather general and refers to the determination of physical parameters that govern the flow. These parameters include temporal quantities, flow velocities, and volumetric flow rates, among others. While some of these quantities can be obtained in vivo by measurements (e.g., image- or catheter-based), others require physiological flow models to be parameterized properly and simulated numerically.

In diagnostic imaging, a variety of modalities are clinically established to evaluate a patient's blood flow. These methods range from ultrasound examinations to more sophisticated imaging techniques such as time-resolved computed tomography angiography (CTA), time-resolved magnetic resonance angiography (MRA), and phase-contrast magnetic resonance imaging (pcMRI) for directly estimating blood flow velocities, among others [21]. These methods either work without the administration of a contrast agent, or some modality-specific contrast agent is injected intravenously. The previously mentioned imaging techniques are thus considered to be noninvasive.

In contrast, approaches toward the interventional (i.e., peritherapeutic) assessment of blood flow primarily cover angiographic imaging using C-arm devices. Additionally, flow catheters and percutaneous Doppler ultrasound measurements are used in clinical practice. Again, we refer to Ref. [21] for a comprehensive overview. Aside from percutaneous Doppler ultrasound, these methods are considered invasive. The acquisition of conventional time-resolved 2D digital subtraction angiography (DSA) image series is based on an intra-arterial injection of contrast agent, which means that a catheter needs to be inserted into the patient's body and advanced to the target region to be examined.

Likewise, the use of flow catheters requires the positioning of the device in the vessel segment to be interrogated. Flow catheters are primarily used in cardiology and interventional radiology, and make use of Doppler ultrasound or the physical principle of thermodilution. The thermodilution method is based on measuring the temperature drop of the bloodstream distal to the injection location of a given amount of a cool fluid such as saline, for instance. See, e.g., the early publication by Ganz and Swan [22] and also Kramme et al. [23]. Due to the requirements of flow catheters regarding the vessel lumen, they are not commonly used in brain vessels. Instead, less invasive image-based methods for assessing cerebral blood flow are desirable.

The Doppler ultrasound in general exploits the fact that the flow velocity of the blood causes a Doppler shift in the frequency of the reflected ultrasound waves. This

[2] As an alternative to iodinated dye in X-ray imaging, CO_2-based contrast media may be used for certain examinations.

shift can be determined and used to estimate the velocity of the blood flow, which may of course be time-varying. Besides its applications in cardiology, percutaneous Doppler ultrasound can also be used to assess the flow in some major cerebral vessel. In this case, it is referred to as a transcranial Doppler (TCD) ultrasound [24]. However, due to the high acoustic impedance of the patient's skull, the appropriate use of TCD is limited to basal intracerebral vessels.

Subsequent sections will concern blood flow measurement and the assessment of cerebral hemodynamics based on angiographic X-ray imagery using C-arm scanners.

2.2 Blood Flow Assessment in 2D

2.2.1 Analysis of Vascular Flow

Generally speaking, the determination of blood flow parameters based on 2D angiographic image series (e.g., 2D DSA series) is a difficult task. 2D X-ray images are projections of the 3D subject onto a plane, i.e., the X-ray detector [21]. Consequently, the 2D images lack depth information and, hence, their interpretation requires both general anatomical knowledge and good 3D understanding of the patient's vascular structures.

This interpretation becomes even more complicated if patient motion occurs during the acquisition of the image series. In this case, the subtraction of the mask image from a fill image may lead to an inaccurate result. Since our major focus is on brain imaging, however, patient motion is commonly less pronounced than it is in cardiac and abdominal imaging. Therefore, algorithmic approaches toward motion artifact correction are not discussed here and we refer to the available literature, cf. [25, 26], for example.

In the following, we distinguish between methods that aim at quantitative flow evaluation in vessels and—as a particular clinical application—approaches that focus on the estimation of flow patterns in cerebral aneurysms in order to assess the efficacy of flow-diverting stents that are deployed to recanalize the bloodstream.

The methods presented in this section have in common that they are all based on 2D DSA image series. Each pixel of a frame of a 2D DSA series is characterized by a time-contrast curve (TCC). For the case of a pixel that corresponds to a blood vessel, its TCC essentially looks like the example curve depicted in Fig. 13.1. The contrast increases until the peak of the curve is reached. The length of the respective time period from the defined starting point until this peak is reached is named time to peak (TTP) opacification. For example, this starting point may correspond to the time point at which the X-ray acquisition starts or at which the TCC reaches a given threshold (e.g., 10% of its peak value). Afterwards, the contrast washes out again. Various curve parameters may be extracted, such as its average wash-in gradient, its average wash-out gradient, and its full width at half maximum (FWHM) [27].

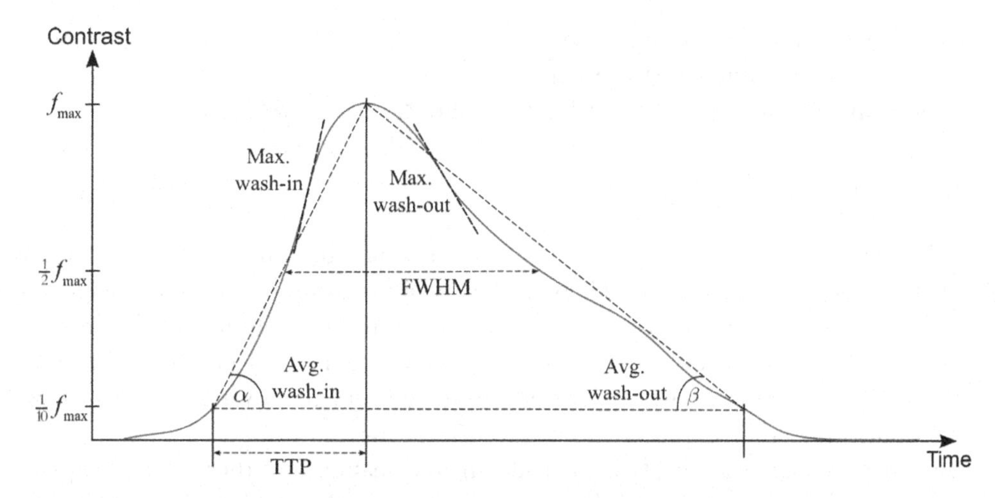

Fig. 13.1 Idealized time-contrast curve (TCC) of a vascular pixel in a 2D DSA image series: once the contrast agent is being injected into the vessel, the contrast in one single vascular pixel increases until the peak of the curve is reached. Afterwards the contrast washes out again until the original pixel intensity value is reached. The length of the respective time period from the defined starting point until the curve's peak is reached is named time to peak (TTP) opacification. Various curve parameters may be extracted such as its average wash-in gradient, its average wash-out gradient, and its full width at half maximum (FWHM).

These curve parameters can be used immediately to quantify temporal characteristics of the patient's blood flow. Color coding may be applied in addition in order to represent these pixel-specific quantities in a single 2D image. Several manufacturers of angiographic C-arm systems offer corresponding software products that generate parametric color images from 2D DSA series, cf. [28].

An example of how TTP measurements may be employed to assess the changes in a patient's cerebral blood flow before and after stenting of his left internal carotid artery (ICA), may be found in Ref. [29]. A similar application of TTP measurements based on 2D DSA image series was presented in Ref. [30]. In this chapter, we focused on the assessment of embolization procedures in patients with carotid-cavernous fistulas (CCFs), which represent a particular form of DAVFs. It was demonstrated that the embolization of these fistulas generally led to normal cerebral circulation times and that TTP measurement tools might therefore be applicable to support the determination of endpoints during embolization procedures.

In order to estimate the velocity of the contrast agent bolus, it is necessary to determine both spatial distances as well as temporal differences. The average velocity $\bar{v}$ of the bolus within a given vessel segment can then be computed as

$$\bar{v} = \frac{\Delta s}{\Delta t}, \tag{13.1}$$

where Δs denotes the length of the vessel segment and Δt represents the time the bolus needs to pass through the vessel segment.

Unfortunately, the estimation of both Δs and Δt can be difficult and thus error-prone. Since 2D DSA images generally lack depth information, Δs can only be approximated, since the 3D structure of the vessel segment under consideration may lead to severe inaccuracies due to foreshortening.

Furthermore, the shape of the contrast agent bolus becomes wider as it travels through the patient's vasculature. There are two major reasons for this bolus widening effect: the dispersion of the contrast agent in blood and the dilution of the mixture of contrast agent and blood due to collateral flows of nonenhanced blood. We refer to Ref. [31] for a comprehensive discussion of these issues and a variety of approaches toward estimating bolus arrival times from 2D DSA series.

In order to overcome the lack of in-depth information and therefore to improve the estimation of Δs, some researchers proposed to use additional vascular 3D images in order to measure the length of the vascular segment under consideration more precisely.

For the sake of simplicity, we generally assume that the intra-arterial injection of the contrast agent does not significantly change the physiological flow of blood in terms of velocity and volumetric flow rate. This may lead to inaccuracies if the location of the injection (i.e., the catheter tip) is close to the locations of the measurements. A physical model of contrast injection into an arterial bloodstream can be found in Ref. [32]. An evaluation of the impact of contrast injection parameters on the resulting TCCs in a canine model was presented in Ref. [33].

Besides velocity estimates, the determination of volumetric flow rates further requires the knowledge of a vessel's cross-sectional area. The volumetric flow rate $Q(l, t)$ at location l along the vessel's centerline and time t is then given by

$$Q(l, t) = \iint_{A(l)} \vec{v}(\vec{x}, t) \times \vec{n}(l) \mathrm{d}\vec{x}, \tag{13.2}$$

where $A(l)$ denotes the cross-sectional area of the vessel at location l along its centerline, $\vec{v}(\vec{x}, t)$ is the flow velocity field, $\vec{n}(l)$ represents the normal vector of $A(l)$, and $\times$ denotes the cross product (vector product) as usual. This situation is illustrated in Fig. 13.2.

Eq. (13.2) can alternatively be written as

$$Q(l, t) = A(l)v(l, t), \tag{13.3}$$

where $v(l, t)$ represents the spatially averaged velocity magnitude of the flow at location l perpendicular to $A(l)$ at time t, i.e.,

$$v(l, t) = \frac{1}{A(l)} \iint_{A(l)} \vec{v}(\vec{x}, t) \times \vec{n}(l) \mathrm{d}\vec{x}. \tag{13.4}$$

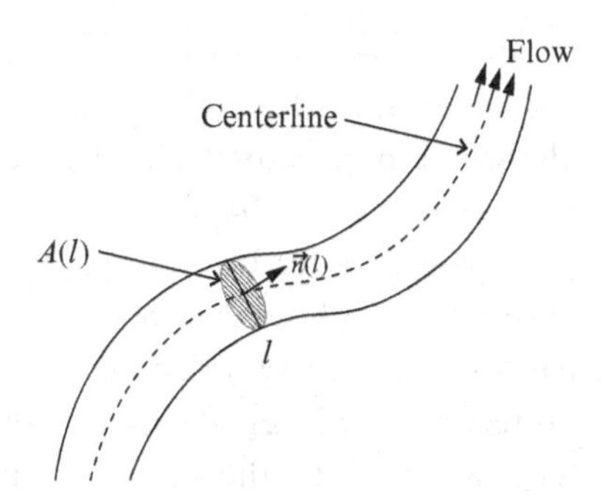

Fig. 13.2 For the calculation of the volumetric flow rate at location l along the vessel's centerline, it is crucial to have knowledge about the cross-sectional area $A(l)$ of the vessel at location l along its centerline and the normal vector of $A(l)$ denoted as $\vec{n}(l)$.

In today's practice, neither $\vec{v}(\vec{x}, t)$ nor $v(l, t)$ can be measured exactly and, therefore, $v(l, t)$ needs to be approximated by an estimate of the temporal and spatial average velocity $\bar{v}$ in the respective vessel segment, see Eq. (13.1). If 2D images are used solely, vascular cross-sections may be estimated based on the assumption of cylindrical vessels [31]. However, if 3D information of the vascular geometry is available as well, cross-sectional areas may be determined more accurately.[3]

The extraction of quantitative flow parameters from 2D DSA image series is highly attractive from a clinical perspective. This is true despite the previously mentioned limitations that may drastically compromise the accuracy of the estimates. The relevant arguments from a practical standpoint are that the acquisition of 2D DSA series is already part of today's clinical routine and that rough estimates of flow parameters or their relative changes may already be sufficient. Hence, the application of flow assessment methods based on such 2D image series fits well into the current clinical workflows and might require neither increased X-ray dose nor increased contrast agent load for the patient.

In contrast, flow assessment techniques that raise the need for additional 3D scans will only be used in practice, if the added clinical value justifies the more complicated and time-consuming workflow as well as—potentially—the additional amounts of X-ray dose and contrast agent.

2.2.2 Analysis of Flow Patterns in Cerebral Aneurysms

Image-Based Metrics for Assessing Flow Diverter Efficacy. A particular application of flow pattern analysis based on 2D DSA refers to the question of efficacy of

[3]Note that the cross-sectional area of a vessel may be time-varying due to cardiac pulsation. Yet, this behavior is difficult to observe using today's medical imaging technology because of limitations in temporal and spatial resolution. This effect is therefore ignored for the sake of simplicity.

flow-diverting devices (also named flow diverters), which represent an endovascular treatment option for cerebral aneurysms [16]. The term "flow diverter" refers to a special stent device with a dense mesh structure that is deployed in the parent artery and placed across the orifice of the aneurysm. Its purpose is to redirect blood flow away from the aneurysm, which physiologically leads to blood clotting and the formation of a thrombus within the aneurysm sac. The flow diverter might further provide a scaffold for neointimal and endothelial tissue overgrowth that eventually separates the aneurysm from the parent vessel and thus restores the original vessel topology. Ideally, the aneurysm finally degenerates completely. We refer to the clinical publications [19, 34–36] for details on endovascular aneurysm treatment using flow diverters. In some patients, several flow diverters need to be deployed in order to treat the aneurysm successfully. An important property of flow diverters is that they can preserve relevant arterial branches that they may cover, instead of blocking them, which might cause severe perfusion deficits in the supplied brain territories [20].

During the endovascular treatment of the aneurysm using flow-diverting stents, the interventionalist needs to decide whether the deployed device (or devices) already redirects the bloodstream sufficiently or whether further measures need to be taken. One further option is the placement of additional coils within the aneurysm sac, for instance. So far, this decision is based on the visual inspection of 2D DSA series that are acquired peritherapeutically.

This evaluation step is highly subjective and depends on the experience of the physician. The physician's decision is thus hard to reproduce, it is not quantifiable, and its predictive value cannot be validated easily. Therefore, the development of image-based quantitative metrics that capture hemodynamic modifications induced by the deployment of flow-diverting stents is desirable for advancing evidence-based medicine and implementing standard clinical guidelines.

Several quantitative metrics for the angiographic assessment of flow diverter efficacy have been proposed so far. They cover visual grading schemes and the extraction of characteristic parameters from TCCs that correspond to user-defined regions of interest (ROIs) placed within the aneurysm, among others. Yet none of these metrics that have been proposed so far can be considered clinically established. Rather, these approaches represent suggestions toward supporting the clinical community with image-based analysis tools that require future in-depth evaluation to determine whether they can add value to peritherapeutic decision making [37]. We refer the reader to Ref. [38], which covers a comprehensive survey of algorithmic approaches that have been proposed so far in order to assess quantitatively the efficacy of flow-diverting stents using angiographic imaging.

In the following, we will present two image-based approaches that aim at quantifying the changes in flow patterns after the endovascular treatment of cerebral aneurysms using flow diverters.

Assessment of Hemodynamics Using Optical Flow. This first approach is motivated by the clinical hypothesis that flow diversion leads to reduced intra-aneurysmal velocity magnitudes. The following description of our method follows the more detailed discussion in Ref. [38].

The fundamental technical idea behind this algorithmic approach is to recover flow patterns in a cerebral aneurysm before and after flow diverter deployment by tracking the motion of a contrast agent. Given a moderate injection rate of iodinated dye into the bloodstream as well as a sufficiently high DSA acquisition frame rate (e.g., 15 frames per second), the resulting mixture exhibits local contrast patterns that can be monitored, e.g., using optical flow methods.

The term "optical flow" refers to a family of methods used in computer vision. These algorithms were introduced about 30 years ago in order to recover the motion of rigid objects from sequences of optical camera images [39]. The aim of optical flow estimators is to determine time-dependent displacement fields, i.e., to estimate the 2D displacement field $\vec{u}(\vec{x}, t_i)$ between any two temporally adjacent 2D frames $I(\vec{x}, t_i)$ and $I(\vec{x}, t_{i+1})$ of the given image series.

The original formulation of the optical flow method is characterized by the assumption of constant brightness of corresponding image pixels, i.e.,

$$I(\vec{x}, t_i) = I(\vec{x} + \vec{u}(\vec{x}, t_i), t_{i+1}). \tag{13.5}$$

Introducing a first-order Taylor series approximation of the image intensities, we obtain

$$I(\vec{x} + \vec{u}(\vec{x}, t_i), t_{i+1}) \approx I(\vec{x}, t_i) + \langle (\nabla I)(\vec{x}, t_i), \vec{u}(\vec{x}, t_i) \rangle + \frac{\delta I}{\delta t}(\vec{x}, t_i), \tag{13.6}$$

where ∇ represents the gradient operator with respect to the spatial dimensions, i.e.,

$$\nabla I = \left(\frac{\partial I}{\partial x}, \frac{\partial I}{\partial y} \right)^{\mathrm{T}}, \tag{13.7}$$

and $\langle .,. \rangle$ denotes the scalar product (inner product). Hence, Eq. (13.5) can be approximated by

$$\frac{\partial I}{\partial t} + (\nabla I) \cdot \vec{u} = 0, \tag{13.8}$$

where we dropped the parameters for the ease of notation.

For the case of motion recovery from X-ray image series, the assumption of constant image brightness is easily violated. Therefore, Wildes et al. proposed replacing Eq. (13.8) with the continuity equation

$$\frac{\partial I}{\partial t} + \nabla(I\vec{u}) = 0, \tag{13.9}$$

which represents the conservation of mass of contrast agent across the images of the DSA series [40]. A comprehensive discussion of the link between the physical properties of fluid flows and their image-based restoration using optical flow methods is given in Ref. [41].

Eq. (13.9) can be shown to be ill-posed. Consequently, further constraints on the estimated displacement fields $\vec{u}(\vec{x}, t_i)$ need to be set. For this purpose, various regularization methods have been proposed [38].

In our approach, we use the second-order div-curl regularizer proposed in Ref. [42]. This means that we compute the solution $\vec{u} = \vec{u}(\vec{x}, t)$ of Eq. (13.9) that minimizes the functional

$$J_u(t) = \int_\Omega \|\nabla \mathrm{div}(\vec{u}(\vec{x}, t))\|^2 + \|\nabla \mathrm{curl}(\vec{u}(\vec{x}, t))\|^2 \mathrm{d}\vec{x} \tag{13.10}$$

on the image domain Ω for all time instances t.

A particular advantage of this regularization approach is that it enforces spatial coherence (i.e., neighboring image pixels will be characterized by similar displacements), while not penalizing nonlaminar flows, which are likely to appear in the context of brain aneurysms. This second property is ensured by considering the gradients of the divergence and curl operators instead of the operators themselves.

Note that Eqs. (13.9), (13.10) are continuous representations of the problem that need to be discretized properly for the sake of their numerical solution. Our implementation uses the numerical scheme proposed in Ref. [42], which is based on a multiresolution finite-difference approximation of the problem. Proper boundary conditions are derived from the physically motivated assumption that there is no flow across the boundaries of the vasculature.

As soon as the time-dependent displacement fields $\vec{u}(\vec{x}, t_i)$ have been computed, various quantities can be derived [38]. One option could be to determine time-averaged magnitudes of projected velocities within user-defined regions of interest (ROIs). In order to account for variations in the patient's cardiac activity during the treatment, the estimated intra-aneurysmal velocities can be normalized by estimated velocities in the parent artery proximal to the flow-diverting device or devices. We refer to this additional correction step as inflow normalization.

Fig. 13.3 shows color-coded (gray scale) representations of the magnitudes of the projected velocities pre- and postflow diverter implantation. The scaling of the left image is the same as the scaling of the right image. Both images were normalized with respect to respective ROIs on the parent artery located proximal to the aneurysm. In this case, the resulting inflow-normalized average velocity magnitude within the aneurysm was reduced by 26% due to flow diverter deployment. Hence, the deployment of flow-diverting stents indeed led to reductions in the estimated velocity magnitudes.

Obviously, such considerations require that the injection and imaging parameters remain constant from pre to post. Injection parameters cover the initial concentration

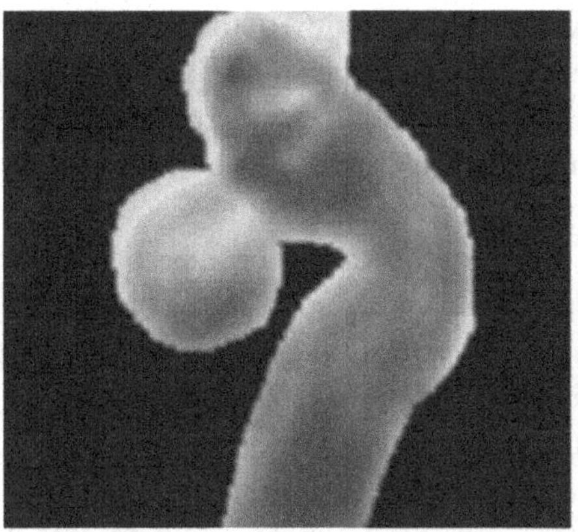
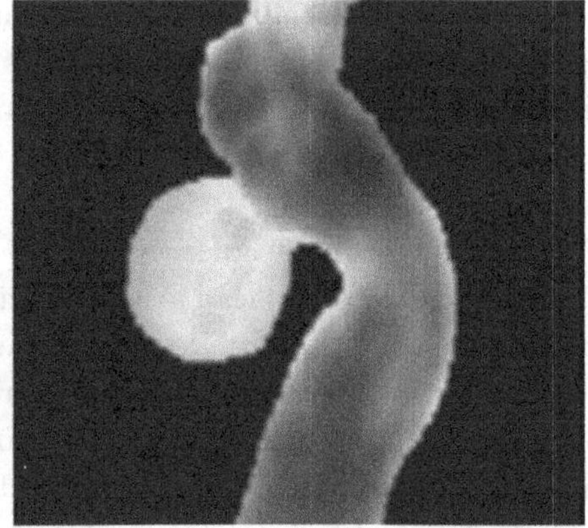

Fig. 13.3 The color-coding (gray scaling) of time-averaged magnitudes of estimated velocities pre- (*left*) and postflow (*right*) diverter treatment reveals that the resulting inflow-normalized average velocity magnitude within the aneurysm was significantly reduced after flow diverter deployment.

of iodinated contrast agent, the amount of contrast agent, and the injection rate, as well as the location of the catheter tip. Imaging parameters cover the angulation of the C-arm, the position of the table and the patient, and the X-ray dose settings, as well as the acquisition frame rate of the DSA series. In addition, the viewing direction (i.e., the angulation of the C-arm) should exhibit as little vascular overlap as possible in order that the projected intra-aneurysmal displacement fields can be estimated as accurately as possible.

A similar method was proposed by Pereira et al. [43], where the induced flow changes in the aneurysm were reduced to a single scalar value that the researchers denoted MAFA (mean aneurysmal flow amplitude) ratio. They also took into account potential flow changes in the parent artery before and after the implantation of the flow diverter. However, their estimation of the volumetric flow rates in the parent artery required 3D vascular datasets, which rendered the clinical workflow more complicated and time-consuming, and potentially led to an increased radiation dose as well as contrast agent load due to the additional scans.

In summary, it needs to be evaluated thoroughly how far the optical flow analysis of 2D DSA image series can be applied in order to assess complicated intra-aneurysmal flow changes in 3D that are caused by the deployment of flow-diverting stents. Different angulations of the C-arm device might lead to different estimates of flow patterns.

However, the analysis of 2D DSA image series yields quantifiable and reproducible quantities. Therefore, this approach already represents an advantage over the current clinical practice, which is only based on the subjective visual inspection of image data.

Fourier-Based Evaluation of Hemodynamics. Due to the patient's cardiac cycle, the physiological blood flow velocities in arteries vary periodically.[4] Consequently, a constant rate of contrast medium injection into the parent artery of the aneurysm leads to periodic brightness patterns in the DSA images, which can also be observed within the aneurysm.

A second approach toward assessing flow diverter efficacy is motivated by the clinical hypothesis that these stents lead to a decoupling of the aneurysm from the bloodstream in the parent artery and therefore to reduced pulsatility patterns of contrast agent within the aneurysm sac. The approach we proposed recently therefore aims at the quantification of pulsatility of intra-aneurysmal hemodynamic brightness patterns. The subsequent summary of this method follows the more detailed presentation in Ref. [38].

In order to quantify the pulsatility exhibited by the intra-aneurysmal flow, a ROI was placed on top of the aneurysm in each frame of the DSA series. The resulting TCC was then transformed into Fourier space by computing its discrete Fourier transform (DFT). After that, its power spectral density (PSD [44]) within a properly selected frequency band around the patient's heart rate was estimated.

In other words, the method is based on determining the energy of the signal given by the TCC that corresponds to a frequency range located around the patient's heart rate. We refer to Ref. [38] for a discussion of appropriate DSA acquisition rates based on the patient's heart rate and Nyquist's sampling theorem [44].

We applied the periodogram estimator [44] to determine the energy of the TCC within the selected frequency range and declared its pulsatility $P = P(f_l, f_h)$ with respect to this range as average quantity as follows:

$$P(f_l, f_h) = \overline{\sum_{k=\lfloor \frac{f_l}{\Delta f} + \frac{1}{2} \rfloor}^{\lfloor \frac{f_h}{\Delta f} + \frac{1}{2} \rfloor} \hat{\Phi}_P(f_k)}, \tag{13.11}$$

where f_l and f_h denote the lower and upper bounds of the selected frequency range, respectively, Δf represents the bin width of the DFT (i.e., $\Delta f = F_s/N$, where F_s is the temporal sampling frequency of the TCC and $N + 1$ is the number of equidistant sampling points of the TCC along its time axis), and $f_k = k\Delta f$.

$\hat{\Phi}_P$ refers to the periodogram estimator of the TCC given by

$$\hat{\Phi}_P(f) = \frac{1}{N} \left| \sum_{n=1}^{N} \mathrm{TCC}_n \mathrm{e}^{-2\pi i f n} \right|^2, \tag{13.12}$$

see Ref. [44].

[4]Venous blood flow is typically much slower due to the significantly lower venous pressure gradients and the relatively large calibers of veins.

Finally, the pulsatility ratio R_P was defined as the ratio of the pulsatility P after (i.e., P^{post}) and before (i.e., P^{pre}) flow diverter deployment, i.e.,

$$R_P = \frac{P^{post}}{P^{pre}}. \quad (13.13)$$

Hence, $R_P < 1$ indicated a decrease in pulsatility. Note that the bounds f_l and f_h of the frequency range were omitted in Eq. (13.13). In our experiments [38], they were chosen to be the same for determining P^{pre} and P^{post}. In general, however, they may vary to account for a change in the patient's heart rate during the intervention.[5] In order to compute the pulsatility ratio R_P according to Eq. (13.13), appropriate ROIs need to be selected in the DSA series acquired before and after flow diverter implantation.

Fig. 13.4 illustrates the steps of the proposed method. Note that the TCCs depicted in Fig. 13.4 were preprocessed properly. This means that the baselines—obtained by low-pass filtering the original TCCs—were subtracted such that only those frequencies attributed to the patient's cardiac cycle were left. Besides, the resulting curves were divided by the mean of the original TCC to account for variations in the amount of dye injected into the parent artery, and a Hamming window function was applied to mitigate spectral leakage effects. Fig. 13.4 illustrates the idealized situation that the frequency components around the patient's heart rate are reduced considerably by the deployment of one or more flow-diverting stents.

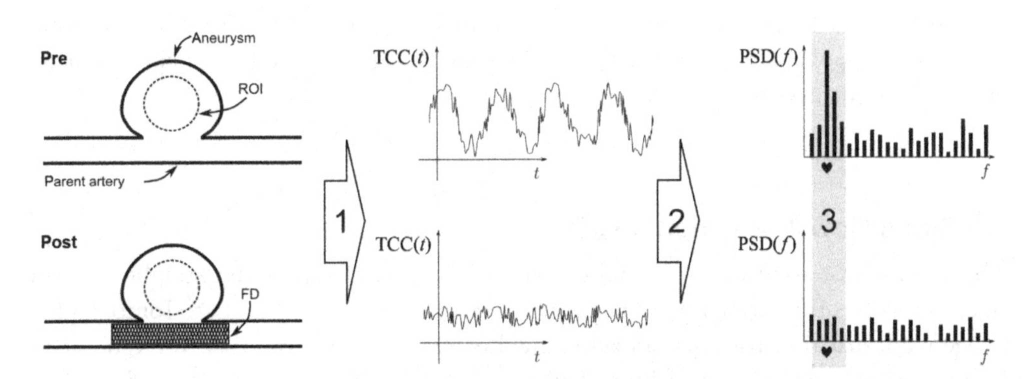

Fig. 13.4 Fourier-based analysis of intra-aneurysmal pulsatility patterns: two regions of interest (ROIs) are delineated in DSA image series acquired pre- and postflow diverter treatment. Time-contrast curves (TCCs) are extracted from both ROIs and preprocessed (1). The power spectral density (PSD) is estimated for each TCC (2). PSD estimates are compared within a certain frequency range around the heart rate (3).

[5] This would require the heart rates to be stored with the DSA series, which is typically not the case in today's clinical practice.

The same process of estimating a pulsatility ratio can be repeated for corresponding ROIs on the parent artery in the pre- and post-DSA series, which should be located proximal to the aneurysm. In order to enhance the accuracy and the predictive value of the proposed method, this proximal pulsatility ratio can then be incorporated into the comparison as a normalization factor to account for variations in the injection profile or the patient's physiological blood flow.

Furthermore, the aneurysmal ROIs may be rasterized in order to eliminate global averaging effects which are caused by phase differences of TCCs corresponding to individual pixels of the selected ROI [38]. As is the case for the optical flow approach, the injection and imaging parameters should of course also remain constant from pre to post.

The pulsatility-based metric for the assessment of flow diverter efficacy was evaluated using 13 pairs of pre-/post-DSA series. The aneurysms that were treated using flow-diverting stents were characterized by a variety of locations, types, and sizes. The results were presented and discussed in detail [38]. In summary, they support the clinical hypothesis that flow diversion leads to a reduction of pulsatility of intra-aneurysmal hemodynamics.

However, it could be observed that these results may be drastically compromised by vascular overlap. Therefore, analogous to the optical flow approach, properly chosen C-arm angulations are mandatory in order to achieve meaningful pulsatility ratio estimates.

Likewise, further clinical studies are required in order to determine whether the proposed pulsatility metric can reliably support clinical decision making during endovascular aneurysm treatment.

2.3 Blood Flow Assessment in 3D

The quantitative assessment of cerebral blood flow in 3D using angiographic C-arm systems involves the tomographic reconstruction of a vascular 3D dataset. Tomographic image reconstruction has been an active field of research for more than three decades. Besides medical imaging modalities that are not based on the application of ionizing radiation (e.g., MRI scanners), a variety of modalities is based on tomographic reconstruction from X-ray projection images, such as CT and CBCT. Typical reconstruction algorithms cover analytic as well as iterative approaches, which can further be subdivided into algebraic and statistical methods. For a more detailed discussion of image reconstruction algorithms, we refer to Ref. [45].

So far, a wide variety of methods toward the interventional reconstruction and the successive assessment of blood flow in 3D using angiographic C-arm devices have been proposed. These approaches may be categorized by the number of individual scans and contrast agent injections that are required, by the complexity of the physical models they are based on, and by the variety of physical flow parameters they can deliver.

The method presented in Ref. [46] is based on the combination of a 2D DSA images series and a static 3D image of the cerebral vasculature. An optical flow algorithm was used to estimate flow velocities in the image plane from the 2D DSA series, which were then backprojected into the 3D representation of the vessel tree. Afterwards, the resulting 3D flow fields were combined with estimates of the cross-sectional areas of the vessels in order to determine volumetric flow rates. Hence, this method required two different scans (i.e., a CBCT scan for the reconstruction of the static 3D image and a 2D DSA series) as well as two contrast agent injections. In addition, an accurate 2D/3D image registration method was needed to align the datasets based on the subsequent acquisitions.

A comparable algorithm was provided in Ref. [47]. This method, however, did not rely on optical flow estimates. Rather, the authors suggested reconstructing bolus arrival times in 3D by reprojecting (i.e., forward projecting) vessel segments onto the detector plane and retrieving the temporal information in 3D from the respective target pixels' TCCs. Likewise, an efficient image registration algorithm was employed to account for inaccuracies due to patient motion.

Another approach similar to the two previous ones was described in Ref. [48]. In this work, projected velocity estimates were determined along the centerlines of vessel segments that had been segmented in the 2D DSA series. A cross-correlation approach was used to robustly determine bolus transit times despite highly pulsatile flow patterns [31]. The projected velocities were then backprojected into a previously reconstructed vascular 3D dataset, where a graph-based approach was used to resolve ambiguities due to vessel overlap. The primary objective of this research on flow velocity estimation was to determine patient-specific boundary conditions for computational fluid dynamics (CFD) simulations.

Another similar approach was presented in Ref. [49]. In this chapter, the authors proposed a method that was based on a 3D image of the brain vasculature as well as a biplane 2D DSA dataset, i.e., two 2D DSA series acquired simultaneously from different viewing directions. The advantage of using biplane DSA over the use of monoplane DSA was that accuracy issues due to vascular overlap could be improved. The time-dependent filling of the vasculature with contrast medium was then determined by minimizing a functional that covered the perspective mappings of the 3D vessel geometry onto the two image planes of the biplane system, along with appropriate modeling assumptions using blood flow and dye propagation. As is the case for the two aforementioned approaches outlined in Refs. [46, 47], two separate acquisitions and two separate contrast agent injections were required. Accurate 2D/3D registration was mandatory once again.

Finally, the method described in Ref. [50] is characterized by a set of mathematical equations that describe the physical and physiological properties of blood flow and contrast agent transport. The parameters governing the underlying model were numerically optimized such that the resulting propagation of the contrast agent through the patient's vasculature matched the acquired projection images as best as possible. The approach is based on subtracted projection images from two successive rotational scans of the C-arm scanner, i.e., two successive CBCT scans. This means that the subtracted projection images were eventually obtained by subtracting the projection images of a native mask scan from the respective projection images based on a contrast-enhanced scan; the so-called fill scan. There are two obvious advantages of this method presented in Ref. [50]. First, only one contrast injection was needed and, second, there was no 2D/3D registration step required. Clinical results based on this method were presented afterwards in Ref. [51].

In comparison to the previously mentioned approaches toward 3D flow reconstruction, the method discussed in Ref. [50] is characterized by a physically and physiologically motivated model of blood flow and dye propagation. Consequently, this method does not only yield the time-dependent concentration of contrast agent within the patient's brain vasculature. It also yields additional flow parameters such as a description of the patient's periodic cardiac activity, flow velocities, and volumetric flow rates within the vessel segments under consideration.

Before elaborating more on model-based flow estimation and dye propagation using virtual angiography, we will first concentrate on another novel flow reconstruction approach we have developed recently, named 4D DSA. This method is purely image-based by construction and does not rely on any simplifying physical and physiological flow model. As a consequence, the 4D DSA method can be considered inherently patient-specific.

2.3.1 4D DSA—Generation of Time-Resolved Vascular Volumes

4D DSA refers to a novel angiographic imaging method for approximately reconstructing time-resolved series of vascular volumes using C-arm scanners. The term "4D" thus refers to 3D plus time. The fundamental idea is that, based on prior knowledge, the time-resolved series of vascular 3D datasets is generated from undersampled input data. The prior knowledge consists of an initially reconstructed vascular 3D dataset, which is then used to constrain the generation of the set of temporal volumes. Therefore, this 3D dataset is also referred to as the constraining volume (or constraint volume). A comprehensive discussion of the 4D DSA method can be found in Refs. [52–54]. Similar approaches were presented for CT imaging as well as for MRI. We again refer to the survey article [52] and to the rich body of literature referenced therein.

Fig. 13.5 shows an illustration of the 4D DSA method. A clear advantage of this novel 4D vascular imaging technology is that it can provide any view at a series of successive time instances. This is particularly true for viewing directions (i.e., angulations of the C-arm) that are mechanically unreachable.

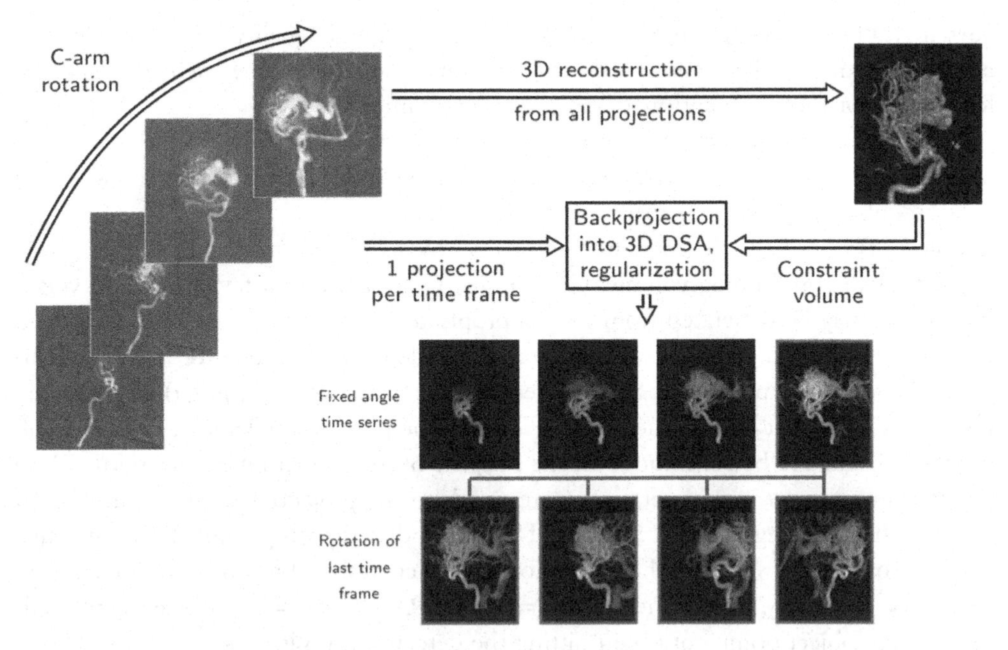

Fig. 13.5 Principle of the 4D DSA method: first, a static 3D vascular dataset is reconstructed. Afterwards, the log-subtracted projection images are revisited, and the time series of vascular volumes is generated.

As a consequence, the 4D DSA method has the potential of reducing the number of 2D DSA acquisitions that are needed to determine the best possible working projection for diagnosis, surgical planning, and treatment. This would lead to reductions in X-ray dose and contrast agent load. However, further clinical studies are required in order to corroborate this hypothesis.

Data Acquisition. In the current 4D DSA implementation, the data acquisition is based on a 3D DSA scan protocol that comprises two rotational runs of the angiographic C-arm device: the mask run and the fill run. The contrast injection is timed such that the inflow of contrast agent can be captured in the projection images of the fill run. Alternative scan protocols that consist solely of a contrast-enhanced rotational run of the C-arm and use suitable bone removal and vessel enhancement techniques in order to represent vascular data only may be used in the future as well.

The angular scan interval is chosen such that a sufficiently large number of projection images can be acquired over an adequate time interval: typically about 260 degree due to mechanical limitations of today's conventional C-arm scanners.[6] As a by-product of our

[6]From the standpoint of CT reconstruction theory, a scan range of 180 degree plus the fan angle of the X-ray beam is sufficient to reconstructed a (2D) object [45]. For today's angiographic C-arm devices, this leads to angular ranges of about 200 degree.

current 4D DSA software prototype implementation, a high-quality 3D vascular dataset is generated during the reconstruction of the time-resolved volumes, which eventually leads to the previously mentioned constraining volume.

Algorithmic Aspects. The basic algorithmic block of the 4D DSA method that generates the time-resolved series of vascular volumes based on an initially reconstructed static vascular volume can formally be written as follows.

Let $I : \mathbf{R}^3 \rightarrow \mathbf{R}$ be the initially reconstructed vascular 3D image and $C : \mathbf{R}^3 \rightarrow \mathbf{R}$ the respective constraining volume. Commonly, C contains even fewer nonzero voxels than I and may be generated from I by appropriate thresholding and noise reduction, for instance, in order to isolate vascular voxels. In addition, we use $p : \mathbf{R}^2 \times \mathbf{N} \rightarrow \mathbf{R}$ to refer to the time-dependent log-subtracted projection images, i.e., $p(\vec{u}, t)$ refers to the projection point that corresponds to the discrete time point t and detector coordinate $\vec{u}$, which is assumed to be continuous for the purpose of this algorithmic description. Note that the time point t in our notation corresponds to the projection angle of the C-arm device. The mapping $A : \mathbf{R}^3 \times \mathbf{N} \rightarrow \mathbf{R}^2$ is introduced to represent the perspective mapping of the 3D space to the projection image corresponding to time point t, i.e., $A(\vec{x}, t) = \vec{u}$. Finally, we define $L(\vec{u}, t) = \{\vec{x} \in \mathbf{R}^3; A(\vec{x}, t) = \vec{u}\}$ as the X-ray path through the object point $\vec{x}$ at time t hitting the detector at position $\vec{u}$.

Using these definitions, the basic scheme of the 4D DSA method for generating a time series of vascular volumes $V(\vec{x}, t)$ is given by

$$V(\vec{x}, t) = C(\vec{x}) \frac{p(A(\vec{x}, t), t)}{\int_{L(A(\vec{x},t),t)} I(y) \mathrm{d}y}. \tag{13.14}$$

Note that Eq. (13.14) corresponds to a multiplicative perspective backprojection of normalized projection values into the constraining image C. The normalization of the projection values $p(A(\vec{x}, t), t)$ is accomplished by introducing the denominator in Eq. (13.14). This normalization step accounts for the proper scaling of the values stored in the constraining image $C(\vec{x})$ independent of the path length of the respective ray through the vasculature from the X-ray focal spot to the detector plane.

Eq. (13.14) further implies that the spatial resolution of the constraining image $C(\vec{x})$ will be preserved by the time series $V(\vec{x}, t)$ of vascular volumes; see again [54].

Obviously, the quality of the time-resolved series of vascular volumes will be compromised due to vascular overlap. The filling states of overlapping vessels can only be restored approximately when using limited angular ranges of projection images. Heuristic approaches based on the interpolation of voxel-specific time-attenuation curves (TACs) may be used in order to mitigate this effect (referred to as regularization in Fig. 13.5). Therefore, the quality of the generated 4D dataset—its temporal resolution, in particular—depends on the sparsity of the vessel tree under consideration [54].

From the standpoint of image reconstruction theory in CT [45, 55], it is important to point out that 4D DSA is an approximative algorithm for generating time-resolved vascular data in 3D. This is for several reasons.

First, due to the circular sampling trajectory, Tuy's data sufficiency condition is violated and the object therefore cannot be reconstructed exactly [45, 55]. According to Tuy's condition, every plane that intersects the irradiated object must contain an X-ray focal point such that the object can be reconstructed exactly.[7] Apparently, this intersection criterion is not true for those planes that are parallel to the mid-plane (i.e., the plane in which the X-ray source rotates along its circular path), but not equal to it. Note that, as a particular consequence of Tuy's condition, the widely used Feldkamp-Davis-Kress (FDK) method does not represent a theoretically exact cone-beam reconstruction algorithm.

Second, due to the propagation of dye during the rotational fill run, the acquired projection images are highly inconsistent. In the early projection images, only arteries are enhanced, while venous structures will typically be filled with contrast agent toward the end of the fill run. Consequently, the initially reconstructed 3D vascular dataset and therefore the constraining volume may be impaired by artifacts such as streaks, for example. This is essentially based on the fact that not all vascular structures are irradiated from all projection directions while they are filled with contrast agent. This is a general issue in X-ray CT of vascular structures based on the administration of iodinated contrast agent, which is not just the case for brain imaging.

Third, the previously mentioned issues due to vascular overlap will compromise the accuracy of the time series of volumes. Sophisticated interpolation schemes in order to close gaps in a vascular voxel's TAC that are characterized by incorrect X-ray attenuation data due to vessel overlap can enhance the accuracy of the results. However, the reconstruction of highly accurate temporal volumes cannot be accomplished for the case of severe vascular overlap using a limited number of projection images only [54].

Clinical Applications. First applications of the 4D DSA method focus on the assessment of cerebral vascular disorders. Ongoing clinical research projects concentrate on the application of 4D DSA for assessing the blood flow in brain AVMs and DAVFs, among others.

For example, in order to guide catheter-based AVM embolization procedures, it is important to understand its angioarchitecture in full detail, which covers the temporal behavior of its arterial feeders and draining veins, as well as the structure of the nidus, e.g., the occurrence of intranidal aneurysms which are associated with enhanced bleeding risk. As an example, Fig. 13.6 illustrates a 4D DSA dataset of a brain AVM at five successive time instances.

See Ref. [54] for further early clinical results. In addition, a recent comment published in a neurosurgery journal also highlights the potential of the 4D DSA method [56].

[7] Intuitively, Tuy's condition ensures that all plane integrals that are needed for the computation of the inverse Radon transform of the irradiated object can be computed from the acquired X-ray projection data. See again [45, 55] for details.

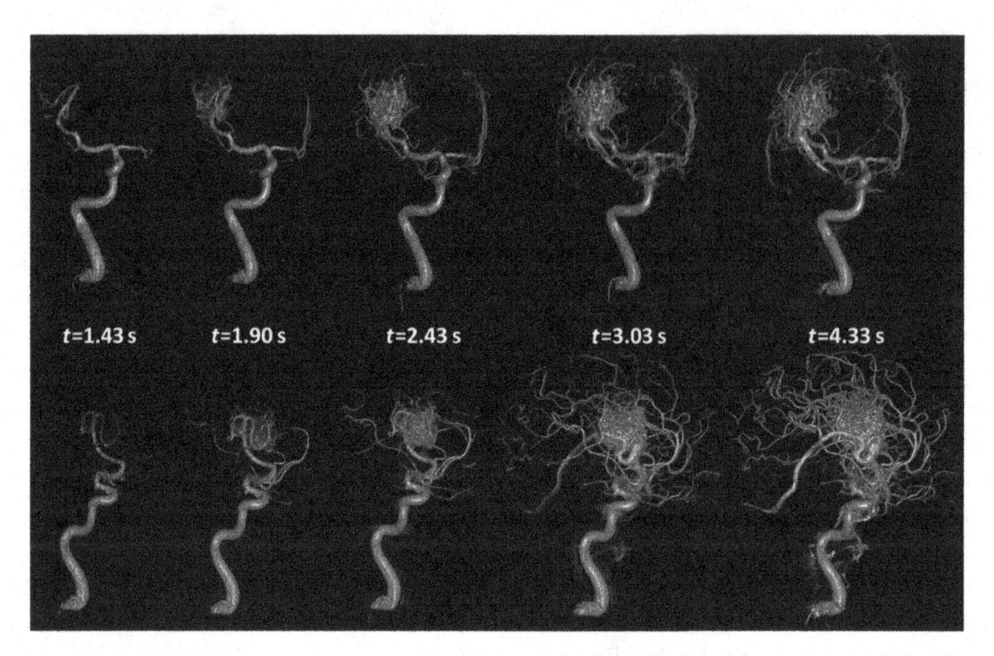

Fig. 13.6 Example of five successive time instances of a 4D DSA dataset of a brain arteriovenous malformation (AVM): the images in the *top row* refer to a different projection direction (anteroposterior view) than the images in the *bottom row* (lateral view).

Parametric Color Coding. 3D parametric datasets can be derived from time-resolved series of vascular volumes. The resulting quantities can be color-coded correspondingly. As an example, Fig. 13.7 shows two views of a color-coded (gray scale) representation of the bolus arrival times of the 4D dataset presented in Fig. 13.6. For each voxel, its bolus arrival time was defined as the first point in time at which its time-attenuation curve (TAC) reaches one-third of its peak. The resulting parametric volume was then rendered appropriately. Fig. 13.7 is based on a ray casting algorithm using a Phong lighting model to improve 3D perception [57]. Empirical vessel segmentation along the rays was implemented in order to ensure that the first vessel hit by a ray determined the color of the respective image pixel, thus reducing artifacts due to vascular overlap. Note that the color allows for easy delineation of the draining veins shown, which may help the physician to identify these important structures and consider them during the planning of the AVM treatment.

Future research on 4D DSA may concentrate on the determination of time-resolved 3D velocity fields. These flow fields can—together with estimates of cross-sectional areas of the respective vessel segments—be used to determine time-resolved volumetric flow rates. Clinically, this may lead to improved efficacy assessment of a variety of

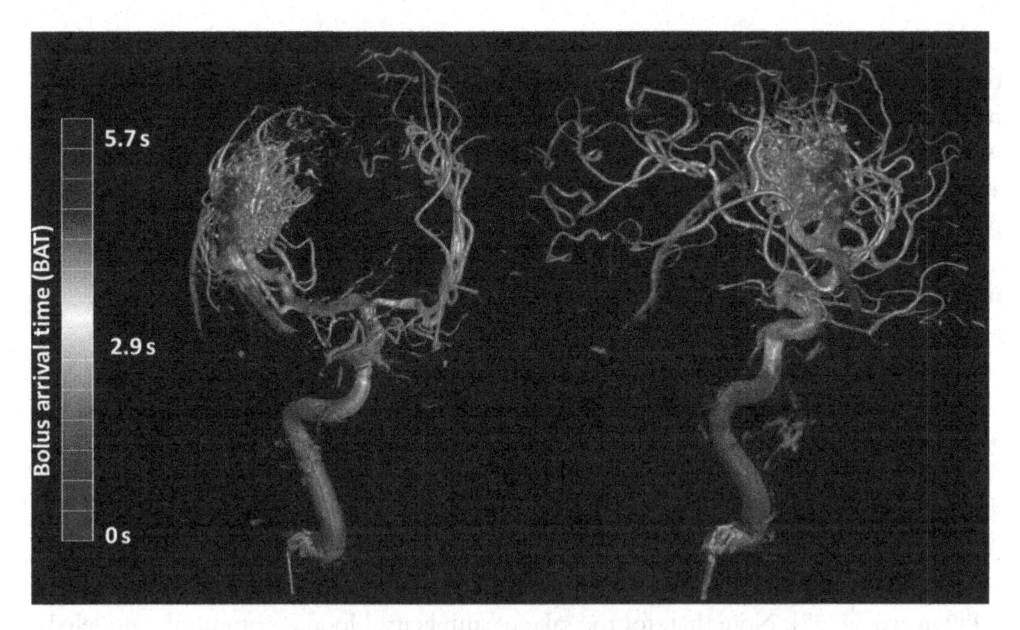

Fig. 13.7 Color-coded (gray scale) representation of bolus arrival times based on the 4D AVM dataset shown in Fig. 13.6: (*left*) anteroposterior view; (*right*) lateral view.

endovascular interventions, e.g., stenting procedures. See Ref. [58] for more details on an approach based on the estimation of dense 3D velocity fields using an optical flow algorithm as well as first results.

2.3.2 Computational Fluid Dynamics (CFD)

Blood flow assessment techniques using numerical CFD simulation represent a family of approaches that are characterized by models of the patient's physiology, blood flow, and contrast agent transport. These models are commonly given by a set of mathematical equations that need to be solved numerically in order to determine the required physical quantities. Depending on the complexity of the flow model, CFD-based approaches yield a variety of parameters such as velocity fields, pressure, wall shear stress, etc.

The simulation of hemodynamics is commonly based on the numerical treatment of the unsteady incompressible Navier-Stokes equations in 3D, given by

$$\rho\left(\frac{\partial \vec{u}}{\partial t} + (\vec{u}\cdot\nabla)\vec{u}\right) = -\nabla p + \mu\Delta\vec{u} + F, \tag{13.15}$$

which is referred to as the momentum equation and essentially relates pressure gradients to velocity fields, and

$$\nabla\cdot\vec{u} = 0, \tag{13.16}$$

which is referred to as the continuity equation for incompressible fluids (i.e., density changes are negligible). As usual, the operator $\cdot$ denotes the scalar product (inner product). Additionally, $\vec{u} = \vec{u}(\vec{x}, t)$ represents the time-varying velocity field, $p = p(\vec{x}, t)$ is the pressure, ρ and μ stand for the density and the viscosity of the blood, respectively, and the term $F = F(\vec{x}, t)$ summarizes all external forces, e.g., gravity as well as other potentially time-dependent effects.

Note that the operators ∇ and Δ in Eqs. (13.15), (13.16) refer to spatial dimensions only, i.e.,

$$\nabla = \left(\frac{\partial}{\partial x}, \frac{\partial}{\partial y}, \frac{\partial}{\partial z}\right)^{\mathrm{T}} \tag{13.17}$$

and

$$\Delta = \frac{\partial^2}{\partial x^2} + \frac{\partial^2}{\partial y^2} + \frac{\partial^2}{\partial z^2}. \tag{13.18}$$

For further details, we refer to the textbook [59] as well as the brief introduction to CFD in article [27]. Note that, for the sake of simplicity, blood is commonly modeled as a continuous Newtonian fluid, cf. [60] for comments on extensions of these simplifying assumptions.

Appropriate boundary conditions are further needed in order to solve Eqs. (13.15), (13.16), e.g., time-varying velocity profiles at the inlets as well as time-varying pressure values at the outlets of the vascular territory under consideration. See again [59] for a general introduction to fluid mechanics and [61] for an introduction to numerical methods for fluid simulation.

The computational domain of the flow simulation is defined by the patient's vascular segment under consideration. Due to the resulting high spatial resolution, CBCT is today's method of choice for imaging a patient's brain vasculature. The boundary conditions for the CFD simulation as well as the density and the viscosity of the blood—cf. Eq. (13.15)—are either taken from the literature or directly measured in the patient. It is obvious that patient-specific simulation parameters will generally lead to more accurate and realistic computational results than average quantities taken from the literature.

CFD Simulations in Brain Aneurysms and Virtual Stenting. The analysis of hemodynamics in cerebral aneurysms and the assessment of their rupture risks represent a major clinical CFD research topic. We refer to the overview provided in Ref. [60] and the references listed therein. An analysis of the statistical correlation between computational results based on CFD and aneurysmal rupture events was recently presented in Ref. [62]. The authors came to the conclusion that ruptured aneurysms tended to be characterized by complex unstable flow patterns, whereas unruptured aneurysms exhibited simple stable flow patterns. However, biological effects, which also play a crucial role in the assessment of aneurysmal rupture, are not yet covered explicitly by today's simulation techniques. For algorithmic approaches toward the visualization of simulated flow data and the respective enhancement of their perception, we refer to the textbook [57] and also the recent publication [63].

Enhancements of CFD methods for cerebral aneurysms cover the simulation of the effect of flow-diverting stents. This approach is often referred to as virtual stenting. It enables the simulation of the treatment of vascular pathologies and, to some extent, the prediction of the treatment success. Using virtual stenting, the physician has the possibility to simulate upfront the treatment of a cerebral aneurysm using a flow-diverting stent in order to figure out what type and size of stent to deploy, and how to implant it properly across the neck of the aneurysm in order to achieve the best possible flow diversion result. We refer to Refs. [64, 65] for descriptions of how to model the individual struts of a flow diverter that is placed within the parent vessel. In contrast, the approach described in Ref. [66] is based on a more macroscopic flow diverter modeling technique, which uses a representation of the stent as a porous medium and thus reduces the complexity of the geometric model and the subsequent numerical simulation.

Note that such modeling and simulation techniques are still far from being applicable in clinical practice. Further validation studies are required in order to establish such an approach in clinical decision making.

Dye Transport Simulation and Virtual Angiography. Available CFD software packages can be used to simulate blood flow, particularly in cerebral aneurysms. The computed velocity fields are then used in order to simulate the propagation of the contrast agent. Afterwards, the resulting time-dependent distributions of the contrast agent are forward projected in order to generate DSA-like image sequences. This approach is commonly referred to as virtual angiography [27]. For the sake of simplicity, the contrast agent is modeled as an ensemble of massless and dimensionless particles that are injected into and advected with the pulsatile bloodstream. In addition to this advective transport of dye, we also model contrast agent diffusion in the blood by estimating continuous concentration gradients from the discrete particle distributions. Based on these concentration gradients and the resulting diffusive forces, the particles are relocated correspondingly. This leads to a numerical time stepping scheme that consists of alternating particle motion steps due to advection and diffusion. In addition to this basic simulation approach, the additional modeling of gravity can further improve the similarity of real (i.e., acquired) and virtual (i.e., simulated) angiographic images. This further allows us to extract patient- and injection-specific parameters (e.g., heart rate, mean inflow velocity, bolus injection profile) from 2D DSA image series that are acquired in addition to the 3D vascular datasets needed to define the computational domains. The use of these parameters generally leads to more accurate simulations of blood flow and dye propagation.

Virtual angiography has various applications. First, it can be used to visualize CFD results in a way that the physician is familiar with, i.e., as DSA-like image series. The comparison of such simulated DSA series with real DSA series represents an important step toward the validation of CFD simulation methods. Fig. 13.8 is taken from Ref. [27] and illustrates the virtual angiography workflow for validating CFD results. In particular, Fig. 13.8 demonstrates what parameters for the CFD simulation and the subsequent virtual angiography stage are retrieved from the input datasets (vascular geometry, heart rate, etc.).

Additionally, once the user has gained confidence in the accuracy of virtual angiography images, an arbitrary number of simulated DSA series can be computed and viewed. In particular, these simulated DSA series may correspond to viewing directions that are mechanically unreachable and, therefore, cannot be acquired as real 2D DSA series.

Second, virtual angiography can be used to generate test datasets for quantitative flow assessment methods such as 4D DSA, for example. The advantage is that, for the case of simulated input data, the ground truth flow is known and can thus be compared against the result that is delivered by the algorithmic flow assessment technique under investigation.

Finally, virtually angiography can play an role in the context of virtual stenting. It may be employed in order to generate DSA-like image sequences that visualize the aneurysmal flow patterns pre- and poststenting such that the interventionalist can get an impression of the flow diversion effect to be expected from the respective treatment

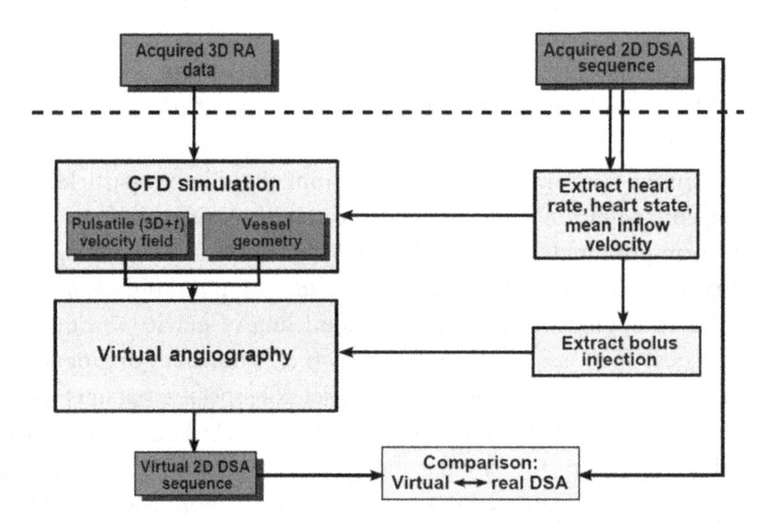

Fig. 13.8 Virtual angiography workflow for the validation of CFD results by comparing virtual angiograms with real 2D DSA series.

option. In particular, methods for assessing the efficacy of flow-diverting stents discussed in previous sections may eventually be applied in order to assess virtual angiograms quantitatively and thus to predict the clinical outcome of the simulated flow diverter treatment.

REFERENCES

[1] Jolesz FA, editor. Intraoperative imaging and image-guided therapy. Berlin, Germany: Springer; 2014.
[2] Peters T, Cleary K, editors. Image-guided interventions: technology and applications. New York: Springer; 2008.
[3] Miles KA, Griffith MR. Perfusion CT: a worthwhile enhancement? Br J Radiol 2003;76(904): 220–31.
[4] Schellinger PD, Fiebach JB. Stellenwert moderner CT-Techniken bei der Diagnostik des akuten Schlaganfalls. Der Radiologe 2004;44:380–8.
[5] Kernan WN, Ovbiagele B, Black HR, Bravata DM, Chimowitz MI, Ezekowitz MD, et al. Guidelines for the prevention of stroke in patients with stroke and transient ischemic attack: a guideline for healthcare professionals from the American Heart Association/American Stroke Association. Stroke 2014;45:2160–236.
[6] Saver JL. Time is brain—quantified. Stroke 2006;37(1):263–6.
[7] Fieselmann A, Kowarschik M, Ganguly A, Hornegger J, Fahrig R. Deconvolution-based CT and MR brain perfusion measurement: theoretical model revisited and practical implementation details. Int J Biomed Imaging 2011;2011 [Article ID 467563].
[8] Manhart M, Kowarschik M, Fieselmann A, Deuerling-Zheng Y, Royalty K, Maier AK, et al. Dynamic iterative reconstruction for interventional 4-D C-arm CT perfusion imaging. IEEE Trans Med Imaging 2013;32(7):1336–48.
[9] Manhart M, Aichert A, Struffert T, Deuerling-Zheng Y, Kowarschik M, Maier AK, et al. Denoising and artefact reduction in dynamic flat detector CT perfusion imaging using high speed acquisition: first experimental and clinical results. Phys Med Biol 2014;59:4505–24.
[10] Al-Shahi R, Warlow C. A systematic review of the frequency and prognosis of arteriovenous malformations of the brain in adults. Brain 2001;124:1900–26.
[11] Spetzler RF, Martin NA. A proposed grading system for arteriovenous malformations. J Neurosurg 1986;65(4):476–83.
[12] van Rooij WJ, Sluzewski M, Beute GN. Brain AVM embolization with onyx. Am J Neuroradiol 2007;28(1):172–7.
[13] Back AG, Zeck O, Shkedy C, Shedden P. Staged embolization with staged gamma knife radiosurgery to treat a large AVM. Can J Neurol Sci 2009;36(4):500–3.
[14] Gupta AK, Periakaruppan AL. Intracranial dural arteriovenous fistulas: a review. Indian J Radiol Imaging 2009;19:43–8.
[15] Cognard C, Gobin YP, Pierot L, Bailly AL, Houdart E, Casasco A, et al. Cerebral dural arteriovenous fistulas: clinical and angiographic correlation with a revised classification of venous drainage. Radiology 1995;194(3):671–80.
[16] Brisman JL, Song JK, Newell DW. Cerebral aneurysms. N Engl J Med 2006;355(9):928–39.
[17] Suarez JI, Tarr RW, Selman WR. Aneurysmal subarachnoid hemorrhage. N Engl J Med 2006;354(4):387–96.
[18] Molyneux AJ, Kerr RS, Birks J, Ramzi N, Yarnold J, Sneade M, et al. Risk of recurrent subarachnoid haemorrhage, death, or dependence and standardised mortality ratios after clipping or coiling of an intracranial aneurysm in the International Subarachnoid Aneurysm Trial (ISAT): long-term follow-up. Lancet Neurol 2009;8(5):427–33.
[19] Bing F, Darsaut T, Salazkin I, Makoyeva A, Gevry G, Raymond J. Stents and flow diverters in the treatment of aneurysms: device deformation in vivo may alter porosity and impact efficacy. Neuroradiology 2013;55(1):85–92.

[20] Darsaut TE, Bing F, Salazkin I, Gevry G, Raymond J. Flow diverters can occlude aneurysms and preserve arterial branches: a new experimental model. Am J Neuroradiol 2012;33(10):2004–9.
[21] Oppelt A, editor. Imaging systems for medical diagnostics: fundamentals, technical solutions and applications for systems applying ionizing radiation, nuclear magnetic resonance and ultrasound. Berlin, Germany: Publicis; 2005.
[22] Ganz W, Swan HJC. Measurement of blood flow by thermodilution. Am J Cardiol 1972;29(2):241–6.
[23] Kramme R, Hoffmann K-P, Pozos R, editors. Springer handbook of medical technology. Berlin: Springer; 2011.
[24] Markus HS. Transcranial Doppler ultrasound. Br Med Bull 2000;56(2):378–88.
[25] Meijering EH, Niesssen WJ, Viergever MA. Retrospective motion correction in digital subtraction angiography: a review. IEEE Trans Med Imaging 1999;18(1):2–21.
[26] Deuerling-Zheng Y, Lell M, Galant A, Hornegger J. Motion compensation in digital subtraction angiography using graphics hardware. Comput Med Imaging Graph 2006;30(5):279–89.
[27] Endres J, Kowarschik M, Redel T, Sharma P, Mihalef V, Hornegger J, et al. A workflow for patient-individualized virtual angiogram generation based on CFD simulation. Comput Math Methods Med 2012; [Article ID 306765].
[28] Strother CM, Bender F, Deuerling-Zheng Y, Royalty K, Pulfer KA, Baumgart J, et al. Parametric color coding of digital subtraction angiography. Am J Neuroradiol 2010;31(5):919–24.
[29] Lin C-J, Hung C-B, Guo W-Y, Chang F-C, Luo C-B, Beilner J, et al. Monitoring peri-therapeutic cerebral circulation time: a feasibility study using color-coded quantitative DSA in patients with steno-occlusive arterial disease. Am J Neuroradiol 2012;33(9):1685–90.
[30] Lin C-J, Luo C-B, Guo W-Y, Chang F-C, Beilner J, Kowarschik M, et al. Application of color-coded digital subtraction angiography in treatment of carotid-cavernous fistulas: initial experience. J Chin Med Assoc 2013;76(4):218–24.
[31] Shpilfoygel SD, Close RA, Valentino DJ, Duckwiler GR. X-ray videodensitometric methods for blood flow and velocity measurement: a critical review of literature. Med Phys 2000;27(9):2008–23.
[32] Lieber BB, Stancampiano AP, Wakhloo AK. Alteration of hemodynamics in aneurysm models by stenting: influence of stent porosity. Ann Biomed Eng 1997;25(3):460–9.
[33] Ahmed AS, Deuerling-Zheng Y, Strother CM, Pulfer KA, Zellerhoff M, Redel T, et al. Impact of intra-arterial injection parameters on arterial, capillary, and venous time-concentration curves in a canine model. Am J Neuroradiol 2009;30(7):1337–41.
[34] Augsburger L, Farhat M, Reymond P, Fonck E, Kulcsar Z, Stergiopulos N, et al. Effect of flow diverter porosity on intraaneurysmal blood flow. Clin Neuroradiol 2009;19(3):204–14.
[35] Pierot L. Flow diverter stents in the treatment of intracranial aneurysms: where are we? J Neuroradiol 2011;38(1):40–6.
[36] Sfyroeras GS, Dalainas I, Giannakopoulos TG, Antonopoulos K, Kakisis JD, Liapis CD. Flow-diverting stents for the treatment of arterial aneurysms. J Vasc Surg 2012;56(3):839–46.
[37] Struffert T, Ott S, Kowarschik M, Bender F, Adamek E, Engelhorn T, et al. Measurement of quantifiable parameters by time-density curves in the elastase-induced aneurysm model: first results in the comparison of a flow diverter and a conventional aneurysm stent. Eur Neuroradiol 2013;23(2):521–7.
[38] Benz T, Kowarschik M, Endres J, Redel T, Demirci S, Navab N. A Fourier-based approach to the angiographic assessment of flow diverter efficacy in the treatment of cerebral aneurysms. IEEE Trans Med Imaging 2014;33(9):1788–802.
[39] Horn BKP, Schunck BG. Determining optical flow. Artif Intell 1981;17(1):185–203.
[40] Wildes RP, Amabile MJ, Lanzillotto AM, Leu TS. Recovering estimates of fluid flow from image sequence data. Comput Vis Image Underst 2000;80(2):246–66.
[41] Liu T, Shen L. Fluid flow and optical flow. J Fluid Mech 2008;614:253–91.
[42] Corpetti T, Mémin E, Pérez P. Dense estimation of fluid flows. IEEE Trans Pattern Anal Mach Intell 2002;24(3):365–80.
[43] Pereira VM, Bonnefous O, Ouared R, Brina O, Stawiaski J, Aerts H, et al. A DSA-based method using contrast-motion estimation for the assessment of the intra-aneurysmal flow changes induced by flow-diverter stents. Am J Neuroradiol 2013;34(4):808–15.

[44] Stoica P, Moses RL. Spectral analysis of signals. Upper Saddle River, NJ: Prentice Hall; 2005.
[45] Zeng GL. Medical image reconstruction: a conceptual tutorial. Berlin: Springer; 2010.
[46] Bonnefous O, Pereira VM, Ouared R, Brina O, Aerts H, Hermans R, et al. Quantification of arterial flow using digital subtraction angiography. Med Phys 2012;30(10):6264–75.
[47] Schmitt H, Grass M, Suurmond R, Köhler T, Rasche V, Hähnel S, et al. Reconstruction of blood propagation in three-dimensional rotational X-ray angiography (3D-RA). Comput Med Imaging Graph 2005;29(7):507–20.
[48] Hentschke CM, Serowy S, Janiga G, Rose G, Tönnies KD. Estimating blood flow velocity in angiographic image data. In: Wong KH, Holmes DR, editors. Medical imaging 2011: visualization, image-guided procedures, and modeling, Proceedings of SPIE medical imaging conference, vol. 7964. International Society for Optics and Photonics; 2011.
[49] Copeland AD, Mangoubi RS, Desai MN, Mitter SK, Malek AM. Spatio-temporal data fusion for 3D+T image reconstruction in cerebral angiography. IEEE Trans Med Imaging 2010;29(6):1238–51.
[50] Waechter I, Bredno J, Hermans R, Weese J, Barratt DC, Hawkes DJ. Model-based blood flow quantification from rotational angiography. Med Image Anal 2008;12(5):586–602.
[51] Groth A, Wächter-Stehle I, Brina O, Perren F, Mendes-Pereira V, Rüfenacht D, et al. Clinical study of model-based blood flow quantification on cerebrovascular data. In: Wong KH, Holmes DR, editors. Medical imaging 2011: visualization, image-guided procedures, and modeling, Proceedings of SPIE medical imaging conference, vol. 7964. International Society for Optics and Photonics; 2011.
[52] Mistretta CA. Sub-Nyquist acquisition and constrained reconstruction in time resolved angiography. Med Phys 2011;38(6):2975–85.
[53] Davis B, Royalty K, Kowarschik M, Rohkohl C, Oberstar E, Aagard-Kienitz E, et al. 4D digital subtraction angiography: implementation and demonstration of feasibility. Am J Neuroradiol 2013;34(10):1914–21.
[54] Royalty K. 4D DSA: new methods and applications for 3D time-resolved angiography for C-Arm CT interventional imaging. Ph.D. thesis, University of Wisconsin-Madison, USA, 2014.
[55] Buzug TM. Computed tomography. Berlin, Germany: Springer; 2008.
[56] Parry PV, Ducruet AF. Four-dimensional digital subtraction angiography: implementation and demonstration of feasibility. World Neurosurg 2014;81(3):454–5.
[57] Preim B, Botha C. Visual computing for medicine. 2nd ed. Waltham, MA: Morgan Kaufmann; 2013.
[58] Maday P, Brosig R, Endres J, Kowarschik M, Navab N. Blood flow quantification using optical flow methods in a body fitted coordinate system. In: Medical imaging 2014: image processing. Proceedings of SPIE, vol. 9034. 2014. p. 90340J.
[59] Munson BR, Rothmayer AP, Okiishi TH, Huebsch WW. Fundamentals of fluid mechanics. 7th ed. Hoboken, NJ: Wiley & Sons; 2012.
[60] Cebral JR, Mut F, Sforza D, Löhner R, Scrivano E, Lylyk P, et al. Clinical application of image-based CFD for cerebral aneurysms. Int J Numer Methods Biomed Eng 2011;27:977–92.
[61] Griebel M, Dornseifer T, Neunhoeffer T. Numerical simulation in fluid dynamics: a practical introduction. Philadelphia, PA: Society for Industrial and Applied Mathematics; 1997.
[62] Byrne G, Mut F, Cebral JR. Quantifying the large-scale hemodynamics of intracranial aneurysms. Am J Neuroradiol 2014;35:333–8.
[63] Oeltze S, Lehmann DJ, Kuhn A, Janiga G, Theisel H, Preim B. Blood flow clustering and applications in virtual stenting of intracranial aneurysms. IEEE Trans Vis Comput Graph 2014;20(5):686–701.
[64] Appanaboyina S, Mut F, Löhner R, Putman CM, Cebral JR. Simulation of intracranial aneurysm stenting: techniques and challenges. Comput Methods Appl Mech Eng 2009;198:3567–82.
[65] Janiga G, Rössl C, Skalej M, Thévenin D. Realistic virtual intracranial stenting and computational fluid dynamics for treatment analysis. J Biomech 2013;46(1):7–12.
[66] Augsburger L, Reymond P, Rüfenacht DA, Stergiopulos N. Intracranial stents being modeled as a porous medium: flow simulation in stented cerebral aneurysms. Ann Biomed Eng 2011;39(2): 850–63.

CHAPTER 14

Virtual Stenting for Intracranial Aneurysms

A Risk-Free, Patient-Specific Treatment Planning Support for Neuroradiologists and Neurosurgeons

P. Berg, L. Daróczy, G. Janiga
University of Magdeburg "Otto von Guericke", Magdeburg, Germany

Chapter Outline

Chapter points

- Virtual stent deployment represents a great opportunity for a risk-free treatment support of intracranial aneurysms.
- Virtual stenting approaches are versatile, covering a broad range of accuracy based on available computational resources.
- In vivo validation of the numerical methodologies is mandatory to accomplish clinical applicability.
- A proof-of-concept for a patient-individualized stent optimization is presented, demonstrating how expert knowledge and automatization can be coupled efficiently.

Computing and Visualization for Intravascular Imaging and Computer-Assisted Stenting
http://dx.doi.org/10.1016/B978-0-12-811018-8.00014-X

1. INTRACRANIAL ANEURYSMS

The prevalence of intracranial aneurysms (IAs), which are permanent dilatations of human brain arteries, is estimated to be 2–5% within the Western population. In contrast to other vessel diseases, e.g., an ischemic stroke, often young people are affected [1]. Further, up to 30% of the patients harbor more than one aneurysm [2], which even increases the number of cases, with a potential risk of rupture. Statistically, 10/10,000 IAs rupture annually, and therefore the rupture rate is relatively small compared to its prevalence [3]. However, in case an aneurysm is detected incidentally, both the physician and the patient have to decide together whether a therapy is required or not. Such a decision should be based not on general and often uncertain criteria (e.g., size, location, shape, or age), but on patient-specific factors such as hemodynamics and its impact on the diseased vessel wall need to be considered. This is of high clinical relevance since not every aneurysm requires treatment. In fact, the risk of rupture has to be carefully weighed against the therapy risk, especially for complex aneurysms. Additionally, the individual growing courses vary significantly and reliable predictions have not been possible until now. In general, two different scenarios exist. Either the diagnosed aneurysm remains constant during long-term follow-ups, or the pathophysiological processes within the vessel wall lead to further growth and consequently to possible rupture. In the case of this severe event, the following clinical course can often hardly be influenced. Despite immediate and optimal medical care, approximately 50% of the patients die within the first 30 days due to complications caused by the bleeding, such as cerebral vasospasm, hydrocephalus, or pneumonia. Among the survivors, a majority suffer from permanent neurological or cognitive deficits. Due to insufficient knowledge about the growth of an individual aneurysm as well as the media presentation of the clinical picture, most patients opt for the treatment.

Over the last decades, several strategies have been developed to treat an IA. In general, these can be classified into invasive and minimal-invasive techniques. The first involves an open skull surgery where a metal clip is placed at the neck of the aneurysm, instantly stopping any blood flow into the sac. However, the clip remains within the head and the risk of injuries during the intervention exists. The second option includes endovascular approaches that are increasingly used, as they are less invasive compared to open surgery [4]. In the most common technique, a platinum coil package is introduced into the IA, reducing its volume by approximately 30%. Hence, the flow velocity of the entering blood is reduced and ideally a thrombotic process is initiated, leading to a complete occlusion. However, due to gravity and dynamic forces, coils may not remain within the dilation, and this can result in a blockage of the parent artery. Therefore, stents are deployed below the ostium (the opening area of an IA) to assist the coiling procedure.

For complexly shaped and large aneurysms, special flow-diverting devices were developed. The low porosity (fraction of void volume to total volume; low porosity

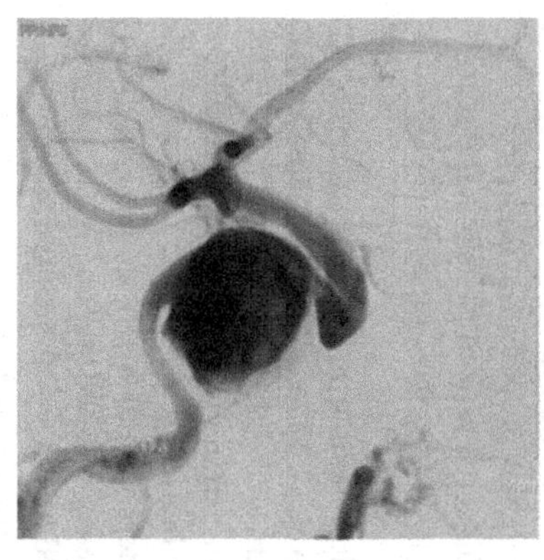
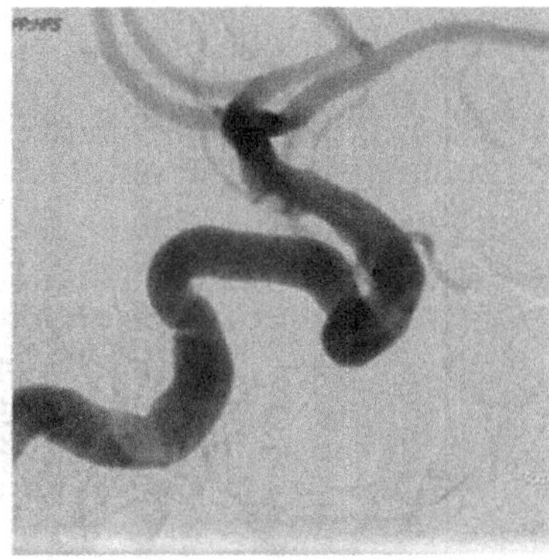

Fig. 14.1 Digital subtraction angiography of an internal carotid artery aneurysm. Pre- (*left*) and posttreatment (*right*) using a flow-diverting device. The absence of a contrast agent within the dilatation indicates full occlusion after 12 months.

also corresponds to high metal coverage) of the flow-diverters reduces the inflow of blood into the aneurysm sac drastically [5] (see Fig. 14.1). An additional advantage includes the reduced risk of perforation, as entering the aneurysm itself with catheters, micro-wires, or coils [6] is not necessary. The initial results of studies, in which flow-diverters were investigated, showed high rates of complete obliteration with a low procedural risk. However, later reports concerning unsuccessful treatments and delayed complications, such as late rupture or in-stent-stenosis [7, 8], raised the question of whether these implants led to the desired effects in all aneurysms or whether their use should be restricted to cases without alternative treatment options. Therefore, virtual stenting approaches, which are presented in the following paragraph, potentially lead to improved treatment strategies and hereafter lower treatment costs due to the avoidance of unnecessary therapies.

2. EXISTING APPROACHES—FROM PRECISE TO PRAGMATIC

While the available stent designs differ only in diameter, length, and porosity, each aneurysm is patient-specific in most aspects, including shape, location, parent artery, branching vessels, and orientation. Therefore, computational fluid dynamics (CFD) is a promising method to study the impact of flow-diverters on intra-aneurysmal hemodynamics with the intention to improve treatment results.

In the following, different approaches to virtually deploy stent or flow-diverting devices within intracranial aneurysm models are presented. Six representative methodologies are chosen, which differ in their complexity and applicability. Since not all existing approaches can be covered within this chapter, an overview table of related works is provided on page 374 (Table 14.1).

Table 14.1 Overview of virtual stenting studies using different deployment approaches for the treatment of intracranial aneurysms (limited to a total number of 50)

First author	Year	Number of aneurysms	Stenting approach	Main findings
Anzai [19]	2012	1[a]	Model with unconnected struts	2D optimization proof-of-concept
Anzai [20]	2014	4[a]	Model with unconnected struts	Optimized stent possesses a nonuniform structure, unique strut placement is required for each aneurysm type
Appanaboyina [21]	2008	1[a]	Hybrid meshing approach	Proof-of-concept of the methodology (body-fitted grids for the vessel, adaptive embedded grids for the stents)
Augsburger [17]	2011	2	Porous medium	Results predicted by the porous medium approach compare well with the real stent geometry model
Auricchio [22]	2011	0	Finite element analysis	Laser-cut closed-cell design provides the highest lumen gain, oversizing is relevant to the stress induced in the vessel wall, stent designs have limited impact on vessel straightening
Babiker [23]	2012	1	CAD-assisted deployment	Highest cross-neck flow reduction due to Y-stenting, good agreement between CFD and experiments
Berg [24]	2016	0	Free-form deformation	The choice of flow-diverting device and the subsequent deployment strategy highly influences the patency of jailed side branches
Bouillot [14]	2016	2	Fast predictive geometrical deployment	Introduction of the deployment method based on geometrical properties of braided stents, measurements showed a very good qualitative and quantitative agreement with the virtual deployments

Cebral [10]	2005	1	Adaptive grid embedding technique	Presentation of the hybrid methodology to simulate endovascular devices
Cebral [8]	2011	7	Adaptive grid embedding technique	Reduction in aneurysm velocity and WSS, increase in pressure for ruptured cases
Cito [25]	2015	1	Deformable simplex models	Good agreement between numerical and experimental results, excellent agreement among the CFD solutions
De Bock [26]	2012	3	Finite element analysis	Open cell designs better cover the aneurysm neck compared to stiff and flexible closed cell stents, but have less struts apposing well to the vessel wall
Janiga [12]	2013	1	Free-form deformation	Flow-diverter deployment leads to increased residence times and significant reductions in the maximum WSS, geometrical validation of the virtual stenting approach
Janiga [27]	2013	1	CAD-assisted deployment	Recommendations regarding geometry, mesh requirements, fluid properties and quantitative analysis are provided for virtual stent computations
Janiga [28]	2015	1	Free-form deformation	Proof-of-concept of an automatic CFD-based flow-diverter optimization principle
Karmonik [29]	2013	8	Porous medium	Velocity and WSS decrease in all cases, pressure increase in one case, pressure at the ostium correlated with pressure changes inside the aneurysm
Kim [30]	2006	1	Asymmetric stent low porosity patch	Asymmetric stent patch blocked strong inflow jet, aneurysmal flow activity was substantially reduced, qualitative agreement with experiments

Continued

Table 14.1 Overview of virtual stenting studies using different deployment approaches for the treatment of intracranial aneurysms (limited to a total number of 50)—contd

First author	Year	Number of aneurysms	Stenting approach	Main findings
Kim [31]	2008	7[a]	CAD-assisted deployment	Aneurysmal stasis and WSS are strongly influenced by stent porosity, damping effect is significantly reduced at high vessel curvatures
Kim [32]	2010	2	Model with unconnected struts	Stent porosity and strut shape highly influence the amount of velocity, vorticity, and shear rate reduction
Kojima [33]	2012	1	CAD-assisted deployment	WSS and pressure decrease in proportion of the stent porosity, flow-diverting stents lead to higher velocity reduction compared to multiple Enterprise stents
Kono [34]	2013	1	Micro-CT scans	Crossing-Y stent leads to highest cycle-average velocity and WSS reduction in an asymmetric bifurcation aneurysm
Kulcsár [35]	2012	8	CAD-assisted deployment only at the ostium	Significant reduction of mean intra-aneurysmal flow velocities and WSS after treatment
Larrabide [11]	2012	1	Deformable simplex models	Validation of the virtual approach succeeded
Larrabide [36]	2013	23	Deformable simplex models	Significant reduction inside the aneurysms for most hemodynamic parameters except intra-aneurysmal pressure, higher reduction in WSS and velocity in fusiform compared to saccular aneurysms
Liou [37]	2008	1[a]	CAD-assisted deployment	Three-layer stents seem not as effective as two-layer stents in reducing the magnitude of aneurysm inflow rate and WSS

Ma [9]	2012	2	Finite element analysis	First mechanical modeling of braided flow-diverting stents in realistic geometries
Meng [38]	2006	4[a]	CAD-assisted deployment	Significant differences between sidewall and curved aneurysm regarding flow stagnancy and impingement
Mut [16]	2012	3	Deployed cylinder approach	Device oversizing leads to a significant reduction of the efficiency
Mut [39]	2014	3	Deployed cylinder approach	Variations in the inflow conditions during the stenting procedure may lead to large deviations in the quantifications of hemodynamic changes
Mut [40]	2015	23	Deployed cylinder approach	Lower mean velocity, inflow rate and shear rate for fast occlusion aneurysms
Ohta [18]	2015	1[a]	Porous medium	Porous media thickness and the ratio of parallel permeability to perpendicular permeability affect the aneurysmal flow pattern, pressure drop in the parallel direction can be determined
Paliwal [41]	2016	2	Simplex mesh structure	Proof-of-concept of the rapid stenting approach, higher flow reduction by the flow-diverter compared to an Enterprise stent
Peach [42]	2014	3	Spring analogy	Large variations in flow reduction between devices and across different aneurysm geometries exist, the industry standard with 70% porosity might not always be optimal
Peach [43]	2016	6	Spring analogy	Increased outlet pressures at jailed daughter vessels have little effect on aneurysm inflow reduction but largely reduce the daughter vessel flow rates

Continued

Table 14.1 Overview of virtual stenting studies using different deployment approaches for the treatment of intracranial aneurysms (limited to a total number of 50)–contd

First author	Year	Number of aneurysms	Stenting approach	Main findings
Radaelli [44]	2008	1	Deformable simplex models	Quantifying the performance of three commercial intracranial stents
Seshadhri [45]	2011	2[a]	CAD-assisted deployment	Intra-aneurysmal residence time increases rapidly with decreasing stent porosity, highest modifications are always obtained for small aneurysm diameters
Shobayashi [46]	2013	1	Finite element analysis	Low velocity reduction by Neuroform stent, high velocity reduction and change of flow pattern by Pipeline device, pressure remains unchanged
Song [47]	2015	5	Porous medium	Good qualitative agreement of posttreatment comparisons between changes after virtual stenting and angiographic changes
Spranger [48]	2014	1	Spring analogy	Proof-of-concept for a spring based, real-time virtual stenting approach for stent grafts and flow-diverters
Spranger [49]	2015	0	Spring analogy + finite element analysis	Optimization of a fast stenting approach by parameter calibration
Srinivas [50]	2010	1[a]	Model with unconnected struts	2D stent strut optimization revealed that placing struts in the proximal region of the neck leads to the best flow-diversion in an aneurysm

Tanemura [51]	2013	2	Micro CT scans	Lower WSS and flow velocity, higher oscillatory shear, WSS gradient, relative residence time
Tremmel [52]	2010	1	CAD-assisted deployment	Reduction of average flow velocity, turnover time and wall shear after deployment of 1–3 flow-diverter
Xu [53]	2013	14	Micro CT scans	Metal coverage correlated differences in hemodynamic parameters between realistic and virtual flow-diverter deployments are confirmed
Zhang [54]	2010	3	Porous medium	Vortex reduction consistent in vitro and in silico, decrease of WSS on the sac
Zhang [55]	2013	2	Porous medium	Complete occlusion of an aneurysm requires higher resistance force compared to dynamic forces
Zhong [56]	2016	6	Fast virtual stenting (active contour model)	Formulation of a novel criterion for terminating virtual stent expansion, efficacy and accuracy of the method is proven by simulations and an experiment

[a] Denotes idealized geometries.

2.1 Explicit Description

To virtually deploy an endovascular device as realistically as possible, it has to be modeled very precisely. This includes the resolving of every individual stent strut and considering its deformation caused by the surrounding vessel section. In this context, explicit approaches can be classified into two categories: (1) finite element analysis (FEA) and (2) fast virtual stenting (FVS) techniques.

2.1.1 Finite Element Analysis (FEA)

One FEA approach was presented by Ma et al. in 2012 [9]. They demonstrated a workflow for simulating mechanical deployments of flow-diverting devices in intracranial aneurysms. Hence mechanical responses, e.g., self-contact within the device, are considered. In order to predict the stent placement as realistic as possible, the complete delivery system was modeled. This includes the microcatheter, the pusher, and the distal coil.

The presented workflow mainly consists of the 3D flow-diverter modeling, the packaging into the deployment system, the actual stent release, and the conformity assessment. Afterwards, precise CFD simulations can be carried out to compare pre- and posttreatment scenarios.

A Pipeline Embolization Device (Covidien, Irvine, USA) consisting of cobalt-chromium-nickel alloy struts served as a model to generate the braided stent geometry. Finite beam elements were used to discretize the stent mesh, which allow the numerical deformations and stress calculations. In this regard, mesh sensitivity tests are crucial to determine the required element size. To realize the packaging process of the undeployed stent, it was compressed in a radial direction by a crimper and expanded until the microcatheter boundaries were reached. For the placement of the flow-diverter below the aneurysm ostium, a delivery pathway was prescribed by a set of reference points along the vessel centerline. This step has to be done with great care because of two contradictory effects. On the one hand, physicians use the ability of flow-diverting devices to produce a lower porosity close to the aneurysm orifice in order to strengthen the efficiency of flow-diversion and, therefore, promote a faster occlusion. However, due to decreased porosity, foreshortening effects are present, which may lead to undesired placements, e.g., ostia that are not fully covered, or insufficient wall apposition. The latter in particular must be prevented since in-stent thrombosis can occur, leading to stenosis or detachments of clots. To account for these effects, Ma et al. [9] formulated a numerical strategy that prevents problematic deployment results. In this regard, the maneuvers are realized by imposing translational as well as rotational boundary conditions. Finally, the whole workflow results in high computational costs and the authors report a duration of approximately 100 hours per virtual stent deployment on a parallelized workstation.

Although the highest reported level of accuracy exists in this FEA approach, assumptions and limitations are still required for the modeling process. The patient-specific vessel wall lumen was assumed to be rigid and model parameters were adopted from previous literature. Furthermore, vessel straightening was not considered and pulsatile effects of the blood flow were neglected, which may influence the actual deployment. Nevertheless, the methodology leads to highly realistic virtual flow-diverter placements that can serve as a basis for precise CFD studies and consequently for real treatment improvements.

2.1.2 Fast Virtual Stenting (FVS)

Fast virtual stenting is the most popular approach for computer-assisted stent deployments, since it considers most relevant device properties but is not limited to enormous numerical efforts. Additionally, FEA models may have difficulties in fitting stents into tortuous vessels with high curvature. Hence contact problems between the stent and the vessel wall can occur, leading to unstable numerical solutions and, therefore, to unrealistic deployment results. In the following, different types of FVS are presented and advantages as well as disadvantages are highlighted to demonstrate their usability in a clinical context.

Cebral and Löhner [10] introduced a robust, simple, and fast method in 2005, where an adaptive grid embedding technique is applied to simulate endovascular devices. They chose a hybrid approach for the modeling of coil packages and stents. The walls of the intracranial vasculature were treated using body-conforming grids, while the endovascular devices were described by an adaptive mesh embedding technique. The use of meshes with faces that match the surrounding vessel surface is well known, but especially for complex geometries, the generation of these can be very challenging. Hence the embedded approach with nonconforming mesh elements allows the consideration of complex objects, when appropriate boundary conditions are applied at the mesh crossings. The whole process can be summarized as follows: first, the anatomic segment of interest has to be reconstructed and spatially discretized using a body-conforming grid. Afterwards, all elements that are intersected by the endovascular device are removed and boundary conditions are applied. Finally, to increase the accuracy of the subsequent hemodynamic simulations, the grid has to be adaptively refined close to the device borders. This is realized by an automated refinement algorithm.

Regarding the device description, Cebral and Löhner suggested the use of a nonuniform rational B-Spline surface (NURBS) to ensure a high level of accuracy. However, since this demands a strong manual effort, they described the endovascular devices using tightly packed spheres. With an increasing number of spheres, the approximation of the device surface becomes more realistic, but clearly at increasing computational costs. To demonstrate the applicability of the proposed method, a stent was modeled using 12 helical structures with a strut thickness of 0.1 mm. In both cases, the stent was placed below the aneurysm orifice to cover the ostium area fully. For the idealized model, a comparison of the hemodynamic predictions pre- and poststenting was shown, demonstrating a clear reduction of velocity and wall shear stress, respectively. However, for the patient-specific configuration the stent geometry does not align appropriately with the surrounding vessel walls, leading to an unrealistic deployment result and consequently to unreliable hemodynamic predictions. Hence improvements were required, particularly for the placement of endovascular devices in individualized vasculature.

To improve the outcome of previous FVS approaches, Larrabide et al. [11] suggested using deformable simplex models. Here, a second-order partial-differential equation is applied to move a mesh (e.g., a virtual stent) under the effect of internal and external forces. In order to apply the method to intracranial stenting, geometrical constrains were taken into account. The authors focused on four key parameters, namely the strut pattern, the strut length, the strut angle, and the deployed stent radius. To deform the simplex mesh, different forces need to be considered, which are referred to as shape-constraining forces. These are divided into strut length forces, angle forces, smoothing forces, and stent expanding forces, respectively. Furthermore, the deformation process is limited by external forces, which are exerted by the vessel walls. The actual stenting algorithm finally consists of the following steps:

1. Simplex point creation around a centerline of the desired vessel segment,
2. Simplex mesh connectivity creation,
3. Stent mesh creation,
4. Iterative computation of simplex and stent mesh forces, and
5. Updating the mesh position until the algorithm converges.

The complete procedure requires very low computational costs and leads to virtual flow-diverter deployments in less than a minute.

Another fast virtual stenting approach was presented by Janiga et al. [12]; this approach uses a nonrigid registration based on a free-form deformation for deployment. The device placement consists of two major steps, which are illustrated in Fig. 14.2. First, a rough, large-scale deformation makes the stent follow a smoothed centerline curve extracted from the vessel data. Second, this initial deformation is refined iteratively such that the stent geometry gets closer and closer to the real vessel boundaries. Both steps rely on free-form deformations based on trivariate triharmonic splines, similar to thin-plate splines employed in 2D. For determining the parameters of the deformation, the stent model is temporarily replaced by a cylindrical geometry that incorporates the convex hull of the stent before deployment. The resulting deformation is finally applied by replacing the cylindrical template with the real stent model. Further details regarding the free-form deformation for virtual stent deployments can be found in the online supplementary of Janiga et al. [12].

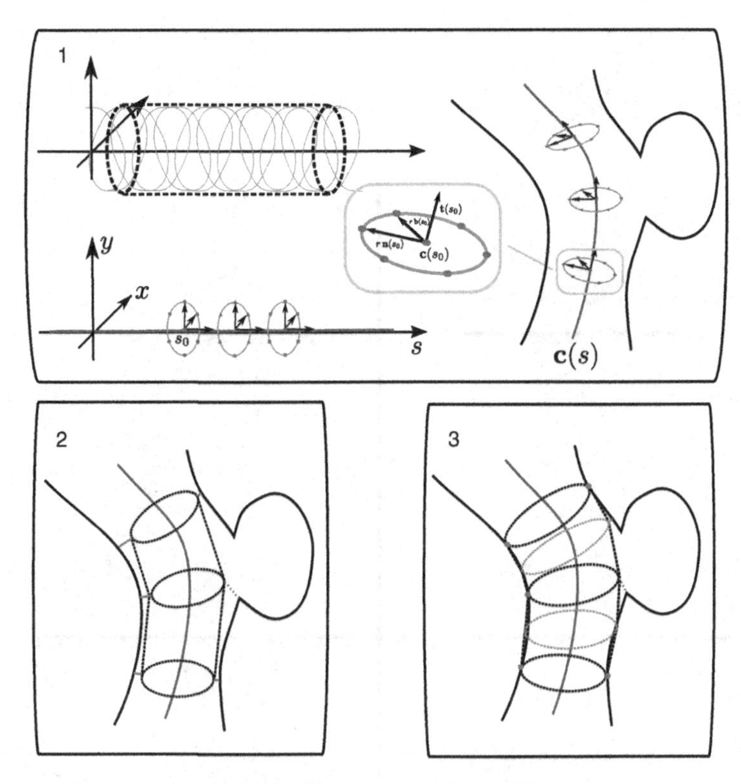

Fig. 14.2 Fast virtual stenting. An initial deformation (1) is obtained from mapping points of the cylinder template to corresponding points in local frames of the centerline curve. This deformation is refined iteratively, starting from a coarse resolution template (2). In each iteration, the deformation is defined by corresponding points on the template and on the vessel walls (3). Corresponding pairs are discarded if a predefined distance or angle threshold is exceeded.

To simplify the usability of this approach, an in-house software (VISCA—VIrtual Stenting for Cerebral Aneurysms) was developed, which enables the interaction through a graphical user interface [13] (Fig. 14.3). Hence users with limited knowledge of the methodology still are able to test different stent placements and evaluate the treatment outcome (see Fig. 14.2 for the required process steps). In this regard, the segmented surface model of the diseased vessel section, the undeployed stent model, and the vessel centerline need to be imported. Furthermore, different parameters such as the axial position of the stent or the radial distance between the stent and the vessel wall need to be defined. Finally, the realistic placement is obtained within seconds on a standard personal computer.

Beside the ability to generate realistic endovascular device deployments, VISCA offers two additional features. First, the user can adjust the local porosity. Ten markings, which are equidistantly distributed along the flow-diverter, can be manually shifted. As

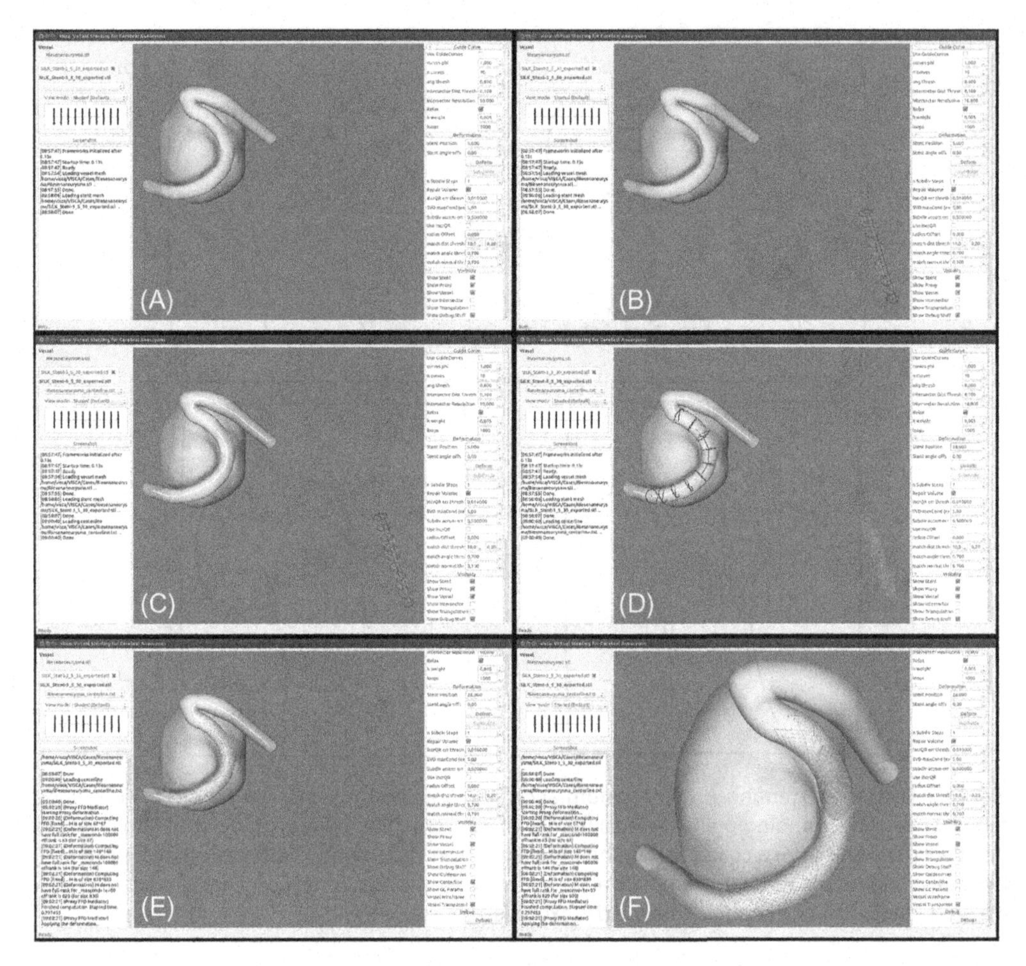

Fig. 14.3 **Illustration of the workflow to virtually deploy a flow-diverter geometry. (A) Import of the three-dimensional aneurysm surface, (B) import of the stent geometry as well as (C) the vessel centerline, (D) setup of the desired deformation parameters, (E) generation of the deployed flow-diverter, and (F) enlarged illustration of the final deformation.**

a result, the local compression can be increased (e.g., below the ostium) and the initial stent length adjusts to the new configuration. Fig. 14.4 illustrates this option for different compression degrees. The second feature demonstrates the option to avoid the use of the graphical user interface. Although it is very comfortable for inexperienced users, the batch-mode allow very rapid stent deployment, which can be easily integrated into an automatized framework.

The last presented example of the FVS representatives was recently published by Bouillot et al. [14]. Their deployment model is based on the geometric properties

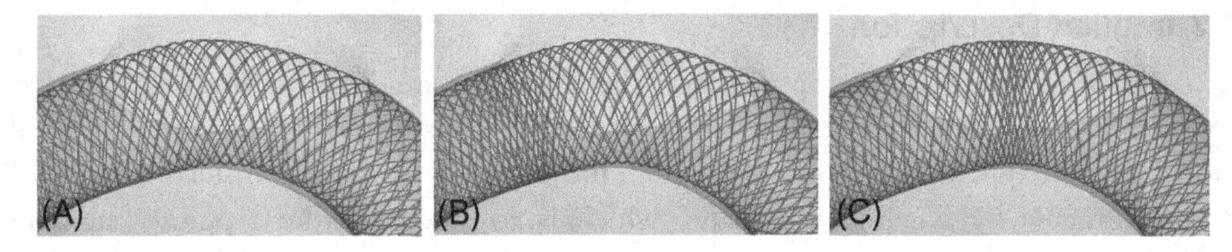

Fig. 14.4 Virtual flow-diverting stent variations to test the influence of different stages of compression. (A) No compression, (B) distal compression, and (C) central compression.

of braided stents and was applied to idealized toroidal vessels and patient-specific vasculature. The following two hypotheses were made by the authors, derived from experimental observations: stent cross sections are assumed to be circular and the distance between the individual struts remains constant. Respecting these assumptions, the model needs a number of cylindrical wires, a total wire length, and a diameter as geometric input parameters. Hence, the flow-diverter porosity depends only on the local strut angle with a maximum value at 45 degree. The complete deployment process is described by the authors as follows:

1. Vascular segmentation: This leads to a realistic three-dimensional representation of the vessel lumen.
2. Vessel centerline and radius computation: The open-source VMTK library (Vascular Modelling Toolkit) [15] (www.vmtk.org) is used for this purpose.
3. Vessel centerline and radius regularization: The calculated local radii are regularized using smoothing splines and limited to the maximal stent radius. Specific care is required between the proximal and distal end of the aneurysm ostium.
4. Meshing of the stent envelope: An envelope representing the stent surface is decomposed into a finite number of torus sections, which are individually treated by the proposed algorithm (see Ref. [14] for a detailed description).
5. Stent weaving: The stent strut path is mapped on the individualized stent envelope.

To test the introduced method, Bouillot et al. [14] virtually reproduced a Pipeline Embolization Device (Covidien, Irvine, USA) consisting of 24 struts with a wire-diameter of 33 μm. They applied their method to idealized as well as patient-specific models and demonstrate the feasibility of their approach. Previous works mostly use an envelope for the fast virtual stenting that closely follows the vessel lumen and may result in incomplete stent appositions. To advance this methodology, the authors consider a tubular envelope inscribed in the vessel lumen. This means that important treatment planning factors such as device sizing or stent positioning can be analyzed in advance. Additionally, the effect of local porosity variations, which clearly influence intra-aneurysmal flow behavior [16], can be evaluated without any risk for the patient.

2.2 Implicit Description

In contrast to the previous approaches, the target of implicit methods is reducing computational efforts and saving simulation times, respectively. In these types of modeling, no explicit description of the stent properties is performed. Instead, the device is approximated by a cylindrical shape, which is considered to be a porous medium following Darcy's law. In the following, two studies are presented that virtually apply implicit approaches to model the effect of flow-diverting devices.

Augsburger et al. [17] based their implementations on a braided SILK stent (Balt International, Montmorency, France), which consists of 48 wires, while 40 struts have a diameter of 30 μm and the remaining 8 have a diameter of 50 μm. To model the resulting porosity, the dense mesh structure of the flow-diverter was assumed to be a porous medium. By adding a momentum source term to the governing flow equations, namely the Navier-Stokes equations, this medium can be considered numerically. This source consists of two terms: a viscous loss term and an inertial loss term, which proportionally influence the pressure gradient with respect to velocity. Eq. (14.1) contains the description of the momentum sink with the three spatial coordinates $i = x, y, z$, the velocity magnitude v, the permeability α, the inertial resistance factor C_2, and the material properties dynamic viscosity μ and density ρ.

$$S_i = -\left(\frac{\mu}{\alpha}v_i + C_2\frac{1}{2}\rho|v|v_i\right) \tag{14.1}$$

The most challenging task when using an implicit method for virtual stent deployments is the acquisition of appropriate values for the coefficients $\frac{1}{\alpha}$ and C_2. In general, such values can be obtained both experimentally and numerically. Augsburger et al. [17] cropped segments of the explicit non-homogenous flow-diverter geometry and placed these sections into test sections. To obtain the perpendicular coefficients, the stent part was put in a long straight pipe perpendicular to the flow. For the tangential coefficients, a long rectangular cuboid was used with the wire segment oriented parallel to the flow direction. Since in reality, flow-diverting devices do perfectly align in a perpendicular or parallel flow direction, each coordinate of the porous medium volume was assigned with a local normal and tangential component. Hence, for curved surfaces the porous medium coefficients vary locally.

To test the efficiency of the reduced model, two patient-specific intracranial aneurysms (one lateral and one terminal) were virtually stented using the porous media and the real flow-diverter geometry, respectively. Further limitations are rigid vessel walls, Newtonian fluid modeling, and boundary conditions based on 1D models. The results demonstrated that the application of a porous medium in the lateral case led to a better agreement with the more realistic stent compared to the bifurcating aneurysm. This is due to the stronger interactions of perpendicular devices with the flow, which makes it more difficult to access these using a reduced model. Therefore, the authors suggest that the model coefficients have to be adapted for each specific stent geometry since stents vary in strut size, cell size, porosity, etc.

This conclusion motivated a study by Ohta et al. [18], which investigates the pressure drop dependency based on different test section geometries and flow scenarios. Specifically, flow-diverter segments were placed perpendicular, parallel, and with inclined angles in a straight and a cylindrical flow domain. Furthermore, CFD computations in an idealized aneurysm model were carried out using a porous medium with different ratios of parallel pressure drops to perpendicular pressure drops to test the impact on the resulting flow patterns.

Again a momentum source term consisting of a viscous resistance term and an inertial resistance term was added to the Navier-Stokes equations. To obtain the relevant permeability and resistance factors (see Eq. 14.1), different test sections with various distances and diameters were generated. For the perpendicular direction, the computations demonstrate that the effect of the stent on the pressure drop becomes negligible for distances over 400 μm. Independent of the inflow velocity, the perpendicular pressure drop was only affected when the strut distances was lower. In contrast, pressure changes approximately linearly over distance in parallel direction. The struts divide the flow into top and bottom regions, leading to an almost constant increase of pressure loss with increasing length. Further, the pressure distribution appears to be constant as well across individual cross sections of the test domain. For the inclining angle configurations, both pressure drops and velocity increase as the stent angle changes from 0 degree (parallel) to 90 degree (perpendicular). Since the number of stent struts exposed to the flow jet increases with higher angles, the authors suggest that the pressure drop depends on the angle at which the stent is inclined. The cylindrical test section revealed that a dramatic pressure drop decrease exists with an increasing cylinder radius. In this regard, pipe frictions was identified to be negligible for pipe diameters higher than 1 mm. The investigation of different parallel permeability to perpendicular permeability ratios demonstrates a huge impact on the resulting flow predictions.

Due to these differences in pressure drops depending on test section geometry and device placement, the authors conclude that it is challenging to acquire appropriate model parameters for Darcy's law. Although the study was limited to straight arteries, idealized aneurysm geometry, and steady-state, Newtonian flow conditions, they suggest using the permeability ratio and the perpendicular permeability to obtain the parallel permeability as well.

3. VALIDATION—THE CURSE OF COMPUTATIONAL PREDICTIONS

The large amount of available methods represented in the related scientific literature raises an important question, whether the applied assumptions still represent a given treatment with an acceptable accuracy. Although the validation of these methods is crucial before applying them in clinical practice, it is rarely discussed and reported.

The validation of a virtual stent treatment technique can be considered in two consecutive steps. First, the geometrical properties of a deployed stent can be compared with in vivo or in vitro deployments. Due to the small strut diameters of the intracranial devices, a high resolution is required, as can be obtained by using, e.g., DynaCT or micro-CT. Often an additional registration is required, because the virtual model is mostly represented in a different coordinate system. Apart from a qualitative comparison, a quantitative analysis is still challenging considering the limited resolution, measurement artifacts, and the complex structure of the flow-diverting devices.

The second step of validation investigates not the geometrical features of the deployed geometry, but rather the flow modification after the implantation of a stent, especially a flow-diverting stent. The flow velocities can be measured in vitro for steady or unsteady configurations in stented phantom models using, e.g., the particle image velocimetry (PIV) method. However, the flow-diverting stents might highly reduce the optical accessibility, especially for patient-specific geometries. In vitro flow velocity measurements using MRI are hardly feasible due to the significant artifacts produced by the flow-diverting devices. Nevertheless, in vivo or in vitro experiments can be used to determine the resistance of a given flow-diverting device and they can be used to prescribe the parameters in a porous medium model. Nevertheless, varying compression is challenging to be incorporated in such a porous medium model; therefore the method cannot be generalized.

Geometrical validation of a detailed virtual deployment technique is demonstrated in Ma et al. [57]. They compared a deployment in a silicone model with their finite element analysis and found a very good agreement. Their method was able to describe the local compression as well; however, their finite element computation might take a few days in order to describe the whole stent opening procedure.

The free-form mesh deformation technique applied for virtual stent deployment by Janiga et al. [12] incorporates a geometrical validation as well. In order to validate the deployment technique, the considered flow-diverter (SILK) deployed in a silicone phantom model based on a real patient geometry has been characterized experimentally. Two different experimental techniques, DynaCT (Biplanar C-Arm System, Axiom Artis, Siemens, Forchheim, Germany) and Micro-CT (Phoenix Nanotom S, GE Sensing & Inspection Technologies GmbH, Wunstorf, Germany), have been tested for this purpose. Even if a full resolution of the finest struts is not possible due to the limited spatial resolution, the position and overall geometry of the deployed stent are extremely well resolved, allowing a direct comparison with the virtual deployment procedure. The geometries obtained experimentally have been reconstructed and registered into a common coordinate system, corresponding to the numerical framework. Fig. 14.5 shows the results of the virtual and real deployments in the same phantom model. Comparing the figures, an excellent agreement can be observed. Hence, the introduced

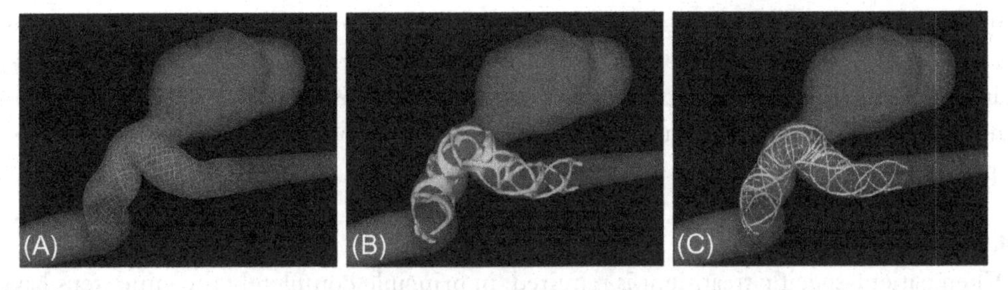

Fig. 14.5 Flow-diverting stent (SILK) in an in-house, silicone phantom model based on a real patient geometry. (A) Virtual stenting based on a free-mesh deformation, (B) deployed SILK stent measured experimentally by DynaCT, and (C) deployed SILK stent measured experimentally by Micro-CT.

virtual stent deployment technique leads to a very faithful representation of the true deployment process.

Apart from the validation based on geometrical comparisons, further validations can be performed to quantify the effect of the flow changes in stented models. However, the wide range of various flow measurement techniques and validation against experiments are not further discussed in this chapter, nevertheless, they are crucial for the clinical applicability.

4. SELECTED APPLICATIONS—HOW NUMERICAL MODELS CAN ASSIST

To demonstrate the applicability of virtual stenting, two advanced examples are chosen. The first combines the numerical stent deployment with optimization techniques in order to improve the individualized outcome of a patient. In the second example, the effects of stent over- and undersizing on side branch occlusion are investigated. In this regard, measurements in swine arteries are compared with numerical observations.

4.1 Individualized Optimization of Intracranial Aneurysm Treatment

Evaluating the flow conditions for a single treated case is very demanding, not only in computational sense but regarding the required expertise and time spent on the computational setup as well. Automating the whole workflow could not only speed up the process drastically, but could also save expensive working time and could enable the optimization of the process as well, enabling a patient-specific treatment.

Despite best efforts, the elimination of human intervention in the process, i.e., full automation, is not yet possible, nor desired. On the one hand, no software can replace the experience and many years of training of medical experts responsible for treatment. On the other hand, cases can vary so much that even the CFD results have to be double-checked, especially when considering that human lives are at risk.

Thus, instead of the full automation, a different approach is chosen. The goal is to speed up the work of the experts as much as possible, i.e., to automate all possible subprocesses, while still giving access for user interventions. When expert intervention is necessary, it should be performed in an optimal way, i.e., requiring minimal time. By choosing an appropriate approach, it is possible to provide patient-specific treatment in this way, while respecting expert decisions. Such an approach is called expert-driven optimization (EDO).

4.1.1 Treatment Process

When patient-specific treatment is required, in principle completely the same steps have to be performed for different geometries, with small variations. Fig. 14.6 illustrates these steps, denoted as "single evaluation." Some steps can be automated and executed with computers (C), if appropriate expertise is available; however, for some steps, input or a decision from medical (M) or technical (T) experts is needed. The steps are as follows:

1. *Angiography (M)*: If, following a medical image acquisition, an aneurysm is identified, the medical expert has to decide if treatment is necessary.
2. *3D vessel reconstruction and segmentation (M&T)*: For flow simulations, the geometry of the aneurysm and the adjacent vessels is required. This could in theory be automated, but due to technological limitations, the methodologies used nowadays produce artifacts. Thus, a highly trained technical expert will remove artifacts, which are deemed to be unrealistic by the medical expert.
3. *Parameters for stent deployment (M&T)*: Following the segmentation, the virtual flow-diverting stent implantation has to be executed. For this step, both technical and medical experts are needed. Only the medical expert can decide which flow-diverter configurations can be adopted, and the software usually requires some expertise as well.
4. *Parameters for meshing and CFD (T)*: Before starting the actual computations, additional decisions have to be met for volume mesh generation and CFD computations. Selecting appropriate values for these parameters is not trivial, as patient-specific geometries exhibit a huge variance, not to mention the very different spatial scales of the flow-diverter struts and the vessel diameter. After this point, all steps could theoretically be automated.
5. *Stent deployment (C)*: Virtual stent deployment is executed based on the fast virtual stenting approach introduced in Section 2.1.2 [12]. The parameters are predefined by the medical and technical experts.
6. *Mesh generation (C)*: The volume between the vessel and flow-diverter is spatially discretized based on the parameters defined by the technical expert using a commercial or in-house mesh generation software.
7. *CFD (C)*: Detailed hemodynamic simulations are executed.
8. *Postprocessing (C)*: Quantities of interest can be analyzed by the simulation software itself or with an adequate postprocessing software.
9. *Evaluation (M)*: The final decision still has to be made by the physician.

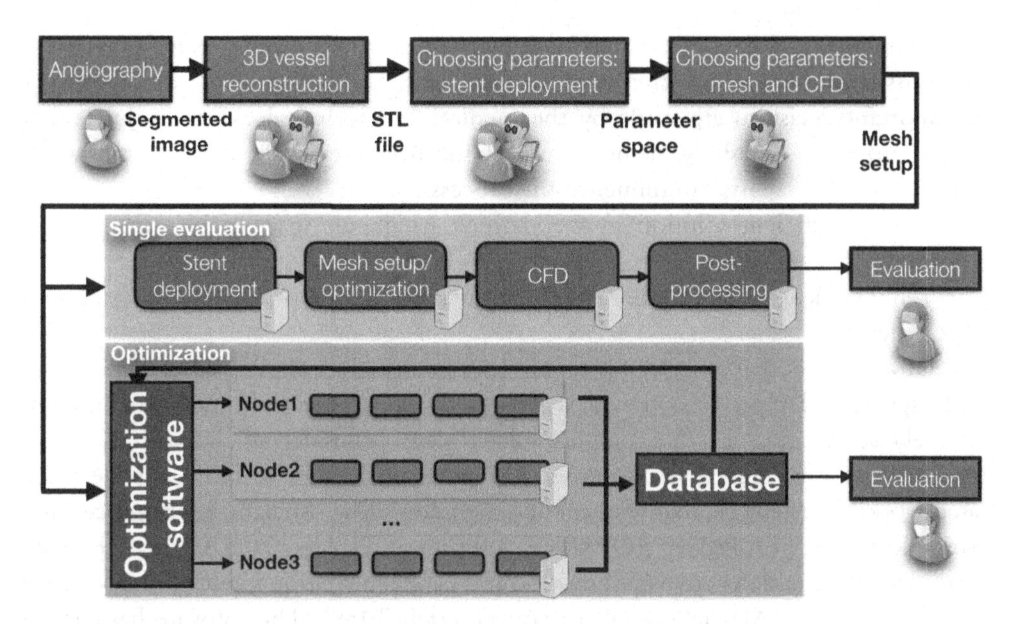

Fig. 14.6 Workflow of a virtual patient specific treatment. A single evaluation involving medical and technical experts.

4.1.2 Expert-Driven Computer Aided Stent Evaluation (ECASE)

In the following, a new method is proposed, which combines fast virtual stent deployment with detailed 3D CFD simulations to support patient-specific treatments. The implemented workflow is fully automated, thus it can be applied to arbitrary patient-specific geometry in a relatively short time, making it compatible with clinical practice.

The idea behind ECASE (Expert-driven Computer Aided Stent Evaluation) is very simple: to achieve maximal efficiency when evaluating a single case, all steps have to be automated, so that an expert intervention can be minimized. However, if these steps are already automated, many different configurations could be evaluated instead of a single configuration, and the results collected in a database. If these evaluations are performed in parallel, the overall computational time will not increase at all. Additionally, instead of just computing a couple of cases, the process can be coupled with an optimization algorithm as well; see Fig. 14.6.

The final evaluation still has to be based on the decision of the physician. However, as more alternatives are presented to the expert, a more well-founded decision can be made.

The most important assumptions and requirements are defined in what follows for ECASE:

- Complete automation of the treatment process is not yet possibly, as many different factors contribute to the efficacy of the treatment.
- Important decisions are made by the medical and technical experts, while other operations (e.g., meshing, hemodynamic simulation) are executed by a computer. If necessary, the experts can influence the process.
- The most efficient solution in the defined parameter space is identified by an optimization algorithm.
- For high-fidelity results, the CFD computations have to resolve the geometry completely.

Even automation of the few selected steps is not trivial, as the coupling of many different, highly specialized types of software is necessary. An automated mesh generation is extremely challenging for the complex treated cases, considering the wide range of length-scales (from cerebral vasculature to fine pores of the stents). The complete automated workflow (stent deployment, mesh generation, flow simulation, postprocessing) was implemented in OPAL++. OPAL++ (OPtimization Algorithm Library++) is an object-oriented multiobjective optimization and parameterization framework developed at the University of Magdeburg "Otto von Guericke" [59]. The software has already been successfully applied to many different problems [59, 60] and is especially focused on CFD-related investigations. In what follows, the two most important steps of the solution will be discussed in detail. Afterwards, a patient-specific aneurysm case will be analyzed.

Mesh and CFD Setup

Based on a segmented 3D surface model of an arbitrary aneurysm, mesh generation is performed with the commercial CFD software package Star-CCM+. This software was chosen as it supports robust polyhedral cell generation for the finite volume discretization. Polyhedral meshes result in better quality compared to, e.g., automatic tetrahedral meshes, as the presence of multiple neighbors leads to a better approximation of the gradients [28].

The automation of the mesh generation was achieved with an intelligent JAVA macro (approximately 3000 lines of code). The macro expects only two input files (vessel and stent geometry), and automatically corrects smaller artifacts, identifies inlets and outlets, and defines the physical setup and mesh generation setup, respectively. Although most parameters for the setup can be defined based on different recommendations [27], due to the widely varying geometry of flow-diverters, vessel, and aneurysm geometries, it is still recommended to check the resulted mesh before proceeding.

The mesh generation is realized using a so-called Script-in-Script (SiS) approach, allowing the mesh generation to occur in a completely automated way during the optimization process. This method calibrates the mesh sizes, but an expert can review the setup and intervene if necessary.

Additionally, if the mesh setup is already parameterized with the mesh sizes, a mesh quality optimization becomes possible. The optimization software can generate different mesh setups by varying the different sizes slightly and choosing the best configuration according to some mesh quality indicator (e.g., minimize skewness angle).

Postprocessing

In order to evaluate the outcome of a given virtual treatment, the efficacy should be analyzed carefully. The most common options for the evaluation include, among others, inflow jet reduction, wall shear stress (WSS), mass inflow through the ostium, maximal velocities, etc. However, up to now, no agreement had been found for the optimal indicator. Increased wall shear stress leads to "mural-cell-mediated destructive remodeling and weakening of the aneurysm wall" [61]. In contrast, low wall shear stress may trigger inflammatory-cell-mediated destructive remodeling, if high oscillatory shear is present [62]. The definition of the right objective to decide whether a treatment is necessary or not is still an active research field.

4.1.3 Exemplary Case

In order to demonstrate the applicability of the presented methodology for clinical practice, a patient-specific geometry was chosen. The geometry, presented schematically in Fig. 14.7, was acquired using CT-angiography of a 53-year-old female patient, who was examined with an acute headache in the emergency department of the University

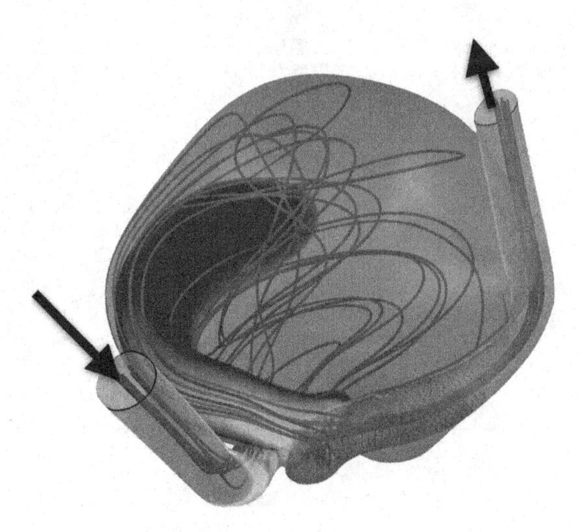

Fig. 14.7 Exemplary patient-specific aneurysm to demonstrate the applicability of the presented automatic virtual stenting approach. Velocity iso-surface (0.04 m/s) as well as streamlines are illustrated for a virtually treated case.

Hospital Magdeburg. The giant aneurysm was treated with a flow-diverter (Silk, Balt, France) and disappeared completely after 6 months, as confirmed by MRI [28].

To provide a proof-of-concept for the proposed automation and optimization process, the effect of compression was analyzed for the present case. For comparison, two reference cases were defined: the untreated case (UN) and the "normal" uncompressed stent (NC); see Fig. 14.8.

Afterwards, the optimal position of a medium compression was analyzed (by maintaining constant total porosity). The local partial stent compression along the centerline was varied to identify the configuration with the smallest inflow through the ostium. As the parameter space of the optimization is small, instead of an actual optimization, a Design-Of-Experiment with seven equidistant locations (C1–C7) suffices.

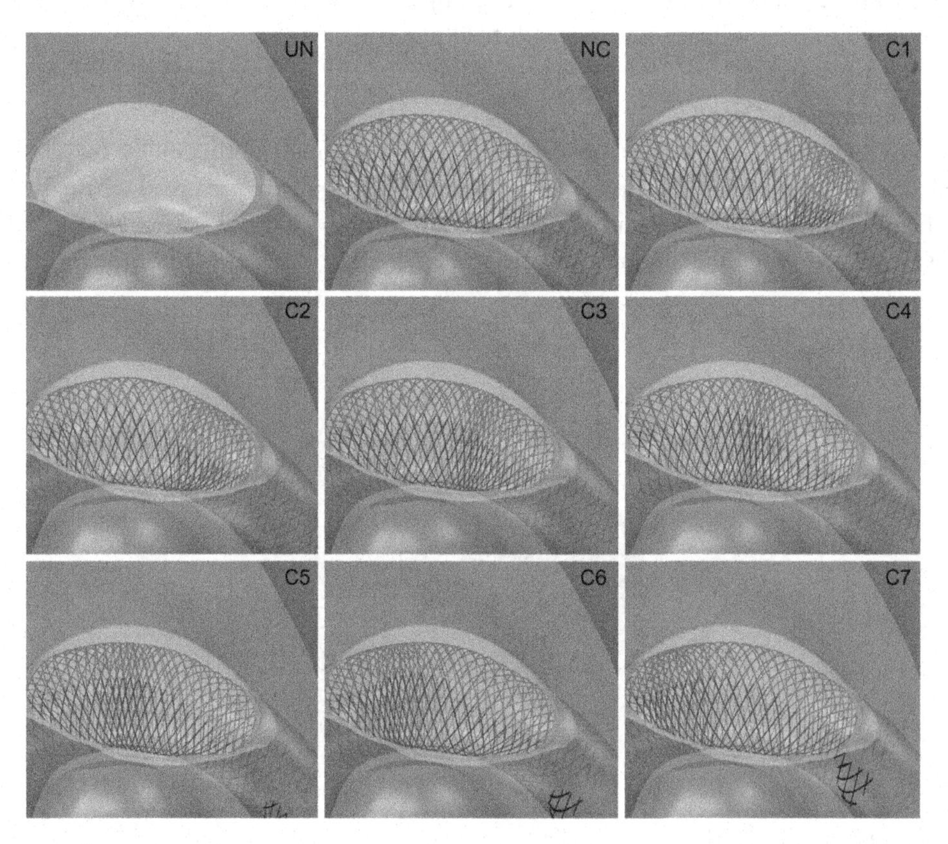

Fig. 14.8 Illustration of the nine considered stent configurations. An untreated (UN) case, one with no compression (NC), seven locally varying compression (C1–C7) are automatically simulated. (View on the aneurysm ostium).

Preparations for the Optimization Process

The process starts with the segmentation of the surface mesh using MeVisLab 2.8 (MeVis Medical Solutions AG, Bremen, Germany). Further, the virtual stent model was created in Creo Parametric 2.0 (Parametric Technology Corporation, Needham, USA), based on the size of the real implant (30 mm length, 3.5 mm diameter, 30 μm mean strut diameter with 44 degree and 136 degree between the struts in the uncompressed regions).

Afterwards, stent deployment and different compressions were obtained by using the in-house software VISCA (see Section 2.1.2), previously validated using DynaCT as well as micro-CT in a patient-specific in vitro silicone phantom model [12].

Mesh Generation and CFD Computations

For discretizing the computational domain, a 0.5 mm mesh resolution was applied for the core region and 0.25 mm for the vessel and aneurysm wall [27]. To resolve the strongly varying flow scales appropriately, additional local mesh refinements were defined around the region compromising the ostium and for the flow-diverter walls, with 0.01–0.02 mm resolution. Additionally, a smooth transition was enforced by applying 0.8 for the mesh density growth rate. This resulted in 1.9 million and 9–11 million polyhedral cells for the untreated and treated cases, respectively.

Mesh generation was performed on workstations (Intel® Xeon E5-2620 (2.10GHz) processor) of a small cluster, relying on the parallelization of OPAL++, requiring 4–6 hours altogether. The final spatial discretization is presented in Fig. 14.9 at different locations for a treated case.

CFD simulations were performed on the same computational cluster. Each simulation required 10–14 hours of computational time with StarCCM+ 9 (CD-adapco, Melville, New York, USA). The evaluations were executed in a parallelized manner. Together with the mesh generation, the complete parameter study did not require more than a single day, thanks to the parallel implementation provided by OPAL++.

The ultimate goal of the study was only to present a proof-of-concept for the implemented expert-driven optimization approach. Thus, several simplifications were made to reduce the runtime significantly:

- Steady flow was assumed with an average inlet velocity of $\overline{u} = 0.5\,\mathrm{m/s}$. Although real hemodynamic flow is transient, steady simulations generate almost equivalent flow patterns compared to cycle-averaged results [63].
- Blood was treated as an incompressible and laminar fluid with constant density ($\rho = 1055\,\mathrm{kg/m^3}$) and dynamic viscosity ($\mu = 0.004\,\mathrm{Pa\,s}$) [64].
- In each simulation, 4000 iterations were performed (ensuring normalized residual values below 10^{-6}).

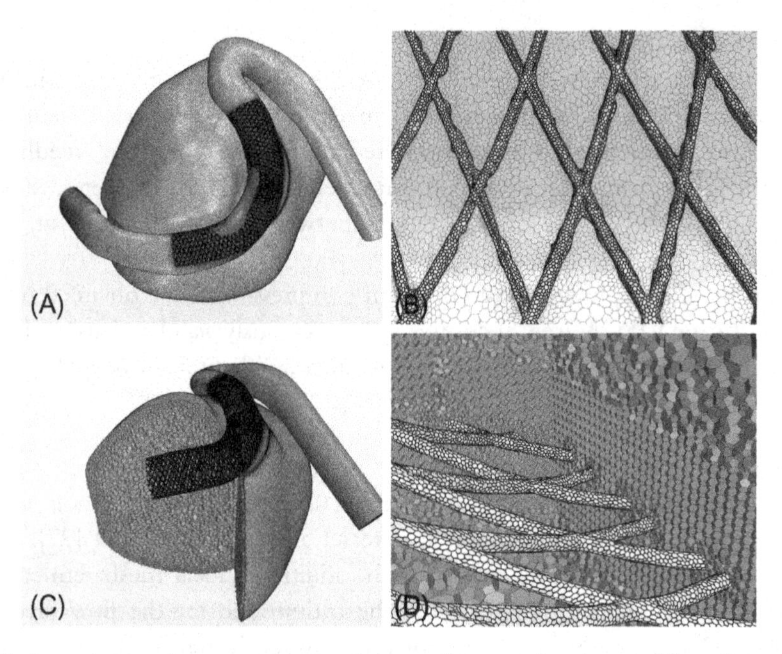

Fig. 14.9 Mesh resolution of an exemplary treated case. Computational mesh (A) on the vessel wall of the treated section, (B) on the strut surface, (C) in the cross-section of the core-region, and (D) in the cross-section of the ostium near the struts.

Optimization Results

The qualitative as well as quantitative analysis was performed by a StarCCM+ script and an additional Ensight 10 (CEI, Apex, USA) script. This can be easily adjusted to automatically quantify a hemodynamic parameter of interest.

- **Qualitative analysis**

Isosurface velocity: Visualizing the flow structures inside an aneurysm is not easy due to the complex three-dimensional flow. One possibility is to use isosurfaces of the velocity, which might be used to represent the impingement jet. Fig. 14.10 shows the iso-surfaces (0.04 m/s) for the untreated (UN), uncompressed (NC) and C5 cases (best configuration according to the quantitative analysis). As it can be seen, blood flow enters the aneurysm sac through the ostium, forming a vortex and developing a stagnation zone. The presence of the flow-diverter leads to a strong reduction of the jet.

In addition to the isosurfaces, cut-planes are shown in Fig. 14.11 near the ostium. The illustration confirms that for the locally varying compressions, configuration C5 leads to the largest improvement, with a much narrower and less intensive jet.

Wall shear stress: The reduction of jet size also influences the wall shear stress distribution on the aneurysm wall. All cases show elevated values around the neck region

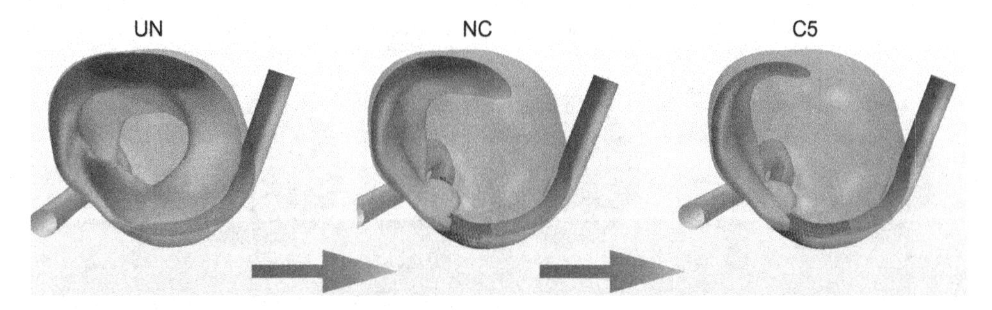

Fig. 14.10 Reduction of the impingement jet. Isosurfaces velocities (0.04 m/s) for the untreated (UN), treated with normal compressed (NC) and optimal medium compressed case (C5).

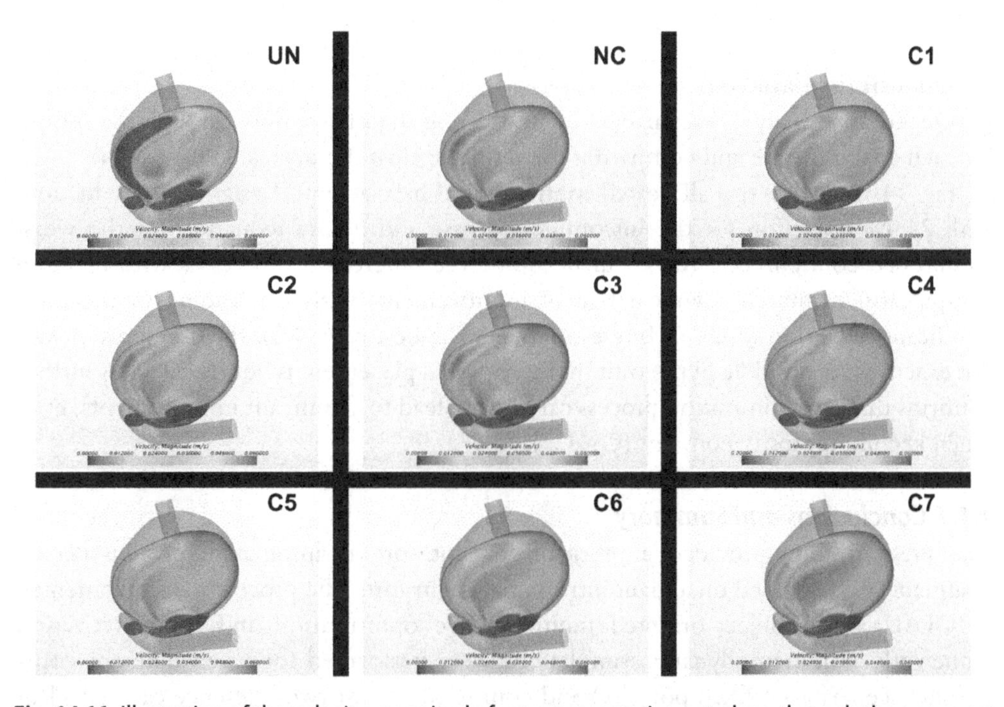

Fig. 14.11 Illustration of the velocity magnitude for a representative cut-plane through the aneurysm sac and ostium for all considered cases. It can be noticed, how different treatment configurations lead to different intensities of velocity reduction compared to the untreated case.

and along the impingement zone, but a wide range of values are identified, with C5 exhibiting the lowest peaks of WSS from the C1–C7 cases.

When illustrating the untreated (UN), uncompressed (NC) and C5 case, the large improvement is clearly visible (see Fig. 14.12).

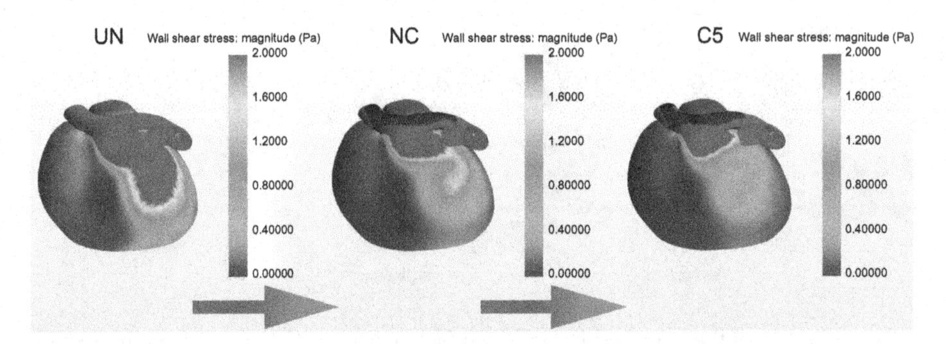

Fig. 14.12 Wall shear stress. Compared to the untreated reference case (UN), the flow-diverter without compression (NC) clearly reduces the WSS in the neck region of the aneurysm. This can be improved even further by choosing an optimized local compression for the stent (C5).

- **Quantitative analysis**

Quantitative analysis was enabled by integrating the mass inflow through the ostium for each analyzed case and comparing the results against the untreated case (UN).

Fig. 14.13 depicts that all treated configurations have achieved a significant reduction, with 24.4% reduction for the uncompressed case and 27.3% reduction for the worst compressed configuration (C1). Furthermore, the different treated cases with medium compression exhibited a wide range of improvement, with C5 leading to the most significant reduction (33.3%). Interestingly, for the best case (C5), the compression was not exactly in the middle of the ostium, i.e., optimal placement is not trivial. This further confirms that an optimization process can indeed lead to significant improvements, even when keeping porosity constant (C1–C7).

4.1.4 Conclusions and Summary

The present study provided a proof-of-concept for an automated, patient-specific treatment process based on a giant intracranial aneurysm. The process was implemented in OPAL++, an object-oriented multiobjective optimization and parameterization framework. The hemodynamic simulations were performed for seven different compression scenarios (at fixed porosity) and compared against two reference cases. A clear reduction was achieved. The method proved to be fast and efficient, and would be compatible to assist physicians during clinical treatment planning.

4.2 Effects of Over- and Undersizing on Jailed Animal Arteries

The previous example of a patient-specific flow-diverting stent deployment demonstrates how the braided wires lead to a clear reduction of velocity and wall shear stress values within the aneurysm. Compared to invasive techniques (e.g., clipping), the endovascular approach requires no opening of the skull and reduces the risk of injuring healthy areas in the vicinity of the dilatation.

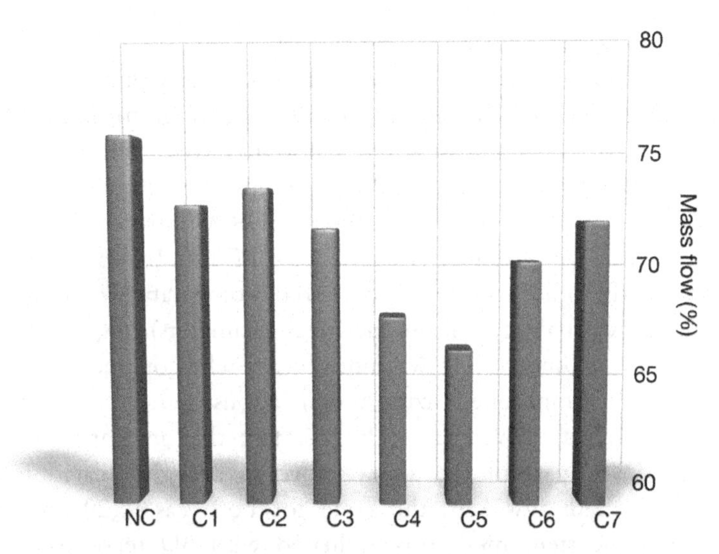

Fig. 14.13 Reduction of the inflow rate. Comparison of the different cases with the untreated case as the reference (100%): treated with normal compressed (NC) and optimal medium compressed cases (C1–C7).

However, complications after flow-diverter placement have still been reported, although high occlusion rates exist [65]. Unsuccessful treatments occur in approximately 3% of the patients: increased pressure might occur and in-stent stenosis were reported, respectively [7, 66]. Furthermore, problems during stent deployment include jailed side branches as well as incomplete wall apposition [67, 68].

To account for one of the described complications, the following study focuses on the effects of over- and undersized flow-diverting stents on covered vessel side branches. Experimental as well as numerical analyses are carried out for two representative vessel bifurcations of large white swine. Hence, comparisons between both approaches allow the derivation of phenomena that might be responsible for the generation of an endothelial layer on the densely braided stent struts.

4.2.1 Experimental Setup

Two large white swine, 3 months old, one male and one female, were used, with a mean weight of 20.05 ± 0.35 kg. The study design and reporting are in accordance with the ARRIVE guidelines [69]. Animals were premedicated with aspirin (10 mg/kg PO) and clopidogrel (10 mg/kg PO), starting 48 hours prior to intervention. This regime was maintained throughout the follow-up period of 3 months. Endovascular procedures were performed under general anesthesia. Each animal was premedicated with intramuscular 20 mg/kg of ketamine and 2 mg/kg of Xylazine, and anesthesia was maintained with propofol and sevoflurane. During femoral puncture, local percutaneous anesthetic was used (lidocaine).

Digital Subtraction Angiography (DSA)

Both swine underwent percutaneous (by right femoral artery puncture) selective digital subtraction angiography (DSA) before and after stenting, including 3D rotational angiography (3DRA). All procedures and follow-ups were performed on a biplane, flat-panel DSA unit (Allura Xper FD20, Philips, Eindhoven, The Netherlands). Arterial diameters for stent size choice were measured by 3DRA (intravenous contrast medium injection rate: 4 mL/s and volume: 16 mL, delay of 1 s). In one case the stent was undersized and in the other case it was oversized, comparatively to the optimal stent choice, by one or two commercial product sizes (diameters), respectively. The reason why a higher over- or undersizing was not chosen relies on the fact that in clinical practice these are the commonest sizing modifications, either chosen deliberately, in order to produce thicker coverage, or by the fact that in important parent-vessel curvatures the convex part of the artery provokes a slight dilatation of the stent pores. The stent length was kept the same for both cases (20 mm). Immediately after deployment, the stent mesh was evaluated with 3D reconstructions of high-resolution CT (HR CT) scans in the angiographic suite. The 3DRA raw data were used a posteriori as an input for the vascular reconstruction, while the 3D HR CT reconstructions of the stents were used only as a means of qualitative evaluation of the virtual stent deployment result.

Four-Dimension (Time Resolved 3D) Phase Contrast MRA

Time-resolved three-dimensional phase-contrast magnetic resonance (PC-MR) data were obtained on a 3.0 Tesla system (Achieva, Philips Healthcare, Best, The Netherlands), using a 16-channel neuro-vascular coil (SENSE-NV-16), with retrospective electrocardiographic gating. The sequence was performed before endovascular stenting and immediately after the procedure for each subject, in intubated swine under general anesthesia, with MRI-compatible anesthesiology apparatus. The scanning parameters were: Field of View (FOV) 230 × 188.5 × 40.5 mm, reconstruction matrix = 352, ACQ voxel MPS (mm) = 0.80/0.86/0.90, REC voxel MPS (mm) = 0.65/0.65/0.90, acquisition matrix MXP = 288 × 219, TR/TE = 8.3 ms/4.5 ms, $\alpha = 15$ degree, NEX = 1 and receiver bandwidth = 64.8 kHz. A flow velocity encoding (venc) = 70 cm/s was chosen along each of the three principal axes. The data was reconstructed at 14 time points evenly spaced over the R-R interval (cardiac cycle). Total acquisition times for 4D flow MR imaging were heart rate-dependent and ranged between 18 and 25 minutes.

Scanning Electron Microscopy (SEM)

The stented arteries were harvested, chemically fixed, and longitudinally opened, in order to expose the inner surface of the stented ascending pharyngeal artery (APhA) ostia. The chemical fixation and scanning electron microscopy preparation protocols are described elsewhere [70]. Inner surfaces of the opened arteries were observed and photographed with SEM (JEOL JSM-7400F). Digital pictures were taken at magnifications 75×, with a voltage of 10 kV. Quantification of the endothelialized ostia surfaces was performed with the image-J1 (NIH, USA) open-source software [71], by hand selection of the nonendothelialized areas. The functional surfaces corresponded to the nonendothelialized openings through the stent struts at 3 months.

4.2.2 Computational Setup

In addition to the experimental investigations, case-specific blood flow simulations were carried out. These included pre- and poststenting computations for each dataset in order to evaluate the impact of over- and undersized stent deployments.

Vascular Reconstruction and Device Modeling

The three-dimensional vessel models were segmented based on the acquired DSA images using MeVisLab 2.3. The final surface models were reviewed, approved by an attending clinician, and are presented in Fig. 14.14. To be able to define appropriate boundary conditions, the in- and outflow regions were truncated orthogonal to the corresponding vessel axis and extruded by at least six mean vessel diameters. Unstructured volume meshes consisting of polyhedral as well as prism elements at the wall were generated using STAR-CCM+ 9.02 (case 1: 0.8 million elements; case 2: 1.5 million elements).

In addition to the untreated configurations, flow-diverting devices have been virtually deployed. Here, an undeformed Pipeline Embolization Device was virtually reproduced based on the implanted stents that were used during each intervention. The stent diameter was selected as 5 mm, whereas for each strut a mean diameter of 30 μm was assumed. The initial length was defined as 200 mm and the angles within the stent pores of the undeformed configurations were chosen to be 44 degree and 136 degree, respectively. Afterwards, an in-house software package VISCA was used to virtually deploy the stents in the corresponding vessel section (see Fig. 14.15) [27]. To reproduce the real stenting with a high accuracy, angle measurements of the deployed flow-diverters were carried out after sacrificing the pigs. This led to the following combinations of acute and blunt angle, respectively: 80 and 100 degree for case 1 (oversizing), 55 and 125 degree for case 2 (undersizing). Due to the fine resolution required to describe the stent pores explicitly, high element numbers resulted for the poststenting cases (case 1: 6.8 million; case 2: 8.1 million).

Hemodynamic Simulations

In all four simulations, blood was treated as a laminar, incompressible ($\rho = 1055\,\mathrm{kg/m^3}$) and Newtonian ($\eta = 0.004\,\mathrm{Pa\,s}$) fluid. Vessel walls as well as the flow-diverting devices were assumed to be rigid. In each subject, time-dependent flow measurements were carried out before and after stent deployment using PC-MRI, which were applied to

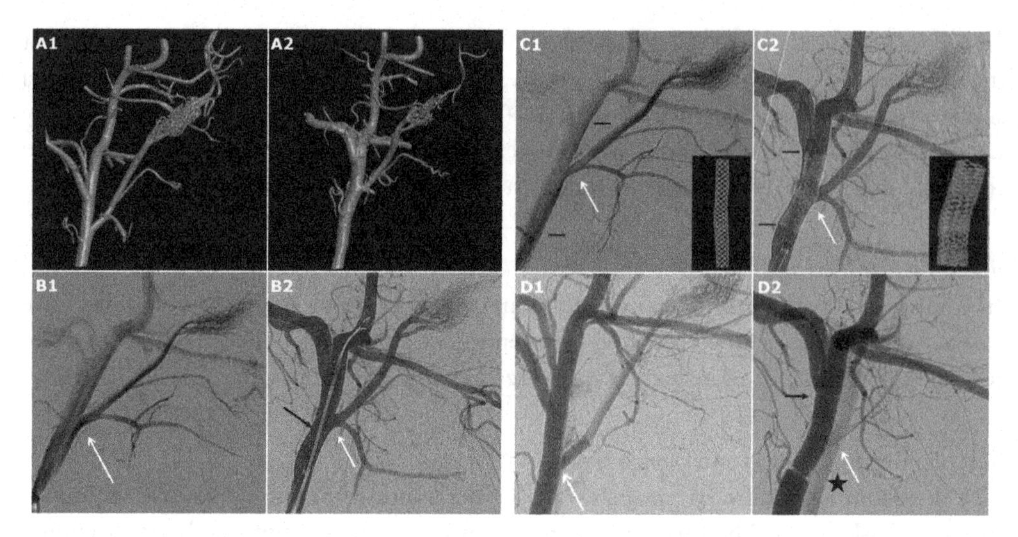

Fig. 14.14 Over- and undersizing of the flow-diverting stent. The numbers correspond to the considered cases. The common carotid artery (CCA)-ascending pharyngeal artery (APhA) bifurcation is marked with a *white arrow* on all DSA figures. (A) Three-dimensional rotational angiographic (3DRA) reconstructions performed with selective contrast medium injection from the right CCA; they were used in order to measure CCA diameters for stent sizing; CCA diameter for case 1 was found to be 4.5 mm and for case 2 was 5.3 mm. The 3DRA raw data were a posteriori used in order to create the 3D anatomic simulations. (B) Selective DSA runs from the right CCA at working projection, before stenting; notice the undeployed flow-diverting stent inside the microcatheter in figure [B2] (*black arrow*). (C) Selective DSA runs from the right CCA at working projection, immediately after stent deployment; the stents were chosen 2 commercial sizes above (case 1) and 1.5 commercial sizes below (case 2) nominal diameter; the stent borders are marked with *black lines*; notice the difference in the deployed stent length among the two cases. The inserted figures on the right show a 3D-rotational CT reconstruction of the deployed stent, showing stent elongation and open pores in case 1 and the shortening of the stent and narrower pores in case 2. (D) Selective DSA runs from the right CCA at working projection, 3 months poststenting; notice the important remodeling with absence of enhancement of the distal part of the right APhA in case 2, comparatively to case 1. The *black arrow* shows parent artery remodeling, and the *asterisk* shows the enhancement of the contralateral CCA during CM injection [24].

the main vessels (common carotid artery (CCA), external carotid artery (ECA), and ascending pharyngeal artery (AphA)). A previous study of Berg et al. [72] demonstrates how hemodynamics CFD simulations based on PC-MRI inflow conditions lead to realistic blood flow predictions. For the remaining outflow side branches, which showed relatively small diameters compared to the three main arteries, a flow splitting was defined that depended on the individual surface areas. For each configuration, three cardiac cycles were simulated using the commercial finite volume CFD solver STAR-CCM+ 9.02. The first two cycles were discarded afterwards, and only the last was considered for analysis.

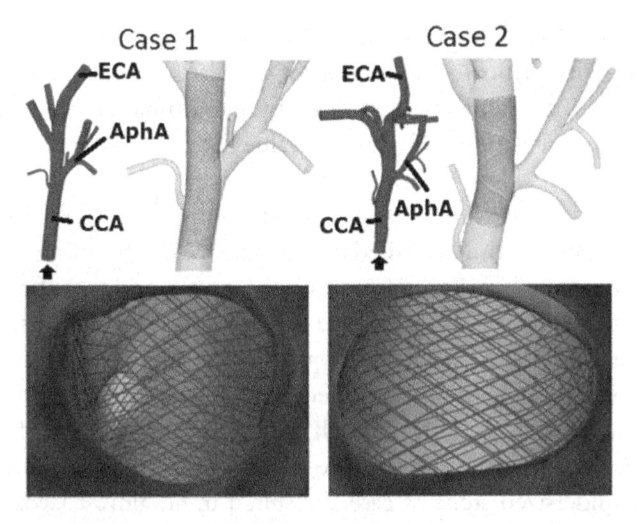

Fig. 14.15 Reconstructed surface models of the investigated arterial sections and the virtual stent deployments for case 1 (*left*) and case 2 (*right*). Opaque representation of the prestented vessel with the flow direction indicated by a *black arrow*, transparent representation after the virtual deployment of the Pipeline Embolization Devices (*top*), luminal perspective from each AphA (*bottom*). Case 1 represents the oversizing and case 2 demonstrates the undersized configuration. The pore angles correspond to the measurements of the real stent deployments acquired after the 3 months of follow-up, by scanning electron microscopy images of the stented ostia [24].

To evaluate the impact of the virtual deployment of a Pipeline Embolization Device (PED) on the local hemodynamics, velocity cut-planes were compared pre- and poststenting in order to describe the flow behavior close to the covered vessel ostium. To analyze the ostium coverage of neointimal cells, the wall shear stress (WSS_{stent}) at the stent was computed. Due to the high spatial resolution of the individual flow-diverter struts, the velocity gradients at the stent pores can be computed. Eq. (14.2) defines the WSS_{stent} with the dynamic viscosity of blood η, the velocity components along the stent u_{stent} and the height above the stent surface n in normal direction.

$$WSS_{\text{stent}} = \eta\sqrt{\left(\frac{\partial u_{x,\text{stent}}}{\partial n_x}\right)^2 + \left(\frac{\partial u_{y,\text{stent}}}{\partial n_y}\right)^2 + \left(\frac{\partial u_{z,\text{stent}}}{\partial n_z}\right)^2} \tag{14.2}$$

4.2.3 Experimental and Numerical Results

No thrombus formation was ascertained for cases 1 and 2 on immediate and 3 months' DSA controls. On the 3-month controls, important remodeling of the APhA and slight intra-stent stenosis were found for case 2, and slight remodeling of the APhA ostium for case 1 (see Fig. 14.14). At 3-month controls, the ostia were successfully harvested. The boundaries of the initial ostia had disappeared due to a coverage by neointimal tissue. Furthermore, a reduction of circulating ostia surfaces was observed. The free segments of the stents were covered by neointimal cells. Ostia surface quantification resulted in circulating ostia surface values of 359,208 μm^2 and 142,937 μm^2, respectively.

The blood flow simulations of the stented configurations reveal the impact of the flow-diverting devices on the jailed vessel branches. The cycle-averaged velocity distributions are illustrated in Fig. 14.16. For the poststenting configurations, highest velocities are present for case 1 at the distal section of the ostia and specifically in the narrow regions between the stent struts. To quantify the effect of both stenting scenarios, the relative flow reductions through the jailed branches are calculated. Here, the ratio of the mean flow rate through the jailed branch with respect to the one of the parent vessel is considered, pre- and poststenting. In case 1, a mean decrease of 14.1% occurred. In comparison, the undersized stent of case 2 resulted in an almost twofold reduction of the mean flow rate (25.5%).

Besides hemodynamic change characterizations, caused by the flow-diverting devices, the reverse effect of the flow on the implants is evaluated. Since the stent struts are spatially discretized with high resolution, the shear stresses can be calculated accordingly. Fig. 14.16 further illustrates the WSS_{stent} distribution along the stent region that covers the ostium of each AphA. Regions that were previously associated with high velocities also experience increased shear stresses. In particular, the distal part of the covered ostium presents with the highest values. For the peak-systolic velocity, this effect even increases but does not change the relative WSS_{stent} distribution. The comparison of cases 1 and 2 showed that the shear stress distribution in the undersized configuration appears to be much more homogeneous. In contrast, case 1 experiences increased shear load especially across the distal area of the jailed side branch.

4.2.4 Conclusions and Summary

Although flow-diverting devices represent a very interesting treatment alternative for complex intracranial aneurysm cases, jailed side branches entail a risk of inadvertent occlusion, even in properly antiaggregated patients. Various reasons may be at the root of this phenomenon, one of which is stent size choice. To evaluate the effect of over- and undersizing in side branch jailing with flow-diverting stents, two large white swine cases were studied with an image-based CFD method and results were confirmed by scanning electron microscopy after 3 months of follow-up. The deployment of an over- and undersized flow-diverter revealed ostium patency for the oversized case, while for the undersized case, significant ostium neointimal formation occurred.

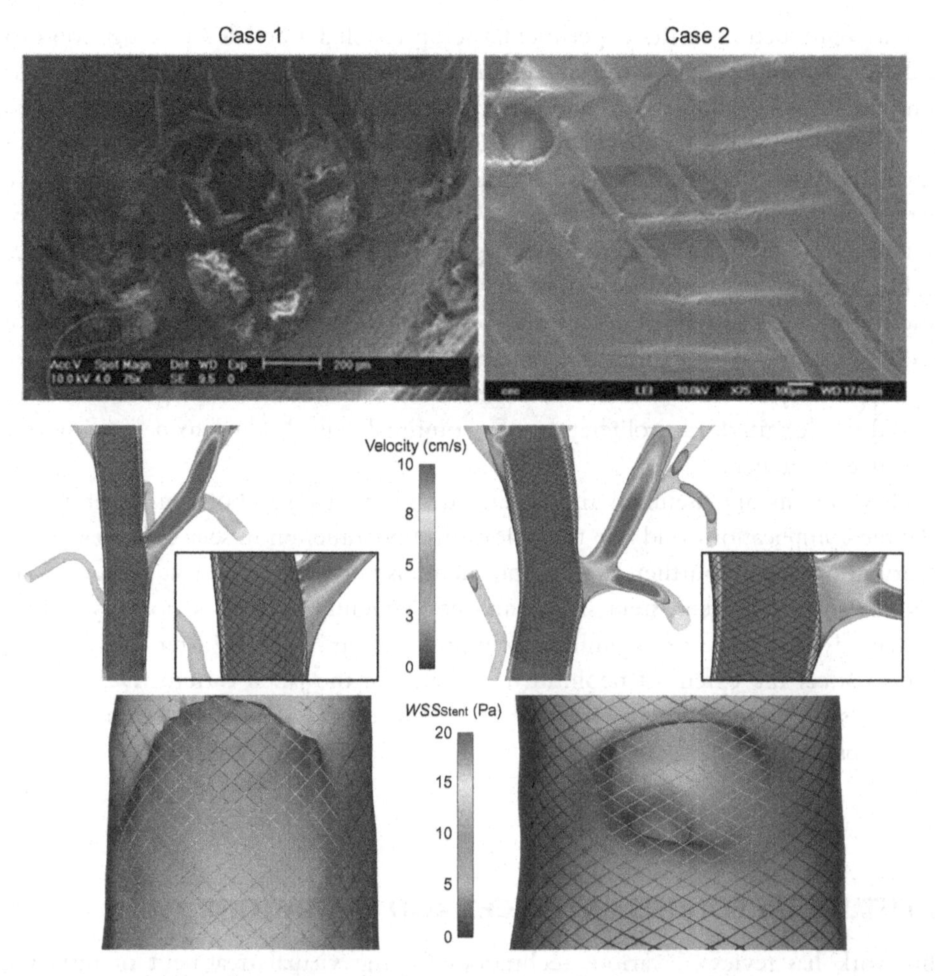

Fig. 14.16 Comparison of experimental and numerical results for case 1 (*left*) and case 2 (*right*). Scanning electron microscopic image of the APhA ostium at 3 months showing the difference in acute stent strut angles and the ostium coverage (*top row*); cycle-averaged cut-plane velocity magnitude for the poststented configurations. The panels contain a magnification of the ostium region to demonstrate the impact of the flow-diverting device on the local hemodynamics (*middle row*); intra-vascular perspective of the stent shear stress. The flow-diverting device of case 1 experiences much higher shear loads, especially at the most distal region of the ostium. In contrast, the shear distribution for case 2 is more homogeneous across the covered ostium and lower WSS_{stent} values are present [24].

The reproduction of the experimental setup enabled a detailed investigation of the local hemodynamics close to the jailed branch ostia. For the oversized stenting scenario, high velocities through the stent pores were still present after treatment. On the contrary, the undersized configuration led to an almost twofold flow rate decrease in the side branch. The values of circulating ostia surfaces seemed to follow the pattern of flow rate decrease at the level of the ostia. It seems that the important flow reduction caused by the interference of a thicker stent mesh triggers neointimal proliferation accordingly.

The identification of shear stress levels along the stent struts provided further indications for the pattern of neointimal formation at the ostium of each case. Increased WSS_{stent} was present in regions that remained patent after 3 months of perfusion, while on the contrary, free stent areas that experienced low shear values right after stenting enabled the longitudinal proliferation of neointimal cells, leading to narrowing of the jailed arterial branch.

These results are useful in therapeutic decision making, both in order to avoid ischemic complications, and also to exploit this phenomenon in selected cases for faster aneurysm occlusion. Further, the findings demonstrate that when a side artery is jailed by a flow-diverting stent, stent sizing, and consequently local stent porosity influence the hemodynamic parameters immediately poststenting inside the artery. These in turn seem to affect the extent of neointimal coverage of the jailed ostium. Depending on the stent strut compression, neointimal formation on the stent struts seems to lead to reduced patency.

5. FUTURE DIRECTIONS—CHANCES AND LIMITATIONS

This work has reviewed various techniques for the virtual treatment of intracranial aneurysms using stents. The presented approaches differ considering their accuracy and the related computational efforts. Careful validation of the methods is absolutely needed before applying them routinely in clinical practice. Despite the large number of published virtual stent deployment techniques, their validation is still rarely discussed. Both geometric as well as flow modification could be examined (Section 3). In vivo validation (Section 4.2) might deliver further insights in the understanding of their working principle and to reduce the complications. With the definition of appropriate objectives, the optimal virtual treatment can be identified for patient-specific cases (Section 4.1).

Further comparisons of the existing methods should be investigated in the future, e.g., as part of virtual stenting challenges [12, 44]. In this regard validation should play an essential role in order to enable the clinical applicability.

ACKNOWLEDGMENTS

The authors thank Dr. Oliver Beuing (Department of Neuroradiology, University Hospital Magdeburg) for fruitful medical discussions and the extensive cooperation. Furthermore, Dr. Christian Rössl (Department of Simulation and Graphics, University of Magdeburg) is warmly acknowledged for his assistance regarding the fast virtual stenting approach. Finally, we thank Dr. Christina Iosif (Jean Minjoz University Hospital, Besançon, France) for a wonderful collaboration regarding stenting experiments in extensive animal studies.

The work is partly funded by the Federal Ministry of Education and Research in Germany within the Research Campus STIMULATE under Grant No. 13GW0095A.

REFERENCES

[1] Bonneville F, Sourour N, Biondi A. Intracranial aneurysms: an overview. Neuroimaging Clin N Am 2006;16(3):371–82, vii. http://dx.doi.org/10.1016/j.nic.2006.05.001.
[2] Juvela S. Risk factors for multiple intracranial aneurysms. Stroke 2000;31(2):392–7. http://dx.doi.org/10.1161/01.STR.31.2.392.
[3] Morita A, Kirino T, Hashi K, Aoki N, Fukuhara S, Hashimoto N, et al. The natural course of unruptured cerebral aneurysms in a Japanese cohort. N Engl J Med 2012;366(26):2474–82. http://dx.doi.org/10.1056/NEJMoa1113260.
[4] Wallace RC, Karis JP, Partovi S, Fiorella D. Noninvasive imaging of treated cerebral aneurysms, Part II: CT angiographic follow-up of surgically clipped aneurysms. Am J Neuroradiol 2007;28(7):1207–12. http://dx.doi.org/10.3174/ajnr.A0664.
[5] Roszelle BN, Gonzalez LF, Babiker MH, Ryan J, Albuquerque FC, Frakes DH. Flow diverter effect on cerebral aneurysm hemodynamics: an in vitro comparison of telescoping stents and the pipeline. Neuroradiology 2013;55(6):751–8. http://dx.doi.org/10.1007/s00234-013-1169-2.
[6] Wallace RC, Karis JP, Partovi S, Fiorella D. Noninvasive imaging of treated cerebral aneurysms, Part I: MR angiographic follow-up of coiled aneurysms. Am J Neuroradiol 2007;28(6):1001–8. http://dx.doi.org/10.3174/ajnr.A0662.
[7] Schneiders JJ, VanBavel E, Majoie CB, Ferns SP, van den Berg R. A flow-diverting stent is not a pressure-diverting stent. Am J Neuroradiol 2013;34(1):E1–4. http://dx.doi.org/10.3174/ajnr.A2613.
[8] Cebral JR, Mut F, Raschi M, Scrivano E, Ceratto R, Lylyk P, et al. Aneurysm rupture following treatment with flow-diverting stents: computational hemodynamics analysis of treatment. Am J Neuroradiol 2011;32(1):27–33. http://dx.doi.org/10.3174/ajnr.A2398.
[9] Ma D, Dargush GF, Natarajan SK, Levy EI, Siddiqui AH, Meng H. Computer modeling of deployment and mechanical expansion of neurovascular flow diverter in patient-specific intracranial aneurysms. J Biomech 2012;45(13):2256–63. http://dx.doi.org/10.1016/j.jbiomech.2012.06.013.
[10] Cebral JR, Löhner R. Efficient simulation of blood flow past complex endovascular devices using an adaptive embedding technique. IEEE Trans Med Imaging 2005;24(4):468–76.
[11] Larrabide I, Kim M, Augsburger L, Villa-Uriol MC, Rüfenacht D, Frangi AF. Fast virtual deployment of self-expandable stents: method and in vitro evaluation for intracranial aneurysmal stenting. Med Image Anal 2012;16(3):721–30. http://dx.doi.org/10.1016/j.media.2010.04.009.
[12] Janiga G, Rössl C, Skalej M, Thévenin D. Realistic virtual intracranial stenting and computational fluid dynamics for treatment analysis. J Biomech 2013;46(1):7–12. http://dx.doi.org/10.1016/j.jbiomech.2012.08.047.
[13] Hann T. Virtuelles Stent-Deployment für zerebrale Aneurysmen. Master's thesis, University of Magdeburg "Otto von Guericke", Germany, 2013.
[14] Bouillot P, Brina O, Ouared R, Yilmaz H, Farhat M, Erceg G, et al. Geometrical deployment for braided stent. Med Image Anal 2016;30:85–94. http://dx.doi.org/10.1016/j.media.2016.01.006.
[15] Antiga L, Piccinelli M, Botti L, Ene-Iordache B, Remuzzi A, Steinman DA. An image-based modeling framework for patient-specific computational hemodynamics. Med Biol Eng Comput 2008;46(11):1097–112. http://dx.doi.org/10.1007/s11517-008-0420-1.

[16] Mut F, Cebral JR. Effects of flow-diverting device oversizing on hemodynamics alteration in cerebral aneurysms. Am J Neuroradiol 2012;33(10):2010–16. http://dx.doi.org/10.3174/ajnr.A3080.
[17] Augsburger L, Reymond P, Rufenacht DA, Stergiopulos N. Intracranial stents being modeled as a porous medium: flow simulation in stented cerebral aneurysms. Ann Biomed Eng 2011;39(2):850–63. http://dx.doi.org/10.1007/s10439-010-0200-6.
[18] Ohta M, Anzai H, Miura Y, Nakayama T. Parametric study of porous media as substitutes for flow-diverter stent. Biomater Biomed Eng 2015;2(2):111–25. http://dx.doi.org/10.12989/bme.2015.2.2.111.
[19] Anzai H, Ohta M, Falcone JL, Chopard B. Optimization of flow diverters for cerebral aneurysms. J Comput Sci 2012;3(1–2):1–7. http://dx.doi.org/10.1016/j.jocs.2011.12.006.
[20] Anzai H, Falcone JL, Chopard B, Hayase T, Ohta M. Optimization of strut placement in flow diverter stents for four different aneurysm configurations. J Biomech Eng 2014;136(6):61006. http://dx.doi.org/10.1115/1.4027411.
[21] Appanaboyina S, Mut F, Löhner R, Putman CM, Cebral JR. Computational fluid dynamics of stented intracranial aneurysms using adaptive embedded unstructured grids. Int J Numer Methods Fluids 2008;57(5):475–93. http://dx.doi.org/10.1002/fld.1590.
[22] Auricchio F, Conti M, de Beule M, de Santis G, Verhegghe B. Carotid artery stenting simulation: from patient-specific images to finite element analysis. Med Eng Phys 2011;33(3):281–9. http://dx.doi.org/10.1016/j.medengphy.2010.10.011.
[23] Babiker MH, Gonzalez LF, Ryan J, Albuquerque F, Collins D, Elvikis A, et al. Influence of stent configuration on cerebral aneurysm fluid dynamics. J Biomech 2012;45(3):440–7. http://dx.doi.org/10.1016/j.jbiomech.2011.12.016.
[24] Berg P, Iosif C, Ponsonnard S, Yardin C, Janiga G, Mounayer C. Endothelialization of over- and undersized flow-diverter stents at covered vessel side branches: an in vivo and in silico study. J Biomech 2016;49(1):4–12. http://dx.doi.org/10.1016/j.jbiomech.2015.10.047.
[25] Cito S, Geers AJ, Arroyo MP, Palero VR, Pallarés J, Vernet A, et al. Accuracy and reproducibility of patient-specific hemodynamic models of stented intracranial aneurysms: report on the virtual intracranial stenting challenge 2011. Ann Biomed Eng 2015;43(1):154–67. http://dx.doi.org/10.1007/s10439-014-1082-9.
[26] de Bock S, Iannaccone F, de Santis G, de Beule M, Mortier P, Verhegghe B, et al. Our capricious vessels: the influence of stent design and vessel geometry on the mechanics of intracranial aneurysm stent deployment. J Biomech 2012;45(8):1353–9. http://dx.doi.org/10.1016/j.jbiomech.2012.03.012.
[27] Janiga G, Berg P, Beuing O, Neugebauer M, Gasteiger R, Preim B, et al. Recommendations for accurate numerical blood flow simulations of stented intracranial aneurysms. Biomed Tech (Berl) 2013;58(3):303–14. http://dx.doi.org/10.1515/bmt-2012-0119.
[28] Janiga G, Daróczy L, Berg P, Thévenin D, Skalej M, Beuing O. An automatic CFD-based flow diverter optimization principle for patient-specific intracranial aneurysms. J Biomech 2015;48(14):3846–52. http://dx.doi.org/10.1016/j.jbiomech.2015.09.039.
[29] Karmonik C, Chintalapani G, Redel T, Zhang YJ, Diaz O, Klucznik R, et al. Hemodynamics at the ostium of cerebral aneurysms with relation to post-treatment changes by a virtual flow diverter: a computational fluid dynamics study. Conf Proc IEEE Eng Med Biol Soc 2013;1895–8. http://dx.doi.org/10.1109/EMBC.2013.6609895.
[30] Kim M, Ionita C, Tranquebar R, Hoffmann KR, Taulbee DB, Meng H, et al. Evaluation of an asymmetric stent patch design for a patient specific intracranial aneurysm using computational fluid dynamic (CFD) calculations in the computed tomography (CT) derived lumen. Proc SPIE Int Soc Opt Eng 2006;6143. http://dx.doi.org/10.1117/12.651773.
[31] Kim M, Taulbee DB, Tremmel M, Meng H. Comparison of two stents in modifying cerebral aneurysm hemodynamics. Ann Biomed Eng 2008;36(5):726–41. http://dx.doi.org/10.1007/s10439-008-9449-4.
[32] Kim YH, Xu X, Lee JS. The effect of stent porosity and strut shape on saccular aneurysm and its numerical analysis with lattice Boltzmann method. Ann Biomed Eng 2010;38(7):2274–92. http://dx.doi.org/10.1007/s10439-010-9994-5.

[33] Kojima M, Irie K, Fukuda T, Arai F, Hirose Y, Negoro M. The study of flow diversion effects on aneurysm using multiple enterprise stents and two flow diverters. Asian J Neurosurg 2012;7(4):159–65. http://dx.doi.org/10.4103/1793-5482.106643.
[34] Kono K, Terada T. Hemodynamics of 8 different configurations of stenting for bifurcation aneurysms. Am J Neuroradiol 2013;34(10):1980–6. http://dx.doi.org/10.3174/ajnr.A3479.
[35] Kulcsár Z, Augsburger L, Reymond P, Pereira VM, Hirsch S, Mallik AS, et al. Flow diversion treatment: intra-aneurismal blood flow velocity and WSS reduction are parameters to predict aneurysm thrombosis. Acta Neurochir 2012;154(10):1827–34. http://dx.doi.org/10.1007/s00701-012-1482-2.
[36] Larrabide I, Aguilar ML, Morales HG, Geers AJ, Kulcsár Z, Rüfenacht D, et al. Intra-aneurysmal pressure and flow changes induced by flow diverters: relation to aneurysm size and shape. Am J Neuroradiol 2013;34(4):816–22. http://dx.doi.org/10.3174/ajnr.A3288.
[37] Liou TM, Li YC. Effects of stent porosity on hemodynamics in a sidewall aneurysm model. J Biomech 2008;41(6):1174–83. http://dx.doi.org/10.1016/j.jbiomech.2008.01.025.
[38] Meng H, Wang Z, Kim M, Ecker RD, Hopkins LN. Saccular aneurysms on straight and curved vessels are subject to different hemodynamics: implications of intravascular stenting. Am J Neuroradiol 2006;27(9):1861–65.
[39] Mut F, Ruijters D, Babic D, Bleise C, Lylyk P, Cebral JR. Effects of changing physiologic conditions on the in vivo quantification of hemodynamic variables in cerebral aneurysms treated with flow diverting devices. Int J Numer Methods Biomed Eng 2014;30(1):135–42. http://dx.doi.org/10.1002/cnm.2594.
[40] Mut F, Raschi M, Scrivano E, Bleise C, Chudyk J, Ceratto R, et al. Association between hemodynamic conditions and occlusion times after flow diversion in cerebral aneurysms. J Neurointerv Surg 2015;7(4):286–90. http://dx.doi.org/10.1136/neurintsurg-2013-011080.
[41] Paliwal N, Yu H, Xu J, Xiang J, Siddiqui AH, Yang X, et al. Virtual stenting workflow with vessel-specific initialization and adaptive expansion for neurovascular stents and flow diverters. Comput Meth Biomech Biomed Eng 2016;1–9. http://dx.doi.org/10.1080/10255842.2016.1149573.
[42] Peach TW, Ngoepe M, Spranger K, Zajarias-Fainsod D, Ventikos Y. Personalizing flow-diverter intervention for cerebral aneurysms: from computational hemodynamics to biochemical modeling. Int J Numer Methods Biomed Eng 2014;30(11):1387–407. http://dx.doi.org/10.1002/cnm.2663.
[43] Peach TW, Spranger K, Ventikos Y. Towards predicting patient-specific flow-diverter treatment outcomes for bifurcation aneurysms: from implantation rehearsal to virtual angiograms. Ann Biomed Eng 2016;44(1):99–111. http://dx.doi.org/10.1007/s10439-015-1395-3.
[44] Radaelli AG, Augsburger L, Cebral JR, Ohta M, Rüfenacht DA, Balossino R, et al. Reproducibility of haemodynamical simulations in a subject-specific stented aneurysm model—a report on the virtual intracranial stenting challenge 2007. J Biomech 2008;41(10):2069–81. http://dx.doi.org/10.1016/j.jbiomech.2008.04.035.
[45] Seshadhri S, Janiga G, Beuing O, Skalej M, Thévenin D. Impact of stents and flow diverters on hemodynamics in idealized aneurysm models. J Biomech Eng 2011;133(7):71005. http://dx.doi.org/10.1115/1.4004410.
[46] Shobayashi Y, Tateshima S, Kakizaki R, Sudo R, Tanishita K, Viñuela F. Intra-aneurysmal hemodynamic alterations by a self-expandable intracranial stent and flow diversion stent: high intra-aneurysmal pressure remains regardless of flow velocity reduction. J Neurointerv Surg 2013;5 Suppl 3:iii38–42. http://dx.doi.org/10.1136/neurintsurg-2012-010488.
[47] Song Y, Choe J, Liu H, Park KJ, Yu H, Lim OK, et al. Virtual stenting of intracranial aneurysms: application of hemodynamic modification analysis. Acta Radiol 2015;57(8):992–7. http://dx.doi.org/10.1177/0284185115613653.
[48] Spranger K, Ventikos Y. Which spring is the best? Comparison of methods for virtual stenting. IEEE Trans Biomed Eng 2014;61(7):1998–2010. http://dx.doi.org/10.1109/TBME.2014.2311856.
[49] Spranger K, Capelli C, Bosi GM, Schievano S, Ventikos Y. Comparison and calibration of a real-time virtual stenting algorithm using finite element analysis and genetic algorithms. Comput Methods Appl Mech Eng 2015;293:462–80. http://dx.doi.org/10.1016/j.cma.2015.03.022.
[50] Srinivas K, Townsend S, Lee CJ, Nakayama T, Ohta M, Obayashi S, et al. Two-dimensional optimization of a stent for an aneurysm. J Med Devices 2010;4(2):21003. http://dx.doi.org/10.1115/1.4001861.

[51] Tanemura H, Ishida F, Miura Y, Umeda Y, Fukazawa K, Suzuki H, et al. Changes in hemodynamics after placing intracranial stents. Neurol Med Chir 2013;53(3):171–8.
[52] Tremmel M, Xiang J, Natarajan SK, Hopkins LN, Siddiqui AH, Levy EI, et al. Alteration of intra-aneurysmal hemodynamics for flow diversion using enterprise and vision stents. World Neurosurg 2010;74(2–3):306–15. http://dx.doi.org/10.1016/j.wneu.2010.05.008.
[53] Xu J, Deng B, Fang Y, Yu Y, Cheng J, Wang S, et al. Hemodynamic changes caused by flow diverters in rabbit aneurysm models: comparison of virtual and realistic FD deployments based on micro-CT reconstruction. PLoS ONE 2013;8(6):e66072. http://dx.doi.org/10.1371/journal.pone.0066072.
[54] Zhang YS, Yang XJ, Wang SZ, Qiao AK, Chen JL, Zhang Ky, et al. Hemodynamic effects of stenting on wide-necked intracranial aneurysms. Chin Med J 2010;123(15):1999–2003.
[55] Zhang Y, Chong W, Qian Y. Investigation of intracranial aneurysm hemodynamics following flow diverter stent treatment. Med Eng Phys 2013;35(5):608–15. http://dx.doi.org/10.1016/j.medengphy.2012.07.005.
[56] Zhong J, Long Y, Yan H, Meng Q, Zhao J, Zhang Y, et al. Fast virtual stenting with active contour models in intracranical aneurysm. Sci Rep 2016;6:21724. http://dx.doi.org/10.1038/srep21724.
[57] Ma D, Dumont TM, Kosukegawa H, Ohta M, Yang X, Siddiqui AH, et al. High fidelity virtual stenting (HiFiVS) for intracranial aneurysm flow diversion: in vitro and in silico. Ann Biomed Eng 2013;41(10):2143–56.
[58] Bouillot P, Brina O, Ouared R, Lovblad KO, Farhat M, Pereira VM. Hemodynamic transition driven by stent porosity in sidewall aneurysms. J Biomech 2015;48(7):1300–09. http://dx.doi.org/10.1016/j.jbiomech.2015.02.020.
[59] Daróczy L, Janiga G, Thévenin D. Systematic analysis of the heat exchanger arrangement problem using multi-objective genetic optimization. Energy 2014;65:364–73.
[60] Daróczy L, Mohamed M, Janiga G, Thévenin D. Analysis of the effect of a slotted flap mechanism on the performance of an H-Darrieus turbine using CFD (GT2014-25250). In: ASME Turbo Expo Conference, Düsseldorf, 2014.
[61] Cebral JR, Mut F, Weir J, Putman C. Quantitative characterization of the hemodynamic environment in ruptured and unruptured brain aneurysms. Am J Neuroradiol 2011;32(1):145–51.
[62] Meng H, Tutino VM, Xiang J, Siddiqui A. High WSS or low WSS? Complex interactions of hemodynamics with intracranial aneurysm initiation, growth, and rupture: toward a unifying hypothesis. Am J Neuroradiol 2014;35(7):1254–62.
[63] Geers AJ, Larrabide I, Morales HG, Frangi AF. Approximating hemodynamics of cerebral aneurysms with steady flow simulations. J Biomech 2014;47(1):178–85.
[64] Castro MA, Olivares MCA, Putman CM, Cebral JR. Unsteady wall shear stress analysis from image-based computational fluid dynamic aneurysm models under Newtonian and Casson rheological models. Med Biol Eng Comput 2014;52(10):827–39.
[65] Brinjikji W, Murad MH, Lanzino G, Cloft HJ, Kallmes DF. Endovascular treatment of intracranial aneurysms with flow diverters: a meta-analysis. Stroke 2013;44(2):442–7. http://dx.doi.org/10.1161/STROKEAHA.112.678151.
[66] Cohen JE, Gomori JM, Moscovici S, Leker RR, Itshayek E. Delayed complications after flow-diverter stenting: reactive in-stent stenosis and creeping stents. J Clin Neurosci 2014;21(7):1116–22. http://dx.doi.org/10.1016/j.jocn.2013.11.010.
[67] Iosif C, Berg P, Ponsonnard S, Carles P, Saleme S, Pedrolo-Silveira E, et al. Role of terminal and anastomotic circulation in the patency of arteries jailed by flow-diverting stents: animal flow model evaluation and preliminary results. J Neurosurg 2016;1–11. http://dx.doi.org/10.3171/2015.8.JNS151296.
[68] Iosif C, Berg P, Ponsonnard S, Carles P, Saleme S, Ponomarjova S, et al. Role of terminal and anastomotic circulation in the patency of arteries jailed by flow-diverting stents: from hemodynamic changes to ostia surface modifications. J Neurosurg 2016;1–12. http://dx.doi.org/10.3171/2016.2.JNS152120 10.3171/2016.2.JNS152120.
[69] Kilkenny C, Browne W, Cuthill IC, Emerson M, Altman DG. Animal research: reporting in vivo experiments—the ARRIVE guidelines. J Cereb Blood Flow Metab 2011;31(4):991–3. http://dx.doi.org/10.1038/jcbfm.2010.220.

[70] Iosif C, Carles P, Trolliard G, Yardin C, Mounayer C. Scanning electron microscopy for flow-diverting stent research: technical tips and tricks. Microscopy 2015;64(3):219–23. http://dx.doi.org/10.1093/jmicro/dfv009.

[71] Schindelin J, Arganda-Carreras I, Frise E, Kaynig V, Longair M, Pietzsch T, et al. Fiji: an open-source platform for biological-image analysis. Nat Methods 2012;9(7):676–82. http://dx.doi.org/10.1038/nmeth.2019.

[72] Berg P, Stucht D, Janiga G, Beuing O, Speck O, Thévenin D. Cerebral blood flow in a healthy Circle of Willis and two intracranial aneurysms: computational fluid dynamics versus four-dimensional phase-contrast magnetic resonance imaging. J Biomech Eng 2014;136(4):041003/1–9. http://dx.doi.org/10.1115/1.4026108.

CHAPTER 15

Preoperative Planning of Endovascular Procedures in Aortic Aneurysms

I. Macía[*,†], J.H. Legarreta[*,†], K. López-Linares[*,†], C. Doblado[*,†], L. Kabongo[*,†]
[*]Vicomtech-IK4 Foundation, San Sebastián, Spain
[†]Biodonostia Health Research Institute, San Sebastián, Spain

Chapter Outline

1. INTRODUCTION

An Abdominal Aortic Aneurysm (AAA) is a condition consisting of a weakening and ballooning of the abdominal aorta, exceeding the normal diameter by more than 50%. If not treated, it tends to grow and rupture with high risk of mortality. The management of an AAA depends on the size or diameter of the aneurysm and the balance between the risk of aneurysm rupture and the operative mortality [1]. When the maximum aortic diameter exceeds 3.0 cm, it is considered an AAA. There is consensus that for very small

Computing and Visualization for Intravascular Imaging and Computer-Assisted Stenting
http://dx.doi.org/10.1016/B978-0-12-811018-8.00015-1

(3.0–3.9 cm) aneurysms, the risk of rupture is negligible. Therefore, these aneurysms do not require surgical intervention and should be kept under surveillance [1].

Endovascular Aneurysm Repair (EVAR) is a minimally invasive alternative to open surgical repair involving the deployment and fixation of a stent graft using a catheter. This procedure excludes the damaged wall from circulation and creates an intraluminal thrombus, which tends to shrink after a successful intervention. EVAR is less aggressive than open surgery and has significantly lower operative mortality, but the 2-year postoperative survival rates of both approaches are almost equivalent [2]. While in open repair the aneurysm wall is removed, and the aortic wall is replaced by a graft, in endovascular repair the aneurysm is not removed, but excluded from blood circulation. In the long term, this may lead to procedure-specific complications known as endoleaks (recurrent flow into the thrombus area), due to deficient stent fixation or device wear. EVAR requires patient-specific preoperative planning to select the endoluminal device that better suits the patient's aorta anatomy. This planning is performed on the basis of Computerized Tomography Angiography (CTA) imaging, where the contrast media enhances the blood flow. Aortic diameters and lengths are then quantified by means of computerized image analysis tools to select the most appropriate device from a catalogue or to design a patient-specific graft in the case of complex anatomies. However, the relative position and deformation introduced by the catheter-aorta and endograft-aorta systems is unknown at the planning and design stage, which may cause postoperative complications. Advanced image analysis tools that systematize measurements, potentially including simulation and prediction of deformations, may lead to better endograft design and a reduction of the long-term post-EVAR mortality.

For more than 60 years, open surgery has been the gold standard for aortic aneurysm repair. However, over the past decades, the management of AAAs has changed dramatically due to the development of the EVAR technique. EVAR is currently the preferred surgical procedure to treat AAAs. The procedure's success is measured in terms of the correct exclusion of the aneurysm from blood circulation; in that case, the pressure exerted by the blood on the aortic wall decreases, leading to an eventual reduction in aneurysm size and decreasing aortic rupture risk. The main advantages of this treatment are surgical time reduction, lower perioperative mortality, faster recovery times and improved postoperative pain control. Nonetheless, the long-term survival rate is almost equivalent to open surgery due to endoleaks [2]. Endoleaks are a major EVAR complications defined as a persistent flow into the excluded aneurysm sac because of incorrect sealing, endograft defects or breakdown, or retrograde blood flow from collateral vessels. Endoleaks may cause aneurysm growth and hence, increased rupture risk that may lead to reintervention.

EVAR requires adequate aortic and iliac fixation sites for effective stent graft sealing. These requirements should be carefully evaluated prior to surgery with adequate aortoiliac imaging (usually using CTA) to assess suitability for EVAR and to configure the most appropriate endograft for each patient [1]. An accurate image-based EVAR planning leads to a personalized and precise endograft design, which may better fit the patient-specific anatomy, and reduces the intervention time and the associated risks. It has been shown [3] that using a 3D workstation for planning reduces the occurrence of endoleaks.

2. OVERVIEW OF ENDOGRAFT SIZING FOR AORTIC ANEURYSMS

EVAR planning requires the evaluation of the anatomy of the patient to select a patient-specific prosthesis. Key planning parameters are usually obtained from the preoperative study, for which the CTA remains the gold standard imaging protocol [1]. Some studies have reported that nonenhanced Magnetic Resonance Imaging (MRI) and Magnetic Resonance Angiography (MRA) could be an alternative to CTA for preoperative planning. Compared to CTA, MRI or MRA eliminate the ionizing radiation source, and reduce the frequency of anaphylactic reaction and contrast material-induced nephropathy [4, 5].

Endograft device sizing is based on the quantification of diameters and lengths along the aorta and its subsidiary branches. A standard endograft is selected in the majority of cases from a set of off-the-shelf models from the manufacturer's catalogue. According to Ref. [5], vessel characterization is more accurate when performed using a 3D workstation, where the measurements are based on a 3D view of the aorta. Thus, the complete isolation of the aortic lumen from the surrounding tissues and the extraction of its centerline are of vital importance for the planning (see Sections 6.2.1 and 6.2.2). Obtaining a segmentation of the aortic tree to display a 3D model of the aorta, or alternatively using volume rendering and specific image reconstructions, usually completed with the vessel centerline tree, is the basis for a good 3D planning. The deformations introduced by the deployment process, which tend to straighten the aorta as the endograft is being released with the catheter, add uncertainty to the procedure outcome.

Planning of fenestrated endografts (those expanding above the renal arteries and provisioning holes for those and other subsidiary arteries) is more complicated, requiring the specification of fenestration parameters, such as height, diameter, and angle. Tools specifically developed for fenestrated Endovascular Aneurysm Repair (f-EVAR) planning are scarce but essential for a successful intervention. Commercial workstations for EVAR planning usually offer a specialized vascular module, allowing segmentation, centerline extraction, and quantification of diameters and lengths with the help of 3D visualizations or planar reformattings for standard, nonfenestrated endograft sizing. Some of these software will be reviewed in Section 7.

3. VASCULAR SEGMENTATION

As stated in Section 1, EVAR has substituted conventional open repair for abdominal aortic aneurysms. An initial preoperative planning based on patient-specific anatomy is required to determine suitability for EVAR and to size the endograft accurately in that case. In this scenario, the segmentation of the vascular structures is of utmost importance, since it is a fundamental step for a precise vessel visualization and the posterior vascular analysis. An aneurysm is usually composed of a luminal area and a thrombotic area: when deciding if an aneurysm is elective for EVAR, the whole aneurysm diameter is measured. Nevertheless, only the luminal area is determining for the planning, since the graft is fixated to the inner aneurysm wall. Hence, this section will only focus on the segmentation of the lumen.

3.1 Challenges When Segmenting the Abdominal Aorta

Vessel segmentation for EVAR planning consists in the isolation of the aortic lumen and branches. In CTA studies these structures are clearly visible due to the high image intensity provided by the contrast agent. Nonetheless, segmenting the lumen of the aorta in a reproducible and robust manner is not a trivial task, and cannot rely only on absolute threshold-based approaches. The abdominal aorta segmentation is challenging due to the need of:

- ***Adapting to intensity inhomogeneities*** along the aorta and acquisition-dependent factors, such as noise or artifacts. The injected contrast is not equally spread along the vessel and thus, the segmentation algorithms need to account for these changes.
- ***Adjusting to vessel shape, scale, and curvature changes.*** The segmentation method needs to deal both with very thin branches of the aorta and large diameter areas in the aneurysm. It should also be flexible enough to adapt to tortuous vessels and to differentiate vessels that are close to each other.
- ***Segmenting branches of the aorta.*** Specific aortic branches are used as reference points, and thus have be reliably delineated. If the aneurysm is close to the mesenteric and renal arteries, the endograft has to be designed taking into account their location.
- ***Avoiding oversegmentation*** due to leaking into adjacent structures, such as vertebrae or other surrounding tissue that have similar intensity values (see Fig. 15.1).

Due to these difficulties, the automatic aortic segmentation is still an active field of research. Techniques based exclusively on intensity contrast thresholding fail to distinguish the lumen from adjacent structures of similar pixel values. Introducing a shape constraint is also a difficult task due to the particular anatomical features each patient may present. Hence, most current aorta segmentation schemes employ semiautomatic methods that include correction steps for an integral vascular tree model.

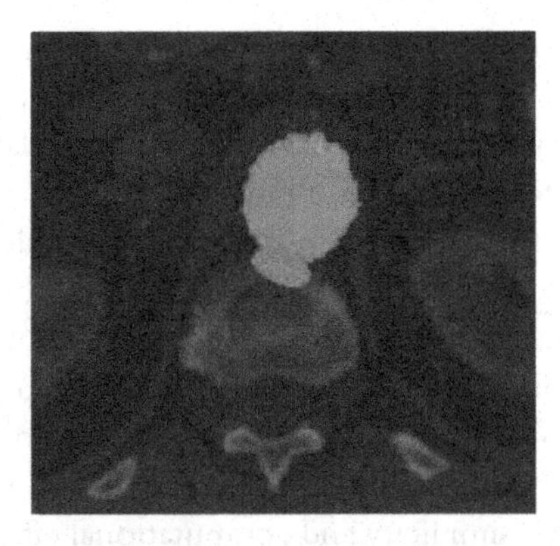

Fig. 15.1 Abdominal aorta oversegmentation. Leakage into vertebrae occurs due to the absence of sharp image intensity contrast in the vessel to bone tissue boundary.

3.2 State of the Art of Abdominal Aorta Segmentation

Aortic lumen segmentation has been recurrently addressed in medical image processing literature in the last decades. Multiple segmentation approaches exist allowing for accurate lumen extraction in a large number of cases. These algorithms can be classified into the following categories:

- region-growing based techniques;
- front propagation based on partial differential equations;
- centerline-based approaches; and
- graph-based methods.

Region growing methods start from a seed point located in the lumen of the aorta and the segmentation is performed by iteratively recruiting neighbor voxels that meet certain inclusion criteria. The success of these algorithms depends on the initialization, usually by manual seed selection, and the criteria employed to control the growing process. Classical region growing approaches relying only on image intensity are sensitive to noise and contrast agent inhomogeneities and are prone to leakage to adjacent structures. Hence, research efforts have been put into adapting the algorithms to take into account intensity inhomogeneities and defining proper growth limiting criteria. Lee et al. [6] proposed to use an empirical stopping criterion based on the mean intensity difference between the luminal and thrombotic surfaces provided by the user. Following this segmentation, a double surface graph search method was utilized to segment the thrombotic area. Almuntashri et al. [7] combined region growing with a logarithmic edge detection algorithm that limits the growth in noncontrasted CT images. Nonetheless, none of these methods deal with bifurcations of the aorta and do not account for contrast inhomogeneities and complex cases where the lumen segmentation can leak into adjacent structures.

Macía et al. [8] developed an adaptive region growing approach that iteratively includes pixels that meet a certain intensity inclusion criterion based on local statistics. The algorithm recomputes the inclusion criterion at regular intervals, based on the statistics of those pixel intensities. This allows to adapt to intensity inhomogeneities along the aorta. However, the algorithm suffers from difficulties in certain cases where structures with similar intensity, such as the vertebrae or the calcifications, provide a continuity in the contrast. Although the algorithm works well in most datasets with default parameters, sometimes a parameter adjustment step may be required to add mis-segmented branches or correct the contours. This algorithm is the basis for the *eVida Vascular* planning software presented in Section 7. The main advantages of region growing approaches are their simplicity and computational efficiency. Nonetheless, these algorithms become less efficient when introducing additional restrictions to prevent leakage. When manual input is required to place the initial seed, the reproducibility of the results may be affected. An automatic seed placement procedure could be provided in order to provide a fully automatic method.

Methods based on front propagation using partial differential equations (PDEs) are commonly used to solve medical image segmentation problems. These algorithms involve solving an initial value problem for some equation for a given time constraint. The solution can be either the image itself at different stages of modification, or some other object (such as a closed curve delineating object boundaries) whose evolution is driven by the image [9]. Wave propagation approaches, active contours (snakes), and level sets [10] with refinements specifically targeting vessel-like objects [11] are widely employed for aortic lumen segmentation. Ayyalasomayajula et al. [12] introduced a 3D active contour method, in which they modified the gradient vector field (GVF) that controls the diffusion of an initial surface. Wang et al. [13] proposed a modified level set method that uses a periodic monotonic speed function to get a coherent propagation of the boundary. This allowed the implementation of a lazy narrow band level set algorithm which prevents repeated computation in points that have reached the vessel border. PDE-based approaches can specifically deal with shape changes and leakage problems by introducing some control over the topology. However, these advantages come at an additional computational cost, a complex parameter setting and special care in ensuring convergence.

The aforementioned methods aim at explicitly extracting the lumen contours by applying intensity and shape related criteria based on edge detection. Some other less widespread approaches have also been proposed. Centerline-based techniques focus on primarily obtaining the vessel centerline, without presegmenting the complete lumen volume. From that centerline, the segmentation of the vessel is obtained. An example of this kind of techniques can be found in Ref. [14], where the aortic lumen is segmented from resampled images orthogonal to a previously extracted centerline.

At each cross-section, the lumen is isolated through a combination of thresholding, morphological operations and connected component analysis. The component that intersects the centerline is then selected and refined by removal of calcium. Apparently, the algorithm does not segment aortic branches and assumes that the intensities along the centerline follow the same Gaussian distribution. Duquette et al. [15] proposed a different approach to both luminal and thrombotic area segmentation based on graph-cuts. The method targets both MRI and contrast-enhanced CT volumes. User interaction is required to initialize the algorithm by selecting a region of interest at every 2 cm, and the graph is constructed from those initializations. Basic graph-cut methods are not well suited for the aorta segmentation problem, so the authors adapted the cost function to deal with low-contrast edges of the aorta and to differentiate the lumen from the aortic wall. The reported segmentation accuracies for both imaging modalities are similar to manual segmentations done by experts. Nonetheless, the method needs a considerable amount of user intervention for initialization and correction.

3.3 Practical Issues

Although the preoperative planning is mainly done on contrast-enhanced CTA images where the lumen exhibits a high image intensity, its segmentation is not straightforward and most segmentation approaches require some editing to include missing branches or to correct the contours that have leaked into adjacent structures. Some commercial software also provide bone-removal tools to eliminate the vertebrae before segmenting the vessel. Including shape and context information improves the results of intensity-based techniques, but introduced constraints should be flexible enough to let the algorithm evolve to the aortic branches without leaking to other structures. Usually, region growing and PDE-based approaches are semiautomatic, starting from a user-selected seed point in the aortic lumen. Nonetheless, some commercial solutions do not require any user interaction for such purpose and provide an automatic lumen segmentation solution, which can then be manually corrected with editing tools. The possibility of interactively refining the segmentations or placing extra seed points may provide an additional degree of robustness in complicated cases.

4. VASCULAR ANALYSIS

Vascular navigation and quantification for endograft sizing requires the extraction of morphological vascular descriptors. These features synthesize the rich geometrical and topological information contained in 3D CT scans and allow for an easier determination of reference points or anatomical landmarks for planning and intervention. Surface-only models that may be obtained in a straightforward manner are difficult to manipulate and are not good descriptors of tubular, elongated and tortuous structures [16]. Centerlines support a more compact representation and hence, are typically used as powerful synthetic descriptors of vascular morphology. According to Ref. [1], the lumen centerline is required for stent graft sizing prior to EVAR, allowing both diameter and length measurements along the aorta and branches, even in the presence of tortuous segments.

4.1 Centerline Extraction and Regularization

The centerline corresponds to the medial line or axis [17] of vessels, whose loci correspond to the centroid of successive cross-sections in the case of CTA volumes. However, the extraction of robust, smooth centerlines is a difficult problem due to the complexity and tortuosity of branched vascular structures. Results may differ depending on the employed technique and are hindered by the presence of different vessel sizes observable in most vessel trees, such as the aorta and its bifurcations. Hence the centerline itself may be considered of multiscale nature [18]. Algorithms dealing with smooth centerline extraction make use, among others, of Voronoi skeletons [19], shock loci of reaction-diffusion equations [20], height ridges of medialness functions [21], and distance transforms [22]. Depending on the extraction and regularization approach, the centerline may be voxel-based, a discrete set of connected linear segments or (typically smooth) mathematical curve approximations, such as polynomials or B-splines.

There are roughly two types of approaches when trying to extract centerlines:

- ***Centerline extraction with prior segmentation***: this approach assumes a prior segmentation or labeling of the vascular tree is available, which is quite feasible usually works well for large vessels. It has the advantage that voxels not corresponding to vascular structures have been previously eliminated. However, the result depends to a large extent on the precision of the segmentation itself.
- ***Direct centerline extraction***: the centerline is estimated with no prior segmentation. This includes extractions based on global metrics, i.e., cost functions or local estimators such as iterative centerline tracking procedures [23].

The large size of the aortic tree as compared to other vessels, whose width may be close to the image resolution, lends itself to a centerline extraction approach based on a prior segmentation. Direct centerline extraction may involve computationally expensive multivoxel algorithms, unless some sort of multiscale coarse-to-fine approach is used for the estimation. In Ref. [8] a segmentation based on a skeletonization of the segmented vessels, which iteratively removes voxels from the external boundaries in a process called thinning [24], is used. Based on predefined voxel template tests, a voxel is marked as nonmeaningful (i.e., it is marked for removal) if it does not alter the topology of the vessel tree. In order to keep centrality, the removal process is driven by the value of a medialness function, which is calculated as a distance map [25] from the boundary inwards. The resulting topological skeleton constitutes a raw voxel-based centerline. One advantage of this algorithm is that parallel versions have been devised to reduce the computation time [24]. The method is sensitive to boundary noise, but in the case of large vessels, such as the aorta, the computed centerline does not deviate importantly from the expected trajectory. Another interesting possibility is to extract the centerline as the ridges of the distance map calculated from the segmentation, as these correspond to the medial line.

The extracted voxel-based centerline cannot be used directly for further vascular analysis, quantification, and navigation. A regularization process is required to avoid jagged voxel-based centerlines and to obtain smooth centerlines and cross-sections whose planes are accurate and vary smoothly. A linear averaging may suffice in large vessels, but excessive smoothing may make the centerline deviate from centrality, as a large averaging may approximate some segments to the boundary in tortuous vessels. Another solution is to use interpolation with smooth curves, such as polynomials or B-splines, as they may be approximated in a piece-wise manner. Medialness cost functions may be introduced in the approximation in order to keep centrality.

Depending on the employed extraction scheme and the presence of spurious structures and local nontubularity, such as in aneurysms, centerlines may be incorrect and spurious or false branches or some disconnections may appear, requiring some correction. Furthermore, navigation and identification of vascular structures requires a topological analysis. In order to provide a computational support for these processes, the vascular tree is organized into a structure called *vessel graph.*

4.2 Vessel Graph Analysis

A vessel graph [8, 26] is a graph-based representation of a vessel tree that may be obtained from the set of extracted centerlines (see Fig. 15.2). This may sound contradictory, as a tree is, indeed, a special type of undirected acyclic graph. However, depending on the centerline extraction algorithm and the presence of spurious or nontubular structures in the original image, the morphological skeleton precursor of the final centerlines may show false cyclic loops. A vessel graph is capable of repressing the topology of the extracted vessels and is suited to store related topological (i.e., connectivity) and geometrical information, such as centerlines corresponding to a vessel segment, and associated cross-sections.

Depending on the centerline extraction approach, the vessel graph may be constructed iteratively, such as in vessel tracking procedures. In this case, a strategy for detecting a bifurcation is required [27]. Global centerline extraction requires parsing the extracted centerlines and performing a topological analysis. Single voxel-width

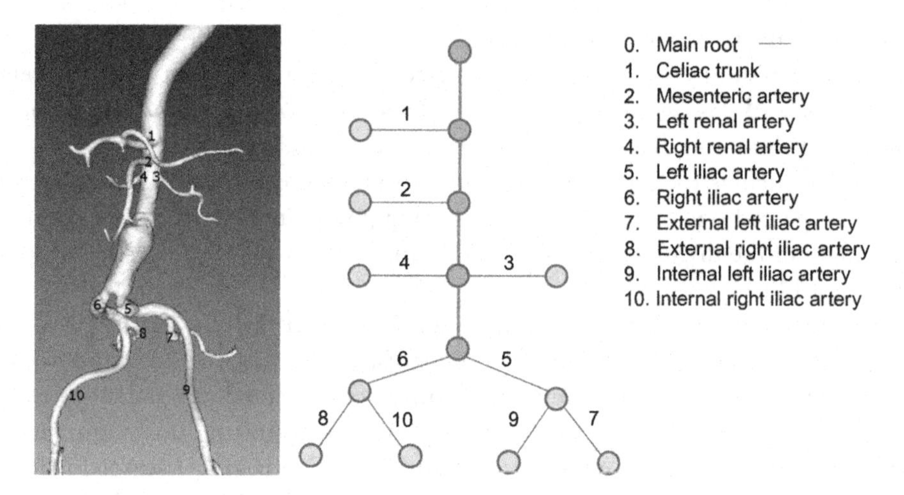

Fig. 15.2 Aortic vessel analysis. Correspondence between the anatomy of the aorta and its topology according to the graph: labeled aorta segmentation (*left*), and target schematic view of the aortic branches for vessel analysis outcome (*right*).

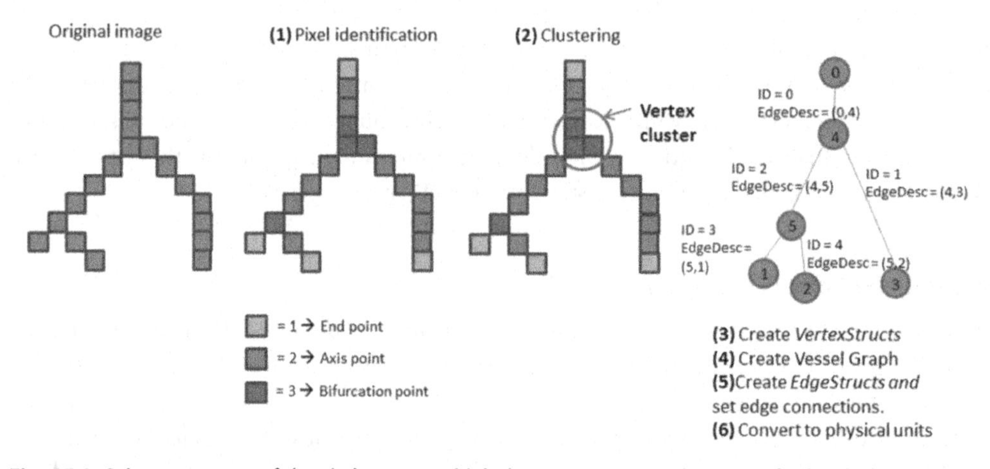

Fig. 15.3 Schematic view of the skeleton voxel labeling process. Voxels are marked as belonging to a bifurcation, some end point, or intermediate edges. Vertex clusters at bifurcations are then identified in order to mark them as belonging to a single bifurcation point. After spurious branch pruning, and loop removal, the final vessel graph contains edges and vertices with unique end points.

centerlines allow distinguishing between end-points, branch-points, and bifurcation-points. In Ref. [8], extracted centerlines' parsing and vessel graph creation is performed iteratively following branch voxels and finding bifurcation and end points through voxel connectivity checking and creating new branches when required (see Fig. 15.3). Knowing the root node is essential in vessel tracking. For the case of the aorta this can be relatively easy, as the graph does not know about anatomical orientation. Macía et al. [8] store centerline information in the edges of the graph, whereas graph nodes represent the bifurcation and associated information.

An advantage of using a vessel graph is that graph-based algorithms may be used to perform different operations, such as detecting and removing loops or spurious branches (branch pruning), searching for minimum paths along the vessel tree, accumulating distances, etc. Furthermore, this allows to smooth branches individually, using some sort of multiscale processing, for example, and taking advantage of the fact that virtually every branch has a different size, as working on a set of clean, true centerlines is preferable. It is desirable that some smoothness is also kept along anatomical centerlines across bifurcation points in the graph.

The vessel graph may be useful also in order to perform other complex tasks such as identification of branches. Note that anatomical branches may consist of more than one connected topological branch in the vessel tree, as it may span more than one bifurcation. For example, the mesenteric bifurcation does not split the aorta, whereas the iliac arteries are a real anatomical bifurcation. Identifying anatomical branches could be relatively straightforward if a clean, error-free set of centerlines and their corresponding graph are available. However, if the presence of errors or spurious branches is suspected or the identification is used purposely for removing or fixing wrong branches, the analysis is more difficult and may require the estimation of diameters and lengths for every topological branch, along with the application of some a priori anatomical knowledge.

From the set of extracted smooth centerlines, tangent and normal vectors may be obtained to perform useful reformattings, such as different sorts of Curved Planar Reformattings (CPR) (see Section 6.1).

4.3 Cross-Section Extraction

Obtaining cross-sections and visualizing them is an important asset for anatomical assessment purposes. Reformation of cross-sections by obtaining local tangent vectors and interpolating normal planes in the acquired imaging volume provides useful visual information by showing the real cross-sections along the vessel path. Cross-sections may be used to estimate vessel diameters relevant to the endograft sizing procedure. If a prior segmentation was present, the extracted plane may provide a segmented cross-section. A simple ray-casting strategy from the central point, corresponding to the centerline, may be used for tracing the (internal) boundaries of the vessel and obtaining a set of diameters. The computation of other vessel cross-section parameters, such as its area, perimeter, and minimum, maximum, and average diameters may also be computed effortlessly. This cross-section information can be stored as part of the centerline structure, since it is associated with every centerline point [28]. If no prior segmentation is available, several methods like simple thresholding (possibly with ray-casting) or region growing may be applied at this stage in order to obtain a segmentation of the section, a strategy that can then be expanded to the whole tree.

5. QUANTITATIVE IMAGE ANALYSIS

Patient-specific planning of an EVAR procedure requires an exhaustive analysis and quantification of the preoperative data in order to (i) select the most suitable endograft based on the patient-specific, image-based measurements, (ii) assess potential intervention risks, and (iii) detect possible complications and measures to prevent them.

Although the preoperative parameters to be explored are well-defined and are usually present in the parameter templates provided by device manufacturers, to the best of our knowledge, there is no standard procedure on how this quantification should be done. Some radiology and vascular surgery specialists approach the planning by manually quantifying required preoperative characteristics from the image, but several studies have reported high inter- and intra-observer variability in these measurements [29, 30]. Thus, efforts have been put in the last two decades into developing semiautomatic and automatic image quantification methods which could lead to more accurate and robust results [31, 32].

5.1 Aortic Characterization for Endograft Sizing

EVAR requires preprocedural planning consisting in image-based patient anatomy assessment for endograft sizing and selection, which may include the design of patient-specific devices in some cases. This process demands the accurate quantification of diameters and lengths along the aorta and its branches [33], as shown in Fig. 15.4.

In the majority of cases, a standard endograft is selected to fit the patient's anatomical characteristics from a set of models in the device manufacturer library or catalogue ensuring the fit. When the healthy aneurysm neck tissue below the renal arteries is not long enough for device fixation, the development of a fenestrated, custom endograft may be necessary. Fenestrated endografts require holes in the device fabric to allow blood flow into the proximal arteries (i.e., celiac trunk, renal, and/or mesenteric arteries), since these devices extend upward beyond the renal arteries. The sizing of this kind of stents is more complicated, requiring the specification of fenestration parameters, such as height, diameter, and angle.

5.1.1 Vessel Diameter, Length, and Area Quantification

In both standard and fenestrated endograft sizing, a precise quantification of the vessel diameters and lengths is crucial. Generally, these measurements are estimated from the segmented lumen and the extracted centerline. Usually, the quantification starts from a user-selected input position in the centerline and hence, user variability is introduced.

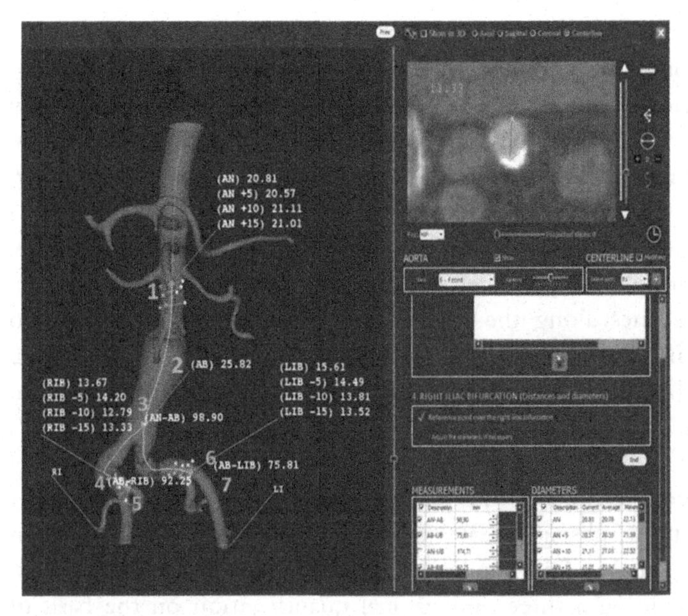

1- Neck diameter distal to the lowest renal artery, at 5 mm, 10 mm and 15 mm
2- Length from the lowest renal artery to the aortic bifurcation
3- Aortic bifurcation diameter
4- Right common iliac artery length
5- Right iliac bifurcation diameter, at 5 mm, 10 mm and 15 mm
6- Left common iliac artery length
7- Left iliac bifurcation diameter, at 5 mm, 10 mm and 15 mm

Fig. 15.4 Endograft sizing. Standard aortic diameter and length measurements for endograft sizing using the eVida Vascular software (see Section 7). f-EVAR endografts require additional measurements to be performed for full aortic branch characterization.

As the aorta is not a regular cylinder, a proper characterization of its shape and diameters must be made at different anatomical points. Since parallax error affects the representation of the true vessel size on axial slices [34], aortic cross-sections are extracted perpendicular to the centerline, allowing real diameters to be quantified [35, 36]. Moreover, aortic sections are not typically circular, so slightly different strategies for diameter quantification are in the literature. Macía et al. [37] proposed an optimization procedure that converts the vessel measurements into a cost function that is optimized with respect to a set of parameters. In order to obtain the section boundaries, a ray-casting strategy from the section center is used, as defined by the centerline, in order to extract the inner wall artery boundaries estimating maximum, minimum and average diameter for the selected section [8]. Another approach, proposed by Martufli et al. [38] utilizes the fluid mechanics definition of hydraulic diameter.

Vessel length quantification is usually assessed by measuring the distance along the centerline between two anatomic landmarks, thus providing a much more realistic model of the aortoiliac segment than directly measuring on the sagittal and coronal planes [37, 39, 40].

Area measurements can be obtained either automatically, based on lumen segmentation, or manually on perpendicular cross-sections, especially when thrombus and vessel wall areas are to be included. Area measurements are usually presented as a profile or contour shapes along the vessel medial axis.

5.1.2 Aortic Neck Angulation

Although measuring the aortic neck angulation is not necessary for endograft sizing, it is a restrictive factor when selecting the prosthesis and it should be qualitatively evaluated. If the neck is very angulated, the upper area of the stent could exert pressure on the aortic wall debilitating that area and causing postoperative complications. Yet, the exact procedure for its quantification is subject to debate among specialists. The most intuitive methodology reproduces some surgeons' manual measurement procedure, consisting in the estimation of the proximal neck angle and the aneurysm angle. It is performed by tracing lines along the medial axis of the vessel between two anatomical points and getting the angle at each intersection just on a single sagittal view (see Fig. 15.5) [31, 41]. This methodology is not robust and reproducible enough, since it is dependent on the user's point of view. van Keulen et al. [42] presented a modification of the former method based on the three-dimensional reconstruction of the aortic lumen. The procedure starts with a manual rotation to get a perpendicular view to the neck flexure and a subsequent 360-degree rotation around that position to find the real maximum angle. The approach, however, is still user-dependent. Diehm et al. [30] suggested using a three-dimensional quantification on the basis of volume reconstructions of the maximum angle between the longitudinal axis of the proximal neck and the longitudinal axis of the aneurysm lumen. In order to get more accurate results and to eliminate completely the subjective interpretation of the angle measurements, Schuurmann et al. [43] proposed to quantify the aortic curvature instead of the angulation, reporting that it is a better descriptor of the aortic shape. The basic

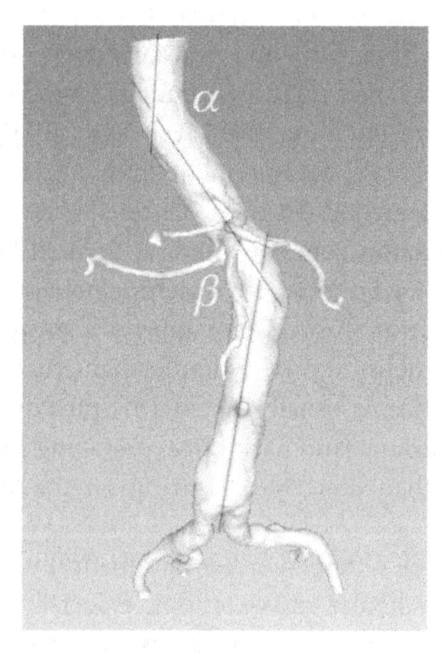

Fig. 15.5 Preoperative aortic angulation measures. Proximal neck angle (α) and aneurysm angle (β) in an abdominal aortic aneurysm.

idea is that curvature takes into account not only the severity of an angle, but also the shape of its trajectory.

Apart from the aneurysm neck angulation and length, the morphology of the neck itself should be considered for endograft planning. However, there seems to be no standard method for evaluating it and only a classification system based on visual appearance has been proposed [44].

5.2 Aortic Characterization for Intervention Risk Assessment

Before proceeding with the intervention, there are additional anatomical characteristics that must be considered because they may cause procedural complications, incrementing the intervention risks. These traits are usually evaluated by visual inspection, but more precise quantification methods are currently being investigated [45].

5.2.1 Iliac Tortuosity

The EVAR procedure consists in deploying the stent graft using fluoroscopic guidance through the iliac arteries. Thus, it is important to evaluate their shape, since a very tortuous iliac artery can hinder the access to the aneurysm. Usually, this characteristic is only qualitatively evaluated and used to decide which path is the most feasible. Boyle et al. [46] highlighted the lack of a good, reliable, and reproducible measure of iliac tortuosity, despite its potential influence in the EVAR outcome. Chaikof et al. [47] proposed to use a tortuosity index defined as the quotient $L1/L2$ between the distance from the femoral artery to the aortic bifurcation along the centerline ($L1$) and the distance on a straight line between those points ($L2$). Taudorf et al. [48] presented a similar approach measuring the ratio of the centerline distance to the straight distance, but they proposed two different indexes: from the pelvic artery to the aortic bifurcation and from the internal iliac artery to the aortic bifurcation. Wolf et al. [49] considered three different measures: (i) a tortuosity index, calculated by adding the inverse of the curvature radius taken at intervals along the centerline from the renal artery origin to the femoral arteries, (ii) the $L1/L2$ ratio, as described in Ref. [47], for both iliac branches from renal arteries to hypogastric arteries, and (iii) the cumulative angulation, which sums all the vessel angles between the renal to the hypogastric arteries from 2D projections of the 3D reconstruction. Finally, Dowson et al. [50] recently proposed a methodology based on the mathematical definition of curvature.

5.2.2 Calcification Score

Arterial calcifications may lead to an unsuccessful sealing of the stent graft, which may result in complications, such as endoleaks [39]. Currently, there is no agreement on a generalized method for quantifying the expanse of calcifications [51]. Arterial calcifications appear as hyperintense structures attached to the vessel inner wall in CTA images. Their segmentation is challenging since due to their intensity, there seems to be a spatial continuity between them and the contrasted lumen. Recently,

Heilmaier et al. [52] have proposed a way to evaluate the calcium score by considering all plaques whose density is higher than 130 Hounsfield units (HU) and with an area higher than 1 mm^2. They register each calcification size and group them according to their thickness (i.e., how deep it extends toward the lumen). Calcium scores are calculated from total plaque size and plaque density is computed as the ratio of calcium score to aortic cross-sectional area. This density is then correlated with a number of cardiovascular risk factors. Finally, An et al. [53] proposed the use of the Agatston method. The method consists in manually selecting regions of interest (ROIs) containing calcified plaques, where the area of pixels above 130 HU is automatically calculated. Then, the calcification score is computed by multiplying each ROI area by a weighting factor related to the highest attenuation value within the ROI.

5.3 Postoperative Control Parameters: AAA Maximum Diameter and Volume

Close follow-up is required after EVAR by performing CTA examinations at least yearly. Correctly excluded aneurysms tend to shrink. Size variations in the aneurysm are usually determined by measuring the largest diameter. Although the aneurysm volume is a better indicator of the changes in the aneurysm, and hence the EVAR procedure outcome [54], it is hardly used in clinical routine. One of the main reasons for this is the difficulty to automatically segment the post-EVAR intra-luminal thrombus in a reliable way. The thrombus appears uncontrasted in the acquisition volume, regardless of the imaging modality, and thus its borders are confounded with adjacent structures with similar image intensity levels. To the best of our knowledge, no method has been proposed yet in literature capable of reliably segmenting the thrombus, and avoiding leakage into adjacent tissues.

Maximum aneurysm diameter estimation is performed similarly to diameter measurement for sizing. Mora et al. [36] proposed a method where the user selects the aortic cross-section that seems to contain the largest cross-section diameter. According to the procedure they described, diameters are computed by a ray-casting strategy in every direction within the selected slice and the highest value is taken. Martufli et al. [38] utilized the aforementioned fluid mechanics definition for lumen characterization to measure the aneurysm diameter, considering "the maximum diameter" the greatest transverse diameter of all cross-sections within the aneurysm sac. Kauffmann et al. [55] used a distance map from a corrected centerline considering the outer wall of the AAA and automatically select the maximum diameter value perpendicular to that central line, reducing the user interaction to just introducing two aneurysm delimiting points. Total aneurysm volume is usually computed adding the volume of the voxels contained within the segmented aneurysm region [56]. While some authors quantify the total volume of the aneurysm (thrombosed and luminal area), others distinguish between the neck volume and total sac volume [45].

Although estimation of maximum aneurysm diameter and volume are the most established parameters predicting rupture and postoperative progression evaluation, other parameters such as wall stress, shape changes and aneurysm content should also be considered [45].

5.4 Conclusions

A precise preoperative planning is crucial for a successful EVAR intervention and a favorable postoperative evolution. EVAR planning involves patient-specific aortic sizing to choose the most suitable endograft through the assessment of lumen diameters, lengths, and angulations. Intervention risk assessment is critical in order to prevent possible complications during the surgical procedure. Literature reports that manual measures are highly variable. Thus, efforts are put into developing robust, fast, and reproducible automatic and semiautomatic approaches. However, this is hindered, among others, by the lack of automatic thrombus and calcification segmentation methods. There is also no standardized procedure for image-based quantification of the aneurysm. Nevertheless, the proliferation of vascular image analysis software is providing more agile and systematic tools for measuring different morphological parameters affecting the planning.

6. VISUALIZATION AND WORKFLOW

6.1 Visualization

Advanced visualization and image reformatting techniques are necessary for preoperative planning and sizing of EVAR stent grafts on the basis of CTA studies. Multimodal visualization of the image-extracted features and models allows to provide better insight at the planning stage. Multiple synchronized views incorporating models, interactive elements, and visual cues are usually employed in order to allow clinicians to validate the planning, providing a better sense of reliability in the measurements.

Based on the features and landmarks extracted from vascular image analysis, visualization procedures render the vascular tree:

- along its central axis;
- in true magnitude; and
- untangled/unhindered by adjacent and/or anterior tissues/structures.

Traditional visualization approaches (most commonly Maximum Intensity Projection (MIP) or Multiplanar Reconstruction (MPR)) only show planar cross-sections through the data volume. However, for vascular disease assessment, this is inefficient as only a small portion of the vessels is revealed in each slice. Specifically, measurements on axial slices do not make use of the three-dimensional, volumetric nature of CT data. Such measurements lack accuracy: they may not be perpendicular to the vessel course most of the times because the aorta is neither a straight tube nor stretches simply in the cranio-caudal direction. The situation is worse for aortic branches and ancillary vessels since they tend to be smaller, tortuous, and oriented in different directions.

Measurements on Multiplanar Reformations (MPR) overcome some of the limitations of axial slices. MPR allows the manual adjustment of the orientation of the three basic imaging planes (axial, sagittal, and coronal). Manual measurements can be performed on the reformatted slabs. However, length measurements are still limited, especially in tortuous vessel segments. The slice thickness may be varied by projecting several slices following some sort of projection algorithm, such as MIP which digitally simulates an X-ray projection and is adequate for contrasted vessels. This allows the visualization of more complete projected vessels, but depth information is still missing.

Addis et al. [57] carried out a comparison of five reconstruction techniques (axial, MIP, MPR, Shaded-Surface Display (SSD), and Volume Rendering (VR)) for vessel lumen diameter and stenosis assessment on CTA images, and they concluded that volume rendering was the most accurate reconstruction technique for measurement purposes. Their research work was anterior to the advent of more advanced techniques, capable of avoiding vessel occlusion by other high-density structures. Specific image reformattings other than traditional MPR have been developed for vascular imaging, the most widespread being Curved Planar Reformation (CPR) (also referred to as Medial Axis Reformation, MAR) [58]. Given a vessel centerline, the CPR works by resampling it and casting lines perpendicular to it, until the whole extent of the centerline is swept. A curvilinear surface is obtained with the resampled pixels. If this surface is flattened and the voxels in the close neighborhood of it are displayed, a 2D image of the vessel is obtained. The CPR method enables the study of peripheral arteries that exhibit one dominant direction. By rotating the orientation of the initial line that cuts every cross-section and defines the CPR, different views of the vessel may be obtained until the 360 degrees are covered, what is call a rotating CPR (rCPR) view. Furthermore, it allows to see the true vessel lumen, which is important to assess calcifications, stenosis, or aneurysms. Desser et al. [59] stressed the value of curved reformattings in the visualization of abdominal pathologies.

Kanitsar et al. [60] proposed multipath CPR (mpCPR) to visualize and assess the entire peripheral tree in CTA by applying an untangling process. The evaluation carried out in Ref. [61] concluded that mpCPR is able to display the peripheral arterial tree over a significantly greater viewing range than MIP. Although mpCPR is reported to cause specific artifacts with certain conditions, such as vessels that run parallel to the horizontal axis, these are predictable.

To analyze the lumen of arbitrarily oriented vessels, which may not necessarily be connected in a tree structure, a technique called Centerline Reformation (CR) has been proposed [62]. Both CPR and CR project the vascular structures into a 2D image space in order to reconstruct the vessel lumen.

Although rotating CPR (rCPR) projections can be produced to explore a vessel around its longitudinal axis completely, Mistelbauer et al. [63] introduced another visualization paradigm called Curvicircular Feature Aggregation (CFA) to project the circular information fully in a single projection and thus, avoiding the need for user interaction. In Ref. [64], Auzinger et al. proposed a technique, called Curved Surface Reformation (CSR), that computed the vessel lumen fully in 3D. According to the authors, this offers high-quality interactive visualizations of vessel lumina and does not suffer from problems of CPR or CR methods such as ambiguous visibility cues or premature discretization of centerline data [64].

Volume rendering produces high quality direct visualization of volumetric data, but one of its main problems is the occlusion of structures of interest. Reducing the opacity of the occluding structures or applying clipping are among the most straightforward solutions to this problem. However, these techniques remove the context of the feature of interest [65]. Another often used possibility is to automatically segment and remove bony structures as they are of high intensity, leaving contrasted vessels almost as the only high-intensity structure. Some authors propose to use volume rendering visualizations instead of planar reformattings to compensate those deficiencies. For example, Chan et al. [66] have proposed to use multidimensional transfer functions [67] to accommodate curvilinear and line structures in angiograms. Sato et al. [68] used filtering techniques to produce classification rules to identify different structures for the enhancement of curvilinear structures in 3D medical images.

The VesselGlyph technique [69] incorporates the CPR slice into the Direct Volume Rendering (DVR) visualization of the context structures in CTA. Thanks to a CPR view presented in the correct context, rendered with DVR, the resulting visualization is able to show interior details of blood vessels in 3D. Glaßer et al. [70] have used specifically designed transfer functions (TFs) for coronary artery evaluation. More recent techniques, such as Exploded Views [71] and the so-called Context-Preserving Exploration of Volume Data technique [72] proposed by Bruckner et al., provide a visualization that enhances interior structures during volume rendering while still preserving the context. Burns et al. [73] also proposed to add contextual information, incorporating mechanisms to discriminate objects according to their relevance. The method incorporates cutaway structures [74, 75] and images from modalities other than CTA, such as US. However, in the context of vascular imaging, these contributions seem to be limited to illustrative rendering.

To obtain a problem-oriented, arbitrarily angled plane, e.g., perpendicular to the aorta at the required site, oblique planes that allow for manual adjustment of the orientation are required. Oblique planes can be defined from either 2D or 3D images.

EVAR stent graft sizing is based on diameter and length measurements. To this end, display of centerline and orthogonal planes is required, a planning technique that has been called "central line technique" by Nyman et al. [76].

As previously described, vessel centerline extraction and analysis is a key element for visualization and quantitative measurements in vascular applications in general and in aortic pathologies in particular. Aortic 3D reconstructions and a Center Lumen

Line (CLL) are necessary for the preoperative investigation of aortic angulations [77]. Extraction of planes normal to the centerline at every point, provide a straightforward mean of measuring diameters in real vessel cross-sections. CPR, in combination with orthogonal planes, segmented lumen surface rendering and/or volume rendering allows for a full featured preoperative planning of endovascular aortic repair.

Centerline analysis gives the opportunity to measure diameters at different positions simply by moving along its length and measuring diameters in extracted cross-sections, without the need of repetitive readjustments, allowing an agile workflow. Compared to MPR, centerline analysis has also the advantage that allows length measurements along its axis, which is important to assess the required length of the device and adequate landing zones, as well as distances to relevant side branches such as the renal arteries. However, the accuracy ultimately depends on an adequate extraction of the centerline, which in turn may depend on a precise and robust segmentation.

Once extracted, the medial axis of vessels can be displayed in 2D views under one of the three CPR modes: either as curved projections (projected or stretched) or as a straightened reformation [58]. Vessel cross-sections perpendicular to the centerline (i.e., to the vessel course or axis of blood flow) are usually indicated over CPR views so that vessel locations and segments can be defined at stent graft sizing.

6.2 Workflow

Most commercial software for endograft sizing and EVAR planning follow some common steps that will be described in the following subsections.

6.2.1 Vessel Segmentation

The vessel tree is segmented (see Section 3) as the first step for the endograft sizing procedure. In contrasted CTA datasets, segmentation algorithms are usually based on techniques that exploit the intensity gradients between vessel lumen, vessel wall and surrounding tissue. Semiautomatic methods relying on front propagation techniques, such as region growing algorithms, employ at least a seed point, whose location may be automatically computed. The number of seed points needed may vary with the image quality, the complexity of the anatomy (e.g., tortuosity), or the presence of adjacent hyperdense structures, like bones or contrast enhanced veins. On one hand, manual selection of seed points poses a problem of reproducibility. However, in some circumstances, it provides more control and an additional degree of robustness.

For the best segmentation results (i.e., no missing arteries or false negatives), sufficient and homogeneous image enhancement of the aorta and its branches should be obtained with as low as reasonably achievable (ALARA) iodinated contrast media dose and optimal timing of contrast agent injection. However, some algorithms are capable of dealing with some degree of nonuniformity of contrast [37].

6.2.2 Vessel Analysis

By employing a skeleton algorithm (see Section 4), the vessel centerline or medial axis (also called Center Lumen Line (CLL), [1]) is extracted based on the lumen mask issued by the segmentation process.

Centerline analysis is prone to errors, particularly in tortuous or ellipsoidal vessel segments, such as aortic dissections. Furthermore, centerline analysis puts high demands on the quality of the image data. High-spatial resolution, with a reconstructed slice thickness not exceeding 2 mm, and overlapping increment should be used to minimize partial volume effects and step artifacts according to Rengier et al. [78].

6.2.3 Verification and Manual Editing

The accuracy of the automatic computation of the segmentation and the vessel centerline should be verified on the source data by a vascular imaging expert. Although the algorithms may be designed to work well in most cases, there may be situations where certain anatomical features (such as the presence of calcifications), imaging artifacts, or procedural pitfalls (such as inhomogeneous contrast) lead to false negatives that cannot be overlooked. Errors in the segmentation process, mostly due to undersegmentation (target branches are not present in the result) or oversegmentation (tissues that do not belong to the vascular tree are included), will prevent correct centerline computation and will therefore lead to incorrect endograft sizing. Likewise, the segmentation mask may be in agreement with the vascular tree, but the skeletonization process may fail at given locations even after applying automatic correction procedures (such as loop reduction). Sections of the aorta with complex anatomy (high angulation, kinking, or tortuosity) are particularly likely to present computational errors in the segmentation and vessel analysis process.

Thus, manual correction and editing should be considered if errors are detected, such as an obvious deviation of the centerline from the visually correct center of the lumen. However, editing should be performed with care both to optimize centerline and to minimize individual errors potentially introduced by the editing process itself. Available software solutions usually provide the possibility for editing the centerline and lumen segmentation. Sometimes a manual editing of the centerline may suffice even in the absence of a correct segmentation.

Any edition procedure should allow the user to either edit incorrect lumen stretches in the segmentation mask or centerline, to add missing branches, or to delete sections where an oversegmentation or centerline loop is observed.

6.2.4 Vessel Identification

Once the vessel tree has been simplified to a number of coherent and anatomically correct medial axes and branching locations, context information may be used to identify relevant branches: it must include, at least, the abdominal aorta or main branch, celiac trunk, superior mesenteric artery, left and right renal arteries, and the left and right iliac arteries. Although vessel identification may be an optional step, it may help automatizing the planning procedure, providing richer information to the vascular surgeon.

6.2.5 Planning (Endograft Sizing)

EVAR planning requires a number of measurements to be taken at various sites along the aorta, depending on the location of the aneurysm and the affected sections. Measurements of aortic diameters and lengths should be taken at reproducible anatomical landmarks and reported in a clear and consistent format. Centerline analysis offers fast and reliable, true-size diameter and length measurements along the centerline for stent graft sizing. Rengier et al. [78] demonstrated the superior planning capabilities offered by the use of the central line technique.

The aortic neck shape or overall configuration of the aneurysm are other aspects to be studied when considering the morphology. Aortic neck characteristics such as diameter, amount of calcification, presence or absence of thrombus, and angulation are important parameters that affect endograft fixation, and they should be evaluated to consider a patient eligible for EVAR [79].

According to Ref. [80], the following morphological features of an aneurysm must be studied in order to assess the feasibility of an EVAR intervention:

- ***Aortic neck scoring***. Identification and characterization of an adequate proximal landing zone for the endovascular graft is of particular importance for an effective sealing [78]. Grading schemes exist to quantify or score the proximal aortic neck diameter, its length, the angle between the flow axis of the suprarenal aorta and the infrarenal neck, the angle between the flow axis of the infrarenal neck and the body of the aneurysm, and the presence of proximal neck calcification or thrombus. A similar scheme exists for the distal aortic neck. Evidence suggests [1] that proximal neck angulation and diameter predispose to device migration, which should be avoided.
- ***Quality of vascular access for device introduction***. Femoral and iliac access must be evaluated for the ability to accommodate stent delivery. In addition, iliac artery morphology is critical to obtain adequate fixation at the distal sealing zone of the stent graft limbs and to maintain limb patency. Thus, the pelvic arteries may be graded based on diameter, calcification, tortuosity, and length [81]. The diameter of the iliac arteries may limit device delivery and will affect the adequacy of the distal attachment sites of the endoprosthesis. Stenoses impact device delivery and outflow patency, whereas an aneurysmatic zone extending along the iliac arteries affects the distal seal zone.
- ***Branch or accessory arteries***. The presence of lumbar arteries has been related to the predisposition of patients to present type II endoleaks. Also, the presence and blood flow access to inferior mesenteric and hypogastric arteries needs to be assessed

[1, 82, 83]: bilateral hypogastric occlusion must be avoided to prevent buttock claudication, erectile dysfunction, and visceral ischemia or even spinal cord ischemia [84].

Thus, the anatomical characteristics that need to be investigated prior to endovascular repair are:

- length, diameter, shape, and wall condition (thrombus, calcifications) of proximal and distal landing zones;
- morphology (angulation, kinking, torsion) of the aorta;
- length and extension of the pathology;
- relation of pathology and landing zones to aortic branches; and
- character of iliac and femoral access arteries (diameter, elongation, calcifications, dissections, etc.).

Workflows for endograft sizing differ depending on whether the distal landing zone (or neck) is above or below the renal arteries, since fenestrations will be needed in the latter case. The sealing (also named "fixation" or "landing") zones must be in healthy sections of the vessel at issue. However, as depicted in Fig. 15.4, there are a number of vessel features at common landmarks to be considered for stent graft placement in elective EVAR [4, 40]:

- aortic neck diameter, usually referring to average diameters;
- aortic neck diameter 15 mm distal to the lowest renal artery;
- aortic neck length;
- aneurysm diameter;
- lowest renal artery to aortic bifurcation length;
- aortic bifurcation diameter;
- common iliac artery diameters;
- external iliac artery diameters;
- lower renal to left and right internal iliac artery lengths; and
- iliac artery sealing lengths.

Other additional measures may be provided at fixed distances (e.g., 5, 10 mm) distal to the above landmarks.

In the case of fenestrated (also referred to as custom) endografts, the additional steps to be completed may be generalized for each fenestration as:

- distance from the lowest renal artery;
- fenestration (ostium) diameters for every subsidiary artery covered by the device; and
- azimuth angle (usually referred to as clock orientation).

Treatment of ruptured AAAs (rAAA) may differ from the one used for unruptured AAAs. Many groups accept the same anatomic criteria for rAAA as in elective EVAR cases [1]. However, more often, since the primary goal of treatment for rAAA is to save the patient's life, more liberal morphologic criteria have also been accepted, particularly concerning the proximal sealing zone length [1]. The hypothesis is that the morbidity/mortality associated with immediate EVAR and eventually delayed conversion to open repair (sometimes referred to as Open Surgery, OS) after EVAR failure is better than that of OS as first option in emergency settings. Modern stent graft systems generally use strong fixation modes, have a wide range of sizes, and can be accommodated in sharp angulations, and thus, a greater number of rAAAs are suitable for EVAR. Both uniliac and bi-iliac device configurations have been successfully used in EVAR for rAAA, without any evidence of significant superiority of one over the other. Aortouniiliac (AUI) stent grafts have the advantages of allowing expeditious introduction and deployment, and rapid control of bleeding [1].

The amount of involved steps or required user interaction depends both on the degree of automatization and the provided visual cues. In a semiautomatic procedure, the user will be required to pick each of the landmarks (either in 3D or 2D, adjusting the point location if necessary) manually.

Some research work [85] also mention other informative quantitative variables, such as the largest and smallest diameters on the first slice distal to the lowermost renal artery, and 15 mm below this landmark.

However, it should be noted that current EVAR procedures do not always ensure the implantation of the endovascular device exactly at the planned landing zones due to aortic shape variations introduced by the sheath and the stent graft itself once fixed to the vessel.

6.2.6 C-Arm Gantry Angle Selection

In order to translate the preoperative planning to the interventional suite, radiologists and vascular surgeons establish a number of landmarks where azimuth and elevation incident angles are defined for automatic C-arm positioning during surgery. Reconstructed images are used to determine preoperatively the most optimal C-arm position during EVAR [86].

6.2.7 Reporting and Device Model Selection

Each AAA patient presents a specific case, due to the anatomical and geometrical specificities of their aorta and the location of the aneurysm. Thus, each patient requires an optimal endograft type.

A correct endograft sizing procedure will yield a number of measurements at standardized, reproducible, anatomical landmarks. This data is transferred to a predefined or templated report so that based on it and the requirements of the procedure, the vascular surgeon chooses the most appropriate device among a variety of available brands and models. The aneurysm geometry needs to be reassessed to check the compatibility with the stent hardware: aneurysm features that suggest a contraindication must be recognized [5].

Bifurcated (or bi-iliac) stent grafts are used in most cases; aortouniiliac grafts may only be used in patients with localized pseudo-aneurysms of the infrarenal aorta [84].

To provide an optimal seal, the stent graft diameter should be oversized by 10–20% according to the aortic diameter at the proximal neck [84]. Oversizing up to 25% has been reported to decrease the risk of proximal endoleaks [1]. Although oversizing has been claimed [87] to be related to late aortic neck dilation and subsequent stent graft migration, to date no study has described a positive relationship between the degree of oversizing and the incidence of endoleaks [1].

A wide variety of endografts is available from manufactures, mostly composed by self-expanding nitinol skeleton covered with a polyester or polytetrafluoroethylene (PTFE) membrane.

6.2.8 Model 3D Printing

EVAR planning involving 3D printing of the aorta model is earning success among radiologists and vascular surgeons. Although the printed models still do not reproduce the properties of the aorta (i.e., its stiffness or compliance properties), they constitute a useful mean to have a better picture of the real dimensions and configuration of the vessel, plan the intervention and foresee possible complications. A recent work by Hoffman et al. [88] highlighted the relevance of 3D printed models for EVAR rehearsal and preoperative stent graft deployment simulation.

7. ENDOGRAFT SIZING SOFTWARE

Currently, endograft sizing is hardly understood without specialized modules incorporated into radiology workstations. From CT scanner manufacturers to specialized radiology and EVAR software companies, endograft sizing software constitutes a highly specialized and competitive market. EVAR planning software is ever providing more powerful and more flexible features, enabling more precise procedures to be performed. The provided list does not pretend to be exhaustive and is based on the information publicly available about the respective software. It only intends to be an overview of the solutions available nowadays. As the software evolve very rapidly, some of the features may vary, so consulting the information of the latest available version is suggested.

Virtually all CT scanner manufacturers provide specialized vascular modules in their radiology workstations:

- ***VesselIQ Xpress*** **(General Electric Healthcare (GE))** (Little Chalfont, Buckinghamshire, UK): it offers a suite made of several components to perform EVAR procedures, from image acquisition, through preoperative planning to surgical interventions. GE's *VesselIQ Xpress* is a postprocessing application for GE's Advantage Workstation (AW) platform, used in the analysis of 2D and 3D CT Angiography images for the purpose of cardiovascular and vascular disease assessment. It is designed to support the physician in performing the vessel analysis, stent planning and directional vessel tortuosity visualization. It can be integrated into their Volume Viewer Innova (VVI) and Synchro3D solutions for the processing, review, analysis, and communication of 3D reconstructed images. Surgery guidance is provided through the registration between the planning volume and the C-arm issued 2D fluoroscopy, with their Innova Vision surgical time software.

- ***syngo.CT Vascular Analysis*** (**Siemens**) (Erlangen, Germany): it provides automatic bone removal, centerline extraction, and main vessel labeling features. It also provides AAA stent planning templates.
- ***IntelliSpace CT Advanced Vessel Analysis (AVA) Stent Planning*** (**Philips Healthcare**) (Best, The Netherlands): it provides, among other features, a customizable workflow so that the endograft sizing can be adapted to the experience of the radiologist or the complexity or nature of the case being considered.
- ***Endovascular Stent Planning (EVSP)*** (**Toshiba Medical Systems**) (Otawara-shi, Tochigi, Japan): it is a planning module integrated into their *Vitrea Imaging* medical imaging suite. One of its most remarkable features is the capability of segmenting aneurysm thrombii, which is still an active field of research.
- ***iNtuition*** (**TeraRecon**) (Foster City, CA, USA): TeraRecon is a workstation manufacturer that provides specialized vascular planning capabilities. Their *iNtuition* product incorporates automatic centerline computation, manufacturer-specific templates, and an internet-based service within their *iNtuition CLOUD* system.

Other companies provide vascular analysis and/or standalone EVAR planning software, without requiring specialized radiology workstations or expensive computing resources. Some of them are:

- ***3mensio Vascular*** (**Pie Medical Imaging**) (Maastricht, The Netherlands): it offers custom reports for specific devices and manufacturers, as well as device order sheets. Its most prominent features are its unwarped fenestration viewer and its surgical procedure simulation. It also provides the ability to plan Thoracic Endovascular Aneurysm Repair (TEVAR) and iliac aneurysm repair.
- ***EndoSize*** (**Therenva SAS**) (Rennes, France): it is a multiplatform solution that provides planning capabilities both for abdominal and thoracic endovascular aneurysm repair, as well as virtual stenting, comprising also fenestrated stent grafts.
- ***aycan workstation*** (**aycan Medical Systems**) (Rochester, NY, USA): it also includes a reporting module for AAA that imports measurements previously performed through the software.
- ***eVida Vascular*** (**eMedica S.L.**) (Derio, Bizkaia, Spain): eMedica markets its specialized EVAR planning module for its all-purpose medical imaging platform *eVida* under the brand *eVida Vascular*. *eVida* makes the core set of features of a radiology workstation available to a general purpose computer. *eVida Vascular* is a

specialized module that offers a full EVAR planning suite workflow with the help of synchronized multiview advanced 2D projections and CPR, together with surface and volume rendering views. It provides semiautomatic vessel tree segmentation, automatic vessel analysis, as well as edition capabilities (add and delete segmentation regions, vessel medial axis edition in 3D, etc.). It also provides annotation capabilities, visual cues, symbolic components (such as time clocks for fenestrations, virtual stent grafts, etc.), C-arm positioning, partial vessel identification, and standard and fenestrated (custom) stent graft design workflows for EVAR and TEVAR. It also provides the possibility to get 3D-printable aorta models.

As medical imaging shifts toward online visualization, solutions for web-based endograft sizing are starting to emerge. *EVARplanning.com* is a web featuring a symbolic representation of the aorta, where off-line measurements can be uploaded to produce an EVAR report. Its loose binding to the off-line planning tool limits its advantages over workstation or computer-based solutions.

Finally, endograft manufacturers, such as Cook Medical (Bloomington, IN, USA), W.L.Gore & Associates, Inc. (Newark, DE, USA) or Medtronic (Minneapolis, MN, USA) also provide their own solutions, specific to the device range they manufacture.

8. CONCLUSIONS AND FUTURE PERSPECTIVES

Image-based EVAR planning is nowadays an indispensable tool for adequate stent sizing and planning of endovascular interventions. The availability of more complex devices and computerized image-analysis software has made it possible to perform EVAR on an evergrowing number of complex cases. Furthermore, less experienced surgeons are now able to successfully perform simpler interventions.

As the technology evolves, there is still much space for improvement at the planning stage. A promising area of research at the preoperative stage is the patient-specific simulation of stent deployment and fixation, which has the promise of training the intervention, specially for complex cases, and preventing possible complications. Some experimental systems already allow to simulate the deployment of the stent and the straightening suffered by the aorta [89], but this is not yet available in any commercial software.

These simulations, in addition to the systematic and reproducible measurements obtained with the available EVAR planning software, may improve the current understanding of some physiopathological processes occurring into the diseased aorta, as well as some of the EVAR complications, such as endoleaks.

Other interesting technologies include the possibility of 3D printing realistic models as well as endovascular devices.

REFERENCES

[1] Moll FL, Powell JT, Fraedrich G, Verzini F, Haulon S, Waltham M, et al. Management of abdominal aortic aneurysms clinical practice guidelines of the European Society for Vascular Surgery. Eur J Vasc Endovasc Surg 2011;41:S1–58.

[2] Stather PW, Sidloff D, Dattani N. Systematic review and meta-analysis of the early and late outcomes of open and endovascular repair of abdominal aortic aneurysm. J Vasc Surg 2013;58(4):1142.

[3] Sobocinski J, Chenorhokian H, Maurel B, Midulla M, Hertault A, Roux M, et al. The benefits of EVAR planning using a 3D workstation. Eur J Vasc Endovasc Surg 2013;46(4):418–23. http://www.sciencedirect.com/science/article/pii/S1078588413005005. http://dx.doi.org/10.1016/j.ejvs.2013.07.018.

[4] Goshima S, Kanematsu M, Kondo H, Kawada H, Kojima T, Sakurai K, et al. Preoperative planning for endovascular aortic repair of abdominal aortic aneurysms: feasibility of nonenhanced MR angiography versus contrast-enhanced CT angiography. Radiology 2013;267(3):948–55. http://dx.doi.org/10.1148/radiol.13121557.

[5] Kicska G, Litt H. Preprocedural planning for endovascular stent-graft placement. Semin Interv Radiol 2009;26(1):44–55.

[6] Lee K, Johnson R, Yin Y, Wahle A, Olszewski M, Scholz T, et al. Three-dimensional thrombus segmentation in abdominal aortic aneurysms using graph search based on a triangular mesh. Comput Biol Med 2010;40(3):271–8.

[7] Almuntashri A, Finol E, Agaian S. Automatic lumen segmentation in CT and PC-MR images of abdominal aortic aneurysm. In: IEEE international conference on systems, man, and cybernetics (SMC), 2012. 2012, p. 2891–6.

[8] Macía I, De Blas M, Legarreta J, Kabongo L, Hernández O, Ega na J, et al. Standard and fenestrated endograft sizing in EVAR planning: description and validation of a semi-automated 3D software. Comput Med Imaging Graph 2015;50:9–23.

[9] Angenent S, Pichon E, Tannenbaum A. Mathematical methods in medical image processing. Bull Am Math Soc 2006;43:365–96.

[10] Sethian JA. Level set methods and fast marching methods: evolving interfaces in computational geometry, fluid mechanics, computer vision, and materials science. Cambridge: Cambridge University Press; 1999.

[11] Lorigo LM, Faugeras O, Grimson WEL, Keriven R, Kikinis R, Nabavi A, et al. Codimension-two geodesic active contours for the segmentation of tubular structures. In: IEEE conference on computer vision and pattern recognition, 2000. Proceedings; vol. 1; 2000. p. 444–51. http://dx.doi.org/10.1109/CVPR.2000.855853.

[12] Ayyalasomayajula A, Polk A, Basudhar A, Missoum S, Nissim L, Vande Geest JP. Three dimensional active contours for the reconstruction of abdominal aortic aneurysms. Ann Biomed Eng 2010;38(1): 164–76.

[13] Wang C, Frimmel H, Smedby Ö. Level-set based vessel segmentation accelerated with periodic monotonic speed function. In: Proc. SPIE 7962, Medical Imaging 2011: Image Processing; vol. 7962; 2011. p. 79621M–79621M-7. http://dx.doi.org/10.1117/12.876704.

[14] Egger J, Freisleben B, Setser R, Renapuraar R, Biermann C, O'Donnell T. Aorta segmentation for stent simulation. CoRR 2011; abs/1103.1773.

[15] Duquette A, Jodoin P, Bouchot O, Lalande A. 3D segmentation of abdominal aorta from CT-scan and MR images. Comput Med Imaging Graph 2012;36(4):294–303.

[16] Antiga L, Ene-Iordache B, Remuzzi A. Computational geometry for patient-specific reconstruction and meshing of blood vessels from MR and CT angiography. IEEE Trans Med Imaging 2003;22(5):674–84. http://dx.doi.org/10.1109/TMI.2003.812261.

[17] Blum H. A transformation for extracting new descriptors of shape. In: Wathen-Dunn W, editor. Models for the perception of speech and visual form. Cambridge, MA: MIT Press; 1967. p. 362–80.

[18] Sironi A, Türetken E, Lepetit V, Fua P. Multiscale centerline detection. IEEE Trans Pattern Anal Mach Intell 2016;38(99):1327–41. http://dx.doi.org/10.1109/TPAMI.2015.2462363.

[19] Näf M. 3D Voronoi skeletons: a semicontinuous implementation of the 'symmetric axis transform' in 3D space. Ph.D. thesis, ETH Zürich; 1996.

[20] Siddiqi K, Kimia BB. A shock grammar for recognition In: IEEE computer vision and pattern recognition (CVPR'96). CVPR'96:507-513, San Francisco, CA, IEEE; 1996. p. 507–13.
[21] Pizer SM, Eberly D, Morse BS, Fritsch DS. Zoom-invariant figural shape: the mathematics of cores. Comput Vis Image Underst 1998;69:55–71.
[22] Bouix S, Siddiqi K, Tannenbaum A. Flux driven automatic centerline extraction. Med Image Anal 2005;9(3):209–21.
[23] Aylward SR, Bullitt E. Initialization, noise, singularities, and scale in height ridge traversal for tubular object centerline extraction. IEEE Trans Med Imaging 2002;21(2):61–75. http://dx.doi.org/10.1109/42.993126.
[24] Ma CM, Sonka M. A fully parallel 3D thinning algorithm and its applications. Comput Vis Image Underst 1996;64(3):420–33. http://www.sciencedirect.com/science/article/pii/S1077314296900697. http://dx.doi.org/10.1006/cviu.1996.0069.
[25] Kimmel R, Kiryati N, Bruckstein AM. Sub-pixel distance maps and weighted distance transforms. J Math Imaging Vis 1996;6(2):223–33. http://dx.doi.org/10.1007/BF00119840.
[26] Selle D, Preim B, Schenk A, Peitgen HO. Analysis of vasculature for liver surgical planning. IEEE Trans Med Imaging 2002;21(11):1344–57. http://dx.doi.org/10.1109/TMI.2002.801166.
[27] Zhang L, Chapman BE, Parker DL, Roberts JA, Guo J, Vemuri P, et al. Automatic detection of three-dimensional vascular tree centerlines and bifurcations in high-resolution magnetic resonance angiography. Invest Radiol 2005;40(10):661–71.
[28] Macía I, Gra na M, Paloc C. Knowledge management in image-based analysis of blood vessel structures. Knowl Inf Syst 2012;30(2):457–91. http://dx.doi.org/10.1007/s10115-010-0377-x.
[29] Diehm N, Baumgartner I, Silvestro A, Herrmann P, Triller J, Schmidli J, et al. Automated software supported versus manual aorto-iliac diameter measurements in CT angiography of patients with abdominal aortic aneurysms: assessment of inter- and intraobserver variation. Eur J Vasc Med 2005;34(4):255–61.
[30] Diehm N, Katzen B, Samuels S, Pena C, Powell A, Dick F. Sixty-four-detector CT angiography of infrarenal aortic neck length and angulation: prospective analysis of interobserver variability. J Vasc Interv Radiol 2008;19(9):1283–8.
[31] Raman B, Raman R, Napel S, Rubin GD. Automated quantification of aortoaortic and aortoiliac angulation for computed tomographic angiography of abdominal aortic aneurysms prior to endovascular repair: preliminary study. J Vasc Interv Radiol 2010;21(11):1746–50.
[32] Parker M, O'Donnell S, Chang A, Johnson C, Gillespie D, Goff J, et al. What imaging studies are necessary for abdominal aortic endograft sizing? A prospective blinded study using conventional computed tomography, aortography, and three-dimensional computed tomography. J Vasc Surg 2005;41(2):199–205.
[33] Kauffmann C, Tang A, Dugas A, Therasse I, Oliva V, Soulez G. Clinical validation of a software for quantitative follow-up of abdominal aortic aneurysm maximal diameter and growth by CT angiography. Eur J Radiol 2011;77(3):502–8.
[34] Krittpracha B, Beebe H, Comerota A. Aortic diameter is an insensitive measurement of early aneurysm expansion after endografting. J Endovasc Ther 2004;11:184–90.
[35] Fairman R, Velazquez O, Carpenter J, Woo E, Baum R, et al. Midterm pivotal trial results of the talent low profile system for repair of abdominal aortic aneurysm: analysis of complicated versus uncomplicated aortic necks. J Vasc Surg 2004;40(6):1074–82.
[36] Mora C, Marcus C, Barbe C, Ecarnot F, Long A. Measurement of maximum diameter of native abdominal aortic aneurysm by angio-CT: reproducibility is better with the semi-automated method. Eur J Vasc Endovasc Surg 2014;47(2):139–50.
[37] Macía I, Legarreta JH, Rajasekharan S, Mu noz E, Hernández Ó, de Blas M, et al. Development of a System for Endovacular Planning of AAA Interventions. In: MICCAI 2013. 16th international conference on medical image computing and computer assisted intervention. MICCAI-STENT'13. The 2nd international MICCAI-workshop on computer assisted stenting. Nagoya, Japan; 2013, p. 45–52.
[38] Martufli G, Di Martino E, Amon C, Muluk S, Finol E. Three-dimensional geometrical characterization of abdominal aortic aneurysms: image-based wall thickness distribution. J Biomed Eng 2009;131(6).

[39] Ghatwary T, Patterson B, Karthikesalingam A, Hinchliffe R, Loftus I, Morgan R, et al. A systematic review of protocols for the three-dimensional morphologic assessment of abdominal aortic aneurysms using computed tomographic angiography. Cardiovasc Intervent Radiol 2013;36:14–24.

[40] Kaladji K, Lucas A, Kervio G, Haigron P, Cardon A. Sizing for endovascular aneurysm repair: clinical evaluation of a new automated three-dimensional software. Ann Vasc Surg 2010;24:912–20.

[41] Martufi G, Auer M, Roy J, Swedenborg J, Sakalihasan N, Panuccio G, et al. Multidimensional growth measurements of abdominal aortic aneurysms. J Vasc Surg 2013;58(3):748–55.

[42] van Keulen J, Moll F, Tolenaar J, Verhagen H, van Herwaarden J. Validation of a new standardized method to measure proximal aneurysm neck angulation. J Vasc Surg 2010;51(4):821–8.

[43] Schuurmann R, Kuster L, Slump C, Vahl A, Van den Heuvel D, Ouriel K, et al. Aortic curvature instead of angulation allows improved estimation of the true aorto-iliac trajectory. Eur J Vasc Endovasc Surg 2016;51(2):216–24.

[44] CO M, M H, A B, SR B. Abdominal aortic aneurysm neck morphology: proposed classification system. J Med Sci 2006;175(3):4–8.

[45] Shum J, Martufi G, Martino ED, Washington CB, Grisafi J, Muluk SC, et al. Quantitative assessment of abdominal aortic aneurysm geometry. Ann Biomed Eng 2011;39(1):277–86.

[46] Boyle J, Thompson M, Vallabhaneni S, Bell R, Bell R J, Browne J, et al. Pragmatic minimum reporting standards for pragmatic minimum reporting standards for endovascular abdominal aortic aneurysm repair. J Endovasc Ther 2011;18(3):263–71.

[47] Chaikof E, Blankensteijn J, Harris P, White G, Zarins C, Bernhard V, et al. Reporting standards for endovascular aortic aneurysm repair. J Vasc Surg 2002;35(5):1048–60.

[48] Taudorf M, Jensen LP, Vogt KC, Gronvall J, Schroeder TV, Lönn L. Endograft limb occlusion in EVAR: Iliac tortuosity quantified by three different indices on the basis of preoperative CTA. Eur J Vasc Endovasc Surg 2014;48(5):527–33.

[49] Wolf YG, Tillich M, Lee WA, Rubin GD, Fogarty TJ, Zarins CK. Impact of aortoiliac tortuosity on endovascular repair of abdominal aortic aneurysms: evaluation of 3D computer-based assessment. J Vasc Surg 2001;34(4):594–9.

[50] Dowson N, Boult M, Cowled P, de Loryn P, Fitridge R. Development of an automated measure of iliac artery tortuosity that successfully predicts early graft-related complications associated with endovascular aneurysm repair. Eur J Vasc Endovasc Surg 2014;48(2):153–60.

[51] De Vries J. The proximal neck: The remaining barrier to a complete EVAR world. Semin Vasc Surg 2012;25(4):182–6.

[52] Heilmaier C, Koester A, Moysidis T, Weishaupt D, Kroger K. Abdominal aortic calcification and its distribution in normal-sized and aneurysmatic abdominal aortas. Vasa 2014;43:132–40.

[53] An C, Lee HJ, Lee HS, Ahn SS, Choi BW, et al. CT-based abdominal aortic calcification score as a surrogate marker for predicting the presence of asymptomatic coronary artery disease. Eur Radiol 2014;24(10):2491–8.

[54] Renapurkar RD, Setser RM, O'Donnell TP, Egger J, Lieber ML, Desai MY, et al. Aortic volume as an indicator of disease progression in patients with untreated infrarenal abdominal aneurysm. Eur J Radiol 2012;81:87–93.

[55] Kauffmann C, Tang A, Therasse E, Giroux MF, Elkouri S, Melanson P, et al. Measurements and detection of abdominal aortic aneurysm growth: accuracy and reproducibility of a segmentation software. Eur J Radiol 2011;81:1688–94.

[56] Van Prehn J, Van der Wal M, Vincken K, Bartels L, Moll F, Van Herwaarden J. Intra- and interobserver variability of aortic aneurysm volume measurement with fast CTA postprocessing software. J Endovasc Ther 2008;15:504–10.

[57] Addis K, Hopper K, Lyriboz A, Liu Y, Wise S, Kasales C, et al. CT angiography: in vitro comparison of five reconstruction methods. Am J Roentgenol 2001;177(5):1171–6. http://dx.doi.org/10.2214/ajr.177.5.1771171.

[58] Kanitsar A, Fleischmann D, Wegenkittl R, Felkel P, Gröller E. CPR—curved planar reformation. In: Visualization, 2002. VIS 2002. IEEE; 2002. p. 37–44.

[59] Desser T, Sommer F, Jeffrey B. Value of curved planar reformations in MDCT of abdominal pathology. Am J Roentgenol 2004;182(6):1477–84. http://dx.doi.org/10.2214/ajr.182.6.1821477.

[60] Kanitsar A, Fleischmann D, Wegenkittl R, Gröller E. Diagnostic relevant visualization of vascular structures. In: Bonneau GP, Ertl T, Nielson GM, editors. Scientific visualization: the visual extraction of knowledge from data. Berlin, Heidelberg: Springer-Verlag; 2006. p. 207–28. http://dx.doi.org/10.1007/3-540-30790-7_13.

[61] Roos JE, Fleischmann D, Koechl A, Rakshe T, Straka M, Napoli A, et al. Multipath curved planar reformation of the peripheral arterial tree in CT angiography. Radiology 2007;244(1):281–90. http://dx.doi.org/10.1148/radiol.2441060976.

[62] Mistelbauer G, Varchola A, Bouzari H, Starinsky J, Köchl A, Schernthaner R, et al. Centerline reformations of complex vascular structures. In: Visualization symposium (PacificVis), 2012 IEEE Pacific; 2012. p. 233–40.

[63] Mistelbauer G, Morar A, Varchola A, Schernthaner R, Baclija I, Köchl A, et al. Vessel visualization using curvicircular feature aggregation. Comput Graph Forum 2013;32(3pt2):231–40. http://dx.doi.org/10.1111/cgf.12110.

[64] Auzinger T, Mistelbauer G, Baclija I, Schernthaner R, Köchl A, Wimmer M, et al. Vessel visualization using curved surface reformation. IEEE Trans Vis Comput Graph 2013;19(12):2858–67. http://dx.doi.org/10.1109/TVCG.2013.215.

[65] Balabanian J. Multi-aspect visualization: going from linked views to integrated views. Ph.D. thesis, Department of Informatics, University of Bergen, Norway; 2009.

[66] Chan MY, Wu Y, Qu H, Chung ACS, Wong WCK. Mip-guided vascular image visualization with multi-dimensional transfer function. In: Nishita T, Peng Q, Seidel HP, editors. CGI'06: Proceedings of the 24th international conference on advances in computer graphics, Hangzhou, China, June 26–28, 2006. Berlin, Heidelberg: Springer-Verlag; 2006. p. 372–384. ISBN 978-3-540-35639-4. http://dx.doi.org/10.1007/11784203_32.

[67] Levoy M. Display of surfaces from volume data. IEEE Comput Graph Appl 1988;8(3):29–37. http://dx.doi.org/10.1109/38.511.

[68] Sato Y, Nakajima S, Shiraga N, Atsumi H, Yoshida S, Koller T, et al. Three-dimensional multi-scale line filter for segmentation and visualization of curvilinear structures in medical images. Med Image Anal 1998;2(2):143–68. http://www.sciencedirect.com/science/article/pii/S1361841598800091. http://dx.doi.org/10.1016/S1361-8415(98)80009-1.

[69] Straka M, Cervenansky M, La Cruz A, Kochl A, Sramek M, Groller E, et al. The vesselglyph: focus & context visualization in ct-angiography. In: Visualization, 2004. IEEE; 2004. p. 385–92. http://dx.doi.org/10.1109/VISUAL.2004.104.

[70] Glaßer S, Oeltze S, Hennemuth A, Wilhelmsen S, Preim B. Adapted transfer function design for coronary artery evaluation. In: Meinzer HP, Deserno TM, Handels H, Tolxdorff T, editors. Bildverarbeitung für die Medizin 2009: Algorithmen, Systeme, Anwendungen. Proceedings des Workshops vom 22. bis 25. März 2009, Heidelberg, Berlin, Heidelberg: Springer-Verlag; 2009. p. 1–5. ISBN 978-3-540-93860-6. http://dx.doi.org/10.1007/978-3-540-93860-6_1.

[71] Bruckner S, Gröller M. Exploded views for volume data. IEEE Trans Vis Comput Graph 2006;12(5):1077–84.

[72] Bruckner S, Grimm S, Kanitsar A, Gröller M. Illustrative context-preserving exploration of volume data. IEEE Trans Vis Comput Graph 2006;12(6):1559–69.

[73] Burns M, Haidacher M, Wein W, Viola I, Gröller M. Feature emphasis and contextual cutaways for multimodal medical visualization. In: Proceedings of the 9th joint Eurographics/IEEE VGTC conference on visualization, EUROVIS'07, Eurographics Association; 2007. p. 275–82. ISBN 978-3-905673-45-6. http://dx.doi.org/10.2312/VisSym/EuroVis07/275-282.

[74] Diepstraten J, Weiskopf D, Ertl T. Interactive cutaway illustrations. Comput Graphics Forum 2003;22(3):523–32. http://dx.doi.org/10.1111/1467-8659.t01-3-00700.

[75] Fischer J, Bartz D, Straßer W. Illustrative display of hidden iso-surface structures. In: Visualization, 2005. VIS 05. IEEE; 2005. p. 663–70.

[76] Nyman R, Eriksson M. The future of imaging in the management of abdominal aortic aneurysm. Scand J Surg 2008;97(2):110–5.

[77] van Keulen J, Moll FL, Tolenaar J, Verhagen H, Herwaarden J. Validation of a new standardized method to measure proximal aneurysm neck angulation. J Vasc Surg 2010;51(4):821–8.

http://www.sciencedirect.com/science/article/pii/S0741521409022629. http://dx.doi.org/10.1016/j.jvs.2009.10.114.

[78] Rengier F, Weber T, Kotelis D, Giesel F, Böckler D, Kauczor H, et al. Three-dimensional preoperative planning of endovascular aortic repair: workflow, clinical benefits and limitations. In: European Congress of Radiology (ECR 2010), European Society of Radiology; 2010.

[79] Tournoij E, Slisatkorn W, Prokop M, Verhagen H, Moll F. Thrombus and calcium in aortic aneurysm necks: validation of a scoring system in a Dutch Cohort Study. Vasc Endovasc Surg 2007;41(2):120–5. http://ves.sagepub.com/content/41/2/120.abstract. http://dx.doi.org/10.1177/1538574406298081.

[80] Walker T, Kalva S, Yeddula K, Wicky S, Kundu S, Drescher P, et al. Clinical practice guidelines for endovascular abdominal aortic aneurysm repair: written by the standards of practice committee for the Society of Interventional Radiology and Endorsed by the Cardiovascular and Interventional Radiological Society of Europe and the Canadian Interventional Radiology Association. J Vasc Interv Radiol 2010;21(11):1632–55. http://www.sciencedirect.com/science/article/pii/S105104431000761X. http://dx.doi.org/10.1016/j.jvir.2010.07.008.

[81] Chaikof E, Fillinger M, Matsumura J, Rutherford R, White G, Blankensteijn J, et al. Identifying and grading factors that modify the outcome of endovascular aortic aneurysm repair. J Vasc Surg 2002;35(5):1061–6. http://www.sciencedirect.com/science/article/pii/S0741521402102552. http://dx.doi.org/10.1067/mva.2002.123991.

[82] Brewster D, Franklin D, Cambria R, Darling R, Moncure A, Lamuraglia G, et al. Intestinal ischemia complicating abdominal aortic surgery. Surgery 1991;109(4):447–54.

[83] Seeger J, Coe D, Kaelin L, Flynn T. Routine reimplantation of patent inferior mesenteric arteries limits colon infarction after aortic reconstruction. J Vasc Surg 1992;15(4):635–41. http://www.sciencedirect.com/science/article/pii/074152149290008V. http://dx.doi.org/10.1016/0741-5214(92)90008-V.

[84] Erbel R, Aboyans V, Boileau C, Bossone E, et al. 2014 ESC Guidelines on the diagnosis and treatment of aortic diseases. Eur Heart J 2014; http://eurheartj.oxfordjournals.org/content/early/2014/08/28/eurheartj.ehu281.

[85] Kaladji A, Lucas A, Kervio G, Haigron P, Cardon A. Sizing for endovascular aneurysm repair: clinical evaluation of a new automated three-dimensional software. Ann Vasc Surg 2010;24(7):912–20. http://www.sciencedirect.com/science/article/pii/S0890509610001846. http://dx.doi.org/10.1016/j.avsg.2010.03.018.

[86] van Keulen J, Moll F, Herwaarden J. Tips and techniques for optimal stent graft placement in angulated aneurysm necks. J Vasc Surg 2010;52(4):1081–6. http://www.sciencedirect.com/science/article/pii/S0741521410003344. http://dx.doi.org/10.1016/j.jvs.2010.02.024.

[87] Resch T, Ivancev K, Brunkwall J, Nyman U, Malina M, Lindblad B. Distal migration of stent-grafts after endovascular repair of abdominal aortic aneurysms. J Vasc Interv Radiol 1999;10(3):257–64. http://www.sciencedirect.com/science/article/pii/S1051044399700278. http://dx.doi.org/10.1016/S1051-0443(99)70027-8.

[88] Hoffman A, Nitecki S, Engel A, Karram T, Leiderman M, Si-On E, et al. In house enhanced 3D printing of complex AAA for EVAR treatment planning and preoperative simulation. J Cardiovasc Dis Diagn 2014; http://dx.doi.org/10.4172/2329-9517.1000181.

[89] Haigron P, Dumenil A, Kaladji A, Rochette M, Said B, et al. Angiovision: aortic stent-graft placement by augmented angionavigation. IRBM 2013;34(2):167–75. http://www.sciencedirect.com/science/article/pii/S1959031813000304. http://dx.doi.org/10.1016/j.irbm.2013.01.016.

INDEX

Note: Page numbers followed by *f* indicate figures and *t* indicate tables.

A

F

G

H

I

J

K

L

M

N

O

P

Q

R

S

T

U

V

W

X

Printed in the United States
By Bookmasters